Differential Equations with Linear Algebra

Differential Equations
with Linear Algebra

Matthew R. Boelkins, J. L. Goldberg, and Merle C. Potter

OXFORD
UNIVERSITY PRESS

2009

OXFORD
UNIVERSITY PRESS

Oxford University Press, Inc., publishes works that further
Oxford University's objective of excellence
in research, scholarship, and education.

Oxford   New York
Auckland   Cape Town   Dar es Salaam   Hong Kong   Karachi
Kuala Lumpur   Madrid   Melbourne   Mexico City   Nairobi
New Delhi   Shanghai   Taipei   Toronto

With offices in
Argentina   Austria   Brazil   Chile   Czech Republic   France   Greece
Guatemala   Hungary   Italy   Japan   Poland   Portugal   Singapore
South Korea   Switzerland   Thailand   Turkey   Ukraine   Vietnam

Published by Oxford University Press, Inc.
198 Madison Avenue, New York, New York 10016
www.oup.com

Oxford is a registered trademark of Oxford University Press

Library of Congress Cataloging-in-Publication Data
Boelkins, Matthew R.
Differential equations with linear algebra / Matthew R. Boelkins, J.L. Goldberg, Merle C. Potter.
    p. cm.
Includes index.
ISBN 978-0-19-538586-1 (cloth)
1. Differential equations, Linear.   2. Algebras, Linear. I. Goldberg, Jack L. (Jack Leonard), 1932–
II. Potter, Merle C. III. Title.
QA372.B657 2009
515′.354–dc22   2008050361

Contents

Introduction

In *Differential Equations with Linear Algebra*, we endeavor to introduce students to two interesting and important areas of mathematics that enjoy powerful interconnections and applications. Assuming that students have completed a semester of multivariable calculus, the text presents an introduction to critical themes and ideas in linear algebra, and then, in its remaining seven chapters, investigates differential equations while highlighting the role that linearity plays in their study. Throughout the text, we strive to reach the following goals:

- To motivate the study of linear algebra and differential equations through interesting applications in order that students may see how theoretical results can answer fundamental questions that arise in physical situations.

- To demonstrate the fact that linear algebra and differential equations can be presented as two parts of a mathematical whole that is coherent and interconnected. Indeed, we regularly discuss how the structure of solutions to linear differential equations and systems of equations exemplify important ideas in linear algebra, and how linear algebra often answers key questions regarding differential equations.

- To present an exposition that is intended to be read and understood by students. While certainly every textbook is written with students in mind, often the rigor and formality of standard mathematical presentation takes over, and books become difficult to read. We employ an examples-first philosophy that uses an intuitive approach as a lead-in to more general, theoretical results.

- To develop in students a deep understanding of what may be their first exposure to post-calculus mathematics. In particular, linear algebra is a fundamental subject that plays a key role in the study of much higher level mathematics; through its study, as well as our investigations of differential equations, we aim to provide a foundation for further study in mathematics for students who are so interested.

Whether designed for mathematics or engineering majors, many universities offer a hybrid course in linear algebra and differential equations, and this text is written for precisely such a class. At other institutions, linear algebra and differential equations are treated in two separate courses; in settings where linear algebra is a prerequisite to the study of differential equations, this text may also be used for the differential equations course, with its first chapter on linear algebra available as a review of previously studied material. More details on the ways the book can be implemented in these courses follows shortly in the section *How to Use this Text*. An overriding theme of the book is that if a differential equation or system of such equations is linear, then we can usually solve it exactly.

Linear algebra and systems first

In most other texts that present the subjects of differential equations and linear algebra, the presentation begins with first-order differential equations, followed by second- and higher order linear differential equations. Following these topics, a modest amount of linear algebra is introduced before beginning to consider systems of linear differential equations. Here, however, we begin on the very first page of the text with an example that shows the natural way that systems of linear differential equations arise, and use this example to motivate the need to study linear algebra. We then embark on a one-chapter introduction to linear algebra that aims not only to introduce such important concepts as linear combinations, linear independence, and the eigenvalue problem, but also to foreshadow the use of such topics in the study of differential equations.

Following chapter 1, we consider first-order differential equations briefly in chapter 2, using the study of linear first-order equations to highlight some of the key ideas already encountered in linear algebra. From there, we quickly proceed to an in-depth presentation of systems of linear differential equations in chapter 3. In that setting, we show how the eigenvalues of an $n \times n$ matrix $\mathbf{A}$ naturally provide the general solution to systems of linear differential equations in the form $\mathbf{x}' = \mathbf{A}\mathbf{x}$. Moreover, we include examples that show how any single higher order linear differential equation may be converted to a system of equations, thus providing further motivation for why we choose to study systems first. Through this approach, we again strive to emphasize critical connections between linear algebra and differential equations and to demonstrate the most important ideas that arise in the study of each. In the remainder of the text, the role of linear algebra is continually emphasized, even in the study of nonlinear equations and systems.

Features of the text

Instructors and students alike will find several consistent features in the presentation.

- Each chapter begins with one or two *motivating problems* that present a natural situation—often a physical application—in which linear algebra or differential equations arises. From such problems, we work to develop related ideas in subsequent sections that enable us to solve the original problem. In discussing the motivating problems, we also endeavor to use our intuition to predict the solution(s) we expect to find, and then later test our results against these predictions.

- In almost every section of the text, we use an *examples-first approach.* By this we mean that we introduce a certain type of problem that we are interested in solving, and then consider a relatively simple one that can be solved by intuition or ideas studied previously. From the solution of an elementary example, we then discuss how this approach can be generalized or modified to solve more complex examples, and then ultimately prove or state theorems that provide general results that enable the solution of a wide range of problems. With this philosophy, we strive to demonstrate how the general theory of mathematics comes from experimenting and investigating through individual examples followed by looking for overall trends. Moreover, we often use this approach to foreshadow upcoming ideas: for example, while studying linear algebra, we look ahead to a handful of fundamental differential equations. Similarly, early on in our investigations of the Laplace transform, we regularly attempt to demonstrate through examples how the transform will be used to solve initial-value problems.

- While there are many formal theoretical results that hold in both linear algebra and differential equations, we have endeavored to *emphasize intuition.* Specifically, we use the aforementioned examples-first approach to solve sample problems and then present evidence as to why the details of the solution process for a small number of examples can be generalized to an overall structure and theory. This is in contrast to many books that first present the overall theory, and then demonstrate the theory at work in a sequence of subsequent examples. In addition, we often eschew formal proofs, choosing instead to present more heuristic or intuitive arguments that offer evidence of the truth of important theorems.

- Wherever possible, we use *visual reasoning* to help explain important ideas. With over 100 graphics included in the text, we have provided figures that help deepen students' understanding and offer additional perspective on essential concepts. By thinking graphically, we often find that an appropriate picture sheds further light on the solution to a problem and how we should expect it to behave, thus adding to our intuition and understanding.

• With computer algebra systems (CASs), such as *Maple* and *Mathematica*, approaching their twentieth year of existence, these technologies are an important part of the landscape of the teaching and learning of mathematics. Especially in more sophisticated subjects with computationally complicated problems, these tools are now indispensable. We have chosen to integrate instructional support for *Maple* directly within the text, while offering similar commentary for *Mathematica*, *MATLAB*, and *SAGE* on our website, www.oup.com/ differentialequations/. For each, students can find directions for how to effectively use computer algebra systems to generate important graphs and execute complicated or tedious calculations. Many sections of the text are followed by a short subsection on "*Using Maple to ....*" Parallel sections for the other CASs, numbered similarly, can be found on the website.

• Each chapter ends with a section titled *For further study*. In this setting, rather than a full exposition, a sequence of leading questions is presented to guide students to discover some key ideas in more advanced problems that arise naturally from the material developed to date. These sections can be used as a basis for instructor-led in-class discussions or as the foundation for student projects or other assignments. Interested students can also pursue these topics on their own.

How to use this text

There are two courses for which this text is well-suited: a hybrid course in linear algebra and differential equations, or a course in differential equations that requires linear algebra as a prerequisite. We address each course separately with some suggestions for instructors.

Linear algebra and differential equations

For a hybrid course in the two subjects, instructors should begin with chapter 1 on linear algebra. There, in addition to an introduction to many essential ideas in the subject, students will encounter a handful of examples on linear differential equations that foreshadow part of the role of linear algebra in the field of differential equations. The goal of the chapter on linear algebra is to introduce important ideas such as linear combinations, linear independence and span, matrix algebra, and the eigenvalue problem. At the close of chapter 1 we also introduce abstract vector spaces in anticipation of the structural role that vector spaces play in solving linear systems of differential equations and higher order linear differential equations. Instructors may choose to move on from chapter 1 upon completing section 1.10 (the eigenvalue problem), as this is the last topic that is absolutely essential for the solution of linear systems of differential equations in chapter 3. Discussion of ideas like basis, dimension, and vector spaces of functions from the final two sections of chapter 1 can occur alongside the development of general solutions to systems of linear differential equations or higher order linear differential equations.

Over the past decade or two, first-order differential equations have become a standard topic that is normally discussed in calculus courses. As such, chapter 2 can be treated lightly at the instructor's discretion. In particular, it is reasonable to expect that students are familiar with direction fields, separable differential equations, Euler's method, and several fundamental applications, such as Newton's law of Cooling and the logistic differential equation. It is less likely that students will have been exposed to integrating factors as a solution technique for linear first-order equations and the solution methods for exact equations. In any case, chapter 2 is not one on which to linger. Instructors can choose to selectively discuss a small number of sections in class, or assign the pages there as a reading assignment or project for independent investigation.

Chapter 3 on systems of linear differential equations is the heart of the text. It can be begun immediately following section 1.10 in chapter 1. Here we find not only a large number of rich ideas that are important throughout the study of differential equations, but also evidence of the essential role that linear algebra plays in the solution of these systems. As is noted on several occasions in chapter 3, any higher order linear differential equation may be converted to a system of first-order equations, and thus an understanding of systems enables one to solve these higher order equations as well. Thus, the material in chapter 4 may be de-emphasized. Instructors may choose to provide a brief overview, in class, of how the ideas in solving linear systems translate naturally to the higher order case, or may choose to have students investigate these details on their own through a sequence of reading and homework assignments or a group project. Section 4.5 on beats and resonance is one to discuss in class as these phenomena are fascinating and important and the perspective of higher order equations is a more natural context in which to consider their solution.

The Laplace transform is a topic that affords discussion of a variety of important ideas: linear transformations, differentiation and integration, direct solution of initial-value problems, discontinuous forcing functions, and more. In addition, it can be viewed as a gateway to more sophisticated mathematical techniques encountered in more advanced courses in mathematics, physics, and engineering. Chapter 5 is written with the goal of introducing students to the Laplace transform from the perspective of how it can be used to solve initial-value problems. This emphasis is present throughout the chapter, and culminates in section 5.5.

Finally, a course in both linear algebra and differential equations should not be considered complete until there has been at least some discussion of nonlinearity. Chapter 6 on nonlinear higher order equations and systems offers an examination of this concept from several perspectives, all of which are related to our previous work with linear differential equations. Direction fields, approximation by linear systems, and an introduction to numerical approximation with Euler's method are natural topics with which to round out the course. Due to the time required to introduce the subject of linear algebra to students, the final two chapters of the text (on numerical methods and series solutions) are ones we would normally not expect to be considered in a hybrid course.

Differential equations with a linear algebra prerequisite

For a differential equations course in which students have already taken linear algebra, chapter 1 may be used as a reference for students, or as a source of review as needed. The comments for the hybrid course above for chapters 2–5 hold for a straight differential equations class as well, and we would expect instructors to use the time not devoted to the study of linear algebra to focus more on the material on nonlinearity in chapter 6, numerical methods in chapter 7, and series solutions in chapter 8. The first several sections of chapter 7 may be treated any time after first-order differential equations have been discussed; only the final section in that chapter is devoted to systems and higher order equations where the methods naturally generalize work with first-order equations.

In addition to spending more time on the final three chapters of the text, instructors of a differential equations-only course can take advantage of the many additional topics for consideration in the *For further study* sections that close each chapter. There is a wide range of subjects from which to choose, both theoretical and applied, including discrete dynamical systems, how raindrops fall, matrix exponentials, companion matrices, Laplace transforms of periodic piecewise continuous forcing functions, and competitive species.

Appendices

Finally, the text closes with five appendices. The first three—on integration techniques, polynomial zeros, and complex numbers—are intended as a review of familiar topics from courses as far back in students' experience as high school algebra. The instructor can refer to these topics as necessary and encourage students to read them for review. Appendix D is different in that it aims to connect some key ideas in linear algebra and differential equations through a more sophisticated viewpoint: linear transformations of vector spaces. Some of the material there is appropriate for consideration following chapter 1, but it is perhaps more suited to discussion after the Laplace transform has been introduced. Finally, appendix E contains answers to nearly all of the odd-numbered exercises in the text.

Acknowledgments

We are grateful to our institutions for the time and support provided to work on this manuscript; to several anonymous reviewers whose comments have improved it; to our students for their feedback in classroom-testing of the text; and to all students and instructors who choose to use this book. We welcome all comments and suggestions for improvement, while taking full responsibility for any errors or omissions in the text.

Matt Boelkins/J. L. Goldberg/Merle Potter

Differential Equations with Linear Algebra

1

Essentials of linear algebra

1.1 Motivating problems

The subjects of differential equations and linear algebra are particularly important because each finds a wide range of applications in fundamental physical problems. We consider two situations that involve *systems of equations* to motivate our work in this chapter and much of the remainder of the text.

The pollution of bodies of water is an important issue for humankind. Environmental scientists are particularly interested in systems of rivers and lakes where they can study the flow of a given pollutant from one body of water to another. For example, there is great concern regarding the presence of a variety of pollutants in the Great Lakes (Lakes Michigan, Superior, Huron, Erie, and Ontario), including salt due to snow melt from highways. Due to the large number of possible ways for salt to enter and exit such a system, as well as the many lakes and rivers involved, this problem is mathematically complicated. But we may gain a feel for how one might proceed by considering a simple system of two tanks, say A and B, where there are independent inflows and outflows from each, as well as two pipes with opposite flows connecting the tanks as pictured in figure 1.1.

We will let x_1 denote the amount of salt (in grams) in A at time t (in minutes). Since water flows into and out of the tank, and each such flow carries salt, the amount of salt x_1 is changing as a function of time. We know from calculus that dx_1/dt measures the rate of change of salt in the tank with respect to time, and is measured in grams per minute. In this basic model, we can see that the rate of change of salt in the tank will be the difference between the net rate of salt flowing in and the net rate of salt flowing out.

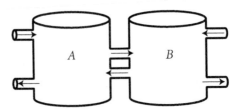

Figure 1.1 Two tanks with inflows, outflows, and connecting pipes.

As a simplifying assumption, we will suppose that the volume of solution in each tank remains constant and all inflows and outflows happen at the identical rate of 5 liters per minute. We will further assume that the tanks are uniformly mixed so that the salt concentration in each is identical throughout the tank at a given time t.

Let us now suppose that the volume of tank A is 200 liters; as we just noted, the pipe flowing into A delivers solution at a rate of 5 liters per minute. Moreover, suppose that this entering water is contaminated with 4 g of salt per liter. An analysis of the units on these quantities shows that the rate of inflow of salt into A is

$$\frac{5 \text{ liters}}{\text{min}} \cdot \frac{4 \text{ g}}{\text{liter}} = 20 \frac{\text{g}}{\text{min}} \tag{1.1.1}$$

There is one other inflow to consider, that being the pipe from B, which we will consider momentarily after first examining the behavior of the outflow.

For the solution exiting the drain from A at a rate of 5 liters/min, observe its concentration is unknown and depends on the amount of salt in the tank at time t. In particular, since there are x_1 g of salt in the tank at time t, and this is distributed over the volume of 200 liters, we can say (using the simplifying assumption that the tank's contents stay uniformly mixed) that the rate of outflow of salt in each of the exiting pipes is

$$\frac{5 \text{ liters}}{\text{min}} \cdot \frac{x_1 \text{ g}}{200 \text{ liters}} = \frac{x_1 \text{ g}}{40 \text{ min}} \tag{1.1.2}$$

Since there are two such exit flows, this means that the combined rate of outflow of salt from A is twice this amount, or $x_1/20$ g/min.

Finally, there is one last inflow to consider. Note that solution from B is entering A at a rate of 5 liters per minute. If we assume that B has a (constant) volume of 400 liters, this flow has a salt concentration of x_2 g/400 liters. Thus the rate of salt entering A from B is

$$\frac{5 \text{ liters}}{\text{min}} \cdot \frac{x_2 \text{ g}}{400 \text{ liters}} = \frac{x_2 \text{ g}}{80 \text{ min}} \tag{1.1.3}$$

Combining the rates of inflow (1.1.1) and (1.1.3) and outflow (1.1.2), where inflows are considered positive and outflows negative, leads us to the *differential equation*

$$\frac{dx_1}{dt} = 20 + \frac{x_2}{80} - \frac{x_1}{20} \tag{1.1.4}$$

Since we have two tanks in the system, there is a second differential equation to consider. Under the assumptions that B has a volume of 400 liters, the pipe entering B carries a concentration of salt of 7 g/liter, and the net rates of inflow and outflow match those into A, a similar analysis to the above reveals that

$$\frac{dx_2}{dt} = 35 + \frac{x_1}{40} - \frac{x_2}{40} \tag{1.1.5}$$

Together, these two DEs form a *system of DEs*, given by

$$\frac{dx_1}{dt} = 20 + \frac{x_2}{80} - \frac{x_1}{20} \tag{1.1.6}$$

$$\frac{dx_2}{dt} = 35 + \frac{x_1}{40} - \frac{x_2}{40}$$

Systems of DEs are therefore, seen to play a key role in environmental processes. Indeed, they find application in studying the vibrations of mechanical systems, the flow of electricity in circuits, the interactions between predators and prey, and much more. We will begin our examination of the mathematics involved with systems of differential equations in chapter 3.

An important question related to the above system of DEs leads us to a more familiar mathematical situation, one that is the foundation of much of the subject of linear algebra. For the system of tanks above, we might ask, "under what circumstances is the amount of salt in the two tanks not changing?" In such a situation, neither x_1 nor x_2 varies, so the rate of change of each is zero, and therefore

$$\frac{dx_1}{dt} = \frac{dx_2}{dt} = 0$$

Substituting these values into the system of DEs, we see that this results in the *system of linear equations*

$$0 = 20 + \frac{x_2}{80} - \frac{x_1}{20} \tag{1.1.7}$$

$$0 = 35 + \frac{x_1}{40} - \frac{x_2}{40}$$

Multiplying both sides of the first equation by eighty and the second by forty and rearranging terms, we find an equivalent system to be

$$4x_1 - x_2 = 1600$$

$$x_1 - x_2 = -1400$$

Geometrically, this system of linear equations represents the set of all points that simultaneously lie on each of the two lines given by the respective equations.

The solution of such 2 × 2 systems is typically discussed in introductory algebra classes where students learn how to solve systems like these with the methods of substitution and elimination. Doing so here leads to the unique solution $x_1 = 1000$, $x_2 = 2400$; one interpretation of this ordered pair is that the system of two tanks has an *equilibrium* state where, if the two tanks ever reach this level of salinity, that salinity will then stay constant. With further study of linear algebra and DEs, we will be able to show that over time, regardless of how much salt is initially in each tank, the amount of salt in A will approach 1000 g, while that in B will approach 2400 g. We will thus call the equilibrium point *stable.*

Electrical circuits are another physical situation where systems of linear equations naturally arise. Flow of electricity through a collection of wires is similar to the flow of water through a sequence of pipes: current measures the flow of electrons (charge carriers) past a given point in the circuit. Typically, we think about a battery as a source that provides a flow of electricity, wires as a collection of paths along which the electricity may flow, and resistors as places in the circuit where electricity is converted to some sort of output such as heat or light. While we will discuss the principles behind the flow of electricity in more detail in section 3.8, for now a basic understanding of Kirchoff's laws enables us to see an important application of linear systems of equations.

In a given loop or branch j of a circuit, current is measured in amperes (A) and is denoted by the symbol I_j. Resistances are measured in ohms (Ω), and the energy produced by the battery is measured in volts. As shown in figure 1.2, we use arrows in the circuit to represent the direction of flow of the current; when

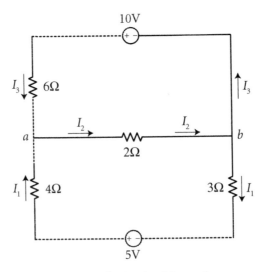

Figure 1.2 A simple circuit with two loops, two batteries, and four resistors.

this flow is away from the positive side of a battery (the circles in the diagram), then the voltage is taken to be positive. Otherwise, the voltage is negative.

Two fundamental laws govern how the currents in various loops of the circuit behave. One is Kirchoff's current law, which is essentially a conservation law. It states that the sum of all current flowing into a node equals the sum of the current flowing out. For example, in figure 1.2 at junction a,

$$I_1 + I_3 = I_2 \tag{1.1.8}$$

Similarly, at junction b, we must have $I_2 = I_1 + I_3$. This equation is identical to (1.1.8) and adds no new information about the currents.

Ohm's law governs the flow of electricity through resistors, and states that the voltage drop across a resistor is proportional to the current. That is, $V = IR$, where R is a constant that is the amount of resistance, measured in ohms. For instance, in the circuit given in figure 1.2, the voltage drop through the 3-Ω resistor on the bottom right is $V = 3\,\Omega$. Kirchoff's voltage law states that, in any closed loop, the sum of the voltage drops must be zero. Since the battery that is present maintains a constant voltage, it follows that in the bottom loop of the given circuit,

$$4I_1 + 2I_2 + 3I_1 = 5 \tag{1.1.9}$$

Similarly, in the upper loop, we have

$$6I_3 + 2I_2 = 10 \tag{1.1.10}$$

Finally, in the outer loop, taking into account the direction of flow of electricity by regarding opposing flows as having opposing signs, we observe

$$6I_3 - 4I_1 - 3I_1 = -5 + 10 \tag{1.1.11}$$

Taking (1.1.8) through (1.1.11), combining like terms, and rearranging each so that indices are in increasing order, we have the system of linear equations

$$\begin{array}{rcl}
I_1 - I_2 + I_3 &=& 0 \\
7I_1 + 2I_2 &=& 5 \\
2I_2 + 6I_3 &=& 10 \\
-7I_1 + 6I_3 &=& 5
\end{array} \tag{1.1.12}$$

We will call the system (1.1.12) a 4×3 system to represent the fact that it is a collection of four linear equations in three unknown variables. Its solution—the set of all possible values of (I_1, I_2, I_3) that make all four equations simultaneously true—provides the current in each loop of the circuit.

In this first chapter, we will develop our understanding of the more general situation of systems of linear equations with m linear equations in n unknown variables. This problem will lead us to consider important ideas from the theory of matrices that play key roles in a variety of applications ranging from computer graphics to population dynamics; related ideas will find further applications in our subsequent study of systems of differential equations.

1.2 Systems of linear equations

Linear equations are the simplest of all possible equations and are involved in many applications of mathematics. In addition, linear equations play a fundamental role in the study of differential equations. As such, the notion of *linearity* will be a theme throughout this book. Formally, a *linear equation* in variables $x_1, \ldots, x_n$ is one having the form

$$a_1 x_1 + a_2 x_2 + \cdots + a_n x_n = b \tag{1.2.1}$$

where the *coefficients* $a_1, \ldots, a_n$ and the value b are real or complex numbers. For example,

$$2x_1 + 3x_2 - 5x_3 = 7$$

is a linear equation, while

$$x_1^2 + \sin x_2 - x_3 \ln x_1 = 5$$

is not. Just as the equation $2x_1 + 3x_2 = 7$ describes a line in the x_1–x_2 plane, the linear equation $2x_1 + 3x_2 - 5x_3 = 7$ determines a plane in three-dimensional space.

A *system of m linear equations in n unknown variables* is a collection of m linear equations in n variables, say $x_1, \ldots, x_n$. We often refer to such a system as an "$m \times n$ system of equations." For example,

$$\begin{aligned} x_1 + 2x_2 + x_3 &= 1 \\ x_1 + x_2 + 2x_3 &= 0 \end{aligned} \tag{1.2.2}$$

is a system of two linear equations in three unknown variables. A *solution* to the system is any point (x_1, x_2, x_3) that makes both equations simultaneously true; the *solution set* for (1.2.2) is the collection of all such solutions. Geometrically, each of these two equations describes a plane in three-dimensional space, as shown in figure 1.3, and hence the solution set consists of all points that lie on both of the planes. Since the planes are not parallel, we expect this solution set to

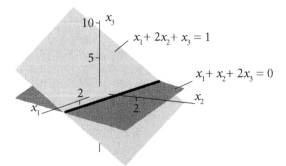

Figure 1.3 The intersection of the planes $x_1 + 2x_2 + x_3 = 1$ and $x_1 + x_2 + 2x_3 = 0$.

form a line in $\mathbb{R}^3$. Note that $\mathbb{R}$ denotes the set of all real numbers; $\mathbb{R}^3$ represents familiar three-dimensional Euclidean space, the set of all ordered triples with real entries.

The solution set for the system (1.2.2) may be determined using elementary algebraic steps. We say that two systems are *equivalent* if they share the same solution set. For example, if we multiply both sides of the first equation by -1 and add this to the second equation, we eliminate x_1 in the second equation and get the equivalent system

$$\begin{aligned} x_1 + 2x_2 + x_3 &= 1 \\ -x_2 + x_3 &= -1 \end{aligned}$$

Next, we multiply both sides of the second equation by -1 to get

$$\begin{aligned} x_1 + 2x_2 + x_3 &= 1 \\ x_2 - x_3 &= 1 \end{aligned}$$

Finally, if we multiply the second equation by -2 and add it to the first equation, it follows that

$$\begin{aligned} x_1 \quad\;\; + 3x_3 &= -1 \\ x_2 - \;\; x_3 &= 1 \end{aligned} \qquad (1.2.3)$$

This shows that any solution (x_1, x_2, x_3) of the original system must satisfy the (simpler) equivalent system of equations $x_1 = -1 - 3x_3$ and $x_2 = 1 + x_3$. Said differently, any point in $\mathbb{R}^3$ of the form $(-1 - 3x_3, 1 + x_3, x_3)$, where $x_3 \in \mathbb{R}$ (here the symbol '$\in$' means *is an element of*), is a solution to the system. Replacing x_3 by the parameter t, we recognize that the solution to the system is the line parameterized by

$$(-1 - 3t, 1 + t, t), \quad t \in \mathbb{R} \qquad (1.2.4)$$

which is the intersection of the two planes with which we began, as seen in figure 1.3. Note that this shows there are infinitely many solutions to the given system of equations; a particular example of such a solution may be found by selecting any value of t (i.e., any point on the line). We can also check that the resulting point makes both of the original equations true.

It is not hard to see in the 2×2 case that any linear system has either no solution (the lines are parallel), a unique solution (the lines intersect once), or infinitely many solutions (the two equations represent the same line). These three options (no solution, exactly one solution, or infinitely many) turn out to be the only possible cases for any $m \times n$ system of linear equations. A system with at least one solution is said to be *consistent*, while a system with no solution is called *inconsistent*.

In our work above from (1.2.2) to (1.2.3) in reducing the given system of equations to a simpler equivalent one, it is evident that the coefficients of the system played the key role, while the variables x_1, x_2, and x_3 (and the equals sign) were essentially placeholders. It proves expedient to therefore change notation and collect all of the coefficients into a rectangular array (called a *matrix*) and eliminate the redundancy of repeatedly writing the variables. Let us reconsider our above work in this light, where we will now refer to *rows* in the *coefficient matrix* rather than *equations* in the original system. When we create a right-most column consisting of the constants from the right-hand side of each equation, we often say we have an *augmented* matrix.

From the 'simplest' version of the system at (1.2.3), the corresponding augmented matrix is

$$\begin{bmatrix} 1 & 0 & 3 & -1 \\ 0 & 1 & -1 & 1 \end{bmatrix}$$

The 0's represent variables that have been eliminated in each equation. From this, we see that our goal in working with a matrix that represents a system of equations is essentially to introduce as many zeros as possible through operations that do not change the solution set of the system. We now repeat the exact same steps we took with the system above, but translate our operations to be on the matrix, rather than the equations themselves.

We begin with the augmented matrix

$$\begin{bmatrix} 1 & 2 & 1 & 1 \\ 1 & 1 & 2 & 0 \end{bmatrix}$$

To introduce a zero in the bottom left corner, we add -1 times the first row to the second row, to yield a new row 2 and the updated matrix

$$\begin{bmatrix} 1 & 2 & 1 & 1 \\ 0 & -1 & 1 & -1 \end{bmatrix}$$

The '0' in the second entry of the first column shows that we have eliminated the presence of the x_1 variable in the second equation. Next, we can multiply row 2 by -1 to obtain an updated row 2 and the augmented matrix

$$\begin{bmatrix} 1 & 2 & 1 & 1 \\ 0 & 1 & -1 & 1 \end{bmatrix}$$

Finally, if we multiply row 2 by -2 and add this to row 1, we find a new row 1 and the matrix

$$\begin{bmatrix} 1 & 0 & 3 & -1 \\ 0 & 1 & -1 & 1 \end{bmatrix}$$

At this point, we have introduced as many zeros as possible[1], and have arrived at our goal of the simplest possible equivalent system. We can reinterpret the matrix as a system of equations: the first row implies that $x_1 + 3x_3 = -1$, while the second row implies $x_2 - x_3 = 1$. This leads us to find, as we did above, that any solution (x_1, x_2, x_3) of the original system must be of the form $(-1 - 3x_3, 1 + x_3, x_3)$, where $x_3 \in \mathbb{R}$.

---

[1] Any additional row operations to introduce zeros in the third or fourth columns will replace the zeros in columns 1 or 2 with nonzero entries.

We will commonly need to refer to the number of rows and columns in a matrix. For example, the matrix

$$\begin{bmatrix} 1 & 0 & 3 & -1 \\ 0 & 1 & -1 & 1 \end{bmatrix}$$

has two rows and four columns; therefore, we say this is a 2×4 matrix. In general, an $m \times n$ matrix has m rows and n columns. Observe that if we have a 2×3 system of equations, its corresponding augmented matrix will be 2×4.

The above example demonstrates the general fact that there are basic operations we can perform on an augmented matrix that, at each stage, result in the matrix representing an equivalent system of equations; that is, these operations do not change the solution to the system, but rather make the solution more easily obtained. In particular, we may

1. *Replace* one row by the sum of itself and a multiple of another row;

2. *Interchange* any two rows; or

3. *Scale* a row by multiplying every entry in a given row by a fixed nonzero constant.

These three types of operations are typically called *elementary row operations.* Two matrices are *row equivalent* if there is a sequence of elementary row operations that transform one matrix into the other. When matrices are used to represent systems of linear equations, as was done above, it is always the case that row-equivalent matrices correspond to equivalent systems.

We desire to use elementary row operations systematically to produce row equivalent matrices from which we may easily interpret the solution to a system of equations. For example, the solution to the system represented by

$$\begin{bmatrix} 1 & 0 & 0 & -5 \\ 0 & 1 & 0 & 6 \\ 0 & 0 & 1 & -3 \end{bmatrix} \tag{1.2.5}$$

is easy to obtain (in particular, $x_1 = -5$, $x_2 = 6$, $x_3 = -3$), while the solution for

$$\begin{bmatrix} 3 & -2 & 4 & -39 \\ -1 & 2 & 7 & -4 \\ 6 & 9 & -3 & 33 \end{bmatrix}$$

is not, even though the two matrices are equivalent. Therefore, we desire each variable in the system to be represented in its corresponding augmented matrix as infrequently as possible. Essentially our goal is to get as many columns of the matrix as possible to have one entry that is 1, while all the rest of the entries in that column are 0.

A matrix is said to be in reduced row echelon form (RREF) if and only if the following characteristics are satisfied:

- All nonzero rows are above any rows with all zeros
- The first nonzero entry (or *leading entry*) in a given row is 1 and is in a column to the right of the first nonzero entry in any row above it
- Every other entry in a column with a leading 1 is 0

For example, the matrix in (1.2.5) is in RREF, while the matrix

$$\begin{bmatrix} 1 & -2 & 4 & -5 \\ 0 & 2 & 7 & 6 \\ 0 & 0 & -3 & -3 \end{bmatrix}$$

is not, since two of the rows lack leading 1's, and columns 2 and 3 lack zeros in the entries above the lowest nonzero locations.

Each leading 1 in RREF is said to be in a *pivot position*, the column in which the 1 lies is termed a *pivot column*, and the leading 1 itself is called a *pivot*. Rows with all zeros do not contain a pivot position. The process by which row operations are applied to a matrix to convert it to RREF is usually called *Gauss–Jordan elimination*. We will also say that we "row-reduced" a given matrix. While this process can be described in a somewhat cumbersome algorithm, it is best demonstrated with a few examples. By working through the details of the following problems (in particular by deciding which elementary row operations were performed at each stage), the reader will not only learn the basics of row reduction, but also will see and understand the key possibilities for the solution set of a system of linear equations.

Example 1.2.1   Solve the system of equations

$$\begin{aligned} 3x_1 + 2x_2 - x_3 &= 8 \\ x_1 - 4x_2 + 2x_3 &= -9 \\ -2x_1 + x_2 + x_3 &= -1 \end{aligned} \qquad (1.2.6)$$

Solution.   We begin with the corresponding augmented matrix

$$\begin{bmatrix} 3 & 2 & -1 & 8 \\ 1 & -4 & 2 & -9 \\ -2 & 1 & 1 & -1 \end{bmatrix}$$

and then perform a sequence of row operations. The arrows below denote the fact that one or more row operations have been performed to produce a row equivalent matrix. We find that

$$\begin{bmatrix} 3 & 2 & -1 & 8 \\ 1 & -4 & 2 & -9 \\ -2 & 1 & 1 & -1 \end{bmatrix} \rightarrow \begin{bmatrix} 1 & -4 & 2 & -9 \\ 3 & 2 & -1 & 8 \\ -2 & 1 & 1 & -1 \end{bmatrix} \rightarrow \begin{bmatrix} 1 & -4 & 2 & -9 \\ 0 & 14 & -7 & 35 \\ 0 & -7 & 5 & -19 \end{bmatrix} \rightarrow$$

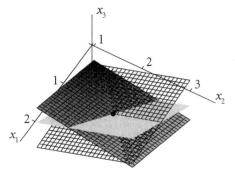

Figure 1.4 The intersection of the three planes given by the linear system (1.2.6).

$$\begin{bmatrix} 1 & -4 & 2 & -9 \\ 0 & 1 & -\frac{1}{2} & \frac{5}{2} \\ 0 & -7 & 5 & -19 \end{bmatrix} \rightarrow \begin{bmatrix} 1 & 0 & 0 & 1 \\ 0 & 1 & -\frac{1}{2} & \frac{5}{2} \\ 0 & 0 & \frac{3}{2} & -\frac{3}{2} \end{bmatrix} \rightarrow \begin{bmatrix} 1 & 0 & 0 & 1 \\ 0 & 1 & -\frac{1}{2} & \frac{5}{2} \\ 0 & 0 & 1 & -1 \end{bmatrix} \rightarrow$$

$$\begin{bmatrix} 1 & 0 & 0 & 1 \\ 0 & 1 & 0 & 2 \\ 0 & 0 & 1 & -1 \end{bmatrix}$$

This shows us that the original 3×3 system has a unique solution, and that this solution is the point $(1, 2, -1)$. Geometrically, this demonstrates that the three planes with equations given by the system (1.2.6) meet in a single point, as we can see in figure 1.4.

Example 1.2.2   Solve the system of equations

$$\begin{aligned} x_1 + 2x_2 - x_3 &= 1 \\ x_1 + x_2 \quad\quad &= 2 \\ 3x_1 + x_2 + 2x_3 &= 8 \end{aligned} \tag{1.2.7}$$

Solution.   We consider the corresponding augmented matrix

$$\begin{bmatrix} 1 & 2 & -1 & 1 \\ 1 & 1 & 0 & 2 \\ 3 & 1 & 2 & 8 \end{bmatrix}$$

and again perform a sequence of row operations:

$$\begin{bmatrix} 1 & 2 & -1 & 1 \\ 1 & 1 & 0 & 2 \\ 3 & 1 & 2 & 8 \end{bmatrix} \rightarrow \begin{bmatrix} 1 & 2 & -1 & 1 \\ 0 & -1 & 1 & 1 \\ 0 & -5 & 5 & 5 \end{bmatrix} \rightarrow \begin{bmatrix} 1 & 2 & -1 & 1 \\ 0 & 1 & -1 & -1 \\ 0 & -5 & 5 & 5 \end{bmatrix} \rightarrow \begin{bmatrix} 1 & 0 & 1 & 3 \\ 0 & 1 & -1 & -1 \\ 0 & 0 & 0 & 0 \end{bmatrix}$$

In this case, we see that one row of the matrix has essentially vanished. This shows that one of the equations in the original system was redundant, and did not contribute any restrictions on the system. Moreover, as the matrix is now in RREF, we can see that the simplest equivalent system is given by the two equations $x_1 + x_3 = 3$ and $x_2 - x_3 = -1$. In other words, $x_1 = 3 - x_3$ and $x_2 = -1 + x_3$. Since the variable x_3 has no restrictions on it, we call x_3 a *free variable*. This implies that the system under consideration has infinitely many solutions, each having the form

$$(3 - t, -1 + t, t), \text{ where } t \in \mathbb{R} \tag{1.2.8}$$

In the next section, we will begin to emphasize the role that vectors play in systems of linear equations. For example, the ordered triple $(3 - t, -1 + t, t)$ in (1.2.8) may be viewed as a vector in $\mathbb{R}^3$. In addition, the representation (1.2.8) of the set of all solutions involving the parameter t is often called the *parametric vector form* of the solution. As we saw in the very first system of equations discussed in this section, example 1.2.2 shows that the three planes given in the system (1.2.7) meet in a line.

Example 1.2.3   Solve the system of equations

$$\begin{aligned}
x_1 + 2x_2 - x_3 &= 1 \\
x_1 + x_2 &= 2 \\
3x_1 + x_2 + 2x_3 &= 7
\end{aligned}$$

Solution.   Observe that the only difference between this example and the previous one is that the "8" in the third equation has been replaced with "7." We proceed with identical row operations to those above and find that

$$\begin{bmatrix} 1 & 2 & -1 & 1 \\ 1 & 1 & 0 & 2 \\ 3 & 1 & 2 & 7 \end{bmatrix} \rightarrow \begin{bmatrix} 1 & 2 & -1 & 1 \\ 0 & -1 & 1 & 1 \\ 0 & -5 & 5 & 4 \end{bmatrix} \rightarrow \begin{bmatrix} 1 & 2 & -1 & 1 \\ 0 & 1 & -1 & -1 \\ 0 & -5 & 5 & 4 \end{bmatrix} \rightarrow \begin{bmatrix} 1 & 0 & 1 & 3 \\ 0 & 1 & -1 & -1 \\ 0 & 0 & 0 & -1 \end{bmatrix}$$

In this case, the final row of the reduced matrix corresponds to the equation $0x_1 + 0x_2 + 0x_3 = -1$. Since there are no points (x_1, x_2, x_3) that make this equation true, it follows that there can be no points which simultaneously satisfy all three equations in the system. Said differently, the three planes given in the original system of equations do not meet at a single point, nor do they meet in a line. Therefore, the system has no solution; recall that we call such a system *inconsistent*.

Note that the only difference between example 1.2.2 and example 1.2.3 is one constant in the righthand side in the equation of one of the planes. This changed the result dramatically, from the case where the system had infinitely many solutions to one where no solutions were present. This is evident geometrically if we think about a situation where three planes meet in a line, and then we alter the equation of one of the planes to shift it to a new plane parallel to its original location: the three planes will no longer have any points in common.

Algebraically, we can see what is so special about the one constant we changed (8 to 7) if we replace this value with an arbitrary constant, say k, and perform row operations:

$$\begin{bmatrix} 1 & 2 & -1 & 1 \\ 1 & 1 & 0 & 2 \\ 3 & 1 & 2 & k \end{bmatrix} \rightarrow \begin{bmatrix} 1 & 2 & -1 & 1 \\ 0 & 1 & -1 & -1 \\ 0 & -5 & 5 & k-3 \end{bmatrix} \rightarrow \begin{bmatrix} 1 & 0 & 1 & 3 \\ 0 & 1 & -1 & -1 \\ 0 & 0 & 0 & k-8 \end{bmatrix}$$

This shows that for any value of k other than 8, the resulting system of linear equations will be inconsistent, therefore having no solutions. In the case that $k = 8$, we see that a free variable arises and then the system has infinitely many solutions.

Overall, the question of consistency is an important one for any linear system of equations. In asking "is this system consistent?" we investigate whether or not the system has at least one solution. Moreover, we are now in a position to understand how RREF determines the answer to this question. We note from considering the RREF of a matrix that there are two overall cases: either the system contains an equation of the form $0x_1 + \cdots + 0x_n = b$, where b is nonzero, or it has no such equation. In the former case, the system is inconsistent and has no solution. In the latter case, it will either be that every variable is uniquely determined, or that there are one or more free variables present, in which case there are infinitely many solutions to the system. This leads us to state the following theorem.

Theorem 1.2.1   For any linear system of equations, there are only three possible cases for the solution set: there are no solutions, there is a unique solution, or there are infinitely many solutions.

This central fact regarding linear systems will play a key role in our studies.

1.2.1 Row-reduction using *Maple*

Obviously one of the problems with the process of row reducing a matrix is the potential for human arithmetic errors. Soon we will learn how to use computer software to execute all of these computations quickly; first, though, we can deepen our understanding of how the process works, and simultaneously eliminate arithmetic mistakes, by using a computer algebra system in a step-by-step fashion. Our software of choice is *Maple*. For now, we only assume that the user is familiar with *Maple*'s interface, and will introduce relevant commands with examples as we go.

We will use the LinearAlgebra package in *Maple*, which is loaded using the command

```
> with(LinearAlgebra):
```

(The symbol '>' is called a *Maple* prompt; the program makes this available to the user automatically, and it should not be entered by the user.) To demonstrate various commands, we will revisit the system from example 1.2.1. The reader should explore this code actively by entering and experimenting on his or her own. Recall that we were interested in row-reducing the augmented matrix

$$\begin{bmatrix} 3 & 2 & -1 & 8 \\ 1 & -4 & 2 & -9 \\ -2 & 1 & 1 & -1 \end{bmatrix}$$

We enter the augmented matrix, say **A**, column-wise in *Maple* with the command

```
> A := <<3,1,-2>|<2,-4,1>|<-1,2,1>|<8,-9,-1>>;
```

We first want to swap rows 1 and 2; this is accomplished by entering

```
> A1 := RowOperation(A,[1,2]);
```

Note that this stores the result of this row operation in the matrix **A1**, which is convenient for use in the next step. After executing the most recent command, the following matrix will appear on the screen:

$$\mathbf{A1} := \begin{bmatrix} 1 & -4 & 2 & -9 \\ 3 & 2 & -1 & 8 \\ -2 & 1 & 1 & -1 \end{bmatrix}$$

To perform row-replacement, our next step is to add $(-3) \cdot R_1$ to R_2 (where rows 1 and 2 are denoted R_1 and R_2) to generate a new second row; similarly, we will add $2 \cdot R_1$ to R_3 for an updated row 3. The commands that accomplish these steps are

```
> A2 := RowOperation(A1,[2,1],-3);
> A3 := RowOperation(A2,[3,1],2);
```

and lead to the following output:

$$\mathbf{A3} := \begin{bmatrix} 1 & -4 & 2 & -9 \\ 0 & 14 & -7 & 35 \\ 0 & -7 & 5 & -19 \end{bmatrix}$$

Next, we will scale row 2 by a factor of $1/14$ using the command

```
> A4 := RowOperation(A3,2,1/14);
```

to find that

$$\mathbf{A4} := \begin{bmatrix} 1 & -4 & 2 & -9 \\ 0 & 1 & -\frac{1}{2} & \frac{5}{2} \\ 0 & -7 & 5 & -19 \end{bmatrix}$$

The remainder of the computations in this example involve slightly modified versions of the three versions of the RowOperation command demonstrated above, and are left as an exercise for the reader. Recall that the unique solution to the original system is $(1, 2, -1)$.

Maple is certainly capable of performing all of these steps at once. After completing each step-by-step command above in the row-reduction process, the result can be checked by executing the command

```
> ReducedRowEchelonForm(A);
```

The corresponding output should be

$$\begin{bmatrix} 1 & 0 & 0 & 1 \\ 0 & 1 & 0 & 2 \\ 0 & 0 & 1 & -1 \end{bmatrix}$$

which clearly reveals the unique solution to the system, $(1, 2, -1)$.

Exercises 1.2   In exercises 1–4, solve each system of equations or explain why no solution exists.

1. $x_1 + 2x_2 = 1$
   $x_1 + x_2 = 0$

2. $\quad x_1 + 2x_2 = \quad 1$
   $-2x_1 - 4x_2 = -2$

3. $\quad x_1 + 2x_2 = \quad 1$
   $-2x_1 - 4x_2 = -3$

4. $4x_1 - 3x_2 = 5$
   $-x_1 + 4x_2 = 2$

In exercises 5–9, for each linear system represented by a given augmented matrix in RREF, decide whether or not the system is consistent or not. If the system is consistent, determine its solution set. For systems with infinitely many solutions, express the solution in parametric vector form.

5. $\begin{bmatrix} 1 & 0 & 0 & 4 \\ 0 & 1 & 0 & -2 \\ 0 & 0 & 1 & 3 \end{bmatrix}$

6. $\begin{bmatrix} 1 & 0 & 0 & 4 \\ 0 & 1 & 1 & -2 \\ 0 & 0 & 0 & 3 \end{bmatrix}$

7. $\begin{bmatrix} 1 & 0 & 2 & -3 \\ 0 & 1 & 1 & -2 \\ 0 & 0 & 0 & 0 \\ 0 & 0 & 0 & 0 \end{bmatrix}$

8. $\begin{bmatrix} 1 & 0 & 0 & -3 & 5 \\ 0 & 0 & 1 & -2 & 4 \\ 0 & 0 & 0 & 0 & 0 \end{bmatrix}$

9. $\begin{bmatrix} 1 & -2 & 0 & 4 & 0 & -1 \\ 0 & 0 & 1 & 3 & 0 & 2 \\ 0 & 0 & 0 & 0 & 1 & -5 \\ 0 & 0 & 0 & 0 & 0 & 0 \end{bmatrix}$

In exercises 10–14, the given augmented matrix represents a system for which some row operations have been performed to partially row-reduce the matrix. By deciding which operations must next be executed, finish row-reducing each matrix. Finally, interpret your results to state the solution set to the system.

10. $\begin{bmatrix} 1 & 3 & 2 & 5 \\ 0 & 1 & -4 & -1 \\ 0 & 0 & 1 & 7 \end{bmatrix}$

11. $\begin{bmatrix} 1 & 0 & 0 & 4 \\ 0 & 0 & 0 & 3 \\ 0 & 1 & 1 & -2 \end{bmatrix}$

12. $\begin{bmatrix} 1 & 0 & 2 & -3 \\ 0 & 1 & 1 & -2 \\ 0 & 3 & 3 & -6 \\ 0 & 2 & 2 & -1 \end{bmatrix}$

13. $\begin{bmatrix} 1 & 0 & 5 & -1 & 6 \\ 0 & 0 & 2 & -8 & 2 \\ 0 & 0 & 0 & 0 & 0 \end{bmatrix}$

14. $\begin{bmatrix} 1 & -3 & 0 & 5 & 0 & -3 \\ 0 & 0 & 1 & 3 & 0 & 4 \\ 0 & 0 & 0 & 1 & 2 & -9 \\ 0 & 0 & 0 & 0 & 1 & 4 \end{bmatrix}$

Determine all value(s) of h that make each augmented matrix in exercises 15–18 correspond to a consistent linear system. For such h, describe the solution set to the system.

15. $\begin{bmatrix} 1 & -2 & 7 \\ -3 & 6 & h \end{bmatrix}$

16. $\begin{bmatrix} 1 & -2 & 7 \\ -3 & h & -21 \end{bmatrix}$

17. $\begin{bmatrix} 1 & h & 3 \\ 2 & h & 6 \end{bmatrix}$

18. $\begin{bmatrix} 1 & 2 & 3 \\ -2 & h & 5 \end{bmatrix}$

Use a computer algebra system to perform step-by-step row operations to solve each of the following linear systems in exercises 19–23. If the system is consistent, determine its solution set. For systems with infinitely many solutions, express the solution in parametric vector form.

19. $\begin{aligned} x_1 - x_2 + x_3 &= 5 \\ 2x_1 - 4x_2 + 3x_3 &= 0 \\ x_1 - 6x_2 + 2x_3 &= 3 \end{aligned}$

20. $\begin{aligned} 4x_1 + 2x_2 - x_3 &= -2 \\ x_1 - x_2 + x_3 &= 6 \\ -3x_1 + x_2 - 4x_3 &= -20 \end{aligned}$

21. $\begin{aligned} 4x_1 + 2x_2 - x_3 &= -2 \\ x_1 - x_2 + x_3 &= 6 \\ -2x_1 - 4x_2 + 3x_3 &= 14 \end{aligned}$

22. $\begin{aligned} 4x_1 + 2x_2 - x_3 &= -2 \\ x_1 - x_2 + x_3 &= 6 \\ -2x_1 - 4x_2 + 3x_3 &= 13 \end{aligned}$

23. $\begin{aligned} 2x_2 + 3x_3 - 4x_4 &= 1 \\ 2x_3 + 3x_4 &= 4 \\ 2x_1 + 2x_2 - 5x_3 + 2x_4 &= 4 \\ 2x_1 \qquad - 6x_3 + 9x_4 &= 7 \end{aligned}$

In exercises 24–27, determine whether or not the given three lines or planes meet in a single point. Justify your answer using appropriate row operations.

24. $x_1 + x_2 = 5,\ 2x_1 - 3x_2 = -5,\ -4x_1 + 2x_2 = -2$

25. $x_1 + x_2 = 5,\ 2x_1 - 3x_2 = -5,\ -4x_1 + 2x_2 = -3$

26. $x_1 + x_2 + x_3 = 5,\ 2x_1 - 3x_2 + x_3 = 1,\ -4x_1 + 2x_2 + 5x_3 = 4$

27. $x_1 + x_2 + x_3 = 5,\ 2x_1 - 3x_2 + x_3 = 3,\ -4x_1 + 2x_2 + 5x_3 = 4$

28. Consider a linear system whose corresponding augmented matrix has all zeros in its final column. Is it ever possible for such a system to be inconsistent? Why or why not?

29. Is it possible for a 2×3 linear system to be inconsistent? Explain.

30. If a 3×4 linear system has three pivot columns in its corresponding augmented matrix, can you determine whether or not the system must be consistent? Explain.

31. A system of linear equations has a unique solution. What can be determined about the relationship between the number of pivot columns in the augmented matrix and the number of variables in the system?

32. Decide whether each of the following sentences is true or false. In every case, write one sentence to support your answer.

   (a) Two lines must either intersect or be parallel.
   (b) A system of three linear equations in three unknown variables can have exactly three solutions.
   (c) If the RREF of a matrix has a row of all zeros, then the corresponding system must have a free variable present.
   (d) If a system has a free variable present, then the system has infinitely many solutions.
   (e) A solution to a 4×3 linear system is a list of four numbers (x_1, x_2, x_3, x_4) that simultaneously makes every equation in the system true.
   (f) A matrix with three columns and four rows is 3×4.
   (g) A consistent system is one with exactly one solution.

33. Suppose that we would like to find a quadratic function $p(t) = a_2 t^2 + a_1 t + a_0$ that passes through the three points $(1, 4), (2, 7)$, and $(3, 6)$. How does this problem lead to a system of linear equations? Find the function $p(t)$. (Hint: $p(1) = 4$ implies that $4 = a_2 1^2 + a_1 1 + a_0$.)

34. Find a quadratic function $p(t) = a_2 t^2 + a_1 t + a_0$ that passes through the three points $(-1, 1), (2, -1)$, and $(5, 4)$. How does this problem involve a system of linear equations?

35. For the circuit shown at the left in figure 1.5, set up and solve a system of linear equations whose solution is the respective currents I_1, I_2, and I_3.

36. For the circuit shown at the right in figure 1.5, set up and solve a system of linear equations whose solution is the respective currents I_1, I_2, and I_3.

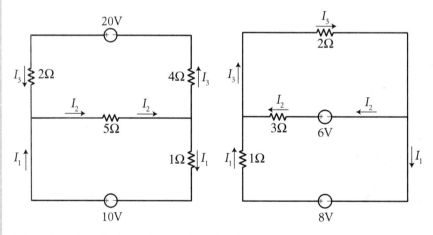

Figure 1.5   Circuits for use in exercises 35 and 36.

1.3 Linear combinations

An important theme in mathematics that is especially present in linear algebra is the value of considering the same idea from a variety of different perspectives. Often, we can make statements that on the surface may seem unrelated, when in fact they ultimately mean the same thing, and one of the statements is most advantageous for solving a particular problem. Throughout our study of linear algebra, we will see that the subject offers a wide variety of perspectives and terminology for addressing the central concept: systems of linear equations. In this section, we take another look at the concept of consistency, but do so in a different, geometric light.

Example 1.3.1   Consider the system of equations

$$\begin{aligned} x_1 - x_2 &= 1 \\ x_1 + x_2 &= 3 \\ x_1 + 2x_2 &= 4 \end{aligned} \qquad (1.3.1)$$

Rewrite the system in vector form and explore how two vectors are being combined to form a third, particularly in terms of the geometry of $\mathbb{R}^3$. Then solve the system.

Solution.   In multivariable calculus, we learn to think of vectors in $\mathbb{R}^3$ very much like we think of points. For example, given the point (a, b, c), we may write $\mathbf{v} = \langle a, b, c \rangle$ or $\mathbf{v} = a\mathbf{i} + b\mathbf{j} + c\mathbf{k}$ to denote the vector $\mathbf{v}$ that emanates from $(0, 0, 0)$ and ends at (a, b, c). (Here $\mathbf{i}$, $\mathbf{j}$, and $\mathbf{k}$ represent the standard unit coordinate vectors: $\mathbf{i}$ is the vector from $(0, 0, 0)$ to $(1, 0, 0)$, $\mathbf{j}$ to $(0, 1, 0)$, and $\mathbf{k}$ to $(0, 0, 1)$.)

In linear algebra, we will prefer to take the perspective of writing such an *ordered triple* as a matrix with only one column, also known as a *column vector*, in the form

$$\mathbf{v} = \begin{bmatrix} a \\ b \\ c \end{bmatrix} \qquad (1.3.2)$$

To save space, we will sometimes use the equivalent notation[2] $\mathbf{v} = [a \ \ b \ \ c]^{\mathrm{T}}$. Recall that two vectors are equal if and only if their corresponding entries are equal, that a vector may be multiplied by a scalar, and that any two vectors of the same size may be added.

---

[2] The 'T' stands for *transpose*, and the transpose of a matrix is achieved by turning every column into a row.

We can now re-examine the system of equations (1.3.1) in the light of equality among vectors. In particular, observe that it is equivalent to say

$$\begin{bmatrix} x_1 - x_2 \\ x_1 + x_2 \\ x_1 + 2x_2 \end{bmatrix} = \begin{bmatrix} 1 \\ 3 \\ 4 \end{bmatrix} \tag{1.3.3}$$

since two vectors are equal if and only if their corresponding entries are equal. Recalling further that vectors are added component-wise, we can rewrite (1.3.3) as

$$\begin{bmatrix} x_1 \\ x_1 \\ x_1 \end{bmatrix} + \begin{bmatrix} -x_2 \\ x_2 \\ 2x_2 \end{bmatrix} = \begin{bmatrix} 1 \\ 3 \\ 4 \end{bmatrix} \tag{1.3.4}$$

Finally, we observe in (1.3.4) that the first vector on the left-hand side has a common factor of x_1 in each component, and the second vector similarly contains x_2. Since a scalar multiple of a vector is computed component-wise, here we can rewrite the equation once more, now in the form

$$x_1 \begin{bmatrix} 1 \\ 1 \\ 1 \end{bmatrix} + x_2 \begin{bmatrix} -1 \\ 1 \\ 2 \end{bmatrix} = \begin{bmatrix} 1 \\ 3 \\ 4 \end{bmatrix} \tag{1.3.5}$$

Equation (1.3.5) is equivalent to the original system (1.3.1), but is now being viewed in a very different way. Specifically, this last equation asks if there are values of x_1 and x_2 for which

$$x_1 \mathbf{v}_1 + x_2 \mathbf{v}_2 = \mathbf{b}$$

where

$$\mathbf{v}_1 = \begin{bmatrix} 1 \\ 1 \\ 1 \end{bmatrix}, \quad \mathbf{v}_2 = \begin{bmatrix} -1 \\ 1 \\ 2 \end{bmatrix}, \quad \text{and } \mathbf{b} = \begin{bmatrix} 1 \\ 3 \\ 4 \end{bmatrix} \tag{1.3.6}$$

If we plot the vectors $\mathbf{v}_1$, $\mathbf{v}_2$, and $\mathbf{b}$, an interesting situation comes to light, as seen in figure 1.6. In particular, it appears as if all three vectors lie in the same plane. Moreover, if we think about the parallelogram law of vector addition and stretch the vector $\mathbf{v}_1$ by a factor of 2, we see the image in figure 1.7. This shows geometrically that it appears $\mathbf{b} = 2\mathbf{v}_1 + \mathbf{v}_2$; a quick check of the vector arithmetic confirms that this is in fact the case. In other words, the unique solution to the system (1.3.1) is $x_1 = 2$ and $x_2 = 1$.

Among the many important ideas in example 1.3.1, perhaps most significant is the way we were able to re-cast a problem about a system of linear equations as a question involving vectors. In particular, we saw that it was equivalent to ask if there exist constants x_1 and x_2 such that

$$x_1 \mathbf{v}_1 + x_2 \mathbf{v}_2 = \mathbf{b} \tag{1.3.7}$$

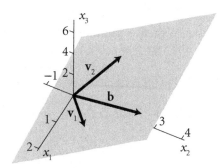

Figure 1.6 The vectors $\mathbf{v}_1, \mathbf{v}_2$, and $\mathbf{b}$ from (1.3.6).

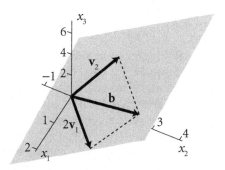

Figure 1.7 The parallelogram formed by the vectors $2\mathbf{v}_1$ and $\mathbf{v}_2$ from (1.3.6).

Note that in (1.3.7), we are only taking scalar multiples of vectors and adding them—computations that are *linear* in nature. We thus naturally come to use the terminology that "$x_1\mathbf{v}_1 + x_2\mathbf{v}_2$ is a *linear combination* of the vectors $\mathbf{v}_1$ and $\mathbf{v}_2$." A more general definition now follows, from which we will be able to widen our perspective on systems of linear equations.

Definition 1.3.1   If $\mathbf{v}_1, \ldots, \mathbf{v}_k$ are vectors in $\mathbb{R}^n$ (that is, each $\mathbf{v}_i$ is a vector with n entries), and $x_1, \ldots, x_k$ are scalars, then the vector $\mathbf{b}$ given by

$$\mathbf{b} = x_1\mathbf{v}_1 + \cdots + x_k\mathbf{v}_k \qquad (1.3.8)$$

is a *linear combination* of the vectors $\mathbf{v}_1, \ldots, \mathbf{v}_k$, with *weights* or *coefficients* $x_1, \ldots, x_k$.

Note the notational convention we use, as in example 1.3.1: a bold, non-italicized, lowercase variable, say $\mathbf{x}$, represents a vector, while a non-bold, italicized, lower-case variable, say c, denotes a scalar. A bold, non-italicized, uppercase variable, say $\mathbf{A}$, will represent a matrix with at least two columns.

In light of this new terminology of linear combinations, in example 1.3.1 we saw that the question "is there a solution to the linear system (1.3.1)?" is equivalent to asking "is the vector **b** a linear combination of the vectors $\mathbf{v}_1$ and $\mathbf{v}_2$?"

If we now consider the more general situation of a system of linear equations, say

$$a_{11}x_1 + a_{12}x_2 + \cdots + a_{1n}x_n = b_1$$

$$a_{21}x_1 + a_{22}x_2 + \cdots + a_{2n}x_n = b_2$$

$$\vdots$$

$$a_{m1}x_1 + a_{m2}x_2 + \cdots + a_{mn}x_n = b_m$$

it follows (as in section 1.2) that we can view this system in terms of the augmented matrix

$$[\mathbf{a}_1 \ \mathbf{a}_2 \ \cdots \ \mathbf{a}_n \ \mathbf{b}]$$

where $\mathbf{a}_1$ is the vector in $\mathbb{R}^m$ representing the first column of the augmented matrix, and so on. Now, however, we have the additional perspective, as in example 1.3.1, that the columns of the augmented matrix **A** are precisely the vectors being used to form a linear combination in an attempt to construct **b**. That is, the general $m \times n$ linear system above asks the question, "is **b** a linear combination of $\mathbf{a}_1, \ldots, \mathbf{a}_n$?"

We make the connection between linear combinations and augmented matrices more explicit by defining *matrix–vector multiplication* in terms of linear combinations.

Definition 1.3.2   Given an $m \times n$ matrix **A** with columns $\mathbf{a}_1, \ldots, \mathbf{a}_n$ that are vectors in $\mathbb{R}^m$, if **x** is a vector in $\mathbb{R}^n$, then we define the *product* **Ax** by the equation

$$\mathbf{Ax} = [\mathbf{a}_1 \ \mathbf{a}_2 \ \cdots \ \mathbf{a}_n] \begin{bmatrix} x_1 \\ x_2 \\ \vdots \\ x_n \end{bmatrix} = x_1\mathbf{a}_1 + x_2\mathbf{a}_2 + \cdots + x_n\mathbf{a}_n \qquad (1.3.9)$$

That is, the matrix–vector product of **A** and **x** is the vector **Ax** obtained by taking the linear combination of the column vectors of **A** according to the weights prescribed by the entries in **x**. Certainly we must have the same number of entries in **x** as columns in **A**, or **Ax** will not be defined. The following example highlights how to compute and interpret matrix–vector products.

Example 1.3.2   Let $\mathbf{a}_1 = [1 \ -4 \ 2]^{\mathrm{T}}$ and $\mathbf{a}_2 = [-3 \ 1 \ 5]^{\mathrm{T}}$, and let **A** be the matrix whose columns are $\mathbf{a}_1$ and $\mathbf{a}_2$. Compute **Ax**, where $\mathbf{x} = [-5 \ 2]^{\mathrm{T}}$, and interpret the result in terms of linear combinations.

Solution.   By definition, we have that

$$\mathbf{Ax} = \begin{bmatrix} 1 & -3 \\ -4 & 1 \\ 2 & 5 \end{bmatrix} \begin{bmatrix} -5 \\ 2 \end{bmatrix} = -5 \begin{bmatrix} 1 \\ -4 \\ 2 \end{bmatrix} + 2 \begin{bmatrix} -3 \\ 1 \\ 5 \end{bmatrix} = \begin{bmatrix} -11 \\ 22 \\ 0 \end{bmatrix}$$

The above computations show clearly that the vector $\mathbf{Ax} = [-11 \ \ 22 \ \ 0]^{\mathrm{T}}$ is a linear combination of $\mathbf{a}_1$ and $\mathbf{a}_2$.

Following a few more computational examples in homework exercises, the reader will quickly see how to compute the product $\mathbf{Ax}$ whenever it is defined; usually we skip past the intermediate stage of writing out the explicit linear combination of the columns and simply write the resulting vector. Matrix–vector multiplication also has several important general properties, some of which will be explored in the exercises. For now, we simply list these properties here for future reference: for any $m \times n$ matrix $\mathbf{A}$, vectors $\mathbf{x}, \mathbf{y} \in \mathbb{R}^n$, and $c \in \mathbb{R}$,

- $\mathbf{A}(\mathbf{x} + \mathbf{y}) = \mathbf{Ax} + \mathbf{Ay}$

- $\mathbf{A}(c\mathbf{x}) = c(\mathbf{Ax})$

The first property shows that matrix multiplication distributes over addition; the second demonstrates that a scalar multiple can be taken either before or after multiplying the vector $\mathbf{x}$ by $\mathbf{A}$. These two properties of matrix multiplication are often referred to as being *properties of linearity*—note the use of only scalar multiplication and vector addition in each, and the *linear* appearance of each equation.[3] Finally, note that it is also the case that $\mathbf{A0}_n = \mathbf{0}_m$, where $\mathbf{0}_n$ is the vector in $\mathbb{R}^n$ with all entries being zero, and $\mathbf{0}_m$ is the corresponding zero vector in $\mathbb{R}^m$.

There is one more important perspective that this new matrix–vector product notation permits. Recall that, in example 1.3.1, we learned that the question "is $\mathbf{b}$ a linear combination of $\mathbf{a}_1$ and $\mathbf{a}_2$?" is equivalent to asking "is there a solution to the system of linear equations whose augmented matrix has columns $\mathbf{a}_1$, $\mathbf{a}_2$, and $\mathbf{b}$?" Now, in light of matrix–vector multiplication, we also see that the question "is $\mathbf{b}$ a linear combination of $\mathbf{a}_1$ and $\mathbf{a}_2$?" may be rephrased as asking "does there exist a vector $\mathbf{x}$ such that $\mathbf{Ax} = \mathbf{b}$?" That is, are there weights x_1 and x_2 (the entries in vector $\mathbf{x}$) such that $\mathbf{b}$ is a linear combination of the columns of $\mathbf{A}$?

In particular, we may now adopt the perspective that we desire to solve the equation $\mathbf{Ax} = \mathbf{b}$ for the unknown vector $\mathbf{x}$, where $\mathbf{A}$ is a matrix whose entries are known, and $\mathbf{b}$ is a vector whose entries are known. This equation is strikingly similar to the most elementary of equations encountered in algebra, ones such as $2x = 7$. Therefore, we see that the linear equation $\mathbf{Ax} = \mathbf{b}$, involving matrices and vectors, is of fundamental importance as it is another way of expressing questions

---

[3] A deeper discussion of the notion of *linear transformations* can be found in appendix D.

regarding linear combinations and solutions of systems of linear equations. In subsequent sections, we will explore this equation from several perspectives.

1.3.1 Markov chains: an application of matrix–vector multiplication

People are often distributed naturally among various groupings. For example, much political discussion in the United States is centered on three classifications of voters: Democrat, Republican, and Independent. A similar situation can be considered with regard to peoples' choices for where to live: urban, suburban, or rural. In each case, the *state* of the population at a given time is its distribution among the relevant categories.

Furthermore, in each of these situations, it is natural to assume that if we consider the state of the system at a given point in time, its state depends on the system's state in the preceding year. For example, the percentage of Democrats, Republicans, and Independents in the year 2020 ought to be connected to the respective percentages in 2019.

Let us assume that a population of voters (of constant size) is considered in which every-one must classified as either D, R, or I (Democrat, Republican, or Independent). Suppose further that a study of voter registrations over many years reveals the following trends: from one year to the next, 95 percent of Democrats keep their registration the same. For the remaining 5 percent who change parties, 2 percent become Republicans and 3 percent become Independents. Similar data for Republicans and Independents is given in the following table.

Future party ($\downarrow$)/current party ($\rightarrow$)	D(%)	R(%)	I(%)
Democrat	95	3	7
Republican	2	90	13
Independent	3	7	80

If we let D_n, R_n, and I_n denote the respective numbers of registered Democrats, Republicans, and Independents in year n, then the table shows us how to determine the respective numbers in year $n+1$. For example,

$$D_{n+1} = 0.95D_n + 0.03R_n + 0.07I_n \qquad (1.3.10)$$

since 95 percent of the Democrats in year n stay registered Democrats, and 3 percent of Republicans and 7 percent of Independents change to Democrats. Similarly, we have

$$R_{n+1} = 0.02D_n + 0.90R_n + 0.13I_n \qquad (1.3.11)$$

$$I_{n+1} = 0.03D_n + 0.07R_n + 0.80I_n \qquad (1.3.12)$$

If we combine (1.3.10), (1.3.11), and (1.3.12) in a single vector equation, then

$$
\begin{bmatrix} D_{n+1} \\ R_{n+1} \\ I_{n+1} \end{bmatrix} = D_n \begin{bmatrix} 0.95 \\ 0.02 \\ 0.03 \end{bmatrix} + R_n \begin{bmatrix} 0.03 \\ 0.90 \\ 0.07 \end{bmatrix} + I_n \begin{bmatrix} 0.07 \\ 0.13 \\ 0.80 \end{bmatrix} \tag{1.3.13}
$$

Here we find that linear combinations of vectors have naturally arisen. Note, for example, that the vector $[0.03\ \ 0.90\ \ 0.07]^T$ is the Republican vector, and represents the likelihood that a Republican in a given year will be in one of the three parties in the following year. More specifically, we observe that probabilities are involved: a Republican has a 3 percent likelihood of registering as a Democrat in the following year, a 90 percent likelihood of staying a Republican, and 7 percent chance of becoming an Independent. The sum of the entries in each column vector is 1.

If we use the vector $\mathbf{x}^{(n)}$ to represent

$$
\mathbf{x}^{(n)} = \begin{bmatrix} D_n \\ R_n \\ I_n \end{bmatrix}
$$

and use matrix–vector multiplication to represent the linear combination of vectors in (1.3.13), then (1.3.13) is equivalently expressed by the equation

$$
\mathbf{x}^{(n+1)} = \mathbf{M}\mathbf{x}^{(n)} \tag{1.3.14}
$$

where $\mathbf{M}$ is the matrix

$$
\mathbf{M} = \begin{bmatrix} 0.95 & 0.03 & 0.07 \\ 0.02 & 0.90 & 0.13 \\ 0.03 & 0.07 & 0.80 \end{bmatrix}
$$

The matrix $\mathbf{M}$ is often called a *transition matrix* since it shows how the population transitions from state n to state $n+1$. We observe that in order for such a matrix to represent the probabilities that groups in a particular set of states will transition to another set of states, the columns of the matrix $\mathbf{M}$ must be non-negative and add to 1. Such a matrix is called a *stochastic matrix* or a *Markov matrix*. Finally, we call any system such as the one with three classifications of voters, where the state of the system in a given observation period results from applying probabilities to a previous state, a *Markov chain* or *Markov process*.

We see, for example, that if we had a group of 250 000 voters that at year $n = 0$ was distributed among Democrats, Republicans, and Independents by the vector (with entries measured in thousands) $\mathbf{x}^{(0)} = [120\ \ 110\ \ 20]^T$ then we can easily compute the projected distribution of voters in subsequent years. In particular, (1.3.14) implies

$$
\mathbf{x}^{(1)} = \mathbf{M}\mathbf{x}^{(0)} = \begin{bmatrix} 118.70 \\ 104 \\ 27.3 \end{bmatrix}, \ \mathbf{x}^{(2)} = \mathbf{M}\mathbf{x}^{(1)} = \begin{bmatrix} 117.80 \\ 99.52 \\ 32.68 \end{bmatrix}, \ \mathbf{x}^{(3)} = \mathbf{M}\mathbf{x}^{(2)} = \begin{bmatrix} 117.18 \\ 96.18 \\ 36.65 \end{bmatrix}
$$

Interestingly, if we continue the sequence, we eventually find that there is very little variation from one vector $\mathbf{x}^{(n)}$ to the next. For example,

$$\mathbf{x}^{(17)} = \begin{bmatrix} 116.67 \\ 85.95 \\ 47.42 \end{bmatrix} \approx \mathbf{x}^{(18)} = \begin{bmatrix} 116.79 \\ 85.76 \\ 47.44 \end{bmatrix}$$

In fact, as we will learn in our later study of eigenvectors, there exists a vector $\mathbf{x}^*$ called the *steady-state vector* for which $\mathbf{x}^* = \mathbf{M}\mathbf{x}^*$. This shows that the system can reach a state in which it does not change from one year to the next.

Another example is instructive.

Example 1.3.3   Geographers studying a metropolitan area have observed a trend that while the population of the area stays roughly constant, people within the city and its suburbs are migrating back and forth. In particular, suppose that 85 percent of people whose homes are in the city keep their residence from one year to the next; the remainder move to the suburbs. Likewise, while 92 percent of people whose homes are in suburbs will live there the next year, the other 8 percent will move into the city.

Assuming that in a given year there are 230 000 people living in the city and 270 000 people in the surrounding suburbs, predict the population distribution over the next 3 years.

Solution.   If we let C_n and S_n denote the populations of the city and suburbs in year n, the given information tells us that the following relationships hold:

$$C_{n+1} = 0.85C_n + 0.08S_n$$
$$S_{n+1} = 0.15C_n + 0.92S_n$$

Using the notation

$$\mathbf{x}^{(n)} = \begin{bmatrix} C_n \\ S_n \end{bmatrix}$$

we can model the changing distribution of the population between the city and suburbs with the Markov process $\mathbf{x}^{(n+1)} = \mathbf{M}\mathbf{x}^{(n)}$, where $\mathbf{M}$ is the Markov matrix

$$\mathbf{M} = \begin{bmatrix} 0.85 & 0.08 \\ 0.15 & 0.92 \end{bmatrix}$$

In particular, starting with $\mathbf{x}^{(0)} = [230 \ 270]^{\mathrm{T}}$, we see that

$$\mathbf{x}^{(1)} = \begin{bmatrix} 217.10 \\ 282.90 \end{bmatrix}, \quad \mathbf{x}^{(2)} = \begin{bmatrix} 207.17 \\ 292.83 \end{bmatrix}, \quad \mathbf{x}^{(3)} = \begin{bmatrix} 199.52 \\ 300.48 \end{bmatrix}$$

As with voter distribution, this example is oversimplified. For instance, we have not taken into account members of the population who move into or away from the metropolitan area. Nonetheless, the basic ideas of Markov processes are important in the study of systems whose current state depends on preceding ones, and we see the key role matrices and matrix multiplication play in representing them.

1.3.2 Matrix products using *Maple*

After becoming comfortable with computing elementary matrix products by hand, it is useful to see how *Maple* can assist us with more complicated computations. Here, we demonstrate the relevant command.

Revisiting example 1.3.2, to compute the product **Ax**, we first enter **A** and **x** using the familiar commands

```
> A := <<1, -4, 2>|<-3, 1, 5>>; x := <<-5,2>>;
```

Next, we use the 'period' symbol to inform *Maple* that we want to multiply. Entering

```
> b := A.x;
```

yields the expected output that

$$b = \begin{bmatrix} -11 \\ 22 \\ 0 \end{bmatrix}$$

Note: *Maple* will obviously only perform the multiplication when it is defined. If, say, we were to attempt to multiply a 2×2 matrix and a 3×1 vector, *Maple* would report the following:

```
Error, (in LinearAlgebra:-MatrixVectorMultiply)
vector dimension (3) must be the same as the
matrix column dimension (2).
```

Exercises 1.3   For exercises 1–4, where a matrix **A** and vector **x** are given, compute the product **Ax** in every case that it is defined. If the product is undefined, explain why.

1. $A = \begin{bmatrix} 1 & -3 & 2 \\ -4 & 1 & 0 \end{bmatrix}, \quad x = \begin{bmatrix} -1 \\ 2 \end{bmatrix}$

2. $A = \begin{bmatrix} 1 & -3 & 2 \\ -4 & 1 & 0 \end{bmatrix}, \quad x = \begin{bmatrix} -1 \\ 2 \\ 4 \end{bmatrix}$

3. $A = \begin{bmatrix} 5 & -2 \\ 1 & -1 \\ -3 & 2 \end{bmatrix}, \quad x = \begin{bmatrix} 3 \\ -2 \end{bmatrix}$

4. $A = \begin{bmatrix} -4 & 2 & 7 \end{bmatrix}, \quad x = \begin{bmatrix} 3 \\ 5 \\ -1 \end{bmatrix}$

5. Recall from multivariable calculus that given vectors $\mathbf{x}, \mathbf{y} \in \mathbb{R}^3$, the *dot product* of $\mathbf{x}$ and $\mathbf{y}$, $\mathbf{x} \cdot \mathbf{y}$, is computed by taking

$$\mathbf{x} \cdot \mathbf{y} = x_1 y_1 + x_2 y_2 + x_3 y_3$$

How can matrix–vector multiplication (when defined) be viewed as the result of computing several appropriate dot products? Explain.

6. For the system of equations given below, determine a vector equation with an equivalent solution. What is the system asking in regard to linear combinations of certain vectors?

$$x_1 + 2x_2 = 1$$
$$x_1 + x_2 = 0$$

In addition, determine a matrix $\mathbf{A}$ and vector $\mathbf{b}$ so that the equation $\mathbf{Ax} = \mathbf{b}$ is equivalent to the given system of equations.

7. For the system of differential equations (1.1.6) (also given below) from the introductory section, how can we rewrite the system in matrix–vector notation?

$$\frac{dx_1}{dt} = 20 - \frac{x_1}{20} + \frac{x_2}{80}$$

$$\frac{dx_2}{dt} = 35 + \frac{x_1}{40} - \frac{x_2}{40}$$

Hint: recall that if $\mathbf{x}(t)$ is a vector function, we write $\mathbf{x}'(t)$ or $d\mathbf{x}/dt$ for the vector $[dx_1/dt \ \ dx_2/dt]^{\mathrm{T}}$.

8. Determine if the vector $\mathbf{b} = [-3 \ 1 \ 5]^{\mathrm{T}}$ is a linear combination of the vectors $\mathbf{a}_1 = [-1 \ 2 \ 1]^{\mathrm{T}}$, $\mathbf{a}_2 = [3 \ 1 \ 1]^{\mathrm{T}}$, and $\mathbf{a}_3 = [1 \ 5 \ 3]^{\mathrm{T}}$. If so, will more than one set of weights work?

9. Determine if the vector $\mathbf{b} = [0 \ 7 \ 4]^{\mathrm{T}}$ is a linear combination of the vectors $\mathbf{a}_1 = [-1 \ 2 \ 1]^{\mathrm{T}}$, $\mathbf{a}_2 = [3 \ 1 \ 1]^{\mathrm{T}}$, and $\mathbf{a}_3 = [1 \ 5 \ 3]^{\mathrm{T}}$. If so, will more than one set of weights work?

10. We know from our work in this section that the matrix equation $\mathbf{Ax} = \mathbf{b}$ corresponds both to a vector equation and a system of linear equations. What is the augmented matrix that represents this system of equations?

In exercises 11–15, let $\mathbf{A}$ be the stated matrix and $\mathbf{b}$ the given vector. Solve the linear equation $\mathbf{Ax} = \mathbf{b}$ by converting the equation to a system of linear equations and row-reducing appropriately. If the system has more than one solution, express the solution in parametric vector form. Finally, write a sentence in each case that explains how the vector $\mathbf{b}$ is related to linear combinations of the columns of $\mathbf{A}$.

11. $\mathbf{A} = \begin{bmatrix} 4 & 5 & -1 \\ 3 & 1 & 2 \end{bmatrix}$, $\mathbf{b} = \begin{bmatrix} 13 \\ -4 \end{bmatrix}$

12. $A = \begin{bmatrix} 2 & 5 \\ -3 & -1 \end{bmatrix}$, $b = \begin{bmatrix} 5 \\ 6 \end{bmatrix}$

13. $A = \begin{bmatrix} 6 & 2 \\ -3 & -1 \end{bmatrix}$, $b = \begin{bmatrix} 7 \\ -1 \end{bmatrix}$

14. $A = \begin{bmatrix} 1 & -3 \\ -2 & 1 \\ 3 & -1 \end{bmatrix}$, $b = \begin{bmatrix} 5 \\ -5 \\ 5 \end{bmatrix}$

15. $A = \begin{bmatrix} 5 & -3 & 1 \\ -2 & 1 & 4 \\ 1 & 0 & -2 \end{bmatrix}$, $b = \begin{bmatrix} 0 \\ 22 \\ -11 \end{bmatrix}$

16. Linear equations of the form $Ax = 0$ are important for a variety of reasons, some of which we will study in the next section. Explain why the system of linear equations corresponding to the equation $Ax = 0$ is always consistent, regardless of the matrix A.

In exercises 17–21, solve the linear equation $Ax = 0$ by row-reducing appropriately. If the system has more than one solution, express the solution in parametric vector form.

17. $A = \begin{bmatrix} 4 & 5 & -1 \\ 3 & 1 & 2 \end{bmatrix}$

18. $A = \begin{bmatrix} 2 & 5 \\ -3 & -1 \end{bmatrix}$

19. $A = \begin{bmatrix} 6 & 2 \\ -3 & -1 \end{bmatrix}$

20. $A = \begin{bmatrix} 1 & -3 \\ -2 & 1 \\ 3 & -1 \end{bmatrix}$

21. $A = \begin{bmatrix} 5 & -3 & 1 \\ -2 & 1 & 4 \\ 1 & 0 & -2 \end{bmatrix}$

22. Let $A = \begin{bmatrix} 3 & -4 \\ -6 & 8 \end{bmatrix}$ and $b = \begin{bmatrix} b_1 \\ b_2 \end{bmatrix}$. Describe the set of all vectors b for which the equation $Ax = b$ is consistent.

23. Let $v_1 = \begin{bmatrix} 3 \\ -6 \end{bmatrix}$, $v_2 = \begin{bmatrix} -4 \\ 8 \end{bmatrix}$, and $b = \begin{bmatrix} b_1 \\ b_2 \end{bmatrix}$. Describe the set of all vectors b for which b is a linear combination of v_1 and v_2.

24. Let $\mathbf{A}$ be an $m \times n$ matrix, $\mathbf{x}$ and $\mathbf{y} \in \mathbb{R}^n$, and $c \in \mathbb{R}$. Show that

   (a) $\mathbf{A}(\mathbf{x} + \mathbf{y}) = \mathbf{A}\mathbf{x} + \mathbf{A}\mathbf{y}$
   (b) $\mathbf{A}(c\mathbf{x}) = c(\mathbf{A}\mathbf{x})$

25. Decide whether each of the following sentences is true or false. In every case, write one sentence to support your answer.

   (a) To compute the product $\mathbf{A}\mathbf{x}$, the vector $\mathbf{x}$ must have the same number of entries as the number of rows in $\mathbf{A}$.
   (b) A linear combination of three vectors in $\mathbb{R}^3$ produces another vector in $\mathbb{R}^3$.
   (c) If $\mathbf{b}$ is a linear combination of $\mathbf{v}_1$ and $\mathbf{v}_2$, then there exist scalars c_1 and c_2 such that $c_1\mathbf{v}_1 + c_2\mathbf{v}_2 = \mathbf{b}$.
   (d) If $\mathbf{A}$ is a matrix and $\mathbf{x}$ and $\mathbf{b}$ are vectors such that $\mathbf{A}\mathbf{x} = \mathbf{b}$, then $\mathbf{x}$ is a linear combination of the columns of $\mathbf{A}$.
   (e) The equation $\mathbf{A}\mathbf{x} = \mathbf{0}$ can be inconsistent.

26. Suppose that for a large population that stays relatively constant, people are classified as living in urban, suburban, or rural settings. Moreover, assume that the probabilities of the various possible transitions are given by the following table:

Future location ($\downarrow$)/current location ($\rightarrow$)	U(%)	S(%)	R(%)
Urban	92	3	2
Suburban	7	96	10
Rural	1	1	88

   Given that the population of 250 million is initially distributed in 100 million urban, 100 million suburban, and fifty million rural, predict the population distribution in each of the following five years.

27. Car-owners can be grouped into classes based on the vehicles they own. A study of owners of sedans, minivans, and sport utility vehicles shows that the likelihood that an owner of one of these automobiles will replace it with another of the same or different type is given by the table

Future vehicle ($\downarrow$)/ current vehicle ($\rightarrow$)	Sedan(%)	Minivan(%)	SUV(%)
Sedan	91	3	2
Minivan	7	95	8
SUV	2	2	90

If there are currently 100 000 sedans, 60 000 minivans, and 80 000 SUVs among the owners being studied, predict the distribution of vehicles among the population after each owner has replaced her vehicle 3 times.

1.4 The span of a set of vectors

In section 1.3, we saw that the question "is **b** a linear combination of $\mathbf{a}_1$ and $\mathbf{a}_2$?" provides an important new perspective on solutions of linear systems of equations. It is natural to slightly rephrase this question and ask more generally "*which* vectors **b** may be written as linear combinations of $\mathbf{a}_1$ and $\mathbf{a}_2$?" We explore this question further through the following sequence of examples.

Example 1.4.1   Describe the set of all vectors in $\mathbb{R}^2$ that may be written as a linear combination of the vector $\mathbf{a}_1 = [2 \ 1]^{\mathrm{T}}$.

Solution.   Since we have just one vector $\mathbf{a}_1$, any linear combination of $\mathbf{a}_1$ has the form $c\mathbf{a}_1$, which of course is a scalar multiple of $\mathbf{a}_1$. Geometrically, the vectors that are linear combinations of $\mathbf{a}_1$ are stretches of $\mathbf{a}_1$, which lie on the line through $(0, 0)$ in the direction of $\mathbf{a}_1$, as shown in figure 1.8.

In this first example, we see a visual way to interpret the question about linear combinations: essentially we want to know "which vectors can we create using only linear combinations of $\mathbf{a}_1$?" The answer is not surprising: only vectors that lie on the line through the origin in the direction of $\mathbf{a}_1$.

Next, we consider how the situation changes when we consider two parallel vectors.

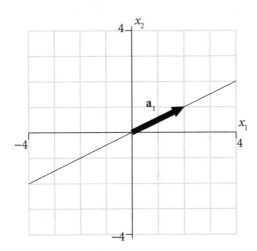

Figure 1.8 The set of all linear combinations of $\mathbf{a}_1$ in example 1.4.1.

Example 1.4.2   Describe the set of all vectors in $\mathbb{R}^2$ that may be written as a linear combination of the vectors $\mathbf{a}_1 = [2\ \ 1]^{\mathrm{T}}$ and $\mathbf{a}_2 = [-1\ \ -\frac{1}{2}]^{\mathrm{T}}$.

Solution.   Observe first that $-\frac{1}{2}\mathbf{a}_1 = \mathbf{a}_2$. Here we are considering the set of all vectors $\mathbf{y}$ of the form

$$\mathbf{y} = c_1 \begin{bmatrix} 2 \\ 1 \end{bmatrix} + c_2 \begin{bmatrix} -1 \\ -\frac{1}{2} \end{bmatrix}$$

In figure 1.9, we observe that the vectors $\mathbf{a}_1$ and $\mathbf{a}_2$ point in opposing directions. When we take a linear combination of these vectors to form $\mathbf{y}$, we are adding a stretch of c_1 units of the first to a stretch of c_2 units of the second. Because the two directions are parallel, this leaves the resulting vector as a stretch of one of the two original vectors, and therefore on the line through the origin in their direction. This may also be seen algebraically since $-\frac{1}{2}\mathbf{a}_1 = \mathbf{a}_2$ implies $\mathbf{y} = c_1\mathbf{a}_1 + c_2\mathbf{a}_2 = c_1\mathbf{a}_1 - \frac{1}{2}c_2\mathbf{a}_1 = (c_1 - \frac{1}{2}c_2)\mathbf{a}_1$.

We note particularly that since the two given vectors $\mathbf{a}_1$ and $\mathbf{a}_2$ are parallel, any linear combination of them is actually a scalar multiple of $\mathbf{a}_1$. Thus, the resulting set of all linear combinations is identical to what we found with the single vector given in example 1.4.1.

Finally, we consider the situation where we consider all linear combinations of two non-parallel vectors.

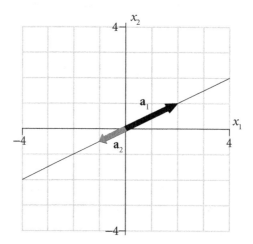

Figure 1.9 The set of all linear combinations of $\mathbf{a}_1$ and $\mathbf{a}_2$ in example 1.4.2.

Example 1.4.3   Describe the set of all vectors in $\mathbb{R}^2$ that may be written as a linear combination of the vectors $\mathbf{a}_1 = [2\ \ 1]^{\mathrm{T}}$ and $\mathbf{a}_2 = [1\ \ 2]^{\mathrm{T}}$.

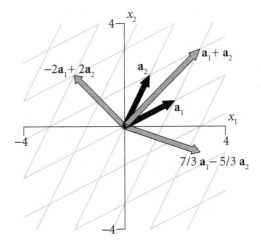

Figure 1.10 Linear combinations of a_1 and a_2 from example 1.4.3.

Solution.   Algebraically, we are again considering the set of all vectors y such that $y = c_1 a_1 + c_2 a_2$. A visual way to think about how the set of all such vectors y looks is found in the question, "which vectors can we create by taking a stretch of a_1 and adding this to a stretch of a_2?"

If we consider a plot of the given two vectors a_1 and a_2 and think of the "grid" that is formed by considering all of their stretches and the sums of their stretches, we have the picture shown in figure 1.10. The fact that a_1 and a_2 are not parallel enables us to "get off the line" that each one generates through the origin. For example, if we simply take the sum of these two vectors and set $y = a_1 + a_2$, by the parallelogram law of vector addition we arrive at the new vector $[3 \ 3]^T$ shown in figure 1.10. Two other linear combinations are shown as well, and from here it is not hard to visualize the fact that we can create any vector in the plane using linear combinations of the non-parallel vectors a_1 and a_2. In other words, the set of all linear combinations of a_1 and a_2 is $\mathbb{R}^2$.

It is also possible to verify our findings in example 1.4.3 algebraically. We will explore this further in the exercises and in section 1.5.

Certainly we are not limited to considering linear combinations of only two vectors. We therefore introduce a more formal perspective and terminology to describe the phenomena examined in the above examples.

Definition 1.4.1   Given a set of vectors $S = \{v_1, \ldots, v_k\}, v_i \in \mathbb{R}^m$, the *span* of S, denoted Span(S) or Span$\{v_1, \ldots, v_k\}$, is the set of all linear combinations of the vectors $v_1, \ldots, v_k$. Equivalently, Span(S) is the set of all vectors y of the form

$$y = c_1 v_1 + \cdots + c_k v_k$$

where $c_1, \ldots, c_k$ are scalars. We also say that Span(S) is *the subset of* $\mathbb{R}^m$ *spanned by the vectors* $\mathbf{v}_1, \ldots, \mathbf{v}_k$.

For any single nonzero vector $\mathbf{v}_1 \in \mathbb{R}^m$, Span$\{\mathbf{v}_1\}$ consists of all vectors that lie on the line through the origin in $\mathbb{R}^m$ in the direction of $\mathbf{v}_1$. For two non-parallel vectors $\mathbf{v}_1, \mathbf{v}_2 \in \mathbb{R}^m$, Span$\{\mathbf{v}_1, \mathbf{v}_2\}$ is the plane through the origin that contains both the vectors $\mathbf{v}_1$ and $\mathbf{v}_2$.

Next, let us recall that our interest in linear combinations was motivated by a desire to look at systems of linear equations from a new perspective. How is the concept of span related to linear systems? We begin to answer this question by considering the special situation where $\mathbf{b} = \mathbf{0}$.

A system of linear equations that can be represented in matrix form by the equation $\mathbf{Ax} = \mathbf{0}$ is said to be *homogeneous*; the case when $\mathbf{b} \neq \mathbf{0}$ is termed *nonhomogeneous*. We also call the equation $\mathbf{Ax} = \mathbf{0}$ a *homogeneous equation*. By the definition of matrix–vector multiplication, it is immediately clear that $\mathbf{A0} = \mathbf{0}$ (note that these two zero vectors may be of different sizes), and thus any homogeneous equation has at least one solution and is guaranteed to be consistent. We will usually call the solution $\mathbf{x} = \mathbf{0}$ the *trivial solution*. Under what circumstances will a homogeneous system have nontrivial solutions? How is this question related to the span of a set of vectors? The following example provides insight into these questions.

Example 1.4.4   Solve the homogeneous system of linear equations given by the equation $\mathbf{Ax} = \mathbf{0}$ where $\mathbf{A}$ is the matrix

$$\mathbf{A} = \begin{bmatrix} 1 & 1 & 1 & 1 \\ 2 & 1 & -1 & 3 \\ 1 & 0 & -2 & 2 \\ 8 & 5 & -1 & 11 \end{bmatrix}$$

If more than one solution exists, express the solution in parametric vector form.

Solution.   To begin, we augment the matrix $\mathbf{A}$ with a column of zeros to represent the vector $\mathbf{0}$ in the system given by $\mathbf{Ax} = \mathbf{0}$. We then row-reduce this augmented matrix to find

$$\begin{bmatrix} 1 & 1 & 1 & 1 & 0 \\ 2 & 1 & -1 & 3 & 0 \\ 1 & 0 & -2 & 2 & 0 \\ 8 & 5 & -1 & 11 & 0 \end{bmatrix} \rightarrow \begin{bmatrix} 1 & 0 & -2 & 2 & 0 \\ 0 & 1 & 3 & -1 & 0 \\ 0 & 0 & 0 & 0 & 0 \\ 0 & 0 & 0 & 0 & 0 \end{bmatrix}$$

We observe that the system has two free variables, and therefore infinitely many solutions. In particular, these solutions must satisfy the equations

$$x_1 - 2x_3 + 2x_4 = 0$$

$$x_2 + 3x_3 - x_4 = 0$$

where x_3 and x_4 are free. Equivalently, using these equations and vector addition and scalar multiplication, it must be the case that any solution $\mathbf{x}$ to $\mathbf{Ax} = \mathbf{0}$ has the form

$$\mathbf{x} = \begin{bmatrix} x_1 \\ x_2 \\ x_3 \\ x_4 \end{bmatrix} = \begin{bmatrix} 2x_3 - 2x_4 \\ -3x_3 + x_4 \\ x_3 \\ x_4 \end{bmatrix} = x_3 \begin{bmatrix} 2 \\ -3 \\ 1 \\ 0 \end{bmatrix} + x_4 \begin{bmatrix} -2 \\ 1 \\ 0 \\ 1 \end{bmatrix} \qquad (1.4.1)$$

where $x_3, x_4 \in \mathbb{R}$. Note particularly that this shows that every solution $\mathbf{x}$ to the original homogeneous equation $\mathbf{Ax} = \mathbf{0}$ can be expressed as a linear combination of the two vectors on the rightmost side of (1.4.1). Moreover, it is also the case that every linear combination of these two vectors is a solution to the equation. In light of the terminology of *span*, we can say that the set of all solutions to the homogeneous equation $\mathbf{Ax} = \mathbf{0}$ is Span$\{\mathbf{v}_1, \mathbf{v}_2\}$, where

$$\mathbf{v}_1 = \begin{bmatrix} 2 \\ -3 \\ 1 \\ 0 \end{bmatrix}, \quad \mathbf{v}_2 = \begin{bmatrix} -2 \\ 1 \\ 0 \\ 1 \end{bmatrix}$$

In this section, we have seen that the set of all linear combinations of a set of vectors can be interpreted geometrically, particularly in the case when we only have one or two vectors present, by thinking about lines and planes. In addition, the span of a set of vectors arises naturally in considering homogeneous equations in which infinitely many solutions are present. In that situation, the set of all solutions can be expressed as the span of a set of k vectors, where k is the number of free variables that arise in row-reducing the augmented matrix.

Exercises 1.4   In exercises 1–6, solve the homogeneous equation $\mathbf{Ax} = \mathbf{0}$, given the matrix $\mathbf{A}$. If infinitely many solutions exist, express the solution set as the span of the smallest possible set of vectors.

1. $\mathbf{A} = \begin{bmatrix} 1 & -3 & 2 \\ -4 & 1 & 0 \end{bmatrix}$

2. $\mathbf{A} = \begin{bmatrix} -4 & 2 \\ 1 & -3 \\ 6 & 5 \end{bmatrix}$

3. $\mathbf{A} = \begin{bmatrix} -5 & 8 \\ 10 & -16 \end{bmatrix}$

4. $\mathbf{A} = \begin{bmatrix} -4 & 2 \\ 2 & -1 \end{bmatrix}$

5. $A = \begin{bmatrix} 3 & 1 & -1 \\ 1 & 3 & 1 \\ -1 & 1 & 3 \end{bmatrix}$

6. $A = \begin{bmatrix} 1 & -1 & 2 \\ 4 & -2 & 6 \\ -7 & 3 & -10 \end{bmatrix}$

7. Let A be an $m \times n$ matrix where $n > m$. Is it possible that $Ax = 0$ has only the trivial solution? Explain why or why not.

8. Let A be an $m \times n$ matrix where $n \leq m$. Is it guaranteed that $Ax = 0$ will have only the trivial solution? Explain why or why not.

9. Determine if the vector $b = [11 \;\; -4]^T$ is in the span of the vectors $a_1 = [3 \;\; -2]^T$ and $a_2 = [-9 \;\; 6]^T$. Justify your answer carefully.

10. Determine if the vector $b = [-17 \;\; 31]^T$ is in the span of the vectors $a_1 = [1 \;\; 0]^T$ and $a_2 = [0 \;\; 1]^T$. What do you observe?

11. Determine if the vector $b = [9 \;\; 17 \;\; 11]^T$ is in the span of the vectors $a_1 = [-1 \;\; 2 \;\; 1]^T$, $a_2 = [3 \;\; 1 \;\; 1]^T$, and $a_3 = [1 \;\; 5 \;\; 3]^T$. Justify your answer.

12. Explain why the vector $b = [3 \;\; 2]^T$ does not lie in the span of the set S, where $S = \{v\}$ and $v = [1 \;\; 1]^T$.

13. Describe geometrically the set $W = \text{Span}\{v_1, v_2\}$, where $v_1 = [1 \;\; 1 \;\; 1]^T$ and $v_2 = [-3 \;\; 0 \;\; 2]^T$.

14. Can every vector $b \in \mathbb{R}^3$ be found in $W = \text{Span}\{v_1, v_2\}$, where $v_1 = [1 \;\; 1 \;\; 1]^T$ and $v_2 = [-3 \;\; 0 \;\; 2]^T$? If so, explain why. If not, find a vector not in W and justify your answer.

15. Show that every point (vector) that lies on the line with equation $2x_1 - 3x_2 = 0$ also lies in the set $W = \text{Span}\{v_1\}$, where $v_1 = [3 \;\; 2]^T$.

16. Show that every point (vector) that lies on the plane with equation $-x + y + z = 0$ also lies in the set $W = \text{Span}\{v_1, v_2\}$, where $v_1 = [1 \;\; -1 \;\; 2]^T$ and $v_2 = [2 \;\; 1 \;\; 1]^T$.

17. Decide whether each of the following sentences is true or false. In every case, write one sentence to support your answer.

   (a) The span of a single nonzero vector in $\mathbb{R}^2$ can be thought of as a line through the origin.
   (b) The span of any two nonzero vectors in $\mathbb{R}^3$ can be viewed as a plane through the origin in $\mathbb{R}^3$.
   (c) If $Ax = b$ holds true for a given matrix A and vectors x and b, then x lies in the span of the columns of A.

(d) It is possible for a homogeneous equation $\mathbf{Ax} = \mathbf{0}$ to be inconsistent.

(e) The number of free variables present in the solution to $\mathbf{Ax} = \mathbf{0}$ is the same as the number of pivot columns in the matrix $\mathbf{A}$.

1.5 Systems of linear equations revisited

From our initial work with row-reducing a system of linear equations to our recent discussions of linear combinations and span, we have seen already that there are several perspectives from which to view a system of linear equations. One is purely algebraic: "is there at least one ordered list $(x_1, \ldots, x_n)$ that makes every equation in a given system true?" Here we are viewing the system in the form

$$a_{11}x_1 + a_{12}x_2 + \cdots + a_{1n}x_n = b_1$$

$$a_{21}x_1 + a_{22}x_2 + \cdots + a_{2n}x_n = b_2$$

$$\vdots = \vdots$$

$$a_{m1}x_1 + a_{m2}x_2 + \cdots + a_{mn}x_n = b_m$$

In light of linear combinations, we can rephrase this question geometrically as "is the vector $\mathbf{b}$ a linear combination of the vectors $\mathbf{a}_1, \ldots, \mathbf{a}_n$?", where $\mathbf{a}_i$ is the ith column of the coefficient matrix of the system. From this standpoint, asking if the system has a solution can be thought of in terms of the question, "does the vector $\mathbf{b}$ belong to the span of the columns of $\mathbf{A}$?" Finally, through matrix multiplication, we can also express this system of equations in its simplest form: $\mathbf{Ax} = \mathbf{b}$. From all of this, we know that the question, "Does $\mathbf{Ax} = \mathbf{b}$ have at least one solution?" is one of fundamental importance.

We have also seen that in the special case of the homogeneous equation $\mathbf{Ax} = \mathbf{0}$, the answer to the above questions is always affirmative, since setting $\mathbf{x} = \mathbf{0}$ guarantees that we have at least one solution. In what follows, we further explore the nonhomogeneous case $\mathbf{Ax} = \mathbf{b}$, with particular emphasis on understanding characteristics of the matrix $\mathbf{A}$ that enable us to answer the questions in the preceding paragraph.

We begin by revisiting example 1.4.2 from a more algebraic perspective.

Example 1.5.1   For which vectors $\mathbf{b}$ is the equation $\mathbf{Ax} = \mathbf{b}$ consistent, if $\mathbf{A}$ is the matrix whose columns are the vectors $\mathbf{a}_1 = [2 \ 1]^{\mathrm{T}}$ and $\mathbf{a}_2 = [-1 \ -\frac{1}{2}]^{\mathrm{T}}$?

Solution.   By the definition of matrix multiplication, this question is equivalent to asking, "which vectors $\mathbf{b}$ are linear combinations of the columns of $\mathbf{A}$?" This question may be equivalently rephrased as "which vectors $\mathbf{b}$ are in the span of the columns of $\mathbf{A}$?" We have already answered this question from a geometric perspective in example 1.4.2, where we saw that since $\mathbf{a}_1$ and $\mathbf{a}_2$ are parallel, it follows that every vector in $\mathbb{R}^2$ that lies on the line through the origin in the direction of $\mathbf{a}_1$ can be written as a linear combination of the two vectors. Nonetheless, it is insightful to explore algebraically why this is the case.

Letting $\mathbf{b}$ be the vector whose entries are b_1 and b_2 and writing the equation $\mathbf{Ax} = \mathbf{b}$ in the form of an augmented matrix, we row-reduce and find that

$$\begin{bmatrix} 2 & -1 & b_1 \\ 1 & -\frac{1}{2} & b_2 \end{bmatrix} \rightarrow \begin{bmatrix} 1 & -\frac{1}{2} & b_2 \\ 0 & 0 & b_1 - 2b_2 \end{bmatrix}$$

The second row in the augmented matrix represents the equation

$$0x_1 + 0x_2 = b_1 - 2b_2$$

Observe that if $b_1 - 2b_2 \neq 0$, this equation cannot possibly be true, and therefore the system would be inconsistent. Said differently, the only way for $\mathbf{Ax} = \mathbf{b}$ to be consistent is for $b_1 - 2b_2 = 0$. That is, if $\mathbf{b}$ is a vector such that $b_1 = 2b_2$, or

$$\mathbf{b} = \begin{bmatrix} 2b_2 \\ b_2 \end{bmatrix}$$

then $\mathbf{Ax} = \mathbf{b}$ is consistent. This makes sense geometrically, since the span of the columns of $\mathbf{A}$ is all the stretches of the vector $\mathbf{a}_1 = [2 \ 1]^{\mathrm{T}}$.

An important lesson to take from example 1.5.1 is that the equation $\mathbf{Ax} = \mathbf{b}$ discussed there is not consistent for every choice of $\mathbf{b}$. In fact, the equation is only consistent for very limited choices of $\mathbf{b}$. For example, if $\mathbf{b} = [6 \ 3]^{\mathrm{T}}$, the equation is consistent, but if $\mathbf{b} = [6 \ k]^{\mathrm{T}}$ for any $k \neq 3$, the equation is inconsistent. Moreover, we should observe that for the matrix in this example, $\mathbf{A}$ does not have a pivot position in every row. This is what ultimately leads to the algebraic equation $0x_1 + 0x_2 = b_1 - 2b_2$, and the potential inconsistency of $\mathbf{Ax} = \mathbf{b}$.

At this point in our work, it is important that we begin to generalize our observations in order to apply them in new, but similar, circumstances. We again emphasize that it is a noteworthy characteristic of linear algebra that the discipline often offers great flexibility through the large number of ways to say the same thing; at times, one way of stating a fact can give more insight than others, and therefore it is important to be well versed in shifting among multiple perspectives. The following theorem is of the form "the following statements are equivalent"; this means that if any one of the statements is true, all the others are as well. Likewise, if any one statement is false, every statement in the theorem must be false.

This theorem formalizes our findings in the example above, and, in some sense, our work in the first several sections of the text.

Theorem 1.5.1   Let $\mathbf{A}$ be an $m \times n$ matrix and $\mathbf{b}$ a vector in $\mathbb{R}^m$ so that the equation $\mathbf{Ax} = \mathbf{b}$ represents a system of m linear equations in n unknown variables. The following statements are equivalent:

a. The equation $\mathbf{Ax} = \mathbf{b}$ is consistent

b. The vector $\mathbf{b}$ is a linear combination of the columns of $\mathbf{A}$

c. The vector **b** is in the span of the columns of **A**

d. When the augmented matrix [**A b**] is row-reduced, there are no rows where the first n entries are zero and the last entry is nonzero.

The following example demonstrates how we can use theorem 1.5.1 to answer questions about span and linear combinations.

Example 1.5.2   Does the vector $\mathbf{b} = [1 \;\; -7 \;\; -14]^{\mathrm{T}}$ belong to the span of the vectors $\mathbf{a}_1 = [1 \;\; 3 \;\; 4]^{\mathrm{T}}$, $\mathbf{a}_2 = [2 \;\; 1 \;\; -1]^{\mathrm{T}}$, and $\mathbf{a}_3 = [0 \;\; 5 \;\; 9]^{\mathrm{T}}$? Does the result change if we ask the same question about the vector $\mathbf{c} = [1 \;\; -7 \;\; -13]^{\mathrm{T}}$?

Solution.   By theorem 1.5.1, we know that it is equivalent to ask if the equation $\mathbf{Ax} = \mathbf{b}$ is consistent, where **b** is the given vector and **A** is the matrix whose columns are $\mathbf{a}_1$, $\mathbf{a}_2$, and $\mathbf{a}_3$. To answer that question, we consider the augmented matrix [**A** | **b**] and row-reduce:

$$\begin{bmatrix} 1 & 2 & 0 & 1 \\ 3 & 1 & 5 & -7 \\ 4 & -1 & 9 & -14 \end{bmatrix} \rightarrow \begin{bmatrix} 1 & 0 & 2 & -3 \\ 0 & 1 & -1 & 2 \\ 0 & 0 & 0 & 0 \end{bmatrix}$$

Because this system of equations is consistent, it follows that **b** is indeed a linear combination of the columns of **A** and therefore **b** lies in the span of $\mathbf{a}_1$, $\mathbf{a}_2$, and $\mathbf{a}_3$.

If we instead consider the vector **c** stated in the example and proceed similarly, row-reduction shows that

$$\begin{bmatrix} 1 & 2 & 0 & 1 \\ 3 & 1 & 5 & -7 \\ 4 & 1 & 9 & -13 \end{bmatrix} \rightarrow \begin{bmatrix} 1 & 0 & 2 & 0 \\ 0 & 1 & -1 & 0 \\ 0 & 0 & 0 & 1 \end{bmatrix}$$

which implies that the system is inconsistent and therefore **c** is not a linear combination of the columns of **A**, or equivalently, **c** does not lie in the span of $\mathbf{a}_1$, $\mathbf{a}_2$, and $\mathbf{a}_3$.

At this point, it is natural to think the situations in examples 1.5.1 and 1.5.2 are somewhat dissatisfying: sometimes $\mathbf{Ax} = \mathbf{b}$ is consistent, and sometimes not, all depending on our choice of **b**. A natural question to ask is, "are there matrices **A** for which $\mathbf{Ax} = \mathbf{b}$ is consistent for every choice of **b**?" With that question, we are certainly interested in the properties of the matrix **A** that make this situation occur. We next revisit example 1.4.3 and explore these issues further.

Example 1.5.3   For which vectors **b** is the equation $\mathbf{Ax} = \mathbf{b}$ consistent, if **A** is the matrix whose columns are the vectors $\mathbf{a}_1 = [2 \;\; 1]^{\mathrm{T}}$ and $\mathbf{a}_2 = [1 \;\; 2]^{\mathrm{T}}$?

Solution. Proceeding as in the previous example, we row reduce the augmented matrix form of the equation and find that

$$\begin{bmatrix} 2 & 1 & b_1 \\ 1 & 2 & b_2 \end{bmatrix} \rightarrow \begin{bmatrix} 1 & 0 & \frac{2}{3}b_1 - \frac{1}{3}b_2 \\ 0 & 1 & -\frac{1}{3}b_1 + \frac{2}{3}b_2 \end{bmatrix}$$

Algebraically, this shows that regardless of the entries we select for the vector **b**, we can always find a solution to the equation $\mathbf{Ax} = \mathbf{b}$. In particular, **x** is the vector in $\mathbb{R}^2$ whose components are $x_1 = \frac{2}{3}b_1 - \frac{1}{3}b_2$ and $x_2 = -\frac{1}{3}b_1 + \frac{2}{3}b_2$. Thus the equation $\mathbf{Ax} = \mathbf{b}$ is consistent for every **b** in $\mathbb{R}^2$. Note that this is not surprising, given our work in example 1.4.3, where we found that from a geometric perspective, every vector $\mathbf{b} \in \mathbb{R}^2$ could be written as a linear combination of $\mathbf{a}_1$ and $\mathbf{a}_2$. This example simply confirms that finding, but now from an algebraic point of view.

In terms of a key property of the matrix in example 1.5.3, we see that **A** has a pivot position in every row. In particular, there is no row in RREF(**A**) where we encounter all zeros, and thus it is impossible to ever encounter an equation of the form $0 = k$, where $k \neq 0$. This is, therefore, one property of the matrix **A** that guarantees consistency for every choice of **b**.

We generalize our findings in this example in the following theorem, which is similar to theorem 1.5.1, but now focuses solely the matrix **A** and no longer requires a vector **b** to be initially chosen.

Theorem 1.5.2   Let **A** be an $m \times n$ matrix. The following statements are equivalent:

   a. The equation $\mathbf{Ax} = \mathbf{b}$ is consistent for every $\mathbf{b} \in \mathbb{R}^m$

   b. Every vector $\mathbf{b} \in \mathbb{R}^m$ is a linear combination of the columns of **A**

   c. The span of the columns of **A** is $\mathbb{R}^m$

   d. **A** has a pivot position in every row. That is, when the matrix **A** is row-reduced, there are no rows of all zeros.

Our next example shows how we can apply theorem 1.5.2 to answer general questions about the span of a set of vectors and the consistency of related systems of equations.

Example 1.5.4   Does the vector $\mathbf{b} = [1 \;\; -7 \;\; -13]^{\mathrm{T}}$ belong to the span of the vectors $\mathbf{a}_1 = [1 \;\; 3 \;\; 4]^{\mathrm{T}}$, $\mathbf{a}_2 = [2 \;\; 1 \;\; -1]^{\mathrm{T}}$, and $\mathbf{a}_3 = [0 \;\; 5 \;\; 10]^{\mathrm{T}}$? Can every vector in $\mathbb{R}^3$ be found in the span of the vectors $\mathbf{a}_1$, $\mathbf{a}_2$, and $\mathbf{a}_3$?

Solution.   Just as in example 1.5.2, we know by theorem 1.5.1 that it is equivalent to ask if the equation $\mathbf{Ax} = \mathbf{b}$ is consistent, where **b** is the given vector and **A** is the matrix whose columns are $\mathbf{a}_1$, $\mathbf{a}_2$, and $\mathbf{a}_3$. We thus consider the augmented matrix [**A** | **b**] and row-reduce:

$$\begin{bmatrix} 1 & 2 & 0 & 1 \\ 3 & 1 & 5 & -7 \\ 4 & -1 & 10 & -13 \end{bmatrix} \rightarrow \begin{bmatrix} 1 & 0 & 0 & -5 \\ 0 & 1 & 0 & 3 \\ 0 & 0 & 1 & 1 \end{bmatrix}$$

Because this system of equations is consistent, it follows that **b** is indeed a linear combination of the columns of **A** and therefore **b** lies in the span of $\mathbf{a}_1$, $\mathbf{a}_2$, and $\mathbf{a}_3$. But by theorem 1.5.2 we can now make a much more general observation. Because we see that the coefficient matrix **A** has a pivot in every row, it follows that regardless of which vector **b** we choose in $\mathbb{R}^3$, we can write that vector as a linear combination of the columns of **A**. That is, the vectors $\mathbf{a}_1$, $\mathbf{a}_2$, and $\mathbf{a}_3$ span all of $\mathbb{R}^3$ and the equation $\mathbf{Ax} = \mathbf{b}$ will be consistent for every choice of **b**.

This example demonstrates that it is in some sense ideal if a matrix **A** has a pivot in every row. As we proceed with further study of linear algebra, we will focus more and more on properties of the coefficient matrix and their implications for related systems of equations. We conclude this section by examining a key link between homogeneous and nonhomogeneous equations in order to foreshadow an essential concept in our pending study of differential equations.

Example 1.5.5   Solve the nonhomogeneous system of linear equations given by the equation $\mathbf{Ax} = \mathbf{b}$ where **A** and **b** are

$$\mathbf{A} = \begin{bmatrix} 1 & 1 & 1 & 1 \\ 2 & 1 & -1 & 3 \\ 1 & 0 & -2 & 2 \\ 8 & 5 & -1 & 11 \end{bmatrix}, \quad \mathbf{b} = \begin{bmatrix} 1 \\ -8 \\ -9 \\ -22 \end{bmatrix}$$

If more than one solution exists, express the solution in parametric vector form.

Solution.   Note that the coefficient matrix **A** is identical to the one in example 1.4.4, so that here we are simply considering a related nonhomogeneous equation. We augment the matrix **A** with **b** and then row reduce to find

$$\begin{bmatrix} 1 & 1 & 1 & 1 & 1 \\ 2 & 1 & -1 & 3 & -8 \\ 1 & 0 & -2 & 2 & -9 \\ 8 & 5 & -1 & 11 & -22 \end{bmatrix} \rightarrow \begin{bmatrix} 1 & 0 & -2 & 2 & -9 \\ 0 & 1 & 3 & -1 & 10 \\ 0 & 0 & 0 & 0 & 0 \\ 0 & 0 & 0 & 0 & 0 \end{bmatrix}$$

As we found with the homogeneous equation, the system is consistent and has two free variables, and therefore infinitely many solutions. These solutions must satisfy the equations

$$x_1 = -9 + 2x_3 - 2x_4$$

$$x_2 = 10 - 3x_3 + x_4$$

where x_3 and x_4 are free. Equivalently, it must be the case that any solution $\mathbf{x}$ has the form

$$\mathbf{x} = \begin{bmatrix} x_1 \\ x_2 \\ x_3 \\ x_4 \end{bmatrix} = \begin{bmatrix} -9 + 2x_3 - 2x_4 \\ 10 - 3x_3 + x_4 \\ x_3 \\ x_4 \end{bmatrix} = \begin{bmatrix} -9 \\ 10 \\ 0 \\ 0 \end{bmatrix} + x_3 \begin{bmatrix} 2 \\ -3 \\ 1 \\ 0 \end{bmatrix} + x_4 \begin{bmatrix} -2 \\ 1 \\ 0 \\ 1 \end{bmatrix}$$

where $x_3, x_4 \in \mathbb{R}$. Observe that if we let $\mathbf{x}_p = [-9\ \ 10\ \ 0\ \ 0]^{\mathrm{T}}$ and let $\mathbf{x}_h$ be any vector of the form

$$\mathbf{x}_h = t \begin{bmatrix} 2 \\ -3 \\ 1 \\ 0 \end{bmatrix} + s \begin{bmatrix} -2 \\ 1 \\ 0 \\ 1 \end{bmatrix}$$

then any solution to the equation $\mathbf{Ax} = \mathbf{b}$ has the form $\mathbf{x} = \mathbf{x}_p + \mathbf{x}_h$. Moreover, it is now apparent that this vector $\mathbf{x}_h$ is the same general solution vector that we found for the corresponding homogeneous equation in example 1.4.4. In addition, it is straightforward to check that $\mathbf{Ax}_p = \mathbf{b}$. Thus, we see that the general solution to the nonhomogeneous equation contains the general solution to the corresponding homogeneous equation.

It appears from example 1.5.5 that if we have a solution, say $\mathbf{x}_p$, to a nonhomogeneous equation $\mathbf{Ax} = \mathbf{b}$, we may add any solution $\mathbf{x}_h$ to the homogeneous equation $\mathbf{Ax} = \mathbf{0}$ to $\mathbf{x}_p$ and still have a solution to $\mathbf{Ax} = \mathbf{b}$. To see why any vector of the form $\mathbf{x}_p + \mathbf{x}_h$ is a solution to $\mathbf{Ax} = \mathbf{b}$, let us assume that $\mathbf{x}_p$ is a solution to $\mathbf{Ax} = \mathbf{b}$, and $\mathbf{x}_h$ is a solution to $\mathbf{Ax} = \mathbf{0}$. We claim that $\mathbf{x} = \mathbf{x}_p + \mathbf{x}_h$ is also a solution to $\mathbf{Ax} = \mathbf{b}$. This holds since

$$\mathbf{Ax} = \mathbf{A}(\mathbf{x}_p + \mathbf{x}_h)$$
$$= \mathbf{Ax}_p + \mathbf{Ax}_h$$
$$= \mathbf{b} + \mathbf{0}$$
$$= \mathbf{b} \tag{1.5.1}$$

Clearly, this shows that the solution to the corresponding homogeneous equation plays a central role in the solution of nonhomogeneous equations. One observation we can make is that in the event we can find a single particular solution $\mathbf{x}_p$ to the nonhomogeneous equation, if the corresponding homogeneous equation has at least one free variable, then we know that there must be infinitely many solutions to the nonhomogeneous equation as well. We could even take the perspective that, in order to solve a nonhomogeneous equation, we simply need to do two things: find one particular solution to $\mathbf{Ax} = \mathbf{b}$, and then combine that particular solution with the general solution to the corresponding homogeneous equation $\mathbf{Ax} = \mathbf{0}$. While this is not so useful with systems of linear algebraic equations, it turns out that this approach of solving the homogeneous equation first is essential in the solution of differential equations.

The following example shows how the same structure is present in a class of differential equations that we will discuss in detail in section 2.3.

Example 1.5.6   Consider the differential equations $y' + 3y = 0$ and $y' + 3y = 6$. Compare and contrast the solutions to these two equations.

Solution.   The first equation, $y' + 3y = 0$, we will call a *homogeneous linear first-order differential equation*. Note that it asks a straightforward question: what function $y(t)$ is such that the function's derivative plus 3 times itself is the zero function? Said differently, we seek a function y such that $y' = -3y$. From our experience with exponential functions in calculus, we know that if $y = e^{-3t}$, then $y' = -3e^{-3t}$. The same is true for functions like $y = 2e^{-3t}$ and $y = -5e^{-3t}$; indeed, we see that for any constant C, the function $y = Ce^{-3t}$ satisfies the differential equation. (It also turns out that these are the only functions that satisfy the differential equation.)

If we next consider the related differential equation $y' + 3y = 6$ – one that we will call a *nonhomogeneous linear first-order differential equation*—we see that there is one obvious solution to the equation. In particular, if we let $y(t)$ be the constant function $y(t) = 2$, then $y'(t) = 0$ and this function clearly makes the differential equation true since $3 \times 2 = 6$.

Now, we should wonder if we have found all of the possible solutions to $y' + 3y = 6$. The answer is no: as we will see in section 2.3, it turns out that the general solution y to this differential equation is

$$y(t) = 2 + Ce^{-3t}$$

We can verify that this is the case by direct substitution. Note that $y' = -3Ce^{-3t}$ and therefore

$$y' + 3y = -3Ce^{-3t} + 3(2 + Ce^{-3t}) = -3Ce^{-3t} + 6 + 3Ce^{-3t} = 6$$

Observe the structure of this solution function: if we let $y_p = 2$, we have a particular solution to the nonhomogeneous equation. Further, letting $y_h = Ce^{-3t}$, this is the general solution to the related homogeneous equation. This demonstrates that the overall solution to the nonhomogeneous equation is

$$y = y_p + y_h = 2 + Ce^{-3t}$$

Exercises 1.5   For each of the following $m \times n$ matrices $\mathbf{A}$ in exercises 1–8, determine whether the equation $\mathbf{Ax} = \mathbf{b}$ is consistent for every choice of $\mathbf{b} \in \mathbb{R}^m$. If not, describe the set of all $\mathbf{b} \in \mathbb{R}^m$ for which the equation is consistent. In each case, explain your reasoning fully.

1. $\mathbf{A} = \begin{bmatrix} 4 & -1 \\ 1 & -4 \end{bmatrix}$

2. $\mathbf{A} = \begin{bmatrix} 4 & -1 \\ -12 & 3 \end{bmatrix}$

3. $A = \begin{bmatrix} 1 & 0 & 2 \\ 0 & 1 & -3 \end{bmatrix}$

4. $A = \begin{bmatrix} 2 & 1 \\ -1 & 3 \\ 4 & -2 \end{bmatrix}$

5. $A = \begin{bmatrix} 1 & 5 & -2 \\ 2 & -1 & 7 \\ -3 & 4 & -14 \end{bmatrix}$

6. $A = \begin{bmatrix} 1 & 5 & -2 \\ 2 & -1 & 7 \\ -3 & 4 & -13 \end{bmatrix}$

7. $A = \begin{bmatrix} 1 & 0 & 0 \\ 0 & 1 & 0 \\ 0 & 0 & 1 \\ 0 & 0 & 0 \end{bmatrix}$

8. $A = \begin{bmatrix} 1 & 0 & 0 & 2 \\ 0 & 1 & 0 & 5 \\ 0 & 0 & 1 & -3 \end{bmatrix}$

9. If A is an $m \times n$ matrix and $m > n$, is it possible for the equation $Ax = b$ to be consistent for every $b \in \mathbb{R}^m$? Explain.

10. If A is an $m \times n$ matrix and $m \le n$, is it guaranteed that the equation $Ax = b$ will be consistent for every $b \in \mathbb{R}^m$? Explain.

In each of exercises 11–16, determine whether the given vector b is in the span of the columns of the given matrix A. If b lies in the span of the columns of A, determine weights that enable you to explicitly write b as a linear combination of the columns of A.

11. $b = \begin{bmatrix} 2 \\ 5 \end{bmatrix}$, $A = \begin{bmatrix} 4 & -1 \\ 1 & -4 \end{bmatrix}$

12. $b = \begin{bmatrix} 6 \\ -20 \end{bmatrix}$, $A = \begin{bmatrix} 4 & -1 \\ -12 & 3 \end{bmatrix}$

13. $b = \begin{bmatrix} 6 \\ -2 \end{bmatrix}$, $A = \begin{bmatrix} 1 & 0 & 2 \\ 0 & 1 & -3 \end{bmatrix}$

14. $b = \begin{bmatrix} 1 \\ -11 \\ 14 \end{bmatrix}$, $A = \begin{bmatrix} 2 & 1 \\ -1 & 3 \\ 4 & -2 \end{bmatrix}$

15. $b = \begin{bmatrix} -4 \\ -2 \\ 1 \end{bmatrix}$, $A = \begin{bmatrix} 1 & 5 & -2 \\ 2 & -1 & 7 \\ -3 & 4 & -14 \end{bmatrix}$

16. $\mathbf{b} = \begin{bmatrix} -4 \\ -2 \\ 1 \end{bmatrix}$, $\quad \mathbf{A} = \begin{bmatrix} 1 & 5 & -2 \\ 2 & -1 & 7 \\ -3 & 4 & -13 \end{bmatrix}$

For each matrix $\mathbf{A}$ given in exercises 17–21, determine the general solution $\mathbf{x}_h$ to the homogeneous equation $\mathbf{Ax} = \mathbf{0}$.

17. $\mathbf{A} = \begin{bmatrix} 1 & -3 & 2 \\ -4 & 1 & 3 \end{bmatrix}$

18. $\mathbf{A} = \begin{bmatrix} 1 & 2 & 0 & 1 \\ 3 & 1 & 5 & -7 \\ 4 & -1 & 10 & -13 \end{bmatrix}$

19. $\mathbf{A} = \begin{bmatrix} -5 & 8 \\ 10 & -16 \end{bmatrix}$

20. $\mathbf{A} = \begin{bmatrix} 3 & 1 & -1 \\ 1 & 3 & 1 \\ -1 & 1 & 3 \end{bmatrix}$

21. $\mathbf{A} = \begin{bmatrix} 1 & -1 & 2 \\ 4 & -2 & 6 \\ -7 & 3 & -10 \end{bmatrix}$

In exercises 22–26, solve the nonhomogeneous equation $\mathbf{Ax} = \mathbf{b}$, given the matrix $\mathbf{A}$ and vector $\mathbf{b}$. Express your solution $\mathbf{x}$ (if one exists) in the form $\mathbf{x} = \mathbf{x}_p + \mathbf{x}_h$, where $\mathbf{x}_p$ is a particular solution to $\mathbf{Ax} = \mathbf{b}$ and $\mathbf{x}_h$ is the solution to the corresponding homogeneous equation $\mathbf{Ax} = \mathbf{0}$. Compare your results to exercises 17–21, respectively.

22. $\mathbf{A} = \begin{bmatrix} 1 & -3 & 2 \\ -4 & 1 & 3 \end{bmatrix}$, $\quad \mathbf{b} = \begin{bmatrix} 5 \\ -9 \end{bmatrix}$

23. $\mathbf{A} = \begin{bmatrix} 1 & 2 & 0 & 1 \\ 3 & 1 & 5 & -7 \\ 4 & -1 & 10 & -13 \end{bmatrix}$, $\quad \mathbf{b} = \begin{bmatrix} 1 \\ 3 \\ 5 \end{bmatrix}$

24. $\mathbf{A} = \begin{bmatrix} -5 & 8 \\ 10 & -16 \end{bmatrix}$, $\quad \mathbf{b} = \begin{bmatrix} -21 \\ 42 \end{bmatrix}$

25. $\mathbf{A} = \begin{bmatrix} 3 & 1 & -1 \\ 1 & 3 & 1 \\ -1 & 1 & 3 \end{bmatrix}$, $\quad \mathbf{b} = \begin{bmatrix} 3 \\ -1 \\ 1 \end{bmatrix}$

26. $\mathbf{A} = \begin{bmatrix} 1 & -1 & 2 \\ 4 & -2 & 6 \\ -7 & 3 & -10 \end{bmatrix}$, $\quad \mathbf{b} = \begin{bmatrix} 5 \\ 16 \\ -27 \end{bmatrix}$

27. Suppose that $\mathbf{A}$ is a 6×9 matrix that has a pivot in every row. What can you say about the consistency of $\mathbf{Ax} = \mathbf{b}$ for every $\mathbf{b} \in \mathbb{R}^6$? Why?

28. Suppose that $\mathbf{A}$ is a 3×4 matrix and that the span of the columns of $\mathbf{A}$ is $\mathbb{R}^3$. What can you say about the consistency of $\mathbf{Ax} = \mathbf{b}$ for every $\mathbf{b} \in \mathbb{R}^3$? Why?

29. If possible, give an example of a 3×2 matrix $\mathbf{A}$ such that the span of the columns of $\mathbf{A}$ is $\mathbb{R}^3$. If finding such a matrix is impossible, explain why.

30. Suppose that $\mathbf{A}$ is a 4×3 matrix for which the homogeneous equation $\mathbf{Ax} = \mathbf{0}$ has only the trivial solution. Will the equation $\mathbf{Ax} = \mathbf{b}$ be consistent for every $\mathbf{b} \in \mathbb{R}^4$? Explain. For the vectors $\mathbf{b}$ for which $\mathbf{Ax} = \mathbf{b}$ is indeed a consistent equation, how many solution vectors $\mathbf{x}$ does each equation have? Why?

31. Suppose that $\mathbf{A}$ is a 3×4 matrix for which the homogeneous equation $\mathbf{Ax} = \mathbf{0}$ has exactly one free variable present. Will the equation $\mathbf{Ax} = \mathbf{b}$ be consistent for every $\mathbf{b} \in \mathbb{R}^3$? Explain. For the vectors $\mathbf{b}$ for which $\mathbf{Ax} = \mathbf{b}$ is indeed a consistent equation, how many solution vectors $\mathbf{x}$ does each equation have? Why?

32. Suppose that $\mathbf{A}$ is a 4×5 matrix for which the homogeneous equation $\mathbf{Ax} = \mathbf{0}$ has exactly two free variables present. Will the equation $\mathbf{Ax} = \mathbf{b}$ be consistent for every $\mathbf{b} \in \mathbb{R}^4$? Explain. For the vectors $\mathbf{b}$ for which $\mathbf{Ax} = \mathbf{b}$ is indeed a consistent equation, how many solution vectors $\mathbf{x}$ does each equation have? Why?

33. Decide whether each of the following sentences is true or false. In every case, write one sentence to support your answer.

    (a) If $\mathbf{Ax} = \mathbf{b}$ is consistent for at least one vector $\mathbf{b}$, then $\mathbf{A}$ has a pivot in every row.
    (b) If $\mathbf{A}$ is a 4×3 matrix, then it is possible for the columns of $\mathbf{A}$ to span $\mathbb{R}^4$.
    (c) If $\mathbf{A}$ is a 3×3 matrix with exactly two pivot columns, then the columns of $\mathbf{A}$ do not span $\mathbb{R}^3$.
    (d) If $\mathbf{A}$ is a 3×4 matrix, then the columns of $\mathbf{A}$ must span $\mathbb{R}^3$.
    (e) If $\mathbf{y}$ and $\mathbf{z}$ are solutions to the equation $\mathbf{Ax} = \mathbf{0}$, then the vector $\mathbf{y} + \mathbf{z}$ is also a solution to $\mathbf{Ax} = \mathbf{0}$.
    (f) If $\mathbf{y}$ and $\mathbf{z}$ are solutions to the equation $\mathbf{Ax} = \mathbf{b}$, where $\mathbf{b} \neq \mathbf{0}$, then the vector $\mathbf{y} + \mathbf{z}$ is also a solution to $\mathbf{Ax} = \mathbf{b}$.

34. Solve the linear first-order differential equation $y' + y = 3$ by first finding all functions y_h that satisfy the homogeneous equation $y' + y = 0$ and then determining a constant function y_p that is a solution to $y' + y = 3$. Verify by direct substitution that $y = y_h + y_p$ is a solution to the given equation.

35. Solve the linear first-order differential equation $y' - 5y = 6$ by first finding all functions y_h that satisfy the homogeneous equation $y' - 5y = 0$ and

then determining a constant function y_p that is a solution to $y' - 5y = 6$. Verify by direct substitution that $y = y_h + y_p$ is a solution to the given equation.

1.6 Linear independence

In theorem 1.5.2, we found that when solving $\mathbf{Ax} = \mathbf{b}$, an ideal situation occurs when $\mathbf{A}$ has a pivot position in every row. Equivalently, this means that the equation $\mathbf{Ax} = \mathbf{b}$ is guaranteed to have at least one solution for every vector $\mathbf{b} \in \mathbb{R}^m$ (when $\mathbf{A}$ is $m \times n$), or that every $\mathbf{b} \in \mathbb{R}^m$ can be written as a linear combination of the columns of $\mathbf{A}$. In other words, regardless of the choice of $\mathbf{b}$, the equation $\mathbf{Ax} = \mathbf{b}$ is always consistent. Because the equation is consistent, we are guaranteed that at least one solution $\mathbf{x}$ exists. In what follows, we explore conditions that imply not only that at least one solution exists, but in fact that only one solution exists. First, we consider the simpler situation of homogeneous equations.

In section 1.4, we discovered that the equation $\mathbf{Ax} = \mathbf{0}$ is always consistent. Because $\mathbf{x} = \mathbf{0}$ always makes this equation true, we know that we at least have the trivial solution present. It is natural to ask: under what conditions on $\mathbf{A}$ is the trivial solution the only solution to the homogeneous equation $\mathbf{Ax} = \mathbf{0}$? Geometrically, we are asking whether or not a nontrivial linear combination of the columns of $\mathbf{A}$ can be formed that leads to the zero vector.

We revisit an earlier example to further explore these issues.

Example 1.6.1   Does the equation $\mathbf{Ax} = \mathbf{0}$ have nontrivial solutions if $\mathbf{A}$ is the matrix whose columns are $\mathbf{a}_1 = [2 \ \ 1]^T$ and $\mathbf{a}_2 = [-1 \ \ -\frac{1}{2}]^T$? Discuss the geometric implications of your conclusions.

Solution.   We first consider the corresponding augmented matrix and row reduce, finding that

$$\begin{bmatrix} 2 & -1 & 0 \\ 1 & -\frac{1}{2} & 0 \end{bmatrix} \rightarrow \begin{bmatrix} 1 & -\frac{1}{2} & 0 \\ 0 & 0 & 0 \end{bmatrix}$$

This shows that any vector $\mathbf{x} = [x_1 \ \ x_2]^T$ that satisfies $x_1 = \frac{1}{2}x_2$ will be a solution to $\mathbf{Ax} = \mathbf{0}$. The presence of the free variable x_2 implies that there are infinitely many nontrivial solutions to this equation.

If we interpret the matrix–vector product $\mathbf{Ax}$ as the linear combination $\mathbf{Ax} = x_1\mathbf{a}_1 + x_2\mathbf{a}_2$, then the equation

$$\frac{1}{2}x_2\mathbf{a}_1 + x_2\mathbf{a}_2 = \mathbf{0}$$

implies geometrically that the zero vector (on the right) may be expressed as a nontrivial linear combination of $\mathbf{a}_1$ and $\mathbf{a}_2$. For example, $\mathbf{a}_1 + 2\mathbf{a}_2 = \mathbf{0}$.

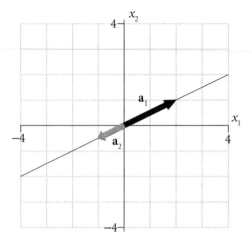

Figure 1.11 Linear combinations of a_1 and a_2 from example 1.6.1.

Indeed, if we consider figure 1.11 this conclusion is evident: if we add one length of $\mathbf{a}_1$ to two lengths of $\mathbf{a}_2$, we end up at $\mathbf{0}$.

Another way to express the equation $\mathbf{a}_1 + 2\mathbf{a}_2 = \mathbf{0}$ is to write $\mathbf{a}_1 = -2\mathbf{a}_2$. In this setting, we can see that $\mathbf{a}_1$ depends on $\mathbf{a}_2$, and that the relationship is given by a linear equation. We hence say that $\mathbf{a}_1$ and $\mathbf{a}_2$ are *linearly dependent* vectors.

The situation in example 1.6.1, where the vectors $\mathbf{a}_1$ and $\mathbf{a}_2$ are parallel is in contrast to that of example 1.4.3, where we instead considered the non-parallel vectors $\mathbf{a}_1 = [2 \ 1]^T$ and $\mathbf{a}_2 = [1 \ 2]^T$; in that setting, if we solve the associated homogeneous equation $\mathbf{Ax} = \mathbf{0}$, we find that

$$\begin{bmatrix} 2 & 1 & 0 \\ 1 & 2 & 0 \end{bmatrix} \rightarrow \begin{bmatrix} 1 & 0 & 0 \\ 0 & 1 & 0 \end{bmatrix}$$

In this case, the only solution to $\mathbf{Ax} = \mathbf{0}$ is the trivial solution, $\mathbf{x} = \mathbf{0}$. The geometry of the situation also informs us: if we desire a linear combination of the vectors $\mathbf{a}_1$ and $\mathbf{a}_2$ (as shown in figure 1.12) that results in the zero vector, we see that the only way to accomplish this is to take $0\mathbf{a}_1 + 0\mathbf{a}_2$. Said differently, if we take any nontrivial linear combination $c_1\mathbf{a}_1 + c_2\mathbf{a}_2$, we end up at a location other than the origin.

When $\mathbf{a}_1$ and $\mathbf{a}_2$ in example 1.6.1 were parallel, we said that $\mathbf{a}_1$ and $\mathbf{a}_2$ were linearly dependent. In the current context, where $\mathbf{a}_1$ and $\mathbf{a}_2$ are not parallel, it makes sense to say that $\mathbf{a}_1$ and $\mathbf{a}_2$ are *linearly independent*, since neither depends on the other.

Of course, in linear algebra we often consider sets of more than two vectors. The next definition formalizes what the terms *linearly dependent* and *linearly independent* mean in a more general context. Observe that the key criterion is

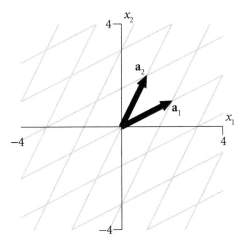

Figure 1.12 Linear combinations of a_1 and a_2 from example 1.4.3.

a geometric one: can we form a nontrivial linear combination of vectors that results in **0**?

Definition 1.6.1   Given a set $S = \{v_1, \ldots, v_k\}$ where each vector $v_i \in \mathbb{R}^m$, the set S is *linearly dependent* if there exists a nontrivial solution x to the vector equation

$$x_1 v_1 + x_2 v_2 + \cdots + x_k v_k = 0 \qquad (1.6.1)$$

If (1.6.1) has only the trivial solution, then we say the set S is *linearly independent*.

Note that (1.6.1) also takes us back to the fundamental questions about any linear system of equations: "does at least one solution exist?" (Yes; the zero vector is always a solution.) And "is that solution unique?" (Maybe; only if the vectors are linearly independent and the zero vector is the only solution.) The latter question addresses the fundamental issue of linear independence. We consider an example to demonstrate how we interpret the language of this most recent definition as well as how we will generally respond to the question of whether or not a set of vectors is linearly independent.

Example 1.6.2   Determine whether the set $S = \{v_1, v_2, v_3\}$ is linearly independent or linearly dependent if

$$v_1 = \begin{bmatrix} 1 \\ 1 \\ -1 \end{bmatrix}, \quad v_2 = \begin{bmatrix} -1 \\ 0 \\ 1 \end{bmatrix}, \quad v_3 = \begin{bmatrix} 0 \\ 1 \\ 1 \end{bmatrix}$$

Solution.   By definition, the linear independence of the set S rests on whether or not nontrivial solutions exist to the vector equation $x_1\mathbf{v}_1 + x_2\mathbf{v}_2 + x_3\mathbf{v}_3 = \mathbf{0}$. Letting $\mathbf{A} = [\mathbf{v}_1\ \mathbf{v}_2\ \mathbf{v}_3]$, we know that this question is equivalent to determining whether or not $\mathbf{A}\mathbf{x} = \mathbf{0}$ has a nontrivial solution. Considering the augmented matrix $[\mathbf{A}\ \mathbf{0}]$ and row-reducing, we find

$$\begin{bmatrix} 1 & -1 & 0 & 0 \\ 1 & 0 & 1 & 0 \\ -1 & 1 & 1 & 0 \end{bmatrix} \rightarrow \begin{bmatrix} 1 & 0 & 0 & 0 \\ 0 & 1 & 0 & 0 \\ 0 & 0 & 1 & 0 \end{bmatrix} \tag{1.6.2}$$

It follows that $\mathbf{A}\mathbf{x} = \mathbf{0}$ has only the trivial solution, and therefore the set S is linearly independent. Geometrically, this means that if we take any nontrivial combination of $\mathbf{v}_1$, $\mathbf{v}_2$, and $\mathbf{v}_3$, the result is a vector that is not the zero vector.

From example 1.6.2, we see how we will normally test a set of vectors for linear independence: we take advantage of our understanding of linear combinations and matrix multiplication and convert the vector equation $x_1\mathbf{v}_1 + x_2\mathbf{v}_2 + \cdots + x_k\mathbf{v}_k = \mathbf{0}$ to the matrix equation $\mathbf{A}\mathbf{x} = \mathbf{0}$, where $\mathbf{A}$ is the matrix with columns $\mathbf{v}_1, \ldots, \mathbf{v}_k$. Row-reducing, we can test whether or not nontrivial solutions exist to $\mathbf{A}\mathbf{x} = \mathbf{0}$ by examining pivot locations in the matrix $\mathbf{A}$.

Several facts about linear dependence and independence will prove to be useful in many aspects of our upcoming work. We simply state them here, and leave their verification to the exercises at the end of this section:

- Any set containing the zero vector is linearly dependent.

- Any set $\{\mathbf{v}_1\}$ consisting of a single nonzero vector is linearly independent.

- Any set of two vectors $\{\mathbf{v}_1, \mathbf{v}_2\}$ is linearly independent whenever $\mathbf{v}_1$ is not a scalar multiple of $\mathbf{v}_2$.

- The columns of a matrix $\mathbf{A}$ are linearly independent if and only if the equation $\mathbf{A}\mathbf{x} = \mathbf{0}$ has only the trivial solution.

The concepts of linear independence and span both involve linear combinations of a set of vectors. Furthermore, there are many important and natural connections between span and linear independence. The next example extends the previous one and lays the foundation for a discussion of several general results.

Example 1.6.3   Let the vectors $\mathbf{v}_1$, $\mathbf{v}_2$, $\mathbf{v}_3$, and $\mathbf{v}_4$ be given by

$$\mathbf{v}_1 = \begin{bmatrix} 1 \\ 1 \\ -1 \end{bmatrix}, \quad \mathbf{v}_2 = \begin{bmatrix} -1 \\ 0 \\ 1 \end{bmatrix}, \quad \mathbf{v}_3 = \begin{bmatrix} 0 \\ 1 \\ 1 \end{bmatrix}, \quad \mathbf{v}_4 = \begin{bmatrix} 5 \\ 6 \\ -1 \end{bmatrix}$$

Let $R = \{\mathbf{v}_1, \mathbf{v}_2\}$, $S = \{\mathbf{v}_1, \mathbf{v}_2, \mathbf{v}_3\}$, and $T = \{\mathbf{v}_1, \mathbf{v}_2, \mathbf{v}_3, \mathbf{v}_4\}$. Which of the sets R, S, and T are linearly independent? Which of the sets R, S, and T span $\mathbb{R}^3$?

Solution.   We have already seen in example 1.6.2 that the set S is linearly independent. Moreover, we saw that when we let $\mathbf{A} = [\mathbf{v}_1 \ \mathbf{v}_2 \ \mathbf{v}_3]$ and row-reduce the augmented matrix for the equation $\mathbf{Ax} = \mathbf{0}$, it follows that

$$\begin{bmatrix} 1 & -1 & 0 & 0 \\ 1 & 0 & 1 & 0 \\ -1 & 1 & 1 & 0 \end{bmatrix} \rightarrow \begin{bmatrix} 1 & 0 & 0 & 0 \\ 0 & 1 & 0 & 0 \\ 0 & 0 & 1 & 0 \end{bmatrix}$$

Not only does this show that the vectors in set S are linearly independent ($\mathbf{Ax} = \mathbf{0}$ has only the trivial solution because $\mathbf{A}$ has a pivot in every column so there are no free variables present), but also, by theorem 1.5.2, the vectors in S span $\mathbb{R}^3$ since $\mathbf{A}$ has a pivot in every row. Since the vectors in S span $\mathbb{R}^3$, this means that we can write every vector in $\mathbb{R}^3$ as a linear combination of the three vectors in S. Moreover, since $\mathbf{A}$ has a pivot in every column, it will also follow that every such linear combination is unique: every vector in $\mathbb{R}^3$ can be written in exactly one way as a linear combination of $\mathbf{v}_1$, $\mathbf{v}_2$, and $\mathbf{v}_3$.

What happens if we remove $\mathbf{v}_3$ from S and instead consider the set $R = \{\mathbf{v}_1, \mathbf{v}_2\}$? To answer the question of linear independence, we ask if there is a nontrivial solution to the vector equation $x_1\mathbf{v}_1 + x_2\mathbf{v}_2 = \mathbf{0}$. Equivalently, we let $\mathbf{B}$ be the 3×2 matrix whose columns are $\mathbf{v}_1$ and $\mathbf{v}_2$ and solve $\mathbf{Bx} = \mathbf{0}$. Doing so, we find that

$$\begin{bmatrix} 1 & -1 & 0 \\ 1 & 0 & 0 \\ -1 & 1 & 0 \end{bmatrix} \rightarrow \begin{bmatrix} 1 & 0 & 0 \\ 0 & 1 & 0 \\ 0 & 0 & 0 \end{bmatrix}$$

so only the trivial solution exists and thus the set R is linearly independent. Note again that this is due to the fact that $\mathbf{B}$ has a pivot in every column. This should not be surprising, since we removed a vector from the linearly independent set S to get the set R: if the vectors in S do not depend on one another, neither should the vectors in R.

On the other hand, we can also say by theorem 1.5.2 that the set R does not span $\mathbb{R}^3$, since $\mathbf{B}$ does not have a pivot position in every row. For example, the vector $\mathbf{b} = [0 \ 1 \ 1]^T$ cannot be written as a linear combination of $\mathbf{v}_1$ and $\mathbf{v}_2$. This can be seen by row-reducing the augmented matrix that represents $\mathbf{Bx} = \mathbf{b}$, where we find that

$$\begin{bmatrix} 1 & -1 & 0 \\ 1 & 0 & 1 \\ -1 & 1 & 1 \end{bmatrix} \rightarrow \begin{bmatrix} 1 & 0 & 0 \\ 0 & 1 & 0 \\ 0 & 0 & 1 \end{bmatrix}$$

The last equation tells us that $0x_1 + 0x_2 = 1$, which is impossible, and thus $\mathbf{b}$ cannot be written as a linear combination of the vectors in R.

Finally, we consider the set $T = \{\mathbf{v}_1, \mathbf{v}_2, \mathbf{v}_3, \mathbf{v}_4\}$. To test if T is linearly independent, we let $\mathbf{C}$ be the matrix whose columns are $\mathbf{v}_1$, $\mathbf{v}_2$, $\mathbf{v}_3$, and $\mathbf{v}_4$, and consider the equation $\mathbf{C}\mathbf{x} = \mathbf{0}$, which corresponds to the equation $x_1\mathbf{v}_1 + x_2\mathbf{v}_2 + x_3\mathbf{v}_3 + x_4\mathbf{v}_4 = 0$. Row-reducing,

$$\begin{bmatrix} 1 & -1 & 0 & 5 & 0 \\ 1 & 0 & 1 & 6 & 0 \\ -1 & 1 & 1 & -1 & 0 \end{bmatrix} \rightarrow \begin{bmatrix} 1 & 0 & 0 & 2 & 0 \\ 0 & 1 & 0 & -3 & 0 \\ 0 & 0 & 1 & 4 & 0 \end{bmatrix}$$

Note that the variable x_4 is free, since $\mathbf{C}$ does not have a pivot in its fourth column. This shows that any vector $\mathbf{x}$ with entries $x_1, x_2, x_3,$ and x_4 such that $x_1 = -2x_4$, $x_2 = 3x_4$, and $x_3 = -4x_4$ will be a solution to the equation $\mathbf{C}\mathbf{x} = \mathbf{0}$. For example, taking $x_4 = 1$, it follows that

$$-2\begin{bmatrix} 1 \\ 1 \\ -1 \end{bmatrix} + 3\begin{bmatrix} -1 \\ 0 \\ 1 \end{bmatrix} - 4\begin{bmatrix} 0 \\ 1 \\ 1 \end{bmatrix} + 1\begin{bmatrix} 5 \\ 6 \\ -1 \end{bmatrix} = \begin{bmatrix} 0 \\ 0 \\ 0 \end{bmatrix}$$

Thus, the set T is linearly dependent. We can also see from our computations that the set T does indeed span $\mathbb{R}^3$, since the matrix $\mathbf{C}$ has a pivot position in every row. This result should be expected: we have already shown that every vector in $\mathbb{R}^3$ can be written as a linear combination of the vectors in S, and the set T contains all three vectors in S.

There are many important generalizations we can make from example 1.6.3. For instance, from an algebraic perspective we see that we can easily answer questions about the linear independence and span of the columns of a matrix simply by considering the location of pivots in the matrix. In particular, the columns of $\mathbf{A}$ are linearly independent if and only if $\mathbf{A}$ has a pivot in every column, while the columns of $\mathbf{A}$ span $\mathbb{R}^m$ if and only if $\mathbf{A}$ has a pivot in every row. We state these results formally in the two following theorems.

Theorem 1.6.1   Let $\mathbf{A}$ be an $m \times n$ matrix. The following statements are equivalent:

a. The columns of $\mathbf{A}$ span $\mathbb{R}^m$.

b. $\mathbf{A}$ has a pivot position in every row.

c. The equation $\mathbf{A}\mathbf{x} = \mathbf{b}$ is consistent for every $\mathbf{b} \in \mathbb{R}^m$.

In the next theorem, note particularly the change in emphasis in statement (b) from rows to columns when considering pivot positions in the matrix.

Theorem 1.6.2   Let $\mathbf{A}$ be an $m \times n$ matrix. The following statements are equivalent:

a. The columns of $\mathbf{A}$ are linearly independent.

b. $\mathbf{A}$ has a pivot position in every column.

c. The equation $\mathbf{A}\mathbf{x} = \mathbf{0}$ has only the trivial solution.

At this point, it appears ideal if a set is linearly independent or spans $\mathbb{R}^m$. The best scenario, then, is the case when a set has both of these properties and forms a linearly independent spanning set. In this case, for the matrix whose columns are the vectors in the set, we need the matrix to have a pivot in every column, as well as in every row. As we saw in example 1.6.3 with the set S and the corresponding matrix $\mathbf{A}$, this can only happen when the number of vectors in the set S matches the number of entries in each vector. In other words, the corresponding matrix $\mathbf{A}$ must be square. Obviously if a square matrix has a pivot in every row, it must also have a pivot in every column, and vice versa. We close our current discussion with an important result that links the concepts of linear independence and span in the columns of a square matrix; theorem 1.6.3 is a consequence of the two preceding ones.

Theorem 1.6.3 Let $\mathbf{A}$ be an $n \times n$ matrix. The following statements are equivalent:

a. The columns of $\mathbf{A}$ are linearly independent.

b. The columns of $\mathbf{A}$ span $\mathbb{R}^n$.

c. $\mathbf{A}$ has a pivot position in every column.

d. $\mathbf{A}$ has a pivot position in every row.

e. For each $\mathbf{b} \in \mathbb{R}^n$, the equation $\mathbf{A}\mathbf{x} = \mathbf{b}$ has a unique solution.

Theorem 1.6.3 shows that square matrices play a particularly important role in linear algebra, an idea that will further demonstrate itself when we study the notion of the *inverse* of a matrix in the following section.

We conclude this section with a look ahead to our study of linear differential equations, in which the concepts of linear independence and span will also find a prominent role.

Example 1.6.4 Consider the differential equation $y'' + y = 0$. Explain why the function $y = c_1 \cos t + c_2 \sin t$ is a solution to the differential equation.

Solution. In our upcoming study of differential equations, we will call the equation $y'' + y = 0$ a linear second-order homogeneous equation with constant coefficients. Equations of this form will be considered in chapter 3 and be the focus of chapter 4.

For now, we can intuitively understand why $y = c_1 \cos t + c_2 \sin t$ is a solution to the equation. Note that in order to solve the equation $y'' + y = 0$, we must find all functions y such that $y'' = -y$. From our experience in calculus, we know that

$$\frac{d}{dt}[\sin t] = \cos t \text{ and } \frac{d}{dt}[\cos t] = -\sin t$$

Furthermore, if we consider second derivatives,

$$\frac{d^2}{dt^2}[\sin t] = \frac{d}{dt}[\cos t] = -\sin t \quad \text{and} \quad \frac{d^2}{dt^2}[\cos t] = \frac{d}{dt}[-\sin t] = -\cos t$$

Hence, the second derivative of each basic trigonometric function is the opposite of itself, which makes both $y = \cos t$ and $y = \sin t$ solutions to the equation $y'' + y = 0$.

Moreover, it is a straightforward exercise to show (using properties of the derivative) that any scalar multiple (such as $y = 3\sin t$) of either function is also a solution to the differential equation, as is any combination of the form $y = 2\cos t + 3\sin t$. More generally, this makes any function

$$y = c_1 \cos t + c_2 \sin t$$

a solution to the differential equation.

If we think about our understanding of linear independence for a set of two vectors, we find an analogy to the two functions $\cos t$ and $\sin t$: since these two functions are not scalar multiples of one another, it makes sense to call these functions *linearly independent*. Moreover, from the form of the function $y = c_1 \cos t + c_2 \sin t$, we are taking *linear combinations* of the basic trigonometric functions to form other solutions to the differential equation. We can even go so far as to say that the solution set to the differential equation is the *span* of the two functions $\cos t$ and $\sin t$.

In future work, we will see that this broader perspective on linear independence and span serves us well in solving linear differential equations. We will gain additional understanding of why the solution set to every second-order linear homogeneous differential equation with constant coefficients demonstrates a similar structure in subsequent work.

Exercises 1.6   In each of exercises 1–8, determine whether the given set S is linearly independent or linearly dependent.

1. $S = \{v_1, v_2\}$ where $v_1 = [3 \ -2]^T$ and $v_2 = [-9 \ 6]^T$

2. $S = \{v_1, v_2\}$ where $v_1 = [1 \ 0]^T$ and $v_2 = [0 \ 1]^T$

3. $S = \{v_1, v_2\}$ where $v_1 = [5 \ -2]^T$ and $v_2 = [5 \ 2]^T$

4. $S = \{v_1, v_2, v_3\}$ where $v_1 = [5 \ -2]^T$, $v_2 = [5 \ 2]^T$, and $v_3 = [11 \ -5]^T$

5. $S = \{v_1, v_2, v_3\}$ where $v_1 = [-1 \ 2 \ 1]^T$, $v_2 = [3 \ 1 \ 1]^T$, and $v_3 = [1 \ 5 \ 3]^T$

6. $S = \{v_1, v_2, v_3\}$ where $v_1 = [-1 \ 2 \ 1]^T$, $v_2 = [3 \ 1 \ 1]^T$, and $v_3 = [1 \ 5 \ 2]^T$

7. $S = \{v_1, v_2\}$ where $v_1 = [1 \ -2 \ 4 \ 3]^T$ and $v_2 = [-3 \ 6 \ -12 \ -9]^T$

8. $S = \{v_1, v_2, v_3, v_4\}$ where $v_1 = [-1 \ 2 \ 1]^T$, $v_2 = [3 \ 1 \ 1]^T$, $v_3 = [1 \ 5 \ 2]^T$, and $v_4 = [1 \ 1 \ 1]^T$

9. For each of the sets S in exercises 1–8, determine whether or not S spans $\mathbb{R}^m$, where m is chosen appropriately.

10. Suppose that S is a set of three vectors in $\mathbb{R}^5$. Is it possible for S to span $\mathbb{R}^5$? Why or why not?

11. Suppose that S is a set of two vectors in $\mathbb{R}^3$. Is S linearly independent, linearly dependent, or not necessarily either? Explain your answer.

12. Let S be a set of four vectors in $\mathbb{R}^3$. Is it possible for S to be linearly independent? Is it possible for S to span $\mathbb{R}^3$? Why or why not?

13. Let S be a set of five vectors in $\mathbb{R}^4$. Must S span $\mathbb{R}^4$? Is it possible for S to be linearly independent? Explain.

14. If A is an $m \times n$ matrix, for what relationship between n and m are the columns of A guaranteed to not span $\mathbb{R}^m$? For what relationship between n and m will the columns have to be linearly dependent?

15. Prove that any set that contains the zero vector must be linearly dependent.

16. Explain why any set consisting of a single nonzero vector must be linearly independent.

17. Show that any set of two vectors, $\{\mathbf{v}_1, \mathbf{v}_2\}$, is linearly independent if and only if $\mathbf{v}_1$ is not a scalar multiple of $\mathbf{v}_2$.

18. Explain why the columns of a matrix A are linearly independent if and only if the equation $A\mathbf{x} = \mathbf{0}$ has only the trivial solution.

19. Let $\mathbf{v}_1 = [-1\ 2\ 1]^T$, $\mathbf{v}_2 = [3\ 1\ 1]^T$, and $\mathbf{v}_3 = [5\ 3\ k]^T$. For what value(s) of k is $\{\mathbf{v}_1, \mathbf{v}_2, \mathbf{v}_3\}$ linearly independent? For what value(s) of k is $\mathbf{v}_3$ in the span of $\{\mathbf{v}_1, \mathbf{v}_2\}$? How are these two questions related?

20. Consider the set $S = \{\mathbf{v}_1, \mathbf{v}_2, \mathbf{v}_3\}$ where $\mathbf{v}_1 = [1\ 0\ 0]^T$, $\mathbf{v}_2 = [0\ 1\ 0]^T$, and $\mathbf{v}_3 = [0\ 0\ 1]^T$. Explain why S spans $\mathbb{R}^3$, and also why S is linearly independent. In addition, determine the weights x_1, x_2, and x_3 that allow you to write the vector $[-27\ 13\ 91]^T$ as a (unique) linear combination of $\mathbf{v}_1, \mathbf{v}_2, \mathbf{v}_3$. What do you observe?

21. Let A be a 4×7 matrix. Suppose that when solving the homogeneous equation $A\mathbf{x} = \mathbf{0}$ there are three free variables present. Do the columns of A span $\mathbb{R}^4$? Explain. Are the columns of A linearly dependent, linearly independent, or is it impossible to say? Justify your answer.

22. Suppose that A is a 9×6 matrix and that A has six pivot columns. Are the columns of A linearly dependent, linearly independent, or is it impossible to say? Do the columns of A span $\mathbb{R}^9$, or is it impossible to tell? Justify your answers.

23. Decide whether each of the following sentences is true or false. In every case, write one sentence to support your answer.

(a) If the system represented by $A\mathbf{x} = \mathbf{0}$ has a free variable present, then the columns of the matrix A are linearly independent vectors.

(b) If a matrix has more columns than rows, then the columns of the matrix must be linearly dependent.

(c) If an $m \times n$ matrix $\mathbf{A}$ has a pivot in every column, then the columns of $\mathbf{A}$ span $\mathbb{R}^m$.

(d) If $\mathbf{A}$ is an $m \times n$ matrix that is not square, it is possible for its columns to be both linearly independent and span $\mathbb{R}^m$.

24. Consider the linear second-order homogeneous differential equation $y'' + y = 0$. Show by direct substitution that $y_1 = e^t$ and $y_2 = e^{-t}$ are solutions to the differential equation. In addition, show by substitution that any linear combination $y = c_1 e^t + c_2 e^{-t}$ is also a solution.

25. We have seen that the general solution to the linear second-order differential equation $y'' + y = 0$ is given by

$$y(t) = c_1 \sin(t) + c_2 \cos(t)$$

Suppose we know initial values for $y(0)$ and $y'(0)$ to be

$$y(0) = 4 \text{ and } y'(0) = -2$$

What are the values of c_1 and c_2? How is a system of linear equations involved?

26. It can be shown that the solution to the linear second-order differential equation $y'' - y = 0$ is given by

$$y(t) = c_1 e^t + c_2 e^{-t}$$

Suppose we know initial values for $y(0)$ and $y'(0)$ to be

$$y(0) = 4 \text{ and } y'(0) = -2$$

What are the values of c_1 and c_2? How is a system of linear equations involved?

1.7 Matrix algebra

For a given system of linear equations, we are now interested in solving the vector equation $\mathbf{Ax} = \mathbf{b}$, where $\mathbf{A}$ is a known $m \times n$ matrix, $\mathbf{b} \in \mathbb{R}^m$ is given, and we seek $\mathbf{x} \in \mathbb{R}^n$. It is natural to compare this equation to an elementary linear equation such as $2x = 7$. The key algebraic step in solving $2x = 7$ is to divide both sides of the equation by 2. Said differently, we multiply both sides by the multiplicative inverse of the number 2. In anticipation of a new approach to solving the vector equation $\mathbf{Ax} = \mathbf{b}$, we carefully state the details required to solve $2x = 7$. In particular, from the equation $2x = 7$, it follows that $\frac{1}{2}(2x) = \frac{1}{2}(7)$, so that $(\frac{1}{2} \cdot 2)x = \frac{7}{2}$. Thus, $1 \cdot x = \frac{7}{2}$, so $x = \frac{7}{2}$. From a sophisticated perspective, to solve the equation $2x = 7$, we need to be able to multiply, to have a multiplicative identity (that is, the number 1), and to be able to compute a multiplicative inverse (here, the number $\frac{1}{2}$).

In this section, we lay the foundation for similar ideas that provide an alternate way to solve the equation $\mathbf{Ax} = \mathbf{b}$: essentially we are interested in determining whether we can find a matrix $\mathbf{B}$ so that when we compute $\mathbf{BA}$ the result is the matrix equivalent of "1". To do this, we will first have to learn what it means to multiply two matrices; a simpler (and still important) place to begin is with the addition of matrices and multiplication of matrices by scalars.

We already know how to add vectors and multiply them by scalars; similar principles hold for matrices. Two matrices can be added (or subtracted) if and only if they have an identical number of rows and columns. When addition (subtraction) is defined, the result is computed component-wise. Furthermore, the multiple of a matrix by a scalar $c \in \mathbb{R}$ is attained by multiplying every entry of the matrix by the same constant c. The following example demonstrates these basic facts.

Example 1.7.1   Let $\mathbf{A}$ and $\mathbf{B}$ be the matrices

$$\mathbf{A} = \begin{bmatrix} 1 & 3 & -4 \\ 0 & -7 & 2 \end{bmatrix}, \quad \mathbf{B} = \begin{bmatrix} -6 & 10 & -1 \\ 3 & 2 & 11 \end{bmatrix}$$

Compute $\mathbf{A} + \mathbf{B}$ and $-3\mathbf{A}$.

Solution.   Since $\mathbf{A}$ and $\mathbf{B}$ are both 2×3, their sum is defined and is given by

$$\mathbf{A} + \mathbf{B} = \begin{bmatrix} 1 & 3 & -4 \\ 0 & -7 & 2 \end{bmatrix} + \begin{bmatrix} -6 & 10 & -1 \\ 3 & 2 & 11 \end{bmatrix} = \begin{bmatrix} -5 & 13 & -5 \\ 3 & -5 & 13 \end{bmatrix}$$

The scalar multiple of a matrix is always defined, and $-3\mathbf{A}$ is given by

$$-3\mathbf{A} = \begin{bmatrix} -3 & -9 & 12 \\ 0 & 21 & -6 \end{bmatrix}$$

Matrix addition, when defined, has all of the expected properties of addition. In particular, $\mathbf{A} + \mathbf{B} = \mathbf{B} + \mathbf{A}$, so order does not matter, and we say matrix addition is *commutative*. Since $\mathbf{A} + (\mathbf{B} + \mathbf{C}) = (\mathbf{A} + \mathbf{B}) + \mathbf{C}$, the way we group more than two matrices to add also does not matter and we say matrix addition is *associative*. There is even a matrix that acts like the number 0. If $\mathbf{Z}$ is a matrix of the same number of rows and columns as $\mathbf{A}$ such that every entry in $\mathbf{Z}$ is zero, then it follows that $\mathbf{A} + \mathbf{Z} = \mathbf{Z} + \mathbf{A} = \mathbf{A}$. We call this zero matrix the *additive identity*.

The next natural operation to consider, of course, is multiplication. What does it mean to multiply two matrices? And when does it even make sense to multiply two matrices? We know for matrix–vector multiplication that the product $\mathbf{Ax}$ computes the vector "$\mathbf{b}$" that is the unique linear combination of the columns of $\mathbf{A}$ having the entries of the vector $\mathbf{x}$ as weights. Moreover, this product is only defined when the number of entries in $\mathbf{x}$ matches the number of columns of $\mathbf{A}$. If we now consider a matrix $\mathbf{B}$, we can naturally think about the matrix product $\mathbf{AB}$ by considering the columns of $\mathbf{B}$, say $\mathbf{b}_1, \ldots, \mathbf{b}_k$. In particular, we make the following definition.

Definition 1.7.1   If **A** is an $m \times n$ matrix, and **B** is a matrix whose columns are $\mathbf{b}_1, \ldots, \mathbf{b}_k$ such that the matrix–vector product $\mathbf{Ab}_j$ is defined for each $j = 1, \ldots, k$, then we define the *matrix product* **AB** by

$$\mathbf{AB} = [\mathbf{Ab}_1 \; \mathbf{Ab}_2 \; \cdots \; \mathbf{Ab}_k] \tag{1.7.1}$$

Note particularly that since **A** has n columns, in order for $\mathbf{Ab}_j$ to be defined each $\mathbf{b}_j$ must belong to $\mathbb{R}^n$. This in turn implies that the matrix **B** must have dimensions $n \times k$. Specifically, the number of rows in **B** must equal the number of columns in **A**. We explore matrix multiplication and its properties in the next example.

Example 1.7.2   Let **A** and **B** be the matrices

$$\mathbf{A} = \begin{bmatrix} 1 & 3 & -4 \\ 0 & -7 & 2 \end{bmatrix}, \quad \mathbf{B} = \begin{bmatrix} -6 & 10 \\ 3 & 2 \end{bmatrix}$$

Compute the matrix products **AB** and **BA**, or explain why they are not defined.

Solution.   First we consider **AB**. To do so, we would have to compute both $\mathbf{Ab}_1$ and $\mathbf{Ab}_2$, where $\mathbf{b}_1$ and $\mathbf{b}_2$ are the columns of **B**. But neither of these products is defined, since **A** has three columns and **B** has just two rows. Thus, **AB** is not defined.

On the other hand, **BA** is defined. For instance, we can compute the first column of **BA** by taking $\mathbf{Ba}_1$, where we see that

$$\mathbf{Ba}_1 = \begin{bmatrix} -6 & 10 \\ 3 & 2 \end{bmatrix} \begin{bmatrix} 1 \\ 0 \end{bmatrix} = \begin{bmatrix} -6 \\ 3 \end{bmatrix}$$

Similar computations for $\mathbf{Ba}_2$ and $\mathbf{Ba}_3$ show that

$$\mathbf{BA} = \begin{bmatrix} -6 & -88 & 44 \\ 3 & -5 & -8 \end{bmatrix}$$

There are several important observations to make based on example 1.7.2. One is that if **A** is $m \times n$ and **B** is $n \times k$ so that the product **AB** is defined, then the resulting matrix **AB** is $m \times k$. This is true since the columns of **AB** are each of the form $\mathbf{Ab}_j$, thus being linear combinations of the columns of **A**, which have m entries, so that **AB** has m rows. Moreover, we have to consider each of the products $\mathbf{Ab}_1, \ldots, \mathbf{Ab}_k$, therefore giving **AB** k columns.

Furthermore, we clearly see that order matters in matrix multiplication. Specifically, given matrices **A** and **B** for which **AB** is defined, it is not even guaranteed that **BA** is defined, much less that **AB** = **BA**. Even when both products are defined, it is possible (even typical) that $\mathbf{AB} \neq \mathbf{BA}$. Formally, we say that matrix multiplication is *not commutative*. This fact will be explored further in the exercises. It is, however, the case that matrix multiplication (for matrices of the appropriate sizes) is both associative and distributive. That is, $\mathbf{A(BC)} = \mathbf{(AB)C}$ and $\mathbf{A(B + C)} = \mathbf{AB} + \mathbf{AC}$, again provided the sizes of the matrices make the relevant products and sums defined.

Now, we should not forget our motivation for considering matrix multiplication: we want to develop an alternative approach to solving equations of the form $\mathbf{Ax} = \mathbf{b}$ by multiplying $\mathbf{A}$ by another matrix $\mathbf{B}$ so that the product $\mathbf{BA}$ is the matrix equivalent of the number 1 (while simultaneously multiplying $\mathbf{b}$ by the same matrix $\mathbf{B}$). What is the matrix equivalent of the number 1? We consider this question and more in the following example.

Example 1.7.3   Consider the matrices

$$\mathbf{A} = \begin{bmatrix} 5 & 11 \\ -3 & -7 \end{bmatrix} \text{ and } \mathbf{I}_2 = \begin{bmatrix} 1 & 0 \\ 0 & 1 \end{bmatrix}$$

Compute $\mathbf{AI}_2$ and $\mathbf{I}_2\mathbf{A}$. What is special about the matrix $\mathbf{I}_2$?

Solution.   Using the rules for matrix multiplication, we observe that

$$\mathbf{AI}_2 = \begin{bmatrix} 5 & 11 \\ -3 & -7 \end{bmatrix}\begin{bmatrix} 1 & 0 \\ 0 & 1 \end{bmatrix} = \begin{bmatrix} 5 & 11 \\ -3 & -7 \end{bmatrix} = \mathbf{A}$$

and similarly

$$\mathbf{I}_2\mathbf{A} = \begin{bmatrix} 1 & 0 \\ 0 & 1 \end{bmatrix}\begin{bmatrix} 5 & 11 \\ -3 & -7 \end{bmatrix} = \begin{bmatrix} 5 & 11 \\ -3 & -7 \end{bmatrix} = \mathbf{A}$$

Thus, we see that multiplying the matrix $\mathbf{A}$ by $\mathbf{I}_2$ has no effect on the matrix $\mathbf{A}$.

The matrix $\mathbf{I}_2$ in example 1.7.3 is important because it has the property that $\mathbf{I}_2\mathbf{A} = \mathbf{A}$ for any matrix $\mathbf{A}$ with two rows (not simply the matrix $\mathbf{A}$ in example 1.7.3) and $\mathbf{AI}_2 = \mathbf{A}$ for any $\mathbf{A}$ with two columns. We can similarly show that if $\mathbf{I}_3$ is the matrix

$$\mathbf{I}_3 = \begin{bmatrix} 1 & 0 & 0 \\ 0 & 1 & 0 \\ 0 & 0 & 1 \end{bmatrix}$$

then $\mathbf{I}_3\mathbf{A} = \mathbf{A}$ for any matrix $\mathbf{A}$ with three rows, and $\mathbf{AI}_3 = \mathbf{A}$ for any matrix $\mathbf{A}$ with three columns. Similar results hold for corresponding matrices $\mathbf{I}_n$ of larger size; each of these matrices acts like the number 1, since multiplying other matrices by $\mathbf{I}_n$ has no effect on the given matrix.

Matrices which when multiplied by other matrices do not change the other matrices, are called *identity matrices*. More formally, the $n \times n$ *identity matrix* $\mathbf{I}_n$ is the square matrix whose diagonal entries all equal 1, and whose off-diagonal entries are all 0. (The *diagonal* entries in a matrix are those whose row and column indices are the same.) Often, when the context is clear, we will write simply $\mathbf{I}$, rather than $\mathbf{I}_n$. We also note that $\mathbf{I}_n$ is the only matrix that is $n \times n$ and acts as a multiplicative identity. Finally, it is evident that for any $m \times n$ matrix $\mathbf{A}$, $\mathbf{I}_m\mathbf{A} = \mathbf{AI}_n = \mathbf{A}$. In the next section, we will explore the notion of the *inverse* of a matrix, and there see that identity matrices play a central role.

One final algebraic operation with matrices merits formal introduction here. Given a matrix $\mathbf{A}$, its *transpose*, denoted $\mathbf{A}^{\mathrm{T}}$, is the matrix whose columns are the rows of **A**. That is, taking the transpose of a matrix replaces its rows with its columns, and vice versa. For example, if **A** is the 2×3 matrix

$$\mathbf{A} = \begin{bmatrix} 1 & 3 & -4 \\ 0 & -7 & 2 \end{bmatrix}$$

then its transpose $\mathbf{A}^{\mathrm{T}}$ is the 3×2 matrix

$$\mathbf{A}^{\mathrm{T}} = \begin{bmatrix} 1 & 0 \\ 3 & -7 \\ -4 & 2 \end{bmatrix}$$

Note that this is the same notation we regularly use to express a column vector in the form $\mathbf{b} = [1\ 2\ 3]^{\mathrm{T}}$. In the case that **A** is a square matrix, taking its transpose results in swapping entries across its diagonal. For example, if

$$\mathbf{A} = \begin{bmatrix} 5 & -2 & 7 \\ 0 & -3 & -1 \\ -4 & 8 & -6 \end{bmatrix}$$

then

$$\mathbf{A}^{\mathrm{T}} = \begin{bmatrix} 5 & 0 & -4 \\ -2 & -3 & 8 \\ 7 & -1 & -6 \end{bmatrix}$$

The transpose operator has several nice algebraic properties, some of which will be explored in the exercises. For example, for matrices for which the appropriate sums and products are defined,

$$(\mathbf{A} + \mathbf{B})^{\mathrm{T}} = \mathbf{A}^{\mathrm{T}} + \mathbf{B}^{\mathrm{T}}$$

and

$$(\mathbf{AB})^{\mathrm{T}} = \mathbf{B}^{\mathrm{T}} \mathbf{A}^{\mathrm{T}}$$

For a square matrix such as

$$\mathbf{A} = \begin{bmatrix} 3 & -1 \\ -1 & 2 \end{bmatrix}$$

it happens that $\mathbf{A}^{\mathrm{T}} = \mathbf{A}$. Any square matrix **A** for which $\mathbf{A}^{\mathrm{T}} = \mathbf{A}$ is said to be *symmetric*. It turns out that symmetric matrices have several especially nice properties in the context of more sophisticated concepts that arise later in the text, and we will revisit them at that time.

1.7.1 Matrix algebra using *Maple*

While it is important that we first learn to add and multiply matrices by hand to understand how these processes work, just like with row-reduction it is reasonable to expect that we will often use available technology to perform tedious computations like multiplying a 4×5 and 5×7 matrix. Moreover, in real-world applications, it is not uncommon to have to deal with matrices that have thousands of rows and thousands of columns, or more. Here we introduce a few *Maple* commands that are useful in performing some of the algebraic manipulations we have studied in this section.

Let us consider some of the matrices defined in earlier examples:

$$\mathbf{A} = \begin{bmatrix} 1 & 3 & -4 \\ 0 & -7 & 2 \end{bmatrix}, \quad \mathbf{B} = \begin{bmatrix} -6 & 10 \\ 3 & 2 \end{bmatrix}, \quad \mathbf{C} = \begin{bmatrix} -6 & 10 & -1 \\ 3 & 2 & 11 \end{bmatrix}$$

After defining each of these three matrices with the usual commands in *Maple*, such as

```
> A := <<1,0>|<3,-7>|<-4,2>>;
```

we can execute the sum of **A** and **C** and the scalar multiple $-3\mathbf{B}$ with the commands

```
> A + C;
> -3*B;
```

for which *Maple* will report the outputs

$$\begin{bmatrix} -5 & 13 & -5 \\ 3 & -5 & 13 \end{bmatrix} \text{ and } \begin{bmatrix} 18 & -30 \\ -9 & -6 \end{bmatrix}$$

We have previously seen that to compute a matrix–vector product, the period is used to indicate multiplication, as in $> \texttt{A.x;}$. The same syntax holds for matrix multiplication, where defined. For example, if we wish to compute the product **BA**, we enter

```
> B.A;
```

which yields the output

$$\begin{bmatrix} -6 & -88 & 44 \\ 3 & -5 & -8 \end{bmatrix}$$

If we try to have *Maple* compute an undefined product, such as **AB** through the command $> \texttt{A.B;}$, we get the error message

```
Error, (in LinearAlgebra:-MatrixMatrixMultiply)
first matrix column dimension (3) <> second matrix
row dimension (2)
```

In the event that we need to execute computations involving an identity matrix, rather than tediously enter all the 1's and 0's, we can use the built-in *Maple* command `IdentityMatrix(n);` where *n* is the number of rows and columns in the matrix. For example, entering

```
> Id := IdentityMatrix(4);
```

results in the output

$$\text{Id} := \begin{bmatrix} 1 & 0 & 0 & 0 \\ 0 & 1 & 0 & 0 \\ 0 & 0 & 1 & 0 \\ 0 & 0 & 0 & 1 \end{bmatrix}$$

Note: **Id** is the name we are using to store this identity matrix. We cannot use the letter I because I is reserved to represent $\sqrt{-1}$ in *Maple* .

Finally, if we desire to compute the transpose of a matrix **A**, such as

$$\mathbf{A} = \begin{bmatrix} 1 & 3 & -4 \\ 0 & -7 & 2 \end{bmatrix}$$

the relevant command is

```
> Transpose(A);
```

which generates the output

$$\mathbf{A}^{\mathrm{T}} = \begin{bmatrix} 1 & 0 \\ 3 & -7 \\ -4 & 2 \end{bmatrix}$$

Exercises 1.7

1. Let **A**, **B**, and **C** be the given matrices. In each of the following problems, compute (by hand) the prescribed algebraic combination of **A**, **B**, and **C** if the operation is defined. If the operation is not defined, explain why.

$$\mathbf{A} = \begin{bmatrix} 3 & -5 & 2 \\ -1 & 5 & -4 \end{bmatrix}, \quad \mathbf{B} = \begin{bmatrix} -6 & 10 \\ 2 & 11 \\ -3 & -2 \end{bmatrix}, \quad \mathbf{C} = \begin{bmatrix} 5 & 3 \\ -1 & 0 \\ 2 & -4 \end{bmatrix}$$

(a) $\mathbf{B} + \mathbf{C}$ (b) $\mathbf{A} + \mathbf{B}$ (c) $-2\mathbf{A}$ (d) $-3\mathbf{B} + 4\mathbf{C}$ (e) $\mathbf{AB}$
(f) $\mathbf{BA}$ (g) $\mathbf{AA}$ (h) $\mathbf{A}(\mathbf{B} + \mathbf{C})$ (i) $\mathbf{CA}$ (j) $\mathbf{C}(\mathbf{A} + \mathbf{B})$
(k) $\mathbf{A}^{\mathrm{T}} + \mathbf{B}$ (l) $(\mathbf{B} + \mathbf{C})^{\mathrm{T}}$ (m) $\mathbf{B}^{\mathrm{T}}\mathbf{C}$ (n) $\mathbf{BC}^{\mathrm{T}}$ (o) $(\mathbf{AB})^{\mathrm{T}}$
(p) $(\mathbf{BA})^{\mathrm{T}}$

2. Let **A**, **B**, and **C** be the given matrices. In each of the following problems, compute (by hand) the prescribed algebraic combination of **A**, **B**, and **C** whenever the operation is defined. If the operation is not defined, explain why.

$$\mathbf{A} = \begin{bmatrix} -5 & 3 \\ 2 & 4 \end{bmatrix}, \quad \mathbf{B} = \begin{bmatrix} 2 & 11 \\ -3 & -2 \end{bmatrix}, \quad \mathbf{C} = \begin{bmatrix} 1 & 0 \\ -5 & 3 \end{bmatrix}$$

(a) $\mathbf{B} + \mathbf{C}$ (b) $\mathbf{A} + \mathbf{B}$ (c) $-2\mathbf{A}$ (d) $-3\mathbf{B} + 4\mathbf{C}$ (e) $\mathbf{AB}$
(f) $\mathbf{BA}$ (g) $\mathbf{AA}$ (h) $\mathbf{A}(\mathbf{B} + \mathbf{C})$ (i) $\mathbf{CA}$ (j) $\mathbf{C}(\mathbf{A} + \mathbf{B})$
(k) $\mathbf{A}^{\mathrm{T}} + \mathbf{B}$ (l) $(\mathbf{B} + \mathbf{C})^{\mathrm{T}}$ (m) $\mathbf{B}^{\mathrm{T}}\mathbf{C}$ (n) $\mathbf{BC}^{\mathrm{T}}$ (o) $(\mathbf{AB})^{\mathrm{T}}$
(p) $(\mathbf{BA})^{\mathrm{T}}$

3. Discuss the differences between multiplying two square matrices versus multiplying non-square matrices. That is, under what circumstances can two square matrices be multiplied? How does the situation change for non-square matrices? In addition, if the product **AB** is defined, is **BA**?

4. Give an example of 2×2 matrices **A** and **B** for which $\mathbf{AB} \neq \mathbf{BA}$.

5. Give an example of 2×2 matrices **A** and **B** for which $\mathbf{AB} = \mathbf{BA}$.

6. If **A** is $m \times n$ and **B** is $n \times k$, and neither **A** nor **B** is square, can **AB** ever equal **BA**? Explain.

In exercises 7–9, let **A** be the given matrix. If possible, find a matrix **B** such that $\mathbf{BA} = \mathbf{I}_2$; if **B** exists, determine whether $\mathbf{BA} = \mathbf{AB}$.

7. $\mathbf{A} = \begin{bmatrix} 2 & 0 \\ 0 & 5 \end{bmatrix}$

8. $\mathbf{A} = \begin{bmatrix} 2 & 4 \\ 0 & 5 \end{bmatrix}$

9. $\mathbf{A} = \begin{bmatrix} 1 & -1 \\ -1 & 2 \end{bmatrix}$

In exercises 10 and 11, for the given matrix **A**, answer each of the following questions:

(a) Are the columns of **A** linearly independent?

(b) Do the columns of **A** span $\mathbb{R}^2$?

(c) How many pivot positions does **A** have?

(d) Solve the equation $\mathbf{Ax} = \mathbf{0}$ by row reducing by hand. Is **A** row equivalent to an important matrix?

(e) If possible, determine a 2×2 matrix **B** such that $\mathbf{BA} = \mathbf{I}_2$.

10. $\mathbf{A} = \begin{bmatrix} -1 & 2 \\ 2 & -3 \end{bmatrix}$

11. $\mathbf{A} = \begin{bmatrix} -1 & 2 \\ 2 & -4 \end{bmatrix}$

12. Decide whether each of the following sentences is true or false. In every case, write one sentence to support your answer.

(a) If **A** and **B** are matrices of the same size, then the products **AB** and **BA** are always defined.

(b) If **A** and **B** are matrices such that the products **AB** and **BA** are both defined, then $\mathbf{AB} = \mathbf{BA}$.

(c) If **A** and **B** are matrices such that **AB** is defined, then $(\mathbf{AB})^{\mathrm{T}} = \mathbf{A}^{\mathrm{T}}\mathbf{B}^{\mathrm{T}}$.

(d) If **A** and **B** are matrices such that **A** + **B** is defined, then
$(\mathbf{A} + \mathbf{B})^{\mathrm{T}} = \mathbf{A}^{\mathrm{T}} + \mathbf{B}^{\mathrm{T}}$.

13. Compute the prescribed algebraic computations in exercise 1 using a computer algebra system.

14. Compute the prescribed algebraic computations in exercise 2 using a computer algebra system.

1.8 The inverse of a matrix

We have observed repeatedly that linear algebra is a subject centered on one idea—systems of linear equations—viewed from several different perspectives. Continuing with this theme, we have recently considered an alternative method for solving the equation **Ax** = **b** by attempting to find a matrix **B** such that **BA** = **I**, where **I** is the appropriate identity matrix. If we can in fact find such a matrix **B**, it follows that

$$\mathbf{B}(\mathbf{Ax}) = \mathbf{Bb} \qquad (1.8.1)$$

By the associativity of matrix multiplication and the defining property of **B**, it follows that

$$\mathbf{B}(\mathbf{Ax}) = (\mathbf{BA})\mathbf{x} = \mathbf{Ix} = \mathbf{x} \qquad (1.8.2)$$

Equations (1.8.1) and (1.8.2) together imply that **x** = **Bb**. Thus, the existence of such a matrix **B** shows us how we can solve **Ax** = **b** by multiplication. It turns out that from a computational point of view, row-reduction is a superior approach to solving **Ax** = **b**; nonetheless, the perspective that it may be possible to solve the equation through the use of a multiplicative inverse has many important theoretical applications. In addition, similar ideas will be encountered in our study of differential equations.

Our work in section 1.7 showed that if **A** and **B** are not square matrices, it is never the case that **AB** and **BA** are equal. Thus it is only possible to find a matrix **B** such that **AB** = **BA** = **I** if **A** is square (though even then it is not always the case that such a matrix **B** exists). Moreover, as we know from theorem 1.6.3, some square matrices have the important property that the equation **Ax** = **b** has a unique solution for every possible choice of **b**.

For the next few sections, we therefore focus our attention almost exclusively on square matrices. Here, our emphasis is on the questions "when does a matrix **B** exist such that **AB** = **BA** = **I**?" and "when such a matrix **B** exists, how can we find it?" The next definition formalizes the notion of the *inverse* of a matrix.

Definition 1.8.1 If **A** is an $n \times n$ matrix, we say that **A** is *invertible* if and only if there exists an $n \times n$ matrix **B** such that

$$\mathbf{AB} = \mathbf{BA} = \mathbf{I}_n \qquad (1.8.3)$$

When **A** is invertible, we call **B** the *inverse of* **A** and write $\mathbf{B} = \mathbf{A}^{-1}$ (read "**B** is A-inverse"). If **A** is not invertible, **A** is often called a *singular* matrix, and thus saying "**A** is invertible" is equivalent to saying "**A** is *nonsingular*."

It can be shown (see exercise 19) that if **A** is an invertible $n \times n$ matrix, then its inverse is unique (i.e., a given matrix cannot have two distinct inverses). In addition, we note from our discussion above in (1.8.1) and (1.8.2) that if **A** is invertible, then the equation $\mathbf{Ax} = \mathbf{b}$ has a solution for every $\mathbf{b} \in \mathbb{R}^n$. In particular, that solution is $\mathbf{x} = \mathbf{A}^{-1}\mathbf{b}$. Moreover, since $\mathbf{Ax} = \mathbf{b}$ has a solution for every $\mathbf{b} \in \mathbb{R}^n$, we know from theorem 1.6.1 that **A** has a pivot position in every row. From this, the fact that **A** is square, and theorem 1.6.3, it follows that $\mathbf{Ax} = \mathbf{b}$ has a unique solution for every $\mathbf{b} \in \mathbb{R}^n$. We state this result formally in the following theorem.

Theorem 1.8.1   If **A** is an $n \times n$ invertible matrix, then the equation $\mathbf{Ax} = \mathbf{b}$ has a unique solution for every $\mathbf{b} \in \mathbb{R}^n$.

Before beginning to explore how to find the inverse of a matrix, as well as when the inverse even exists, we consider an example to see how we may check if two matrices are inverses and how to apply an inverse to solve a related equation.

Example 1.8.1   Let **A** and **B** be the matrices

$$\mathbf{A} = \begin{bmatrix} 4 & 5 \\ 1 & 2 \end{bmatrix}, \quad \mathbf{B} = \begin{bmatrix} 2/3 & -5/3 \\ -1/3 & 4/3 \end{bmatrix}$$

Show that **A** and **B** are inverses, and then use this fact to solve $\mathbf{Ax} = \mathbf{b}$, where $\mathbf{b} = [-7 \ 3]^{\mathrm{T}}$, without using row reduction.

Solution.   The reader should verify that the following matrix products indeed hold:

$$\mathbf{AB} = \begin{bmatrix} 4 & 5 \\ 1 & 2 \end{bmatrix} \begin{bmatrix} 2/3 & -5/3 \\ -1/3 & 4/3 \end{bmatrix} = \begin{bmatrix} 1 & 0 \\ 0 & 1 \end{bmatrix}$$

and

$$\mathbf{BA} = \begin{bmatrix} 2/3 & -5/3 \\ -1/3 & 4/3 \end{bmatrix} \begin{bmatrix} 4 & 5 \\ 1 & 2 \end{bmatrix} = \begin{bmatrix} 1 & 0 \\ 0 & 1 \end{bmatrix}$$

This shows that indeed $\mathbf{B} = \mathbf{A}^{-1}$. Note, equivalently, that $\mathbf{A} = \mathbf{B}^{-1}$. Now, we can easily solve the equation $\mathbf{Ax} = \mathbf{b}$ where **b** is the given vector:

$$\mathbf{x} = \mathbf{A}^{-1}\mathbf{b} = \begin{bmatrix} 2/3 & -5/3 \\ -1/3 & 4/3 \end{bmatrix} \begin{bmatrix} -7 \\ 3 \end{bmatrix} = \begin{bmatrix} -29/3 \\ 19/3 \end{bmatrix}$$

Of course, what is not clear in example 1.8.1 is how, given the matrix **A**, one might determine the entries in the inverse matrix $\mathbf{B} = \mathbf{A}^{-1}$. We now explore this in the 3×3 case for a general matrix **A**, and along the way learn conditions that guarantee that $\mathbf{A}^{-1}$ exists.

Given a 3×3 matrix $\mathbf{A}$, we seek a matrix $\mathbf{B}$ such that $\mathbf{AB} = \mathbf{I}_3$. Let the columns of $\mathbf{B}$ be $\mathbf{b}_1$, $\mathbf{b}_2$, and $\mathbf{b}_3$, and the columns of $\mathbf{I}_3$ be $\mathbf{e}_1$, $\mathbf{e}_2$, and $\mathbf{e}_3$. The column-wise definition of matrix multiplication then tells us that the following three vector equations must hold:

$$\mathbf{Ab}_1 = \mathbf{e}_1, \quad \mathbf{Ab}_2 = \mathbf{e}_2, \quad \text{and} \quad \mathbf{Ab}_3 = \mathbf{e}_3 \tag{1.8.4}$$

For the unique inverse matrix $\mathbf{B}$ to exist, it follows that each of these equations must have a unique solution. Clearly if $\mathbf{A}$ has a pivot position in every row (or, equivalently, the columns of $\mathbf{A}$ span $\mathbb{R}^3$), then by theorem 1.6.3 it follows that we can find unique vectors $\mathbf{b}_1$, $\mathbf{b}_2$, and $\mathbf{b}_3$ that make these three equations hold. Thus, any one of the conditions in theorem 1.6.3 will guarantee that $\mathbf{B} = \mathbf{A}^{-1}$ exists. Moreover, if $\mathbf{A}^{-1}$ exists, we know from theorem 1.8.1 that every condition in theorem 1.6.3 also holds.

Momentarily, let us assume that $\mathbf{A}$ is indeed invertible. If we proceed to find the matrix $\mathbf{B}$ by solving the three equations in (1.8.4), we see that row-reduction provides an approach for producing all three vectors at once. To find these vectors one at a time, it would be necessary to row-reduce each of the three augmented matrices

$$[\mathbf{A} \; \mathbf{e}_1], \quad [\mathbf{A} \; \mathbf{e}_2], \quad \text{and} \quad [\mathbf{A} \; \mathbf{e}_3] \tag{1.8.5}$$

In each case, the exact same elementary row operations will be applied to $\mathbf{A}$ and thus be applied, respectively, to the vectors $\mathbf{e}_1$, $\mathbf{e}_2$, and $\mathbf{e}_3$. As such, we may do all of them at once by considering the augmented matrix

$$[\mathbf{A} \; \mathbf{e}_1 \; \mathbf{e}_2 \; \mathbf{e}_3] \tag{1.8.6}$$

Note particularly that the form of the augmented matrix in (1.8.6) is $[\mathbf{A} \; \mathbf{I}_3]$. If we now row-reduce this matrix, and $\mathbf{A}$ has a pivot in every row, it follows that we will be able to read the coefficients of $\mathbf{A}^{-1}$ from the result. This process is best illuminated by an example, so we now explore how these computations lead us to $\mathbf{A}^{-1}$ in a concrete situation.

Example 1.8.2   Find the inverse of the matrix

$$\mathbf{A} = \begin{bmatrix} 2 & 1 & -2 \\ 1 & 1 & -1 \\ -2 & -1 & 3 \end{bmatrix}$$

Solution.   Following the discussion above, we augment $\mathbf{A}$ with the 3×3 identity matrix and row-reduce. It follows that

$$\begin{bmatrix} 2 & 1 & -2 & 1 & 0 & 0 \\ 1 & 1 & -1 & 0 & 1 & 0 \\ -2 & -1 & 3 & 0 & 0 & 1 \end{bmatrix} \rightarrow \begin{bmatrix} 1 & 0 & 0 & 2 & -1 & 1 \\ 0 & 1 & 0 & -1 & 2 & 0 \\ 0 & 0 & 1 & 1 & 0 & 1 \end{bmatrix}$$

These computations demonstrate two important things. The first is that the row reduction of $\mathbf{A}$ in the first three columns of the augmented matrix shows that $\mathbf{A}$ has a pivot position in every row, and therefore $\mathbf{A}$ is invertible. Moreover, the row-reduced form of $[\mathbf{A}\ \mathbf{I}_3]$ tells us that $\mathbf{A}^{-1}$ is the matrix

$$\mathbf{A}^{-1} = \begin{bmatrix} 2 & -1 & 1 \\ -1 & 2 & 0 \\ 1 & 0 & 1 \end{bmatrix}$$

Again, we observe from our preceding discussion and example 1.8.2 that we have found an algorithm for finding the inverse of a square matrix $\mathbf{A}$. We augment $\mathbf{A}$ with the corresponding identity matrix and row-reduce. Provided that $\mathbf{A}$ has a pivot in every row, we find by row-reducing that

$$[\mathbf{A}\ \mathbf{I}] \rightarrow [\mathbf{I}\ \mathbf{A}^{-1}]$$

That is, row-reduction of an invertible matrix $\mathbf{A}$ augmented with the identity matrix leads us directly to the inverse, $\mathbf{A}^{-1}$.

Next, we examine what happens in the event that a square matrix is not invertible.

Example 1.8.3   Find the inverse of the matrix

$$\mathbf{A} = \begin{bmatrix} 2 & 1 \\ -6 & -3 \end{bmatrix}$$

provided the inverse exists. If the inverse does not exist, explain why.

Solution.   We augment $\mathbf{A}$ with the 2×2 identity matrix and row-reduce, finding that

$$\begin{bmatrix} 2 & 1 & 1 & 0 \\ -6 & -3 & 0 & 1 \end{bmatrix} \rightarrow \begin{bmatrix} 1 & \frac{1}{2} & 0 & -\frac{1}{6} \\ 0 & 0 & 1 & \frac{1}{3} \end{bmatrix}$$

Again, we see at least two key facts from these computations: $\mathbf{A}$ does not have a pivot position in every row, and thus $\mathbf{A}$ is not invertible. In particular, recall that we are solving two vector equations simultaneously in these computations: $\mathbf{A}\mathbf{b}_1 = \mathbf{e}_1$ and $\mathbf{A}\mathbf{b}_2 = \mathbf{e}_2$. If we consider the first of these and observe the row-reduction

$$\begin{bmatrix} 2 & 1 & 1 \\ -6 & -3 & 0 \end{bmatrix} \rightarrow \begin{bmatrix} 1 & \frac{1}{2} & 0 \\ 0 & 0 & 1 \end{bmatrix}$$

we see that this system of equations is inconsistent—the last row of the augmented matrix is equivalent to the equation $0b_{11} + 0b_{12} = 1$, where $\mathbf{b} = [b_{11}\ b_{12}]^T$. This is yet another way of saying that $\mathbf{A}$ does not have an inverse.

The above two examples together show us, in general, how we answer two questions at once: does the square matrix $\mathbf{A}$ have an inverse? And if so, what is $\mathbf{A}^{-1}$? In a computational sense, we can simply row-reduce $\mathbf{A}$ augmented with the appropriate identity matrix and then observe if $\mathbf{A}$ has a pivot position in every row. If $\mathbf{A}$ is row equivalent to the appropriately sized identity matrix, then $\mathbf{A}$ is invertible and $\mathbf{A}^{-1}$ will be revealed through the row-reduction.

We close this section with a formal statement of a theorem that summarizes our discussion. Note particularly how this result extends theorem 1.6.3 and demonstrates the theme of linear algebra: one idea from several perspectives. We will refer to this result as *The Invertible Matrix Theorem.*

Theorem 1.8.2   (The Invertible Matrix Theorem) Let $\mathbf{A}$ be an $n \times n$ matrix. The following statements are equivalent:

a. $\mathbf{A}$ is invertible.

b. The columns of $\mathbf{A}$ are linearly independent.

c. The columns of $\mathbf{A}$ span $\mathbb{R}^n$.

d. $\mathbf{A}$ has a pivot position in every column.

e. $\mathbf{A}$ has a pivot position in every row.

f. $\mathbf{A}$ is row equivalent to $\mathbf{I}_n$.

g. For each $\mathbf{b} \in \mathbb{R}^n$, the equation $\mathbf{A}\mathbf{x} = \mathbf{b}$ has a unique solution.

In addition to being of great theoretical significance, inverse matrices find many key applications. We investigate one such use in the following subsection.

1.8.1 Computer graphics

Linear algebra is the engine that drives computer animations. While animated movies originally were constructed by artists hand-drawing thousands of similar sketches that were photographed and played in sequence, today such films are created entirely with computers. Once a figure has been constructed, moving the image around the screen is essentially an exercise in matrix multiplication.

Every pixel in an image on a computer screen can be represented through coordinates. For an elementary example, consider an animated figure which, at a given point in time, has its hand located at the point $(3, 4)$. To see how a basic animation can be built, assume further that the figure's elbow is at the origin $(0, 0)$, and that an animator wishes to make the hand wave back and forth. This enables us to represent the forearm of the figure with the vector $\mathbf{v} = [3\ 4]^{\mathrm{T}}$.

If we now consider the matrix

$$\mathbf{R} = \begin{bmatrix} \sqrt{3}/2 & -1/2 \\ 1/2 & \sqrt{3}/2 \end{bmatrix}$$

and apply the matrix $\mathbf{R}$ to the vector $\mathbf{v}$, we see that the product is

$$\mathbf{R}\mathbf{v} = \begin{bmatrix} \sqrt{3}/2 & -1/2 \\ 1/2 & \sqrt{3}/2 \end{bmatrix} \begin{bmatrix} 3 \\ 4 \end{bmatrix} = \begin{bmatrix} 3\sqrt{3} - 4/2 \\ 3 + 4\sqrt{3}/2 \end{bmatrix} \approx \begin{bmatrix} 0.598 \\ 4.964 \end{bmatrix}$$

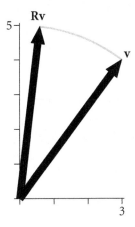

Figure 1.13 The vectors
$\mathbf{v} = [3 \quad 4]$ and $\mathbf{Rv} =$
$[0.598 \ 4.964]^{\mathrm{T}}$.

Thus, the figure's hand is now located at the point $(0.598, 4.964)$. In fact, the hand has been rotated $30°$ counterclockwise about the origin, as shown in figure 1.13.

The matrix $\mathbf{R}$ is known as a *rotation matrix*; its impact on any vector is to rotate the vector $30°$ counterclockwise about the origin. One way to see why this is so is to compute the vectors $\mathbf{Re}_1$ and $\mathbf{Re}_2$, where $\mathbf{e}_1$ and $\mathbf{e}_2$ are the columns of the 2×2 identity matrix. Since each of those two vectors is rotated $30°$ when multiplied by $\mathbf{R}$, the same thing happens to any vector in $\mathbb{R}^2$, because any such vector may be written as a linear combination of $\mathbf{e}_1$ and $\mathbf{e}_2$.

Not only do computer animations show one application of matrix–vector multiplication, but they also demonstrate the need for inverse matrices. For instance, suppose we knew that the matrix $\mathbf{R}$ had been applied to some unknown vector $\mathbf{v}$ and that the result was

$$\mathbf{Rv} = \begin{bmatrix} 2 \\ 5 \end{bmatrix}$$

That is, a hand located at some unknown point $\mathbf{v}$ was waved and had been moved to the new point $(2, 5)$. An animator might want to wave the hand back so that it ended up at its original location, which is again represented by the vector $\mathbf{v}$. To do so, he must answer the question "for which vector $\mathbf{v}$ is $\mathbf{Rv} = [2 \ 5]^{\mathrm{T}}$?"

We now know that one way to solve for $\mathbf{v}$ is to use the inverse of $\mathbf{R}$. The matrix $\mathbf{R}$ is clearly invertible because its columns are linearly independent; we can compute $\mathbf{R}^{-1}$ in the standard way to find that

$$\mathbf{R}^{-1} = \begin{bmatrix} \sqrt{3}/2 & 1/2 \\ -1/2 & \sqrt{3}/2 \end{bmatrix}$$

We can solve for **v** by computing

$$\mathbf{v} = \mathbf{R}^{-1}(\mathbf{Rv}) = \mathbf{R}^{-1}\begin{bmatrix} 2 \\ 5 \end{bmatrix}$$

so that

$$\mathbf{v} = \mathbf{R}^{-1}\begin{bmatrix} 2 \\ 5 \end{bmatrix} = \begin{bmatrix} \sqrt{3}/2 & 1/2 \\ -1/2 & \sqrt{3}/2 \end{bmatrix}\begin{bmatrix} 2 \\ 5 \end{bmatrix} \approx \begin{bmatrix} 4.232 \\ 3.330 \end{bmatrix}$$

Of course, in actual animations, we would not wave the hand by a single 30° rotation, but rather through a sequence of consecutive small rotations, for instance, 1-degree rotations. Again, computers enable us to do thousands of such computations almost instantly and make amazing animations possible.

We consider an additional example to see the role of matrices to store data as well as matrices and their inverses to transform the data.

Example 1.8.4   Consider the matrix

$$\mathbf{B} = \begin{bmatrix} 0 & 1 \\ 1 & 0 \end{bmatrix}$$

Let $\mathbf{v}_1 = [2\ 1]^T$, $\mathbf{v}_2 = [3\ 3]^T$, and $\mathbf{v}_3 = [4\ 0]^T$ be the vertices of a triangle in the plane. Compute $\mathbf{Bv}_1$, $\mathbf{Bv}_2$, and $\mathbf{Bv}_3$. Sketch a picture of the new triangle that has resulted from applying the matrix **B** to the vertices $(2, 1)$, $(3, 3)$, and $(4, 0)$. What is the impact of the matrix **B** on each point? Finally, determine the inverse of **B**. What do you observe?

Solution.   We observe first that

$$\mathbf{Bv}_1 = \begin{bmatrix} 0 & 1 \\ 1 & 0 \end{bmatrix}\begin{bmatrix} 2 \\ 1 \end{bmatrix} = \begin{bmatrix} 1 \\ 2 \end{bmatrix}, \quad \mathbf{Bv}_2 = \begin{bmatrix} 0 & 1 \\ 1 & 0 \end{bmatrix}\begin{bmatrix} 3 \\ 3 \end{bmatrix} = \begin{bmatrix} 3 \\ 3 \end{bmatrix}, \quad \text{and}$$

$$\mathbf{Bv}_1 = \begin{bmatrix} 0 & 1 \\ 1 & 0 \end{bmatrix}\begin{bmatrix} 4 \\ 0 \end{bmatrix} = \begin{bmatrix} 0 \\ 4 \end{bmatrix}$$

From these calculations, we see that multiplying by **B** moves a given point to a new point that corresponds to the one found by switching the coordinates of the given point. Geometrically, the matrix **B** accomplishes a reflection across the line $y = x$ in the plane, as we can see in figure 1.14.

Moreover, if we think about how we might undo reflection across the line $y = x$, it is clear that to restore a point to its original location, we need to reflect the point back across the line. Said differently, the inverse of the matrix **B** must be the matrix itself. We can confirm that $\mathbf{B}^{-1} = \mathbf{B}$ by computing the product

$$\mathbf{BB} = \begin{bmatrix} 0 & 1 \\ 1 & 0 \end{bmatrix}\begin{bmatrix} 0 & 1 \\ 1 & 0 \end{bmatrix} = \mathbf{I}$$

It is noteworthy that the calculations of $\mathbf{Bv}_1$, $\mathbf{Bv}_2$, and $\mathbf{Bv}_3$ can be simplified into a single matrix product if we let $\mathbf{T} = [\mathbf{v}_1\ \mathbf{v}_2\ \mathbf{v}_3]$. That is, the matrix **T** holds the

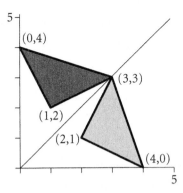

Figure 1.14 The triangle with vertices $\mathbf{v}_1 = [2 \ \ 1]^T$, $\mathbf{v}_2 = [3 \ \ 3]^T$, and $\mathbf{v}_3 = [4 \ \ 0]^T$ and its image under multiplication by the matrix $\mathbf{B}$.

coordinates of the three points in the given triangle; the product $\mathbf{BT}$ is then the image of the triangle under multiplication by the matrix $\mathbf{B}$. A more complicated polygonal figure than a triangle would be stored in a matrix with additional columns.

Of course, the actual work of computer animations is much more complicated than what we have presented here. Nonetheless, matrix multiplication is the platform on which the entire enterprise of animated films is built. In addition to achieving rotations and reflections, matrices can be used to dilate (or magnify) images, to shear images, and even to translate them (provided that we are clever about the coordinate system we use to represent points). Finally, matrices are even essential to the storage of images, as each column of a matrix can be viewed as a data point in an image. More about the application of matrices and their inverses to computer graphics can be learned in one of the projects found at the end of this chapter. In addition, a deeper discussion of the notion of *linear transformations* (of which reflection and rotation matrices are a part) can be found in appendix D.

1.8.2 Matrix inverses using *Maple*

Certainly we can use *Maple*'s row-reduction commands to find inverses of matrices. However, an even simpler command exists that enables us to avoid having to enter the corresponding identity matrix. Let us consider the two matrices from examples 1.8.2 and 1.8.3. Let

$$\mathbf{A} = \begin{bmatrix} 2 & 1 & -2 \\ 1 & 1 & -1 \\ -2 & -1 & 3 \end{bmatrix}$$

If we enter the command

```
> MatrixInverse(A);
```

we see the resulting output which is indeed $\mathbf{A}^{-1}$,

$$\begin{bmatrix} 2 & -1 & 1 \\ -1 & 2 & 0 \\ 1 & 0 & 1 \end{bmatrix}$$

For the matrix

$$\mathbf{A} = \begin{bmatrix} 2 & 1 \\ -6 & -3 \end{bmatrix}$$

executing the command > `MatrixInverse(A);` produces the output

```
Error, (in LinearAlgebra:-LA_Main:-MatrixInverse)
singular matrix
```

which is *Maple's* way of saying "$\mathbf{A}$ is not invertible."

Exercises 1.8   In exercises 1–5, find the inverse of each matrix (doing the computations by hand), or show that the inverse does not exist.

1. $\begin{bmatrix} 2 & 1 \\ 2 & 2 \end{bmatrix}$

2. $\begin{bmatrix} 5 & 0 \\ 0 & -3 \end{bmatrix}$

3. $\begin{bmatrix} 2 & -1 \\ -4 & 2 \end{bmatrix}$

4. $\begin{bmatrix} 1 & 2 & -1 \\ 0 & 1 & 3 \\ 0 & 0 & 2 \end{bmatrix}$

5. $\begin{bmatrix} 1 & -2 & -1 \\ -1 & 1 & 0 \\ 1 & 3 & 4 \end{bmatrix}$

6. Let $\mathbf{A} = \begin{bmatrix} 1 & 3 \\ 1 & 4 \end{bmatrix}$ and $\mathbf{b}_1 = \begin{bmatrix} -3 \\ 5 \end{bmatrix}$, $\mathbf{b}_2 = \begin{bmatrix} 2 \\ -7 \end{bmatrix}$, $\mathbf{b}_3 = \begin{bmatrix} 11 \\ 4 \end{bmatrix}$. Find $\mathbf{A}^{-1}$ and use it to solve the equations $\mathbf{Ax} = \mathbf{b}_1$, $\mathbf{Ax} = \mathbf{b}_2$, and $\mathbf{Ax} = \mathbf{b}_3$. In addition, show how you can use row reduction to solve all three of these equations simultaneously.

7. Let $\mathbf{A} = \begin{bmatrix} 1 & -3 \\ -2 & 6 \end{bmatrix}$ and $\mathbf{b}_1 = \begin{bmatrix} 10 \\ -20 \end{bmatrix}$, $\mathbf{b}_2 = \begin{bmatrix} -1/2 \\ 1 \end{bmatrix}$, $\mathbf{b}_3 = \begin{bmatrix} 2 \\ 1 \end{bmatrix}$. Solve the equations $\mathbf{Ax} = \mathbf{b}_1$, $\mathbf{Ax} = \mathbf{b}_2$, and $\mathbf{Ax} = \mathbf{b}_3$. What do you observe about the matrix $\mathbf{A}$?

8. Let $\mathbf{A} = \begin{bmatrix} 1 & -2 \\ 1 & 2 \end{bmatrix}$ and $\mathbf{b} = \begin{bmatrix} 3 \\ 5 \end{bmatrix}$. Without doing any computations, explain why $\mathbf{b}$ may be written as a linear combination of the columns of $\mathbf{A}$.

Then execute computations to find the explicit weights by which **b** is a linear combination of the columns of **A**.

9. Let **E** be the *elementary matrix* given by $\mathbf{E} = \begin{bmatrix} 1 & 0 & 0 \\ 0 & 0 & 1 \\ 0 & 1 & 0 \end{bmatrix}$. Note that **E** is obtained by interchanging rows 2 and 3 of the 3×3 identity matrix. Choose a 3×3 matrix **A**, and compute **EA**. What is the effect on **A** of multiplication by **E**?

10. Without doing any row-reduction, determine $\mathbf{E}^{-1}$ where **E** is the matrix defined in exercise 9. (Hint: $\mathbf{E}^{-1}\mathbf{EI} = \mathbf{I}$. Think about the impact that **E** has on **I**, and then what $\mathbf{E}^{-1}$ must accomplish.)

11. Let **E** be the elementary matrix given by $\mathbf{E} = \begin{bmatrix} 1 & 0 & 0 \\ 0 & c & 0 \\ 0 & 0 & 1 \end{bmatrix}$. Note that **E** is obtained by scaling the second row of the 3×3 identity matrix by the constant c. Choose a 3×3 matrix **A**, and compute **EA**. What is the effect on **A** of multiplication by **E**?

12. Without doing any row reduction, determine $\mathbf{E}^{-1}$ where **E** is the matrix defined in exercise 11. What do you observe?

13. Let **E** be the elementary matrix given by $\mathbf{E} = \begin{bmatrix} 1 & 0 & 0 \\ 0 & 1 & 0 \\ a & 0 & 1 \end{bmatrix}$. Note that **E** is obtained by applying the row operation of taking a times row 1 of the 3×3 identity matrix and adding it to row 3 to form a new row 3. Choose a 3×3 matrix **A**, and compute **EA**. What is the effect on **A** of multiplication by **E**?

14. Without doing any row reduction, determine $\mathbf{E}^{-1}$ where **E** is the matrix defined in exercise 13. (Hint: $\mathbf{E}^{-1}\mathbf{EI} = \mathbf{I}$. Think about the impact that **E** has on **I**, and then what $\mathbf{E}^{-1}$ must accomplish.)

15. Let $\mathbf{A} = \begin{bmatrix} 1/\sqrt{2} & -1/\sqrt{2} \\ 1/\sqrt{2} & 1/\sqrt{2} \end{bmatrix}$. Compute $\mathbf{A}^{-1}$. What do you observe about the relationship between **A** and $\mathbf{A}^{-1}$?

16. Let θ be any real number and $\mathbf{A} = \begin{bmatrix} \cos\theta & -\sin\theta \\ \sin\theta & \cos\theta \end{bmatrix}$. Compute $\mathbf{A}^{T}$ and $\mathbf{A}^{T}\mathbf{A}$. What do you observe about the relationship between **A** and $\mathbf{A}^{T}$?

17. Let **A** and **B** be invertible $n \times n$ matrices with inverses $\mathbf{A}^{-1}$ and $\mathbf{B}^{-1}$, respectively. Show that **AB** is also an invertible matrix by finding $(\mathbf{AB})^{-1}$ in terms of $\mathbf{A}^{-1}$ and $\mathbf{B}^{-1}$.

18. Let **A** be an invertible matrix. Explain why $\mathbf{A}^{-1}$ is also invertible, and find $(\mathbf{A}^{-1})^{-1}$.

19. Show that if **A** is an invertible $n \times n$ matrix, then its inverse is unique. (Hint: suppose that both **B** and **C** are inverses of **A**. What can you say about **AB** and **AC**?)

20. For real numbers a and b, the Zero Product Property states that "if $a \cdot b = 0$, then $a = 0$ or $b = 0$." Said differently, if $a \neq 0$ and $b \neq 0$, then $a \cdot b \neq 0$. Let $\mathbf{0}$ be the 2×2 zero matrix (i.e., all entries are zero). Does the Zero Product Property hold for matrices? That is, can you find two nonzero matrices $\mathbf{A}$ and $\mathbf{B}$ such that $\mathbf{AB} = \mathbf{0}$? Can you find such matrices where *none* of the entries in $\mathbf{A}$ or $\mathbf{B}$ are zero? If so, what kind of matrices are $\mathbf{A}$ and $\mathbf{B}$?

21. Does there exist a 2×2 matrix $\mathbf{A}$, none of whose entries are zero, such that $\mathbf{A}^2 = \mathbf{0}$?

22. Does there exist a 2×2 matrix $\mathbf{A}$ other than the identity matrix such that $\mathbf{A}^2 = \mathbf{I}$? What is special about such a matrix?

23. Let $\mathbf{D}$ be a diagonal matrix, $\mathbf{P}$ an invertible matrix, and $\mathbf{A} = \mathbf{PDP}^{-1}$. Using the expression $\mathbf{PDP}^{-1}$ for $\mathbf{A}$, compute and simplify the matrix $\mathbf{A}^2 = \mathbf{A} \cdot \mathbf{A}$. Do likewise for $\mathbf{A}^3 = \mathbf{A} \cdot \mathbf{A} \cdot \mathbf{A}$. What will be the simplified form of $\mathbf{A}^n$ in terms of $\mathbf{P}$, $\mathbf{D}$, and $\mathbf{P}^{-1}$?

24. Let $\mathbf{A}$ be the matrix $\begin{bmatrix} a & b \\ c & d \end{bmatrix}$. Find conditions on a, b, c, and d that guarantee that $\mathbf{Ax} = \mathbf{0}$ has infinitely many solutions. What must therefore be true about a, b, c, and d in order for $\mathbf{A}$ to be invertible?

25. Let $\mathbf{A} = \begin{bmatrix} 1/2 & \sqrt{3}/2 \\ -\sqrt{3}/2 & 1/2 \end{bmatrix}$ and $\mathbf{v}_1, \mathbf{v}_2, \mathbf{v}_3$ be the vectors that emanate from the origin to the vertices of the triangle given by $(2, 1)$, $(3, 3)$, and $(4, 0)$. Compute the new triangle that results from applying the matrix $\mathbf{A}$ to the given vertices, and sketch a picture of the original triangle and the resulting image. What is the effect of multiplying by $\mathbf{A}$?

26. Suppose that $\mathbf{A}$ in exercise 25 was applied to a different set of three unknown vectors $\mathbf{x}_1$, $\mathbf{x}_2$, and $\mathbf{x}_3$. The resulting output from these products is

$$\mathbf{Ax}_1 = \begin{bmatrix} -4 \\ 2 \end{bmatrix}, \quad \mathbf{Ax}_2 = \begin{bmatrix} 0 \\ 3 \end{bmatrix}, \quad \text{and } \mathbf{Ax}_3 = \begin{bmatrix} 2 \\ 1 \end{bmatrix}$$

In other words, the new image after multiplying by $\mathbf{A}$ is the triangle whose vertices are $(-4, 2)$, $(0, 3)$, and $(2, 1)$.

Determine the exact vectors $\mathbf{x}_1$, $\mathbf{x}_2$, and $\mathbf{x}_3$ and sketch the original triangle that was mapped to the triangle with vertices $(-4, 2)$, $(0, 3)$, and $(2, 1)$.

27. Consider the matrix

$$\mathbf{B} = \begin{bmatrix} 0 & -1 \\ 1 & 0 \end{bmatrix}$$

Let $\mathbf{v}_1 = [2 \ 1]^T$, $\mathbf{v}_2 = [3 \ 3]^T$, and $\mathbf{v}_3 = [4 \ 0]^T$. Compute $\mathbf{Bv}_1$, $\mathbf{Bv}_2$, and $\mathbf{Bv}_3$. Sketch a picture of the new triangle that has resulted from applying the matrix $\mathbf{B}$ to the vertices $(1, 1)$, $(2, 3)$, and $(4, 0)$. What is the geometric effect of the matrix $\mathbf{B}$ on each point?

28. Determine the inverse of **B** in exercise 27. What do you observe?

29. An unknown 2×2 matrix **C** is applied to the two vectors $\mathbf{v}_1 = [1\ \ 1]^T$ and $\mathbf{v}_2 = [2\ \ 3]^T$, and the results are $\mathbf{Cv}_1 = [0.1\ \ 0.7]^T$ and $\mathbf{Cv}_2 = [-0.1\ \ 1.8]^T$. Determine the entries in the matrix **C**.

30. Suppose that a computer graphics programmer decides to use the matrix

$$\mathbf{A} = \begin{bmatrix} 1/\sqrt{2} & 1/\sqrt{2} \\ 1/\sqrt{2} & 1/\sqrt{2} \end{bmatrix}$$

Why is the programmer's choice a bad one? What will be the result of applying this matrix to any collection of points?

31. Suppose that for a large population that stays relatively constant, people are classified as living in urban, suburban, or rural settings. Moreover, assume that the probabilities of the various possible transitions are given by the following table:

Future location ($\downarrow$)/current location ($\rightarrow$)	U(%)	S(%)	R(%)
Urban	92	3	2
Suburban	7	96	10
Rural	3	1	88

Given that the population of 250 million in a certain year is distributed among 100 million urban, 100 million suburban, and 50 million rural, determine the population distribution in each of the preceding two years.

32. Car-owners can be grouped into classes based on the vehicles they own. A study of owners of sedans, minivans, and sport-utility vehicles shows that the likelihood that an owner of one of these automobiles will replace it with another of the same or different type is given by the table

Future vehicle ($\downarrow$)/ current vehicle ($\rightarrow$)	Sedan(%)	Minivan(%)	SUV(%)
Sedan	91	3	2
Minivan	7	95	8
sUV	2	2	90

If there are currently 100 000 sedans, 60 000 minivans, and 80 000 SUVs among the owners being studied, determine the distribution of vehicles among the population before each current owner replaced his or her previous vehicle.

33. Decide whether each of the following sentences is true or false. In every case, write one sentence to support your answer.

(a) If $\mathbf{A}$ is a matrix with a pivot in every row, then $\mathbf{A}$ is invertible.

(b) If $\mathbf{A}$ is an invertible matrix, then its columns are linearly independent.

(c) If $\mathbf{Ax} = \mathbf{b}$ has a unique solution, then $\mathbf{A}$ is an invertible matrix.

(d) If $\mathbf{A}$ and $\mathbf{B}$ are invertible matrices, then $(\mathbf{AB})^{-1}$ exists and $(\mathbf{AB})^{-1} = \mathbf{A}^{-1}\mathbf{B}^{-1}$.

(e) If $\mathbf{A}$ is a square matrix row equivalent to the identity matrix, then $\mathbf{A}$ is invertible.

(f) If $\mathbf{A}$ is a square matrix and $\mathbf{Ax} = \mathbf{b}$ has a solution for a given vector $\mathbf{b}$, then $\mathbf{Ax} = \mathbf{c}$ has a solution for every choice of $\mathbf{c}$.

(g) If $\mathbf{R}$ is a matrix that reflects points across a line through the origin, then $\mathbf{R}^{-1} = \mathbf{R}$.

(h) If $\mathbf{A}$ and $\mathbf{B}$ are 2×2 matrices with all nonzero entries, then $\mathbf{AB}$ cannot equal the 2×2 zero matrix.

1.9 The determinant of a matrix

The Invertible Matrix Theorem (theorem 1.8.2) tells us that there are several different ways to determine whether or not a matrix is invertible, and hence whether or not an $n \times n$ system of linear equations has a unique solution. There is at least one more useful way to characterize invertibility, and that is through the concept of a determinant. As seen in exercise 24 of section 1.8, it may be shown through row-reduction that the general 2×2 matrix

$$\begin{bmatrix} a & b \\ c & d \end{bmatrix}$$

is invertible if and only if $ad - bc \neq 0$. We call the quantity $(ad - bc)$ the *determinant* of the matrix $\mathbf{A}$, and write[4] $\det(\mathbf{A}) = ad - bc$. Note that this expression provides a condition on the entries of matrix $\mathbf{A}$ that determines whether or not $\mathbf{A}$ is invertible.

We can explore similar ideas for larger matrices. For example, if we take an arbitrary 3×3 matrix

$$\mathbf{A} = \begin{bmatrix} a_{11} & a_{12} & a_{13} \\ a_{21} & a_{22} & a_{23} \\ a_{31} & a_{32} & a_{33} \end{bmatrix}$$

and row-reduce in order to explore conditions under which the matrix has a pivot position in every row, it turns out to be necessary that the quantity

$$D = a_{11}a_{22}a_{33} - a_{11}a_{23}a_{32} - a_{12}a_{21}a_{33} + a_{12}a_{23}a_{31} + a_{13}a_{21}a_{32} - a_{13}a_{22}a_{31}$$

---

[4] Some authors use the notation $|\mathbf{A}|$ instead of $\det(\mathbf{A})$.

is nonzero. Grouping and factoring, we see that D may be rewritten in the form

$$D = a_{11}(a_{22}a_{33} - a_{23}a_{32}) - a_{12}(a_{21}a_{33} - a_{23}a_{31}) + a_{13}(a_{21}a_{32} - a_{22}a_{31}) \quad (1.9.1)$$

We again call this quantity D the *determinant* of the matrix $\mathbf{A}$. In (1.9.1) we see evidence of the fact that determinants of larger matrices can be defined recursively in terms of smaller matrices found within the original matrix $\mathbf{A}$. For example, letting

$$\mathbf{A}_{11} = \begin{bmatrix} a_{22} & a_{23} \\ a_{32} & a_{33} \end{bmatrix}$$

it follows that $\det(\mathbf{A}_{11}) = a_{22}a_{33} - a_{23}a_{32}$, which is the expression multiplied by a_{11} in (1.9.1). More generally, if we let $\mathbf{A}_{ij}$ be the *submatrix* defined by deleting row i and column j of the original matrix $\mathbf{A}$, then we see from (1.9.1) that

$$D = a_{11}\det(\mathbf{A}_{11}) - a_{12}\det(\mathbf{A}_{12}) + a_{13}\det(\mathbf{A}_{13})$$

The formal definition of the determinant of an $n \times n$ matrix is given through a similar recursive process.

Definition 1.9.1   The *determinant* of an $n \times n$ matrix $\mathbf{A}$ with entries a_{ij} is defined to be the number given by

$$\det(\mathbf{A}) = a_{11}\det(\mathbf{A}_{11}) - a_{12}\det(\mathbf{A}_{12}) + \cdots + (-1)^{n+1}a_{1n}\det(\mathbf{A}_{1n}) \quad (1.9.2)$$

where $\mathbf{A}_{ij}$ is the matrix found by deleting row i and column j of $\mathbf{A}$.

We next consider an example to see some concrete computations.

Example 1.9.1   Compute the determinant of the matrix

$$\mathbf{A} = \begin{bmatrix} 2 & -1 & 1 \\ 1 & 1 & 2 \\ -3 & 0 & -3 \end{bmatrix}$$

In addition, determine if $\mathbf{A}$ is invertible.

Solution.   By definition,

$$\det \begin{bmatrix} 2 & -1 & 1 \\ 1 & 1 & 2 \\ -3 & 0 & -3 \end{bmatrix} = 2\det \begin{bmatrix} 1 & 2 \\ 0 & -3 \end{bmatrix} - (-1)\det \begin{bmatrix} 1 & 2 \\ -3 & -3 \end{bmatrix} + 1\det \begin{bmatrix} 1 & 1 \\ -3 & 0 \end{bmatrix}$$

$$= 2(-3 - 0) + 1(-3 - (-6)) + 1(0 - (-3))$$

$$= -6 + 3 + 3$$

$$= 0$$

Next, to determine whether or not **A** is invertible, we row-reduce **A** to see if **A** has a pivot position in every row. Doing so, we find that

$$
\begin{bmatrix} 2 & -1 & 1 \\ 1 & 1 & 2 \\ -3 & 0 & -3 \end{bmatrix} \rightarrow \begin{bmatrix} 1 & 0 & 1 \\ 0 & 1 & 1 \\ 0 & 0 & 0 \end{bmatrix}
$$

Thus, we see that **A** does not have a pivot in every row, and therefore **A** is not invertible.

Of course, we should note that the primary motivation for the concept of the determinant comes from the question, "is **A** invertible?" Indeed, one reason the 3×3 matrix in the above example is not invertible is precisely because its determinant is zero. Later in this section, we will formally establish the connection between the value of the determinant and the invertibility of a general $n \times n$ matrix.

It is clear at this point that determinants of most $n \times n$ matrices with $n \geq 3$ require a substantial number of computations. Certain matrices, however, have particularly simple determinants to calculate, as the following example demonstrates.

Example 1.9.2   Compute the determinant of the matrix

$$
A = \begin{bmatrix} 2 & -2 & 7 \\ 0 & -5 & 3 \\ 0 & 0 & 4 \end{bmatrix}
$$

In addition, determine if **A** is invertible.

Solution.   Again using the definition, we see that

$$
\det(A) = 2 \det \begin{bmatrix} -5 & 3 \\ 0 & 4 \end{bmatrix} - (-2) \det \begin{bmatrix} 0 & 3 \\ 0 & 4 \end{bmatrix} + 7 \det \begin{bmatrix} 0 & -5 \\ 0 & 0 \end{bmatrix}
$$

$$
= 2(-5 \cdot 4 - 2 \cdot 0) + 2(0 - 0) + 7(0 - 0)
$$

$$
= 2(-5)(4) = -40
$$

Note particularly that the determinant of **A** is the product of its diagonal entries. Moreover, **A** clearly has a pivot position in every row, and so by this fact (or equivalently by the nonzero determinant of **A**) we see that **A** is invertible.

In general, the determinant of any triangular matrix (one where all entries either below or above the diagonal are zero) is simply the product of its diagonal entries. There are other interesting properties that the determinant has, several of which are explored in the next example for the 2×2 case.

Example 1.9.3   Let

$$
A = \begin{bmatrix} a & b \\ c & d \end{bmatrix}
$$

be an arbitrary 2×2 matrix. Explore the effect of elementary row operations on the determinant of **A**.

Solution.   First, let us consider a row swap, calling $\mathbf{A}_1$ the matrix

$$\mathbf{A}_1 = \begin{bmatrix} c & d \\ a & b \end{bmatrix}$$

We observe immediately that $\det(\mathbf{A}) = ad - bc$ and $\det(\mathbf{A}_1) = cb - ad = -\det(\mathbf{A})$.

We next consider scaling; let $\mathbf{A}_2$ be the matrix whose first row is $[ka \ \ kb]$, a scaled version of row 1 in **A**. We see that $\det(\mathbf{A}_2) = kad - kbc = k(ad - bc) = k \cdot \det(\mathbf{A})$.

Finally, replacing, say, row 2 of **A** by the sum of k times row 1 with itself, we arrive at the matrix

$$\mathbf{A}_3 = \begin{bmatrix} a & b \\ c + ka & d + kb \end{bmatrix}$$

Then $\det(\mathbf{A}_3) = a(d + kb) - b(c + ka) = ad + kab - bc - kab = ad - bc = \det(\mathbf{A})$.

Thus, we see that for the 2×2 case, swapping rows in a matrix changes only the sign of the determinant, scaling a row by a nonzero constant scales the determinant by the same constant, and executing a row replacement does not change the value of the determinant at all. These demonstrate the effect that the three elementary row operations from the process of row-reduction have on a 2×2 matrix **A**.

Given that the general definition of the determinant is recursive, it should not be surprising that the properties witnessed in example 1.9.3 can be shown to hold for $n \times n$ matrices. We state this result formally as our next theorem.

Theorem 1.9.1   Let **A** be an $n \times n$ matrix and k a nonzero constant. Then

a. If two rows of **A** are exchanged to produce matrix **B**, then $\det(\mathbf{B}) = -\det(\mathbf{A})$.

b. If one row of **A** is multiplied by k to produce **B**, then $\det(\mathbf{B}) = k \det(\mathbf{A})$.

c. If **B** results from a row replacement in **A**, then $\det(\mathbf{B}) = \det(\mathbf{A})$.

Theorem 1.9.1 enables us to more clearly see the link between invertibility and determinants. Through a finite number of row interchanges and row replacements, any square matrix **A** may be row-reduced to upper triangular form **U** (where we have all subdiagonal zeros, but we do not necessarily scale to get 1's on the diagonal). It follows from theorem 1.9.1 that

$$\det(\mathbf{A}) = (-1)^k \det(\mathbf{U}),$$

where k is the number of row interchanges needed. Note that since **U** is triangular, its determinant is the product of its diagonal entries, and these entries lie in the pivot locations of **A**. Thus, **A** has a pivot in every row if and only if this determinant is nonzero. Specifically, we have shown that **A** is invertible if and only if $\det(\mathbf{A}) \neq 0$.

To conclude this section, we note that linear algebra has once again afforded an alternate perspective on the problem of solving an $n \times n$ system of linear equations, and we can now add an additional statement involving determinants to the Invertible Matrix Theorem.

Theorem 1.9.2   (Invertible Matrix Theorem) Let **A** be an $n \times n$ matrix. The following statements are equivalent:

a. **A** is invertible.
b. The columns of **A** are linearly independent.
c. The columns of **A** span $\mathbb{R}^n$.
d. **A** has a pivot position in every column.
e. **A** has a pivot position in every row.
f. **A** is row equivalent to $\mathbf{I}_n$.
g. For each $\mathbf{b} \in \mathbb{R}^n$, the equation $\mathbf{Ax} = \mathbf{b}$ has a unique solution.
h. $\det(\mathbf{A}) \neq 0$.

1.9.1 Determinants using *Maple*

Obviously for most square matrices of size greater than 3×3, the computations necessary to find determinants are tedious and present potential for error. As with other concepts that require large numbers of arithmetic operations, *Maple* offers a single command that enables us to take advantage of the program's computational powers. Given a square matrix **A** of any size, we simply enter

```
> Determinant(A);
```

As we explore properties of determinants in the exercises of this section, it will prove useful to be able to generate random matrices. Within the LinearAlgebra package in *Maple*, one accomplishes this for a 3×3 matrix with the command

```
> RandomMatrix(3);
```

For example, if we wanted to consider the determinant of a random matrix **A** we could enter the code

```
> A := RandomMatrix(3);
> det(A);
```

See exercise 11 for a particular instance where this code will be useful.

Exercises 1.9   Compute (by hand) the determinant of each of the following matrices in exercises 1–7, and hence state whether or not the matrix is invertible.

1. $A = \begin{bmatrix} 2 & 1 \\ 2 & 2 \end{bmatrix}$

2. $A = \begin{bmatrix} 2 & 4 \\ 1 & 2 \end{bmatrix}$

3. $A = \begin{bmatrix} 2 & 1 & -3 \\ 2 & 2 & 5 \\ 2 & 3 & -1 \end{bmatrix}$

4. $A = \begin{bmatrix} 2 & 1 & 3 \\ 2 & 2 & 4 \\ 2 & 3 & 5 \end{bmatrix}$

5. $A = \begin{bmatrix} -3 & 1 & 0 & 5 \\ 0 & 2 & -4 & 0 \\ 0 & 0 & -7 & 11 \\ 0 & 0 & 0 & 6 \end{bmatrix}$

6. $A = \begin{bmatrix} a & a & d \\ b & b & e \\ c & c & f \end{bmatrix}$

7. I_n, where I_n is the $n \times n$ identity matrix.

8. For which value(s) of h is the matrix $\begin{bmatrix} 1 & 2 \\ -3 & h \end{bmatrix}$ invertible? Explain your answer in at least two different ways.

9. For which value(s) of z is the matrix $\begin{bmatrix} 2-z & 1 \\ 1 & 2-z \end{bmatrix}$ invertible? Why?

10. For which value(s) of z do nontrivial solutions $\mathbf{x}$ to the equation $\begin{bmatrix} 2-z & 1 \\ 1 & 2-z \end{bmatrix} \mathbf{x} = \mathbf{0}$ exist? For one such value of z, determine a nontrivial solution $\mathbf{x}$ to the equation.

11. In a computer algebra system, devise code that will generate two random 3×3 matrices $\mathbf{A}$ and $\mathbf{B}$, and that subsequently computes $\det(\mathbf{A})$, $\det(\mathbf{B})$, and $\det(\mathbf{AB})$. What theorem do you conjecture is true about the relationship between $\det(\mathbf{AB})$ and the individual determinants $\det(\mathbf{A})$ and $\det(\mathbf{B})$?

12. In a computer algebra system, devise code that will generate a random 3×3 matrix $\mathbf{A}$ and that subsequently computes its transpose $\mathbf{A}^\mathrm{T}$, as well as $\det(\mathbf{A})$ and $\det(\mathbf{A}^\mathrm{T})$. What theorem do you conjecture is true about the relationship between $\det(\mathbf{A})$ and $\det(\mathbf{A}^\mathrm{T})$?

13. Use the formula conjectured in exercise 11 above to show that if $\mathbf{A}$ is invertible, then $\det(\mathbf{A}^{-1}) = \dfrac{1}{\det(\mathbf{A})}$. (Hint: $\mathbf{A}\mathbf{A}^{-1} = \mathbf{I}$.)

14. What can you say about the determinant of any square matrix in which one of the columns (or rows) is zero? Why?

15. What can you say about the determinant of any square matrix where one of the columns (or rows) is repeated in the matrix? Why?

16. Suppose that $\mathbf{A}$ is a $n \times n$ matrix and that $\mathbf{Ax} = \mathbf{0}$ has infinitely many solutions. What can you say about $\det(\mathbf{A})$? Why?

17. Suppose that $\mathbf{A}^2$ is not invertible. Can you determine if $\mathbf{A}$ is invertible or not? Explain.

18. Two matrices $\mathbf{A}$ and $\mathbf{B}$ are said to be *similar* if there exists an invertible matrix $\mathbf{P}$ such that $\mathbf{A} = \mathbf{PBP}^{-1}$. What can you say about the determinants of similar matrices?

19. Let $\mathbf{A}$ be an arbitrary 2×2 matrix of the form

$$\begin{bmatrix} a & b \\ c & d \end{bmatrix}$$

where $a \neq 0$ and $\mathbf{A}$ is assumed to be invertible. Working by hand, row reduce the augmented matrix $[\mathbf{A} \ \ \mathbf{I}_2]$ and hence determine a formula for $\mathbf{A}^{-1}$ in terms of the entries of $\mathbf{A}$. What role does $\det(\mathbf{A})$ play in the formula for $\mathbf{A}^{-1}$?

20. Decide whether each of the following sentences is true or false. In every case, write one sentence to support your answer.

   (a) Swapping the rows in a square matrix $\mathbf{A}$ does not change the value of $\det(\mathbf{A})$.
   (b) If $\mathbf{A}$ is a square matrix with a pivot in every column, then $\det(\mathbf{A}) = 0$.
   (c) The determinant of any diagonal matrix is the product of its diagonal entries.
   (d) If $\mathbf{A}$ is an $n \times n$ matrix and $\mathbf{Ax} = \mathbf{b}$ has a unique solution for every $\mathbf{b} \in \mathbb{R}^n$, then $\det(\mathbf{A}) \neq 0$.

1.10 The eigenvalue problem

Another powerful characteristic of linear algebra is the way the subject often allows us to better understand an infinite collection of objects in terms of the properties of a small, finite number of elements in the set. For example, if we have a set of three linearly independent vectors that spans $\mathbb{R}^3$, then every vector in $\mathbb{R}^3$ may be understood as a unique linear combination of the three special vectors in the linearly independent spanning set. Thus, in some ways it is sufficient to understand these three vectors, and to use that knowledge to better understand the rest of the vectors in $\mathbb{R}^3$. In a similar way, as we will see in this section, for an $n \times n$ matrix $\mathbf{A}$ there are up to n important vectors (called *eigenvectors*) that enable us to better understand a variety of properties of the matrix.

The process of matrix multiplication enables us to associate a function with any given matrix $\mathbf{A}$. For example, if $\mathbf{A}$ is a 2×2 matrix, then we may define a function T by the formula

$$T(\mathbf{x}) = \mathbf{Ax} \qquad (1.10.1)$$

Note that the domain of the function T is $\mathbb{R}^2$, the set of all vectors with two entries. Moreover, note that every output of the function T is also a vector in $\mathbb{R}^2$. We therefore use the notation $T : \mathbb{R}^2 \to \mathbb{R}^2$. This is analogous to familiar functions like $f(x) = x^2$, where for every real number input we obtain a real number output ($f : \mathbb{R} \to \mathbb{R}$); the difference here is that for the function T, for every vector input we get a vector output. In what follows, we go in search of special input vectors to the function T for which the corresponding output is particularly simple to compute. The next example will highlight the properties of the vector(s) we seek.

Example 1.10.1 Explore the geometric effect of the matrix

$$\mathbf{A} = \begin{bmatrix} 2 & 1 \\ 1 & 2 \end{bmatrix}$$

on the vectors $\mathbf{u} = [1 \ 0]^\mathsf{T}$ and $\mathbf{v} = [1 \ 1]^\mathsf{T}$ from the perspective of the function $T(\mathbf{x}) = \mathbf{Ax}$.

Solution. We first compute $T(\mathbf{u}) = \mathbf{Au} = [2 \ 1]^\mathsf{T}$. In figure 1.15, we see a plot of the vector $\mathbf{u}$ on the left, and $T(\mathbf{u})$ on the right. This shows that the geometric effect of T on $\mathbf{u}$ is to rotate $\mathbf{u}$ and stretch it. For the vector $\mathbf{v}$, we observe that $T(\mathbf{v}) = \mathbf{Av} = [3 \ 3]^\mathsf{T}$. Graphically, as shown in figure 1.16, it is clear that $T(\mathbf{v})$ is simply a stretch of $\mathbf{v}$ by a factor of 3. Said slightly differently, we might write that

$$T(\mathbf{v}) = \mathbf{Av} = \begin{bmatrix} 3 \\ 3 \end{bmatrix} = 3 \begin{bmatrix} 1 \\ 1 \end{bmatrix} = 3\mathbf{v}$$

This shows that the result of the function T (and hence the matrix $\mathbf{A}$) being applied to the vector $\mathbf{v}$ is particularly simple: $\mathbf{v}$ is only stretched by T.

For any $n \times n$ matrix $\mathbf{A}$, there is an associated function $T : \mathbb{R}^n \to \mathbb{R}^n$ defined by $T(\mathbf{x}) = \mathbf{Ax}$. This function takes a given vector in $\mathbb{R}^n$ and maps it to a corresponding vector in $\mathbb{R}^n$; in every case, we may view this output as resulting from the input vector being stretched and/or rotated. Input vectors that are

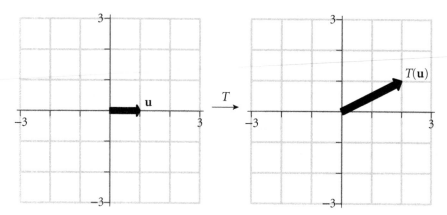

Figure 1.15 The vectors **u** and $T(\mathbf{u})$ in example 1.10.1.

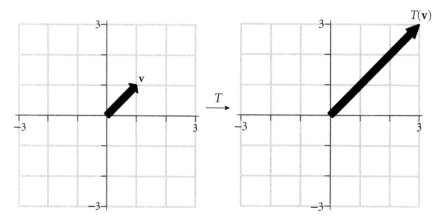

Figure 1.16 The vectors **v** and $T(\mathbf{v})$.

only stretched have corresponding outputs that are simplest to determine: the input vector is simply multiplied by a scalar. To put this another way, for these stretched-only vectors, multiplying them by **A** is equivalent to multiplying them by a constant. Such vectors prove to be important for a host of reasons, and are called the *eigenvectors* of a matrix **A**.

Definition 1.10.1   For a given $n \times n$ matrix **A**, a nonzero vector **v** is said to be an *eigenvector* of **A** if and only if there exists a scalar λ such that

$$\mathbf{A}\mathbf{v} = \lambda \mathbf{v} \tag{1.10.2}$$

The scalar λ is called the *eigenvalue* corresponding to the eigenvector **v**.

In example 1.10.1, we found that the vector $\mathbf{v} = [1 \ 1]^{\mathrm{T}}$ is an eigenvector of the given matrix $\mathbf{A}$ with corresponding eigenvalue 3 since $\mathbf{Av} = 3\mathbf{v}$. What is not yet clear is how we even begin to find eigenvectors and eigenvalues. We will soon see that some of the many different perspectives we can take on systems of linear equations will help us solve this problem.

In general, given an $n \times n$ matrix $\mathbf{A}$, we seek eigenvectors $\mathbf{v}$ that are, by definition, nonzero and satisfy the equation $\mathbf{Av} = \lambda \mathbf{v}$. In one sense, what makes this problem challenging is that neither $\mathbf{v}$ nor λ is initially known. We thus explore some different perspectives on the problem to see if we can highlight the role of either $\mathbf{v}$ or λ. Early in this chapter, we spent significant effort studying homogeneous equations and the circumstances under which they have nontrivial solutions. Here, the eigenvector problem can be rephrased in a similar light. Subtracting $\lambda \mathbf{v}$ from both sides of (1.10.2), we equivalently seek λ and $\mathbf{v}$ such that

$$\mathbf{Av} - \lambda \mathbf{v} = \mathbf{0} \tag{1.10.3}$$

Viewing $\lambda \mathbf{v}$ as $(\lambda \mathbf{I})\mathbf{v}$, we can factor (1.10.3) and write

$$(\mathbf{A} - \lambda \mathbf{I})\mathbf{v} = \mathbf{0} \tag{1.10.4}$$

Now the question becomes, "for which values of λ does (1.10.4) have a nontrivial solution?" At this point, we recall theorem 1.6.2, which tells us that the equation $\mathbf{Bx} = \mathbf{0}$ has only the trivial solution if and only if the matrix $\mathbf{B}$ has a pivot in every column. To have a nontrivial solution, we therefore want $\mathbf{A} - \lambda \mathbf{I}$ to not have a pivot in every column. In (1.10.4), the matrix $\mathbf{A} - \lambda \mathbf{I}$ is square, so by the Invertible Matrix Theorem such a nontrivial solution exists if and only if $\mathbf{A} - \lambda \mathbf{I}$ is not invertible.

This last observation brings us, finally, to determinants. As we saw in Section 1.9, a matrix is invertible if and only if its determinant is nonzero. Therefore, a nontrivial solution to (1.10.4) exists whenever λ is such that $\det(\mathbf{A} - \lambda \mathbf{I}) = 0$. In the next example, we explore how this equation enables us to find the eigenvalues of a matrix $\mathbf{A}$, and hence the eigenvectors as well.

Example 1.10.2   Find the eigenvalues and eigenvectors of the matrix

$$\mathbf{A} = \begin{bmatrix} 2 & 1 \\ 1 & 2 \end{bmatrix}$$

Solution.   As seen in our preceding discussion, by the definition of eigenvalues and eigenvectors, λ is an eigenvalue of $\mathbf{A}$ if and only if the equation $(\mathbf{A} - \lambda \mathbf{I})\mathbf{v} = \mathbf{0}$ has a nontrivial solution. Note first that $\mathbf{A} - \lambda \mathbf{I}$ is the matrix $\mathbf{A}$ with the scalar λ subtracted from each diagonal entry since

$$\mathbf{A} - \lambda \mathbf{I} = \begin{bmatrix} 2 & 1 \\ 1 & 2 \end{bmatrix} - \begin{bmatrix} \lambda & 0 \\ 0 & \lambda \end{bmatrix} = \begin{bmatrix} 2-\lambda & 1 \\ 1 & 2-\lambda \end{bmatrix}$$

We next compute $\det(\mathbf{A} - \lambda \mathbf{I})$ so that we can see which values of λ make this determinant zero. In particular, we have

$$\det(\mathbf{A} - \lambda \mathbf{I}) = \det \begin{bmatrix} 2 - \lambda & 1 \\ 1 & 2 - \lambda \end{bmatrix}$$

$$= (2 - \lambda)^2 - 1$$

$$= \lambda^2 - 4\lambda + 3 \tag{1.10.5}$$

Thus, in order for $\det(\mathbf{A} - \lambda \mathbf{I}) = 0$, λ must satisfy the equation $\lambda^2 - 4\lambda + 3 = 0$. Factoring, $(\lambda - 3)(\lambda - 1) = 0$, and therefore $\lambda = 3$ and $\lambda = 1$ are eigenvalues of $\mathbf{A}$. The value $\lambda = 3$ is not surprising, given our earlier discoveries in example 1.10.1.

Next, we proceed to find the eigenvectors that correspond to each eigenvalue. Beginning with $\lambda = 3$, we seek nonzero vectors $\mathbf{v}$ that satisfy $\mathbf{Av} = 3\mathbf{v}$, or equivalently

$$(\mathbf{A} - 3\mathbf{I})\mathbf{v} = \mathbf{0}$$

This problem is a familiar one: solving a homogeneous system of linear equations for which infinitely many solutions exist. Augmenting $\mathbf{A} - 3\mathbf{I}$ with a column of zeros and row-reducing, we find that

$$\begin{bmatrix} -1 & 1 & 0 \\ 1 & -1 & 0 \end{bmatrix} \rightarrow \begin{bmatrix} 1 & -1 & 0 \\ 0 & 0 & 0 \end{bmatrix}$$

Note that from the very definition of an eigenvector, by which we seek a nontrivial solution to $(\mathbf{A} - \lambda \mathbf{I})\mathbf{v} = \mathbf{0}$, it must be the case at this point that the matrix $\mathbf{A} - \lambda \mathbf{I}$ does not have a pivot in every row. Interpreting the row-reduced matrix with the free variable v_2, we find that the vector $\mathbf{v} = [v_1 \ v_2]^\mathsf{T}$ must satisfy $v_1 - v_2 = 0$. Thus, any vector $\mathbf{v}$ of the form

$$\mathbf{v} = \begin{bmatrix} v_2 \\ v_2 \end{bmatrix} = v_2 \begin{bmatrix} 1 \\ 1 \end{bmatrix}$$

is an eigenvector of $\mathbf{A}$ that corresponds to the eigenvalue $\lambda = 3$. In particular, we observe that any scalar multiple of the vector $\mathbf{v} = [1 \ 1]^\mathsf{T}$ is an eigenvector of $\mathbf{A}$ with associated eigenvalue 3. We say that the set of all eigenvectors associated with eigenvalue 3 is the *eigenspace* corresponding to $\lambda = 3$.

It now only remains to find the eigenvectors associated with $\lambda = 1$. We proceed in the same manner as above, now solving the homogeneous equation $(\mathbf{A} - 1\mathbf{I})\mathbf{v} = \mathbf{0}$. Row-reducing, we find that

$$\begin{bmatrix} 1 & 1 & 0 \\ 1 & 1 & 0 \end{bmatrix} \rightarrow \begin{bmatrix} 1 & 1 & 0 \\ 0 & 0 & 0 \end{bmatrix}$$

and therefore the eigenvector $\mathbf{v}$ must satisfy $v_1 + v_2 = 0$ and have the form

$$\mathbf{v} = \begin{bmatrix} -v_2 \\ v_2 \end{bmatrix} = v_2 \begin{bmatrix} -1 \\ 1 \end{bmatrix}$$

Here, any scalar multiple of $\mathbf{v} = [-1 \ 1]^{\mathrm{T}}$ is an eigenvector of $\mathbf{A}$ corresponding to $\lambda = 1$.

There are several important general observations to be made from example 1.10.2. One is that for any 2×2 matrix, the matrix will have 0, 1, or 2 real eigenvalues. This comes from the fact that $\det(\mathbf{A} - \lambda \mathbf{I})$ is a quadratic function in the variable λ, and therefore can have up to two real zeros. While it is possible to consider complex eigenvalues, we will wait until these arise in our study of systems of differential equations to address them in detail. In addition, we note that there are infinitely many eigenvectors associated with each eigenvalue. Often we will be interested in finding *representative* eigenvectors— ones for which all others with the same eigenvalue are linear combinations. Finally, it is worthwhile to note that the two representative eigenvectors found in example 1.10.2, corresponding respectively to the two distinct eigenvalues, are linearly independent. More on why this is important will be discussed at the end of this section; for now, we remark that it is possible to show that eigenvectors corresponding to distinct eigenvalues are always linearly independent. This fact will be proved in exercise 16.

The observations in the preceding paragraph generalize to the case of $n \times n$ matrices. It may be shown that $\det(\mathbf{A} - \lambda \mathbf{I})$ is a polynomial of degree n in λ. This function is usually called the *characteristic polynomial*; the equation $\det(\mathbf{A} - \lambda \mathbf{I}) = 0$ is typically referred to as the *characteristic equation*. Because the characteristic polynomial has degree n, it follows that $\mathbf{A}$ has up to n real eigenvalues[5].

Next we consider two additional examples that demonstrate some more of the possibilities and important ideas that arise in trying to find the eigenvalues and eigenvectors of a given matrix.

Example 1.10.3   Determine the eigenvalues and eigenvectors of the matrix

$$\mathbf{R} = \begin{bmatrix} 1/\sqrt{2} & -1/\sqrt{2} \\ 1/\sqrt{2} & 1/\sqrt{2} \end{bmatrix}$$

In addition, explore the geometric effect of the function $T(\mathbf{v}) = \mathbf{R}\mathbf{v}$ on vectors in $\mathbb{R}^2$.

---

[5]   See appendix C for a review and discussion of important properties of roots of polynomial equations.

Solution.   We consider the characteristic equation $\det(\mathbf{R} - \lambda \mathbf{I}) = 0$ and hence solve

$$0 = \det \begin{bmatrix} \frac{1}{\sqrt{2}} - \lambda & -\frac{1}{\sqrt{2}} \\ \frac{1}{\sqrt{2}} & \frac{1}{\sqrt{2}} - \lambda \end{bmatrix}$$

$$= \left(\frac{1}{\sqrt{2}} - \lambda \right)^2 + \frac{1}{2}$$

$$= \lambda^2 - \sqrt{2}\lambda + 1$$

By the quadratic formula, it follows that

$$\lambda = \frac{\sqrt{2} \pm \sqrt{2-4}}{2} = \frac{\sqrt{2} \pm i\sqrt{2}}{2}$$

which shows that $\mathbf{R}$ does not have any real eigenvalues. If we explore the geometric effect of $T(\mathbf{v}) = \mathbf{R}\mathbf{v}$ graphically, we can better understand why this is the case. Beginning with the vector $\mathbf{e}_1 = [1 \quad 0]^{\mathrm{T}}$ and computing $\mathbf{R}\mathbf{e}_1 = [1/\sqrt{2} \ 1/\sqrt{2}]^{\mathrm{T}}$, as seen in figure 1.17, we see that the function $T(\mathbf{x}) = \mathbf{R}\mathbf{x}$ rotates the vector $\mathbf{e}_1$ counterclockwise by $\pi/4$ radians, and (as computing the length of each vector shows) there is no stretching involved. Similarly, for the vector $\mathbf{e}_2 = [0 \ 1]^{\mathrm{T}}$, we can see that $\mathbf{R}\mathbf{e}_2 = [-1/\sqrt{2} \ 1/\sqrt{2}]^{\mathrm{T}}$. Just as with the previous vector $\mathbf{e}_1$, we see that the function $T(\mathbf{v}) = \mathbf{R}\mathbf{v}$ simply rotates the vector $\mathbf{e}_2$ counterclockwise by $\pi/4$ radians.

In fact, since every vector in $\mathbb{R}^2$ can be written as a linear combination of $\mathbf{e}_1$ and $\mathbf{e}_2$, it follows that the image $\mathbf{R}\mathbf{v}$ of any vector $\mathbf{v}$ is simply the original vector rotated counterclockwise $\pi/4$ radians. This shows that no vector in $\mathbb{R}^2$ is simply stretched under multiplication by $\mathbf{R}$, and therefore $\mathbf{R}$ has no real eigenvectors.

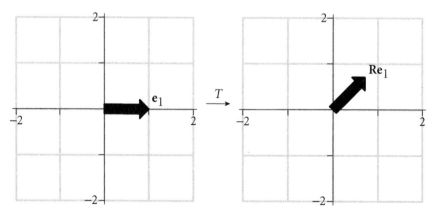

Figure 1.17   The vectors $\mathbf{e}_1$ and $T(\mathbf{e}_1) = \mathbf{R}\mathbf{e}_1$.

Matrices such as **R** in example 1.10.3 with the property that they rotate every vector by a fixed angle (with no stretching factor) are usually called *rotation matrices*.

Other interesting cases arise in the search for eigenvectors when some of the eigenvalues are repeated. That is, when a value λ is a multiple root of the characteristic equation $\det(\mathbf{A} - \lambda\mathbf{I}) = 0$. We explore this further in the next example.

Example 1.10.4 Determine all eigenvalues and eigenvectors of the matrix

$$\mathbf{A} = \begin{bmatrix} 5 & 6 & 2 \\ 0 & -1 & -8 \\ 1 & 0 & -2 \end{bmatrix}$$

Solution. As in previous examples, we first compute $\det(\mathbf{A} - \lambda\mathbf{I})$. Doing so and simplifying yields

$$\det(\mathbf{A} - \lambda\mathbf{I}) = -36 + 15\lambda + 2\lambda^2 - \lambda^3$$

Factoring, it follows that

$$\det(\mathbf{A} - \lambda\mathbf{I}) = -(\lambda + 4)(\lambda - 3)^2$$

Setting the characteristic polynomial equal to zero, it is required that $-(\lambda + 4)(\lambda - 3)^2 = 0$. This shows that **A** has two distinct eigenvalues; moreover, just as with zeros of polynomials, we say that $\lambda = -4$ has multiplicity 1, while $\lambda = 3$ has multiplicity 2.

We now find the eigenvectors corresponding to each eigenvalue. For $\lambda = -4$, we solve the equation $(\mathbf{A} + 4\mathbf{I})\mathbf{v} = \mathbf{0}$, and see by row-reducing that

$$\begin{bmatrix} 9 & 6 & 2 & 0 \\ 0 & 3 & -8 & 0 \\ 1 & 0 & 2 & 0 \end{bmatrix} \rightarrow \begin{bmatrix} 1 & 0 & 2 & 0 \\ 0 & 1 & -\frac{8}{3} & 0 \\ 0 & 0 & 0 & 0 \end{bmatrix}$$

Note that v_3 is a free variable, and that the corresponding eigenvector **v** must have components which satisfy $v_1 + 2v_3 = 0$ and $v_2 - \frac{8}{3}v_3 = 0$, which shows that **v** has form

$$\mathbf{v} = v_3 \begin{bmatrix} -2 \\ \frac{8}{3} \\ 1 \end{bmatrix}$$

Likewise, for $\lambda = 3$, we consider $(\mathbf{A} - 3\mathbf{I})\mathbf{v} = \mathbf{0}$, and row-reduce to find that

$$\begin{bmatrix} 2 & 6 & 2 \\ 0 & -4 & -8 \\ 1 & 0 & -5 \end{bmatrix} \rightarrow \begin{bmatrix} 1 & 0 & -5 \\ 0 & 1 & 2 \\ 0 & 0 & 0 \end{bmatrix}$$

This leads us to see that the corresponding eigenvector has form

$$\mathbf{v} = v_3 \begin{bmatrix} 5 \\ -2 \\ 1 \end{bmatrix}$$

Therefore, we see that for this matrix $\mathbf{A}$, the matrix has two distinct eigenvalues (-4 and 3), and each of these eigenvalues has only one associated linearly independent eigenvector. That is, every eigenvector of $\mathbf{A}$ associated with $\lambda = -4$ is a scalar multiple of $[-2 \ \frac{8}{3} \ 1]^\mathrm{T}$ while every eigenvector associated with $\lambda = 3$ is a scalar multiple of $[5 \ -2 \ 1]^\mathrm{T}$.

In the three preceding examples, we have seen that an $n \times n$ matrix has up to n real eigenvalues. It turns out that there are also up to n linearly independent eigenvectors of the matrix. For many reasons, the best possible scenario is when a matrix has n linearly independent eigenvectors, such as the matrix $\mathbf{A}$ in example 1.10.2. In that 2×2 situation, $\mathbf{A}$ had two distinct real eigenvalues, and two corresponding linearly independent eigenvectors. One reason that this is so useful is that the eigenvectors are not only linearly independent, but also span $\mathbb{R}^2$. If we call the two eigenvectors found in example 1.10.2 $\mathbf{u}$ and $\mathbf{v}$, corresponding to $\lambda = 3$ and $\mu = 1$, respectively, then, since these two vectors are linearly independent in $\mathbb{R}^2$ and span $\mathbb{R}^2$, we can write every vector in $\mathbb{R}^2$ uniquely as a linear combination of $\mathbf{u}$ and $\mathbf{v}$.

In particular, given a vector $\mathbf{x}$, there exist coefficients α and β such that

$$\mathbf{x} = \alpha\mathbf{u} + \beta\mathbf{v}$$

If we are interested in computing $\mathbf{Ax}$, we can do so now solely by knowing how $\mathbf{A}$ acts on the eigenvectors. Specifically, if we apply the linearity of matrix multiplication and the definition of eigenvectors, we have

$$\mathbf{Ax} = \mathbf{A}(\alpha\mathbf{u} + \beta\mathbf{v})$$
$$= \alpha\mathbf{Au} + \beta\mathbf{Av}$$
$$= \alpha\lambda\mathbf{u} + \beta\mu\mathbf{v}$$

This then reduces matrix multiplication essentially to scalar multiplication.

In conclusion, we have seen in this section that via matrix multiplication, every matrix can be viewed as a function in the way that, through multiplication, it stretches and rotates vectors. Those vectors that are only stretched are called eigenvectors, and the factor by which the matrix stretches them are called eigenvalues. By knowing the eigenvalues and eigenvectors, we can better understand how $\mathbf{A}$ acts on an arbitrary vector, and, with some more sophisticated approaches, even further understand key properties of the matrix. Some of these properties will be studied in detail later in this text when we consider systems of differential equations.

1.10.1 Markov chains, eigenvectors, and Google

In a Markov process such as the one discussed in subsection 1.3.1 that represents the transition of voters from one classification to another, it is natural to wonder whether or not there is a distribution of voters for which the total number in each category will remain constant from one year to the next. For example, for the Markov process represented by

$$\mathbf{x}^{(n+1)} = \mathbf{M}\mathbf{x}^{(n)} \tag{1.10.6}$$

where **M** is the matrix

$$\mathbf{M} = \begin{bmatrix} 0.95 & 0.03 & 0.07 \\ 0.02 & 0.90 & 0.13 \\ 0.03 & 0.07 & 0.80 \end{bmatrix}$$

we can ask: is there a voter distribution **x** such that $\mathbf{M}\mathbf{x} = \mathbf{x}$? In light of our most recent work with eigenvalues and eigenvectors, we see that this question is equivalent to asking if the matrix **M** has $\lambda = 1$ as an eigenvalue with some corresponding eigenvector that can represent a voter distribution.

If we compute the eigenvalues and eigenvectors of **M**, we find that the eigenvalues are $\lambda = 1.000, 0.911, 0.739$. The eigenvector corresponding to $\lambda = 1$ is $\mathbf{v} = [0.770 \ \ 0.558 \ \ 0.311]^{\mathrm{T}}$. Scaling **v** so that the sum of its entries is 250, we see that the eigenvector

$$\mathbf{v} = [117.450 \ \ 85.113 \ \ 47.437]^{\mathrm{T}}$$

represents the distribution of a population of 250 000 people in such a way that the total number of Democrats, Republicans, and Independents does not change from one year to the next, under the hypothesis that voters change categories annually according to the likelihoods expressed in the Markov matrix **M**. This eigenvector is sometimes also called a *stationary vector*.

Remarkably, we can also note that in our earlier computations in subsection 1.3.1 for this Markov chain, we observed that the sequence of vectors $\mathbf{x}^{(1)}, \mathbf{x}^{(2)}, \ldots, \mathbf{x}^{(20)}, \ldots$ was approaching a single vector. In fact, the limiting value of this sequence is the eigenvector $\mathbf{v} = [117.450 \ \ 85.113 \ \ 47.437]^{\mathrm{T}}$. That this phenomenon occurs is the result of the so-called *Power method*, a rudimentary numerical technique for computing an eigenvalue–eigenvector pair of a matrix. More about this concept can be studied in the project on discrete dynamical systems found in section 1.13.3.

Example 1.10.5   Find the stationary vector from the matrix in example 1.3.3.

Solution.   Under the assumptions stated in example 1.3.3, we saw that the migration of citizens from urban to suburban areas of a metropolitan area, or vice versa, were modeled by the Markov process $\mathbf{x}^{(n+1)} = \mathbf{M}\mathbf{x}^{(n)}$ where **M** is the matrix

$$\mathbf{M} = \begin{bmatrix} 0.85 & 0.08 \\ 0.15 & 0.92 \end{bmatrix}$$

Solving the equation $\mathbf{x} = \mathbf{Mx}$ by writing $(\mathbf{M} - \mathbf{I})\mathbf{x} = \mathbf{0}$, we see that we need to find the eigenvector of $\mathbf{x}$ that corresponds to $\lambda = 1$. Doing so, we find that the eigenvector is

$$\mathbf{v} = \left[\begin{array}{c} 0.4706 \\ 0.8824 \end{array} \right]$$

Scaling this vector so that the sum of its entries is one, we see that the population stabilizes when it is distributed with 34.78 percent in the city and 65.22 percent in the suburbs, in accordance with the vector $[0.3478 \;\; 0.6522]^{\mathrm{T}}$.

One of the most stunning applications of eigenvalues and eigenvectors can be found on the World Wide Web. In particular, the idea of finding a stationary vector that satisfies $\mathbf{Mx} = \mathbf{x}$ is at the center of Google's Page Rank Algorithm that it uses to index the importance of billions of pages on the Internet. What is particularly challenging about this problem is the fact that the stochastic matrix $\mathbf{M}$ used by the algorithm is a square matrix that has one column for every page on the World Wide Web that is indexed by Google! In early 2007, this meant that $\mathbf{M}$ was a matrix with 25 billion columns. Nonetheless, properties of the matrix $\mathbf{M}$ and sophisticated numerical algorithms make it possible for modern computers to quickly find the stationary vector of $\mathbf{M}$ and hence provide the user with the results we have all grown accustomed to in using Google.[6]

1.10.2 Using *Maple* to find eigenvalues and eigenvectors

Due to its reliance upon determinants and the solution of polynomial equations, the eigenvalue problem is computationally difficult for any case larger than 3×3. Sophisticated algorithms have been developed to compute eigenvalues and eigenvectors efficiently and accurately. One of these is the so-called *QR* algorithm, which through an iterative technique produces excellent approximations to eigenvalues and eigenvectors simultaneously.

While *Maple* implements these algorithms and can find both eigenvalues and eigenvectors, it is essential that we not only understand what the program is attempting to compute, but also how to interpret the resulting output. As always, in what follows we are working within the `LinearAlgebra` package.

Given an $n \times n$ matrix $\mathbf{A}$, we can compute the eigenvalues of $\mathbf{A}$ with the command

```
> Eigenvalues(A);
```

---

[6] A detailed description of how the Page Rank Algorithm works and the role that eigenvectors play may be read at `http://www.ams.org/featurecolumn/archive/pagerank.html`.

Doing so for the matrix

$$\mathbf{A} = \begin{bmatrix} 2 & 1 \\ 1 & 2 \end{bmatrix}$$

from example 1.10.2 yields the *Maple* output

$$\begin{bmatrix} 3 \\ 1 \end{bmatrix}$$

Despite the vector format, the program is telling us that the two eigenvalues of the matrix **A** are 3 and 1. If we desire the eigenvectors, too, we can use the command

```
> Eigenvectors(A);
```

which leads to the output

$$\begin{bmatrix} 3 \\ 1 \end{bmatrix}, \begin{bmatrix} 1 & -1 \\ 1 & 1 \end{bmatrix}$$

Here, the first vector tells us the eigenvalues of **A**. The following matrix holds the corresponding eigenvectors in its columns; the vector $[1\ \ 1]^{\mathrm{T}}$ is the eigenvector corresponding to $\lambda = 3$ and $[-1\ \ 1]^{\mathrm{T}}$ corresponds to $\lambda = 1$.

Maple is extremely powerful. It is not at all bothered by complex numbers. So, if we enter a matrix like the one in example 1.10.3 that has no real eigenvalues, *Maple* will find complex eigenvalues and eigenvectors. To see how this appears, we enter the matrix

$$\mathbf{R} = \begin{bmatrix} 1/\sqrt{2} & -1/\sqrt{2} \\ 1/\sqrt{2} & 1/\sqrt{2} \end{bmatrix}$$

and execute the command

```
> Eigenvectors(R);
```

The resulting output is

$$\begin{bmatrix} \frac{1}{2}\sqrt{2} + \frac{1}{2}I\sqrt{2} \\ \frac{1}{2}\sqrt{2} - \frac{1}{2}I\sqrt{2} \end{bmatrix}, \begin{bmatrix} I & -I \\ 1 & 1 \end{bmatrix}$$

Note that here *Maple* is using '*I*' to denote not the identity matrix, but rather $\sqrt{-1}$. Just as we saw in example 1.10.3, **R** does not have any real eigenvalues. We can use familiar properties of complex numbers (most importantly, $I^2 = 1$) to actually check that the equation $\mathbf{Ax} = \lambda\mathbf{x}$ holds for the listed complex eigenvalues and complex eigenvectors above. However, at this point in our study, these complex eigenvectors are of less importance, so we defer further details on them until later work with systems of differential equations.

One final example is relevant here to see how *Maple* deals with repeated eigenvalues and missing eigenvectors. If we enter the 3×3 matrix **A** from example 1.10.4 and execute the `Eigenvectors` command, we receive the output

$$\begin{bmatrix} 3 \\ 3 \\ -4 \end{bmatrix}, \begin{bmatrix} 5 & 0 & -2 \\ -2 & 0 & \frac{8}{3} \\ 1 & 0 & 1 \end{bmatrix}$$

Here we see that 3 is a repeated eigenvalue of **A** with multiplicity 2. The first two columns of the matrix in the output contain the (potentially) linearly independent eigenvectors which correspond to this eigenvalue. The second column of all zeros indicates that **A** has only one linearly independent eigenvector corresponding to this particular eigenvalue. The third column, of course, is the eigenvector associated with the eigenvalue $\lambda = -4$. The column of all zeros also demonstrates that $\mathbb{R}^3$ does not have a linearly independent spanning set that consists of eigenvectors of **A**.

Exercises 1.10   In exercises 1–8, compute (by hand) the eigenvalues and any corresponding real eigenvectors of the given matrix **A**.

1. $A = \begin{bmatrix} 5 & 1 \\ 0 & 3 \end{bmatrix}$

2. $A = \begin{bmatrix} 3 & -1 \\ -1 & 3 \end{bmatrix}$

3. $A = \begin{bmatrix} 3 & 4 \\ -5 & -5 \end{bmatrix}$

4. $A = \begin{bmatrix} 1 & 4 \\ 1 & 4 \end{bmatrix}$

5. $A = \begin{bmatrix} 2 & 1 & 0 \\ 0 & 2 & 1 \\ 0 & 0 & 2 \end{bmatrix}$

6. $A = \begin{bmatrix} 2 & 1 & 0 \\ 0 & 2 & 0 \\ 0 & 0 & 2 \end{bmatrix}$

7. $A = \begin{bmatrix} 2 & 0 & 0 \\ 0 & 2 & 0 \\ 0 & 0 & 2 \end{bmatrix}$

8. $A = \begin{bmatrix} -3 & 2 & 5 \\ 0 & 6 & -2 \\ 0 & 0 & 5 \end{bmatrix}$

9. A 2×2 matrix $\mathbf{A}$ has eigenvalues 5 and -1 and corresponding eigenvectors $\mathbf{u} = [0 \ 1]^{T}$ and $\mathbf{v} = [1 \ 0]^{T}$. Use this information to compute $\mathbf{Ax}$, where $\mathbf{x}$ is the vector $\mathbf{x} = [-5 \ 4]^{T}$.

10. A 2×2 matrix $\mathbf{A}$ has eigenvalues -3 and -2 and corresponding eigenvectors $\mathbf{u} = [-1 \ 1]^{T}$ and $\mathbf{v} = [1 \ 1]^{T}$. Use this information to compute $\mathbf{Ax}$, where $\mathbf{x}$ is the vector $\mathbf{x} = [-3 \ 5]^{T}$.

11. Consider the matrix

$$\mathbf{A} = \begin{bmatrix} -2 & 1 & 1 \\ 1 & -2 & 1 \\ 1 & 1 & -2 \end{bmatrix}$$

(a) Determine the eigenvalues and eigenvectors of $\mathbf{A}$.
(b) Does $\mathbb{R}^3$ have a linearly independent spanning set that consists of eigenvectors of $\mathbf{A}$?

12. Consider the matrix

$$\mathbf{A} = \begin{bmatrix} 3 & -1 \\ -1 & 3 \end{bmatrix}$$

(a) Determine the eigenvalues and eigenvectors of $\mathbf{A}$, and show that $\mathbf{A}$ has two linearly independent eigenvectors.
(b) Let $\mathbf{P}$ be the matrix whose columns are two linearly independent eigenvectors of $\mathbf{A}$. Why is $\mathbf{P}$ invertible?
(c) Let $\mathbf{D}$ be the diagonal matrix whose diagonal entries are the eigenvalues of $\mathbf{A}$; place the eigenvalues on the diagonal in an order corresponding to the order of the eigenvectors in the columns of $\mathbf{P}$, where $\mathbf{P}$ is the matrix defined in (b) above. Compute $\mathbf{AP}$ and $\mathbf{PD}$. What do you observe?
(d) Explain why $\mathbf{A} = \mathbf{PDP}^{-1}$. Use this factorization to compute $\mathbf{A}^2$, $\mathbf{A}^3$, and $\mathbf{A}^{10}$ in terms of $\mathbf{P}$, $\mathbf{D}$, and $\mathbf{P}^{-1}$. In particular, explain how $\mathbf{A}^{10}$ can be easily computed by using the diagonal matrix $\mathbf{D}$ along with $\mathbf{P}$ and $\mathbf{P}^{-1}$.

13. Consider the matrix

$$\mathbf{A} = \begin{bmatrix} 3 & -1 & 1 \\ -1 & 3 & -1 \\ 1 & -1 & 3 \end{bmatrix}$$

(a) Determine the eigenvalues and eigenvectors of $\mathbf{A}$, and show that $\mathbf{A}$ has three linearly independent eigenvectors.
(b) Let $\mathbf{P}$ be the matrix whose columns are three linearly independent eigenvectors of $\mathbf{A}$. Why is $\mathbf{P}$ invertible?
(c) Let $\mathbf{D}$ be the diagonal matrix whose diagonal entries are the eigenvalues of $\mathbf{A}$; place the eigenvalues on the diagonal in an order corresponding to the order of the eigenvectors in the columns of $\mathbf{P}$,

where **P** is the matrix defined in (b) above. Compute **AP** and **PD**. What do you observe?

(d) Explain why $\mathbf{A} = \mathbf{PDP}^{-1}$. Use this factorization to compute $\mathbf{A}^2$, $\mathbf{A}^3$, and $\mathbf{A}^{10}$ in terms of **P**, **D**, and $\mathbf{P}^{-1}$.

14. Prove that an $n \times n$ matrix **A** is invertible if and only if **A** has no eigenvalue equal to zero.

15. Show that if **A**, **B**, and **P** are square matrices (with **P** invertible) such that $\mathbf{B} = \mathbf{PAP}^{-1}$, then **A** and **B** have the same eigenvalues. (Hint: consider the characteristic equation for $\mathbf{PAP}^{-1}$.)

16. Prove that if **A** is a 2×2 matrix and **v** and **u** are eigenvectors of **A** corresponding to distinct eigenvalues λ and μ, then **v** and **u** are linearly independent. (Hint: suppose to the contrary that **v** and **u** are linearly dependent.)

17. For a differentiable function y, denote the derivative of y with respect to x by $D(y)$. Now consider the function $y = e^{7x}$, and compute $D(y)$. For what value of λ is $D(y) = \lambda y$? Explain how this value behaves like an eigenvalue of the operator D. What is the corresponding eigenvector? How does the problem change if we consider $y = e^{rx}$ for any other real value of r?

18. For a vector-valued function $\mathbf{x}(t)$, let the derivative of **x** with respect to t be denoted by $D(\mathbf{x})$. For the function

$$\mathbf{x}(t) = \begin{bmatrix} e^{-2t} \\ -3e^{-2t} \end{bmatrix}$$

compute $D(\mathbf{x})$. For what value(s) of λ is $D(\mathbf{x}) = \lambda\mathbf{x}$? Explain how it appears from your work that the operator D has an eigenvalue-eigenvector pair.

19. Suppose that for a large population that stays relatively constant, people are classified as living in urban, suburban, or rural settings. Moreover, assume that the probabilities of the various possible transitions are given by the following table:

Future location ($\downarrow$)/current location ($\rightarrow$)	U(%)	S(%)	R(%)
Urban	90	3	2
Suburban	7	96	10
Rural	3	1	88

Given that a population of 250 million is present, is there a stationary vector that reveals a population which does not change from year to year?

20. Car-owners can be grouped into classes based on the vehicles they own. A study of owners of sedans, minivans, and sport utility vehicles shows

that the likelihood that an owner of one of these automobiles will replace it with another of the same or different type is given by the table

Future vehicle ($\downarrow$)/ current vehicle ($\rightarrow$)	Sedan(%)	Minivan(%)	SUV(%)
Sedan	91	3	2
Minivan	7	95	8
SUV	2	2	90

If there are currently 100 000 vehicles in the population under study, is there a stationary vector that represents a distribution in which the number of owners of each type of vehicle will not change as they replace their vehicles?

21. Decide whether each of the following sentences is true or false. In every case, write one sentence to support your answer.

(a) If $\mathbf{x}$ is any vector and λ is a constant such that $\mathbf{Ax} = \lambda\mathbf{x}$, then $\mathbf{x}$ is an eigenvector of $\mathbf{A}$.

(b) If $\mathbf{Ax} = \mathbf{0}$ has nontrivial solutions, then $\lambda = 0$ is an eigenvalue of $\mathbf{A}$.

(c) Every 3×3 matrix has three real eigenvalues.

(d) If $\mathbf{A}$ is a 2×2 matrix, then $\mathbf{A}$ can have up to two real linearly independent eigenvectors.

1.11 Generalized vectors

Throughout our work with vectors in $\mathbb{R}^n$, we have regularly used several key algebraic properties they possess. For example, any two vectors $\mathbf{u}$ and $\mathbf{v}$ can be added to form a new vector $\mathbf{u} + \mathbf{v}$, any single vector can be multiplied by a scalar to determine a new vector $c\mathbf{u}$, and there is a zero vector $\mathbf{0}$ with the property that for any vector $\mathbf{v}$, $\mathbf{v} + \mathbf{0} = \mathbf{v}$. Of course, we use other algebraic properties of vectors as well, often implicitly.

Other sets of mathematical objects behave in ways that are algebraically similar to vectors. The purpose of this section is to expand our perspective on what familiar mathematical entities might also reasonably be called vectors; much of this expanded perspective is in anticipation of our pending work with differential equations and their solutions. We motivate our study with several familiar examples, and then summarize a collection of formal properties that all these examples share.

Example 1.11.1 Let $\mathcal{M}_{2\times 2}$ denote the collection of all 2×2 matrices with real entries. Show that if $\mathbf{A}$ and $\mathbf{B}$ are any 2×2 matrices and $c \in \mathbb{R}$, then $\mathbf{A} + \mathbf{B}$ and $c\mathbf{A}$ are also 2×2 matrices. In addition, show that there exists a "zero matrix" $\mathbf{Z}$ such that $\mathbf{A} + \mathbf{Z} = \mathbf{A}$ for every matrix $\mathbf{A}$.

Solution. Let

$$\mathbf{A} = \begin{bmatrix} a_{11} & a_{12} \\ a_{21} & a_{22} \end{bmatrix} \text{ and } \mathbf{B} = \begin{bmatrix} b_{11} & b_{12} \\ b_{21} & b_{22} \end{bmatrix}$$

By the definition of matrix addition,

$$\mathbf{A} + \mathbf{B} = \begin{bmatrix} a_{11} + b_{11} & a_{12} + b_{12} \\ a_{21} + b_{21} & a_{22} + b_{22} \end{bmatrix}$$

and thus we see that $\mathbf{A} + \mathbf{B}$ is also a 2×2 matrix. Recall that it only makes sense for matrices of the same size to be added; here we are simply pointing out the obvious fact that the sum of two matrices of the same size is yet another matrix of the same size. In the same way,

$$c\mathbf{A} = \begin{bmatrix} ca_{11} & ca_{12} \\ ca_{21} & ca_{22} \end{bmatrix}$$

which shows that not only is the scalar multiple defined, but also that $c\mathbf{A}$ is a 2×2 matrix. Finally, if we let $\mathbf{Z}$ be the 2×2 matrix all of whose entries are zero,

$$\mathbf{Z} = \begin{bmatrix} 0 & 0 \\ 0 & 0 \end{bmatrix}$$

then our work with matrix sums shows us immediately that $\mathbf{A} + \mathbf{Z} = \mathbf{A}$ for every possible 2×2 matrix $\mathbf{A}$.

Certainly, we can see that there is nothing particularly special about the 2×2 case in this example; the same properties will hold for $\mathcal{M}_{m \times n}$ for any positive integer values of m and n.

Mathematicians often use the language "$\mathcal{M}_{2 \times 2}$ is *closed under addition and scalar multiplication*" and "$\mathcal{M}_{2 \times 2}$ *contains a zero element*" to describe the observations we made in example 1.11.1. Specifically, to say that a set is closed under an operation means simply that if we perform the operation on an appropriate number of elements from the set, the result is another element in the set. We next consider several more examples of sets that demonstrate the properties of being closed and having a zero element.

Example 1.11.2 Let $\mathbb{P}_2$ denote the set of all polynomials of degree 2 or less. That is, $\mathbb{P}_2$ is the set of all functions of the form

$$p(x) = a_2 x^2 + a_1 x + a_0$$

where $a_0, a_1, a_2 \in \mathbb{R}$. Show that $\mathbb{P}_2$ is closed under addition and scalar multiplication, and that $\mathbb{P}_2$ contains a zero element.

Solution. Before we formally address the stated tasks, let us remind ourselves how we add polynomial functions. If we are given, say, $f(x) = 2x^2 - 5x + 11$ and $g(x) = 4x - 3$, we compute $(f + g)(x) = f(x) + g(x) = 2x^2 - 5x + 11 + 4x - 3$. We can then add like terms to simplify and find that $(f + g)(x) = 2x^2 - x + 8$.

Similarly, if we wanted to compute $(-3f)(x)$, we have $(-3f)(x) = -3f(x) = -3(2x^2 - 5x + 11) = -6x^2 + 15x - 33$.

We now show that $\mathbb{P}_2$ is indeed closed under the operations of addition and scalar multiplication. Given two arbitrary elements of $\mathbb{P}_2$, say $f(x) = a_2x^2 + a_1x + a_0$ and $g(x) = b_2x^2 + b_1x + b_0$, it follows upon adding and combining like terms that

$$(f + g)(x) = (a_2 + b_2)x^2 + (a_1 + b_1)x + (a_0 + b_0)$$

which is obviously a polynomial of degree 2 or lower, and thus $f + g$ is an element of $\mathbb{P}_2$. In the same way, for any real value c,

$$(cf)(x) = ca_2x^2 + ca_1x + ca_0$$

which also belongs to $\mathbb{P}_2$. Finally, it is evident that if we let $z(x) = 0x^2 + 0x + 0$ (i.e., $z(x)$ is the zero function), then $(f + z)(x) = f(x)$ for any choice of f in $\mathbb{P}_2$.

Here, too, we should observe that while these properties hold for $\mathbb{P}_2$, there is nothing special about the 2. In fact, $\mathbb{P}_n$ (the set of all polynomials of degree n or less) has the exact same properties. Even $\mathbb{P}$, the set of all polynomials, behaves in the same manner.

Example 1.11.3   From calculus, consider the set $C[-1, 1]$ of all continuous functions on the interval $[-1, 1]$. That is,

$$C[-1, 1] = \{f \mid f \text{ is continuous on } [-1, 1]\}.$$

Show that $C[-1, 1]$ is closed under addition and scalar multiplication, and also that $C[-1, 1]$ contains a zero element.

Solution.   Two standard facts from calculus tell us that the sum of any two continuous functions is also a continuous function and that a constant multiple of a continuous function is also a continuous function. Thus $C[-1, 1]$ is closed under addition and scalar multiplication. Furthermore, the zero function $z(x) = 0$ is itself continuous, which shows that $C[-1, 1]$ indeed has a zero element.

One of the principal reasons that we are shifting our attention from vectors in $\mathbb{R}^n$ to this more generalized concept of vector where the objects under consideration are often functions is the fact that our focus in subsequent chapters will be solving differential equations. The solution to a differential equation is a function that makes the equation true. Moreover, we will also see that for certain important classes of differential equations, there are multiple solutions to the equation and that often these solution sets are closed under addition and scalar multiplication and also contain the zero function.

From each of the above examples, we see that $\mathbb{R}^n$ has many important properties that we can consider in a broader context. We therefore introduce the notion of a *vector space*, which is a set of objects that have defined operations of addition and scalar multiplication that satisfy the list of ten rules below. The concept of a vector space is a generalization of $\mathbb{R}^n$.

While many of the rules are technical in nature, the most important ones to verify turn out to be the three that we have focused on so far: being closed under addition, closed under scalar multiplication, and having a zero element. All three sets described in the above examples are vector spaces, as is $\mathbb{R}^n$.

Definition 1.11.1   A *vector space* is a nonempty set V of objects, on which operations of addition and scalar multiplication are defined, where the objects in V (called *vectors*) adhere to the following ten rules:

1. For every $\mathbf{u}$ and $\mathbf{v}$ in V, the sum $\mathbf{u} + \mathbf{v}$ is in V (V is "closed under vector addition")

2. For every $\mathbf{u}$ and $\mathbf{v}$ in V, $\mathbf{u} + \mathbf{v} = \mathbf{v} + \mathbf{u}$ ("vector addition is commutative")

3. For every $\mathbf{u}, \mathbf{v}, \mathbf{w}$ in V, $(\mathbf{u} + \mathbf{v}) + \mathbf{w} = \mathbf{v} + (\mathbf{u} + \mathbf{w})$ ("vector addition is associative")

4. There exists a zero vector $\mathbf{0}$ in V such that $\mathbf{u} + \mathbf{0} = \mathbf{u}$ for every $\mathbf{u} \in V$ ($\mathbf{0}$ is called the *additive identity* of V)

5. For every $\mathbf{u} \in V$, there is a vector $-\mathbf{u}$ such that $\mathbf{u} + (-\mathbf{u}) = \mathbf{0}$ ($-\mathbf{u}$ is called the *additive inverse* of $\mathbf{u}$)

6. For every $\mathbf{u} \in V$ and every scalar c, the scalar multiple $c\mathbf{u} \in V$ (V is "closed under scalar multiplication")

7. For every $\mathbf{u}$ and $\mathbf{v}$ in V and every scalar c, $c(\mathbf{u} + \mathbf{v}) = c\mathbf{u} + c\mathbf{v}$ ("scalar multiplication is distributive over vector addition")

8. For every $\mathbf{u} \in V$ and scalars c and d, $(c + d)\mathbf{u} = c\mathbf{u} + d\mathbf{u}$

9. For every $\mathbf{u} \in V$ and scalars c and d, $c(d\mathbf{u}) = (cd)\mathbf{u}$

10. For every $\mathbf{u} \in V$, $1\mathbf{u} = \mathbf{u}$

Sometimes we can take a sub-collection (i.e., a subset) of the vectors in a vector space, and that smaller set itself acts like a vector space. For example, the set of all polynomial functions is a vector space. If we take just the polynomials of degree 2 or less (as in example 1.11.2 above), that subset is itself a vector space. This leads us to introduce the notion of a *subspace*.

Definition 1.11.2   Given a vector space V, let H be a subset of V (i.e., every object in H is also in V.) There are then operations of addition and scalar multiplication on objects in H: specifically, the same addition and scalar multiplication as on the objects in V. We say H is a *subspace* of V if and only if all three of the following conditions hold:

1. H is closed under addition

2. H is closed under scalar multiplication

3. H contains the zero element of V

We close this section with two important examples of subspaces. The first is a subspace of $\mathbb{R}^n$ associated with a given matrix $\mathbf{A}$. The second is a subspace of the set of all continuous functions on $[-1, 1]$.

Example 1.11.4   Recall the matrix $\mathbf{A}$ from example 1.10.4 in section 1.10,

$$\mathbf{A} = \begin{bmatrix} 5 & 6 & 2 \\ 0 & -1 & -8 \\ 1 & 0 & -2 \end{bmatrix}$$

Show that the set of all eigenvectors that correspond to a given eigenvalue of $\mathbf{A}$ forms a subspace of $\mathbb{R}^3$.

Solution.   In example 1.10.4, we saw that the eigenvalues of $\mathbf{A}$ are $\lambda = -4$ (with multiplicity 1) and $\lambda = 3$ (with multiplicity 2). In addition, the corresponding eigenvectors are $\mathbf{v} = [-2 \quad \frac{8}{3} \quad 1]^T$ for $\lambda = -4$ and $\mathbf{v} = [5 \quad -2 \quad 1]^T$ for $\lambda = 3$. In particular, recall that every scalar multiple of $\mathbf{v}_{\lambda = -4}$ is also an eigenvector of $\mathbf{A}$ corresponding to $\lambda = -4$. We now show that the set of all these eigenvectors corresponding to $\lambda = -4$ is a subspace of $\mathbb{R}^3$.

Let $\mathbb{E}_{\lambda = -4}$ denote the set of all vectors $\mathbf{v}$ such that $\mathbf{Av} = -4\mathbf{v}$. First, certainly it is the case that $\mathbf{A0} = -4\mathbf{0}$. This shows that the zero element of $\mathbb{R}^3$ is an element of $\mathbb{E}_{\lambda = -4}$. Furthermore, we have already seen that every scalar multiple of an eigenvector is itself an eigenvector, and thus $\mathbb{E}_{\lambda = -4}$ is closed under scalar multiplication. Finally, suppose we have two vectors $\mathbf{x}$ and $\mathbf{y}$ such that $\mathbf{Ax} = -4\mathbf{x}$ and $\mathbf{Ay} = -4\mathbf{y}$. Observe that by properties of linearity,

$$\mathbf{A}(\mathbf{x} + \mathbf{y}) = \mathbf{Ax} + \mathbf{Ay}$$
$$= -4\mathbf{x} - 4\mathbf{y}$$
$$= -4(\mathbf{x} + \mathbf{y})$$

which shows that $(\mathbf{x} + \mathbf{y})$ is also an eigenvector of $\mathbf{A}$ corresponding to $\lambda = -4$. Therefore, $\mathbb{E}_{\lambda = -4}$ is closed under addition.

This shows that $\mathbb{E}_{\lambda = -4}$ is indeed a subspace of $\mathbb{R}^3$. In a similar fashion, $\mathbb{E}_{\lambda = 3}$ is also a subspace of $\mathbb{R}^3$.

Our observations for the eigenspaces of the 2×2 matrix $\mathbf{A}$ in example 1.11.4 hold in general for any $n \times n$ matrix $\mathbf{A}$: the set of all eigenvectors corresponding to a given eigenvalue of $\mathbf{A}$ forms a subspace of $\mathbb{R}^n$.

Example 1.11.5   Show that the set of all linear combinations of the sine and cosine functions is a subspace of the vector space $\mathcal{C}$ of all continuous functions.

Solution.   We let $\mathcal{C}$ denote the vector space of all continuous functions, and now let H be the subset of $\mathcal{C}$ which is defined to be all functions that are linear combinations of $\sin t$ and $\cos t$. That is, a typical element of H is a function f of the form

$$f(t) = c_1 \sin t + c_2 \cos t$$

where c_1 and c_2 are any real scalars. We need to show that the set H contains the zero function from $\mathcal{C}$, that H is closed under scalar multiplication, and that H is closed under addition.

First, if we choose $c_1 = c_2 = 0$, the function $z(t) = 0 \sin t + 0 \cos t = 0$ is the function that is identically zero, which is the (continuous) zero function from $\mathcal{C}$. Next, if we take a function from H, say $f(t) = c_1 \sin t + c_2 \cos t$, and multiply it by a scalar k, we get

$$kf(t) = k(c_1 \sin t + c_2 \cos t) = (kc_1) \sin t + (kc_2) \cos t$$

which is of course another element in H, so H is closed under scalar multiplication. Finally, if we consider two elements f and g in H, given by $f(t) = c_1 \sin t + c_2 \cos t$ and $g(t) = d_1 \sin t + d_2 \cos t$, then it follows that

$$f(t) + g(t) = (c_1 \sin t + c_2 \cos t) + (d_1 \sin t + d_2 \cos t)$$
$$= (c_1 + d_1) \sin t + (c_2 + d_2) \cos t$$

so that H is closed under addition, too. Thus, H is a subspace of $\mathcal{C}$.

In fact, it turns out that the subspace considered in example 1.11.5 contains all of the solutions to a familiar differential equation. We will revisit this issue in example 1.11.7. It is also instructive to consider an example of a set that is not a subspace.

Example 1.11.6   Consider the vector space $\mathcal{C}[-1, 1]$ of all continuous functions on the interval $[-1, 1]$. Let H be the set of all functions with the property that $f(-1) = f(1) = 2$. Determine whether or not H is a subspace of $\mathcal{C}[-1, 1]$.

Solution.   The set H does not satisfy any of the three required properties of subspaces, so any one of these suffices to show that H is not a subspace. In particular, the zero function $z(t) = 0$ does not have the property that $z(-1) = 2$, and thus the zero function from $\mathcal{C}[-1, 1]$ does not lie in H, so H is not a subspace.

We could also observe that any scalar multiple of a function whose value at $t = -1$ and $t = 1$ is 2 will result in a new function whose value at these points is not 2; similarly, the sum of two functions whose values at $t = -1$ and $t = 1$ are 2 will lead to a new function whose values at these points is 4. These facts together show that H is not closed under scalar multiplication, nor under addition.

As we have already mentioned, we are considering this generalization of the term vector to include mathematical objects like functions because this structure underlies the study of differential equations, and this vector space perspective will help us to better understand a variety of key ideas when we are solving important problems later on. To foreshadow these coming ideas, we present an example of an elementary differential equation that shows how the set of solutions to the equation is in fact the subspace of continuous functions considered in example 1.11.5.

Example 1.11.7   Consider the differential equation

$$y'' + y = 0$$

Show that $y_1 = \sin t$ and $y_2 = \cos t$ are solutions to this differential equation, and that every function of the form $y = c_1 y_1 + c_2 y_2$ is a solution as well.

Solution.   This example is very similar to example 1.6.4. Because of its importance, we discuss the current problem in full detail here as well.

For any equation, a *solution* is an object that makes the equation true. In the above differential equation, y represents a function. The equation asks "for which functions y is the sum of y and its second derivative equal to zero?"

Observe first that if we let $y_1 = \sin t$, then $y_1' = \cos t$, so $y_1'' = -\sin t$, and therefore $y_1'' + y_1 = -\sin t + \sin t = 0$. In other words, y_1 is a solution to the differential equation. Similarly, for $y_2 = \cos t$, $y_2' = -\sin t$ and $y_2'' = -\cos t$, so that $y_2'' + y_2 = -\cos t + \cos t = 0$. Thus, y_2 is also a solution to the differential equation.

Now, consider any function y of the form $y = c_1 y_1 + c_2 y_2$. That is, let y be any linear combination of the two solutions we have already found. We then have

$$y = c_1 \sin t + c_2 \cos t$$

so that, using standard properties of the derivative (properties which are *linear* in nature), it follows that

$$y' = c_1 \cos t - c_2 \sin t$$

and

$$y'' = -c_1 \sin t - c_2 \cos t$$

We, therefore see that

$$y'' + y = (-c_1 \sin t - c_2 \cos t) + (c_1 \sin t + c_2 \cos t)$$
$$= -c_1 \sin t + c_1 \sin t - c_2 \cos t + c_2 \cos t$$
$$= 0$$

so that y is indeed also a solution of $y'' + y = 0$.

In example 1.11.7, we find a large number of connections to our work in systems of linear equations and linear algebra: properties of linearity, linear combinations of vectors, homogeneous equations, infinitely many solutions, and more. In particular, the set of all solutions to the differential equation in example 1.11.7 is precisely the subspace of continuous functions examined in example 1.11.5. Certainly, we will revisit these topics in greater detail as we progress in our study of differential equations.

Exercises 1.11   In exercises 1–16, determine whether or not the set H is a subspace of the given vector space V. If H is a subspace, show that it satisfies the three required properties stipulated by the definition; if not, show at least one example of why at least one of the properties does not hold.

1. $V = \mathbb{R}^2$, $H = \left\{ \begin{bmatrix} x \\ y \end{bmatrix} : x \geq 0, y \geq 0 \right\}$

2. $V = \mathbb{R}^2$, $H = \left\{ \begin{bmatrix} x \\ y \end{bmatrix} : x \cdot y \geq 0 \right\}$

3. $V = \mathbb{R}^3$, $H = \left\{ t \begin{bmatrix} 2 \\ 0 \\ -1 \end{bmatrix} : t \in \mathbb{R} \right\}$

4. $V = \mathbb{R}^3$, $H = \left\{ t \begin{bmatrix} 2 \\ 0 \\ -1 \end{bmatrix} + \begin{bmatrix} 1 \\ 1 \\ 1 \end{bmatrix} : t \in \mathbb{R} \right\}$

5. $V = \mathbb{P}_2$, $H = \left\{ at^2 : a \in \mathbb{R} \right\}$

6. $V = \mathbb{P}_2$, $H = \left\{ at^2 + 1 : a \in \mathbb{R} \right\}$

7. $V = \mathbb{R}^2$, $H = \left\{ \mathbf{x} : \mathbf{Ax} = \mathbf{b} \text{ where } \mathbf{A} = \begin{bmatrix} 2 & -1 \\ -6 & 3 \end{bmatrix} \text{ and } \mathbf{b} = \begin{bmatrix} 5 \\ -15 \end{bmatrix} \right\}$

8. $V = \mathbb{R}^2$, $H = \left\{ \mathbf{x} : \mathbf{Ax} = \mathbf{b} \text{ where } \mathbf{A} = \begin{bmatrix} 2 & -1 \\ -6 & 3 \end{bmatrix} \text{ and } \mathbf{b} = \begin{bmatrix} 0 \\ 0 \end{bmatrix} \right\}$

9. $V = \mathcal{M}_{2 \times 2}$, $H = \{ \mathbf{A} \in \mathcal{M}_{2 \times 2} : \mathbf{A} \text{ is invertible} \}$

10. $V = \mathcal{M}_{2 \times 2}$, $H = \{ \mathbf{A} \in \mathcal{M}_{2 \times 2} : \mathbf{A} \text{ is not invertible} \}$

11. $V = \mathcal{M}_{2 \times 2}$, $H = \left\{ \mathbf{A} \in \mathcal{M}_{2 \times 2} : \mathbf{A} = \begin{bmatrix} a & 0 \\ b & c \end{bmatrix} \right\}$

12. $V = \mathcal{M}_{2 \times 2}$, $H = \left\{ \mathbf{A} \in \mathcal{M}_{2 \times 2} : \mathbf{A} = \begin{bmatrix} a & 1 \\ b & c \end{bmatrix} \right\}$

13. $V = \mathcal{C}[-1, 1]$, $H = \{ f \in \mathcal{C}[-1, 1] : f(-1) = 0 \}$

14. $V = \mathcal{C}[-1, 1]$, $H = \{ f \in \mathcal{C}[-1, 1] : f(-1) = 5 \}$

15. $V = \mathcal{C}[-1, 1]$, $H = \{ f \in \mathcal{C}[-1, 1] : f' + f = 0 \}$

16. $V = \mathcal{C}[-1, 1]$, $H = \{ f \in \mathcal{C}[-1, 1] : f' + f = 1 \}$

17. Recall that for a given eigenvalue λ of a matrix $\mathbf{A}$, the eigenspace associated to that eigenvalue is the set of all eigenvectors that correspond to λ. For the matrix $\mathbf{A} = \begin{bmatrix} 2 & -1 \\ -1 & 2 \end{bmatrix}$, describe all of the eigenspaces of $\mathbf{A}$.

18. For the matrix $\mathbf{A} = \begin{bmatrix} 2 & 1 \\ 0 & 2 \end{bmatrix}$, describe all of the eigenspaces of $\mathbf{A}$.

19. Explain why for any set of vectors $\{\mathbf{u}, \mathbf{v}\}$ in $\mathbb{R}^n$, Span$\{\mathbf{u}, \mathbf{v}\}$ is a subspace of $\mathbb{R}^n$. Similarly, explain why Span $\{\mathbf{v}_1, \ldots, \mathbf{v}_k\}$ is a subspace of $\mathbb{R}^n$ for any set $\{\mathbf{v}_1, \ldots, \mathbf{v}_k\}$.

20. Let $V = \mathbb{R}^3$ and $H = \left\{ \begin{bmatrix} 2a+b \\ a-b \\ 3a+5b \end{bmatrix} : a, b \in \mathbb{R} \right\}$. Determine vectors $\mathbf{u}$ and $\mathbf{v}$ so that H can be expressed as the set Span$\{\mathbf{u}, \mathbf{v}\}$, and hence explain why H is a subspace of $\mathbb{R}^3$.

21. Let $V = \mathbb{R}^3$ and $H = \left\{ \begin{bmatrix} 2a+b \\ -2 \\ 3a+5b \end{bmatrix} : a, b \in \mathbb{R} \right\}$. Explain why H is not a subspace of $\mathbb{R}^3$.

22. Let $\mathbf{A}$ be an $m \times n$ matrix. The *null space* of the matrix $\mathbf{A}$, denoted Nul$(\mathbf{A})$ is the set of all solutions to the equation $\mathbf{Ax} = \mathbf{0}$. Explain why Nul$(\mathbf{A})$ is a subspace of $\mathbb{R}^n$.

23. Let $\mathbf{A}$ be an $m \times n$ matrix. The *column space* of the matrix $\mathbf{A}$, denoted Col$(\mathbf{A})$ is the set of all linear combinations of the columns of $\mathbf{A}$. Explain why Col$(\mathbf{A})$ is a subspace of $\mathbb{R}^m$.

In exercises 24–27, use the definitions of the null space Nul$(\mathbf{A})$ and column space Col$(\mathbf{A})$ of a matrix given in exercises 22 and 23.

24. Let $\mathbf{A} = \begin{bmatrix} 2 & 1 & -1 \\ 1 & 3 & 4 \end{bmatrix}$. Is the vector $\mathbf{v} = [-2 \ 1 \ 1]^T$ in Nul$(\mathbf{A})$? Justify your answer clearly. In addition, describe all vectors that belong to Nul$(\mathbf{A})$ as the span of a finite set of vectors.

25. Let $\mathbf{A} = \begin{bmatrix} 1 & -2 \\ 3 & 1 \\ -4 & 0 \end{bmatrix}$. Is the vector $\mathbf{v} = [-2 \ 1 \ 1]^T$ in Col$(\mathbf{A})$? Justify your answer. Is the vector $\mathbf{u} = [-1 \ 4 \ -4]^T$ in Col$(\mathbf{A})$? In addition, describe all vectors that belong to Col$(\mathbf{A})$ as the span of a finite set of vectors.

26. Given a matrix $\mathbf{A}$ and a vector $\mathbf{v}$, is it easier to determine whether $\mathbf{v}$ lies in Nul$(\mathbf{A})$ or Col$(\mathbf{A})$? Why?

27. Given a matrix $\mathbf{A}$ and a vector $\mathbf{v}$, is it easier to describe Nul$(\mathbf{A})$ or Col$(\mathbf{A})$ as the span of a finite set of vectors? Why?

28. Consider the differential equation $y' = 3y$. Explain why any function of the form $y = Ce^{3t}$ is a solution to this equation. Is the set of all these solutions a subspace of the vector space of continuous functions?

29. Consider the differential equation $y' = 3y - 3$. Explain why any function of the form $y = Ce^{3t} + 1$ is a solution to this equation. Is the set of all these solutions a subspace of the vector space of continuous functions?

30. Decide whether each of the following sentences is true or false. In every case, write one sentence to support your answer.

    (a) If H is a subspace of a vector space V, then H is itself a vector space.

    (b) If H is a subset of a vector space V, then H is a subspace of V.

    (c) The set of all linear combinations of any two vectors in $\mathbb{R}^3$ is a subspace of $\mathbb{R}^3$.

    (d) Every nontrivial subspace of a vector space has infinitely many elements.

1.12 Bases and dimension in vector spaces

In section 1.11, we saw that some common sets we encounter in mathematics are very similar to $\mathbb{R}^n$. For instance, the set $\mathcal{M}_{2\times2}$ of all 2×2 matrices, the set $\mathbb{P}_2$ of all polynomials of degree 2 or less, and the set $\mathcal{C}[-1, 1]$ of all continuous functions on $[-1, 1]$ are sets that contain a zero element, are closed under addition, and are closed under scalar multiplication. In addition, because they each satisfy the other required seven characteristics we noted, these sets are all vector spaces. We specifically observe that this enables us to take *linear combinations* of elements of a vector space, because addition and scalar multiplication are defined and closed in these collections of objects.

Every vector space has further characteristics that are similar to $\mathbb{R}^n$. For example, it is natural to discuss now-familiar concepts such as linear independence and span in the context of the more generalized notion of vector. As we will see, the definitions of these terms in the setting of vector spaces are almost identical to those we encountered earlier in $\mathbb{R}^n$. Moreover, just as we can frequently describe sets in $\mathbb{R}^n$ in terms of a small number of special vectors, we will find that this often occurs in general vector spaces.

We begin by updating two key definitions.

Definition 1.12.1   In a vector space V, given a set $S = \{\mathbf{v}_1, \ldots, \mathbf{v}_k\}$ where each vector $\mathbf{v}_i \in V$, the set S is *linearly dependent* if there exists a nontrivial solution to the vector equation

$$x_1\mathbf{v}_1 + x_2\mathbf{v}_2 + \cdots + x_k\mathbf{v}_k = \mathbf{0} \qquad (1.12.1)$$

If (1.12.1) has only the trivial solution ($x_1 = \cdots = x_k = 0$), then we say the set S is *linearly independent*.

The only difference between this definition and definition 1.6.1 that we encountered in section 1.6 is that $\mathbb{R}^n$ has been replaced by V. Just as with vectors in $\mathbb{R}^n$, it is an equivalent formulation to say that a set S in a vector space V is linearly independent if and only if no vector in the set may be written as a linear combination of the other vectors in the set.

We can also define the span of a set of vectors in a vector space V.

Definition 1.12.2   In a vector space V, given a set of vectors $S = \{\mathbf{v}_1, \ldots, \mathbf{v}_k\}$, $\mathbf{v}_i \in V$, the *span* of S, denoted Span(S) or Span$\{\mathbf{v}_1, \ldots, \mathbf{v}_k\}$, is the set of all linear combinations of the vectors $\mathbf{v}_1, \ldots, \mathbf{v}_k$. Equivalently, Span($S$) is the set of all vectors $\mathbf{y}$ of the form

$$\mathbf{y} = c_1 \mathbf{v}_1 + \cdots + c_k \mathbf{v}_k,$$

where $c_1, \ldots, c_k$ are scalars. We also say that Span(S) is *the subset of V spanned by the vectors* $\mathbf{v}_1, \ldots, \mathbf{v}_k$.

In example 1.6.3 in section 1.6, we studied three sets R, S, and T in $\mathbb{R}^3$. R contained two vectors and was linearly independent but did not span $\mathbb{R}^3$; S contained three vectors, was linearly independent, and spanned $\mathbb{R}^3$; and T consisted of four vectors, was linearly dependent, and spanned $\mathbb{R}^3$. In that setting, we came to see that the set S was in some ways the best of the three: it had both key properties of being linearly independent and a spanning set. In other words, the set had enough vectors to span $\mathbb{R}^3$, but not so many vectors as to generate redundancy by being linearly dependent.

Through the next definition, we will now call such a set a *basis*, even in the generalized setting of vector spaces and subspaces.

Definition 1.12.3   Let V be a vector space and H a subspace of V. A set $\mathcal{B} = \{\mathbf{v}_1, \mathbf{v}_2, \ldots, \mathbf{v}_k\}$ of vectors in H is called a *basis* of H if and only if $\mathcal{B}$ is linearly independent and Span($\mathcal{B}$) $= H$. That is, $\mathcal{B}$ is a basis of H if and only if it is a linearly independent spanning set.

Several examples now follow that use the terminology of linear independence, span, and basis in the context of different vector spaces.

Example 1.12.1   In the vector space $\mathbb{P}$ of all polynomials, consider the subspace $H = \mathbb{P}_2$ of all polynomials of degree 2 or less. Show that the set $\mathcal{B} = \{1, t, t^2\}$ is a basis for H. Is the set $\{1, t, t^2, 4 - 3t\}$ also a basis for H?

Solution.   To begin, we observe that every element of $H = \mathbb{P}_2$ is a polynomial function of the form $p(t) = a_0 + a_1 t + a_2 t^2$. In particular, every element of $\mathbb{P}_2$ is a linear combination of the functions 1, t, and t^2, and therefore the set $\mathcal{B} = \{1, t, t^2\}$ spans H.

In addition, to determine whether the set $\mathcal{B}$ is linearly independent, we consider the equation

$$c_0 + c_1 t + c_2 t^2 = 0 \tag{1.12.2}$$

and ask whether or not this equation has a nontrivial solution. Keeping in mind that the '0' on the right-hand side represents the zero function in $\mathbb{P}_2$, the function that is everywhere equal to zero, we can see that if at least one of c_0, c_1, or c_2 is nonzero, we will be guaranteed to have either a nonzero constant function, a linear function, or a quadratic function, thus making $c_0 + c_1 t + c_2 t^2$

not identically zero. This shows that (1.12.2) has only the trivial solution, and therefore the set $\mathcal{B} = \{1, t, t^2\}$ is linearly independent. Having shown that $\mathcal{B}$ is a linearly independent spanning set for $H = \mathbb{P}_2$, we can conclude that $\mathcal{B}$ is a basis for H.

On the other hand, the set $\{1, t, t^2, 4 - 3t\}$ is not a basis for H since we can observe that the element $4 - 3t$ is a linear combination of the elements 1 and t: $4 - 3t = 4 \cdot 1 - 3 \cdot t$. This shows that the set $\{1, t, t^2, 4 - 3t\}$ is linearly dependent and thus cannot be a basis.

Example 1.12.2   Consider the set H of all functions of the form $y = c_1 \sin t + c_2 \cos t$. In the vector space $\mathcal{C}$ of all continuous functions, explain why the set $\mathcal{B} = \{\sin t, \cos t\}$ is a basis for the subspace H.

Solution.   First, we recall that H is indeed a subspace of $\mathcal{C}[-1, 1]$ due to our work in example 1.11.5.

By the definition of H (the set of all functions of the form $y = c_1 \sin t + c_2 \cos t$), we see immediately that $\mathcal{B}$ is a spanning set for H. In addition, it is clear that the functions $\sin t$ and $\cos t$ are not scalar multiples of one another: any scalar multiple of $\sin t$ is simply a vertical stretch of the function, which cannot result in $\cos t$. This tells us that the set $\mathcal{B} = \{\sin t, \cos t\}$ is also linearly independent, and therefore is a basis for H.

Example 1.12.3   In $\mathbb{R}^3$, consider the set $\mathcal{B} = \{\mathbf{e}_1, \mathbf{e}_2, \mathbf{e}_3\}$, where $\mathbf{e}_1 = [1 \ 0 \ 0]^T$, $\mathbf{e}_2 = [0 \ 1 \ 0]^T$, and $\mathbf{e}_3 = [0 \ 0 \ 1]^T$. Explain why $\mathcal{B}$ is a basis for $\mathbb{R}^3$.
Is the set $\mathcal{S} = \{\mathbf{v}_1, \mathbf{v}_2, \mathbf{v}_3\}$, where $\mathbf{v}_1 = [1 \ 2 \ -1]^T$, $\mathbf{v}_2 = [-1 \ 1 \ 3]^T$, and $\mathbf{v}_3 = [0 \ 3 \ 1]^T$ also a basis for $\mathbb{R}^3$?

Solution.   First, we observe that while the formal definition of a basis refers to the basis of a subspace H of a vector space V, since every vector space is a subspace of itself, it follows that we can also discuss a basis for a vector space.

Considering the set $\mathcal{B} = \{\mathbf{e}_1, \mathbf{e}_2, \mathbf{e}_3\}$, we observe that the vectors in this set are the columns of the 3×3 identity matrix. By the Invertible Matrix Theorem, it follows that the set $\mathcal{B}$ is linearly independent because $\mathbf{I}_3$ has a pivot in every column. Likewise, the set $\mathcal{B}$ spans $\mathbb{R}^3$ since $\mathbf{I}_3$ has a pivot in every row. As a linearly independent spanning set in $\mathbb{R}^3$, $\mathcal{B}$ is indeed a basis.

For the set $\mathcal{S}$ whose elements are the columns of the matrix

$$\mathbf{A} = \begin{bmatrix} 1 & -1 & 0 \\ 2 & 1 & 3 \\ -1 & 3 & 1 \end{bmatrix}$$

we again use the Invertible Matrix Theorem to determine whether or not S is a basis for $\mathbb{R}^3$. Row-reducing $\mathbf{A}$, it is straightforward to see that $\mathbf{A}$ is row equivalent to the identity matrix, and therefore is invertible. In particular, $\mathbf{A}$ has a pivot in every column and every row, and thus the columns of $\mathbf{A}$ are linearly independent and span $\mathbb{R}^3$. It follows that S is also a basis for $\mathbb{R}^3$.

The basis $\mathcal{B} = \{\mathbf{e}_1, \mathbf{e}_2, \mathbf{e}_3\}$ consisting of the columns of the 3×3 identity matrix is often referred to as the "standard basis of $\mathbb{R}^3$." In addition, by our work in example 1.12.3, we can see the role that the Invertible Matrix Theorem plays in determining whether a set of vectors in $\mathbb{R}^n$ is a basis or not. Specifically, since we know that it is logically equivalent for the columns of a square matrix $\mathbf{A}$ to be linearly independent and to be a spanning set for $\mathbb{R}^n$, it follows that a matrix $\mathbf{A}$ is invertible if and only if its columns form a basis for $\mathbb{R}^n$. We therefore update the Invertible Matrix Theorem with an additional statement as follows.

Theorem 1.12.1   (Invertible Matrix Theorem) Let $\mathbf{A}$ be an $n \times n$ matrix. The following statements are equivalent:

 a. $\mathbf{A}$ is invertible.

 b. The columns of $\mathbf{A}$ are linearly independent.

 c. The columns of $\mathbf{A}$ span $\mathbb{R}^n$.

 d. $\mathbf{A}$ has a pivot position in every column.

 e. $\mathbf{A}$ has a pivot position in every row.

 f. $\mathbf{A}$ is row equivalent to $\mathbf{I}_n$.

 g. For each $\mathbf{b} \in \mathbb{R}^n$, the equation $\mathbf{Ax} = \mathbf{b}$ has a unique solution.

 h. $\det(\mathbf{A}) \neq 0$.

 i. The columns of $\mathbf{A}$ form a basis for $\mathbb{R}^n$.

Our next example demonstrates how certain families of vectors naturally form subspaces of $\mathbb{R}^n$ and how vector arithmetic can be used to determine a basis for the subspace they form.

Example 1.12.4   Consider the set W of all vectors of the form $\begin{bmatrix} 3a + b - c \\ 4a - 5b + c \\ a + 2b - 3c \\ a - b \end{bmatrix}$.

Show that W is a subspace of $\mathbb{R}^4$ and determine a basis for this subspace.

Solution.   First, we observe that a typical element $\mathbf{v}$ of W is a vector of the form

$$\mathbf{v} = \begin{bmatrix} 3a + b - c \\ 4a - 5b + c \\ a + 2b - 3c \\ a - b \end{bmatrix}$$

Using properties of vector addition and scalar multiplication, we can write

$$\mathbf{v} = a \begin{bmatrix} 3 \\ 4 \\ 1 \\ 1 \end{bmatrix} + b \begin{bmatrix} 1 \\ -5 \\ 2 \\ -1 \end{bmatrix} + c \begin{bmatrix} -1 \\ 1 \\ -3 \\ -1 \end{bmatrix}$$

From this, we observe that W may be viewed as the span of the set $S = \{\mathbf{w}_1, \mathbf{w}_2, \mathbf{w}_3\}$, where

$$\mathbf{w}_1 = \begin{bmatrix} 3 \\ 4 \\ 1 \\ 1 \end{bmatrix}, \quad \mathbf{w}_2 = \begin{bmatrix} 1 \\ -5 \\ 2 \\ -1 \end{bmatrix}, \quad \mathbf{w}_3 = \begin{bmatrix} -1 \\ 1 \\ -3 \\ -1 \end{bmatrix}'$$

As seen in exercise 19 in section 1.11, the span of any set of vectors in $\mathbb{R}^n$ generates a subspace of $\mathbb{R}^n$; it follows that W is a subspace of $\mathbb{R}^4$. Moreover, we can observe that $S = \{\mathbf{w}_1, \mathbf{w}_2, \mathbf{w}_3\}$ is a linearly independent set since

$$\begin{bmatrix} 3 & 1 & -1 \\ 4 & -5 & 1 \\ 1 & 2 & -3 \\ 1 & -1 & -1 \end{bmatrix} \rightarrow \begin{bmatrix} 1 & 0 & 0 \\ 0 & 1 & 0 \\ 0 & 0 & 1 \\ 0 & 0 & 0 \end{bmatrix}$$

Since S both spans the subspace W and is linearly independent, it follows that S is a basis for W.

In example 1.12.4 we used the fact that the span of any set in $\mathbb{R}^n$ is a subspace of $\mathbb{R}^n$. This result extends to general vector spaces and is stated formally in the following theorem.

Theorem 1.12.2   In any vector space V, the span of any set of vectors forms a subspace of V.

It is not hard to prove this result. Since the span of a set contains all linear combinations of the set, it must contain the zero combination and be closed under both vector addition and scalar multiplication.

One of the reasons that a basis for a subspace is important is that a basis tells us the minimum number of vectors needed to fully describe every element of the subspace. More specifically, given a basis $\mathcal{B}$ for a subspace W, we know that we can write every element of W uniquely as a linear combination of the elements in the basis. Note that a subspace does not have a unique basis; for example, in example 1.12.3, we saw two different bases for $\mathbb{R}^3$.

Furthermore, in $\mathbb{R}^3$ we have seen that the standard basis (and one example of another basis) has three elements. By the Invertible Matrix Theorem, it is clear that every basis of $\mathbb{R}^3$ consists of three vectors since we are required to have a set that is both linearly independent and spans $\mathbb{R}^3$. Likewise, any basis of $\mathbb{R}^n$ will have n elements. It can be shown that even in vector spaces other than $\mathbb{R}^n$, any two bases of a subspace are guaranteed to have the same number of elements. Therefore, this number of elements in a basis can be used to identify a fundamental property of any subspace: the minimum number of elements needed to describe all of the elements in the space. We call this number the *dimension* of the subspace.

Definition 1.12.4   Given a subspace W in a vector space V and a basis $\mathcal{B}$ for W, the number of elements in $\mathcal{B}$ is the *dimension* of W. Equivalently, if $\mathcal{B}$ has k elements, we write $dim(W) = k$.

Thus we naturally use the language that "$\mathbb{R}^3$ is three-dimensional" and similarly that "$\mathbb{R}^n$ has dimension n." Similarly, we can say $dim(\mathbb{P}_2) = 3$ (see example 1.12.1), and that the dimension of the vector space of all linear combinations of the functions $\sin t$ and $\cos t$ is two (see example 1.12.2).

In closing, it is worth recalling example 1.6.3 in section 1.6, where we considered three sets R, S, and T in $\mathbb{R}^3$. R contained two vectors and was linearly independent but did not span $\mathbb{R}^3$; S contained three vectors, was linearly independent, and spanned $\mathbb{R}^3$; and T consisted of four vectors, was linearly dependent, and spanned $\mathbb{R}^3$. Since the set S has both key properties of being linearly independent and a spanning set, we can say that the set S is a basis for $\mathbb{R}^3$, which further reflects the fact that $dim(\mathbb{R}^3) = 3$.

Exercises 1.12   In the vector space V given in each of exercises 1–7, determine a basis for the subspace H and hence state the dimension of H.

1. $V = \mathbb{R}^3$, $H = \left\{ t \begin{bmatrix} 2 \\ 0 \\ -1 \end{bmatrix} : t \in \mathbb{R} \right\}$

2. $V = \mathbb{P}_2$, $H = \left\{ at^2 : a \in \mathbb{R} \right\}$

3. $V = \mathbb{R}^4$, $H = \left\{ \begin{bmatrix} 2a + 3b \\ a - 4b \\ -3a + 2b \\ a - b \end{bmatrix} : a, b \in \mathbb{R} \right\}$

4. $V = \mathbb{P}$ (the vector space of all polynomials), $H = \mathbb{P}_n$ (the subspace of all polynomials of degree n or less)

5. $V = \mathbb{R}^2$, $H = \left\{ \mathbf{x} : \mathbf{Ax} = \mathbf{0} \text{ where } \mathbf{A} = \begin{bmatrix} 2 & -1 \\ -6 & 3 \end{bmatrix} \right\}$

6. $V = \mathbb{R}^4$, $H = \left\{ \mathbf{x} : \mathbf{Ax} = \mathbf{0} \text{ where } \mathbf{A} = \begin{bmatrix} 1 & -3 & 2 & -1 \\ -2 & 5 & 0 & 4 \end{bmatrix} \right\}$

7. $V = \mathcal{M}_{2\times2}$, $H = \left\{ \mathbf{A} \in \mathcal{M}_{2\times2} : \mathbf{A} = \begin{bmatrix} a & 0 \\ b & c \end{bmatrix} \right\}$

8. Determine whether or not the following set S is a basis for $\mathbb{R}^3$. If not, is some subset of S a basis for $\mathbb{R}^3$? Explain.

$$S = \left\{ \begin{bmatrix} 1 \\ 0 \\ 1 \end{bmatrix}, \begin{bmatrix} 0 \\ 1 \\ 1 \end{bmatrix}, \begin{bmatrix} 1 \\ 1 \\ 1 \end{bmatrix}, \begin{bmatrix} 2 \\ 1 \\ 3 \end{bmatrix} \right\}$$

9. Is the set $S = \{[1\ 2]^T, [2\ 1]^T\}$ a basis for $\mathbb{R}^2$? Justify your answer.

10. Is the set $S = \{[1\ 2]^T, [-4\ -8]^T\}$ a basis for $\mathbb{R}^2$? Justify your answer.

11. Is the set $S = \{[1\ 2\ 1\ 1]^T, [2\ 1\ 1\ -1]^T, [-1\ 1\ 3\ 1]^T, [2\ 4\ 5\ 1]^T\}$ a basis for $\mathbb{R}^4$? Justify your answer.

12. Is the set $S = \{[1\ 2\ 1\ 1]^T, [2\ 1\ 1\ -1]^T, [-1\ 1\ 3\ 1]^T, [2\ 4\ 5\ 0]^T\}$ a basis for $\mathbb{R}^4$? Justify your answer.

13. Can a set with three vectors be a basis for $\mathbb{R}^4$? Why or why not?

14. Can a set with seven vectors be a basis for $\mathbb{R}^6$? Why or why not?

15. Not every vector space has a basis with finitely many elements. If there is not a finite basis, then we say that the vector space is *infinite dimensional*. Explain why the vector space $\mathbb{P}$ of all polynomial functions is an infinite dimensional vector space.

16. Let V be the vector space $V = \mathcal{C}[-1, 1]$ and H the subset defined by

$$H = \{f \in \mathcal{C}[-1, 1] : f \text{ is differentiable}\}$$

Explain why H is an infinite dimensional subspace of V and why we cannot explicitly write down the elements in a basis for H.

17. Recall from exercises 22 and 23 in section 1.11 that the null space of a matrix $\mathbf{A}$ is the subspace of all solutions to the equation $\mathbf{Ax} = \mathbf{0}$ and that the column space of $\mathbf{A}$ is the space spanned by the columns of $\mathbf{A}$. By exploring several different examples of matrices $\mathbf{A}$ of your choice, discuss how the dimensions of the null and column spaces are related to the number of pivot columns in the matrix. In particular, explain what you can say about the relationship between the sum of the dimensions of the null and column spaces and the number of columns in the matrix $\mathbf{A}$.

18. Decide whether each of the following sentences is true or false. In every case, write one sentence to support your answer.

(a) Any set of five vectors is a basis for $\mathbb{R}^5$.
(b) If S is a linearly independent set of six vectors in $\mathbb{R}^6$, then S is a basis for $\mathbb{R}^6$.
(c) If the determinant of a 3×3 matrix $\mathbf{A}$ is zero, then the columns of $\mathbf{A}$ form a basis for $\mathbb{R}^3$.
(d) If $\mathbf{A}$ is an $n \times n$ matrix whose columns span $\mathbb{R}^n$, then the columns of $\mathbf{A}$ form a basis for $\mathbb{R}^n$.

1.13 For further study

1.13.1 Computer graphics: geometry and linear algebra at work

In modern computer graphics, images consisting of sets of pixels are moved around the screen through mathematical computations that rely on linear algebra. If we focus on two-dimensional objects, there are several basic moves that we must be able to perform: translation, rotation, reflection, and dilation. In what follows, we explore the role that linear algebra plays in the geometry of linear transformations and computer graphics.

(a) In section 1.8.1 we began to develop an understanding of how matrix multiplication can be used to move a two-dimensional image around the plane. If you have not already read this section, do so now.

   If we take the perspective that a given point in the plane is stored in the vector $\mathbf{v}$, then for any 2×2 matrix $\mathbf{A}$, the matrix $\mathbf{A}$ moves the vector via multiplication to the new location $\mathbf{Av}$. If we have a finite set of points (which together constitute an image), we can store the points in a matrix $\mathbf{M}$ whose columns represent the individual points), and the new image which results from multiplication by $\mathbf{A}$ is given by $\mathbf{AM}$.

   Consider the triangle with vertices $(0, 0)$, $(3, 1)$, and $(2, 2)$, stored in the matrix

   $$\mathbf{M} = \begin{bmatrix} 0 & 3 & 2 \\ 0 & 1 & 2 \end{bmatrix}$$

   Choose three different matrices $\mathbf{A}$ and compute $\mathbf{AM}$. Then explain why it is impossible to use multiplication by a 2×2 matrix to translate the triangle so that all three of its vertices appear in new locations.

(b) Due to our discovery in (a) that a simple translation is impossible using 2×2 matrices, we introduce the notion of *homogeneous coordinates*; instead of representing points in the two-dimensional plane as $[x \ y]^\mathrm{T}$, we move to a plane in three-dimensional space where the third coordinate is always 1. That is, instead of $[x \ y]^\mathrm{T}$ we use $[x \ y \ 1]^\mathrm{T}$.

   Consider the matrix $\mathbf{A}$ given by

   $$\mathbf{A} = \begin{bmatrix} 1 & 0 & a \\ 0 & 1 & b \\ 0 & 0 & 1 \end{bmatrix} \qquad (1.13.1)$$

   and the triangle from (a) which can be represented in homogeneous coordinates by the matrix

   $$\mathbf{M} = \begin{bmatrix} 0 & 3 & 2 \\ 0 & 1 & 2 \\ 1 & 1 & 1 \end{bmatrix}$$

Compute **AM**. What has happened to each vertex of the triangle represented by **M**? Explain in terms of the parameters a and b in **A**.

(c) Using $a = 2$ and $b = -1$ in (1.13.1) along with the triangle **M** from above, compute **AM** in order to determine the translation of the triangle 2 units in the x-direction and -1 units in the y-direction. Sketch both the original triangle and its image under this translation.

(d) In order to view some more sophisticated graphics, we use *Maple* in our computations that follow. Rather than performing operations on a triangle, we will use the syntax

```
> with(plots): with(LinearAlgebra):
> setoptions(scaling=constrained, axes=boxed,
  tickmarks=[5,5]):
> X := cos(t)*(1+sin(t))*(1+0.3*cos(8*t))*
  (1+0.1*cos(24*t)):
> Y := sin(t)*(1+sin(t))*(1+0.3*cos(8*t))*
  (1+0.1*cos(24*t)):
> plot([X,Y,t=0..2*Pi], color=blue,
  thickness=3);
```

which generates a parametric curve whose plot is the leaf shown in figure 1.18. Input these commands in *Maple*, as well as the syntax

```
> leaf := plot([X,Y,t=0..2*Pi], color=grey,
  thickness=1):
```

to store the image of the original leaf in `leaf`.

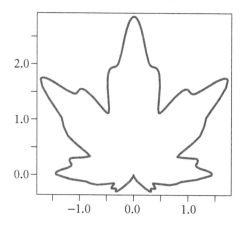

Figure 1.18 A *Maple* leaf.

Finally, for a given matrix **A** of the form

$$\mathbf{A} = \begin{bmatrix} a_{11} & a_{12} & a_{13} \\ a_{21} & a_{22} & a_{23} \\ 0 & 0 & 1 \end{bmatrix}$$

and a vector $\mathbf{Z} = [X \ Y \ 1]$, compute **AZ** (by hand) to show how **AZ** depends on the entries in **A**.

(e) By our work in (c) and (d), if we now let

$$\mathbf{A} = \begin{bmatrix} 1 & 0 & 2 \\ 0 & 1 & -1 \\ 0 & 0 & 1 \end{bmatrix}$$

the product **AZ** should result in translation of the leaf by the vector $[2 \ -1]^{\mathrm{T}}$. To test this, we define the matrix **A** in *Maple* by

```
> A := <<1,0,0>|<0,1,0>|<2,-1,1>>;
```

and compute the coordinates in the new image by

```
> Xnew := A[1,1]*X + A[1,2]*Y + A[1,3]*1:
> Ynew := A[2,1]*X + A[2,2]*Y + A[2,3]*1:
> image1 := plot([Xnew,Ynew,t=0..2*Pi],
    thickness=3, color=blue):
```

The last command above plots the resulting image and stores it in `image1`. Display both the original leaf and the new image with the command

```
> display(leaf, image1);
```

and show that this results indeed in the translated leaf as shown in figure 1.19.

(f) In section 1.8.1, we learned that a matrix of the form

$$\mathbf{R} = \begin{bmatrix} \cos\theta & -\sin\theta \\ \sin\theta & \cos\theta \end{bmatrix}$$

is known as a rotation matrix and, through multiplication, rotates any vector by θ radians counterclockwise about the origin. To work with a rotation matrix in homogeneous coordinates, we update the matrix as follows:

$$\mathbf{R} = \begin{bmatrix} \cos\theta & -\sin\theta & 0 \\ \sin\theta & \cos\theta & 0 \\ 0 & 0 & 1 \end{bmatrix}$$

Let us say that we wanted to perform two operations on the leaf. First, we wish to translate the leaf as above along the vector $[2 \ -1]^{\mathrm{T}}$, and then we

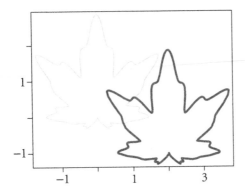

Figure 1.19 The original leaf and its translation by $[2 \ -1]^{\mathrm{T}}$.

want to rotate the resulting image $\pi/4$ radians clockwise about the origin. We can accomplish this through two matrices by computing their product, as the following discussion shows.

From (e), we know that using the matrix

```
> Translation := <<1,0,0>|<0,1,0>|<2,-1,1>>;
```

leads to the desired translation. Likewise, the matrix

```
> Rotation := <<1/sqrt(2),-1/sqrt(2),
  0>|<1/sqrt(2),1/sqrt(2),0>|<0,0,1>>;
```

will produce the sought rotation.

Explain why the matrix

```
> A := Rotation.Translation;
```

will produce the combined translation and rotation, and plot the resulting figure by updating your computations for Xnew and Ynew and using the syntax

```
> image2 := plot([Xnew,Ynew,t=0..2*Pi],
  thickness=4, color=black):
> display(leaf, image1, image2);
```

(g) What is the result of applying the matrix

$$\mathbf{A} = \begin{bmatrix} 1 & 0 & 0 \\ 0 & \frac{1}{2} & 0 \\ 0 & 0 & 1 \end{bmatrix}$$

on the leaf? What kind of geometric transformation is performed by this matrix? What matrix would keep the height of the leaf constant but stretch its width by a factor of 2?

(h) It can be shown that to reflect an image across a line through the origin that forms an angle α with the positive x-axis, the necessary matrix is

$$\mathbf{A} = \begin{bmatrix} \cos 2\alpha & \sin 2\alpha & 0 \\ \sin 2\alpha & -\cos 2\alpha & 0 \\ 0 & 0 & 1 \end{bmatrix}$$

By finding the appropriate value of α, find the matrix that will reflect an image across the line $y = x$ and compute and plot the image of the original leaf under this reflection.

(i) **Exercises for further practice and investigation:**

1. Find the image of the original leaf under rotation about the origin by $2\pi/3$ radians, followed by a reflection across the y-axis.
2. Find the image of the original leaf under rotation about the point $(-3, 1)$ by $-\pi/6$ radians. (**Hint:** To rotate about a point other than the origin, first translate that point to the origin, then rotate, then translate back.)
3. Find the image of the original leaf under translation along the vector $[3 \ 2]^{\mathrm{T}}$, followed by reflection across the line $y = x/2$.

1.13.2 Bézier curves

In what follows[7], we explore the use of a specific type of parametric curves, called *Bézier curves* (pronounced "bezzy-eh"), which have a variety of important applications. These curves were originally developed by two automobile engineers in France in the 1960s, P. Bézier and P. de Casteljau, who were working to develop mathematical formulas to graph the smooth, wiggle-free curves that formed the shape of a car's body. Today, Bézier curves find their way into our lives every day: they are used to create the letters that appear in typeset fonts. The principles that govern these curves involve fundamental mathematics from linear algebra and calculus.

(a) In calculus, we study parametric curves given in the form $x = f(t)$, $y = g(t)$, where f and g are each functions of the parameter t. Another way to denote this situation is to write

$$P(t) = (f(t), g(t))$$

where t belongs to some interval of real numbers. Note that $P(t)$ is essentially a vector; the graph of $P(t)$ is the parametric curve traced out by

---

[7] The material in this project has been adapted from Steven Janke's chapter "Designer Curves" in *Applications of Calculus*, MAA Notes Number 29, Philip Straffin, Ed.

the vector over time. It will be most convenient if we simply write this as $P(t) = (x(t), y(t))$ in what follows. In this problem we begin to consider some special formulas for $x(t)$ and $y(t)$.

To parameterize the line between the points $P_0(1, 3)$ and $P_1(3, 7)$, we can think about wanting to make x go from 1 to 3, and y go from 4 to 7. Indeed, we want these to occur simultaneously as t goes from 0 to 1. Consider the parameterization:

$$x = x(t) = 1 + t(3 - 1) = t \cdot 3 + (1 - t) \cdot 1$$

$$y = y(t) = 3 + t(7 - 3) = t \cdot 7 + (1 - t) \cdot 3$$

$$0 \le t \le 1$$

Observe that when $t = 0$, $x = 1$ and $y = 3$, and when $t = 1$, $x = 3$ and $y = 7$.

Show that the curve parameterized by these two equations is indeed the line segment between P_0 and P_1. For instance, you might use algebra to eliminate the variable t, thereby deducing a relationship between x and y.

(b) We can think about the equations for x and y in (a) in a more compact manner. Consider the following vector notation to replace the previous equations:

$$P(t) = \begin{bmatrix} x(t) \\ y(t) \end{bmatrix} = t \begin{bmatrix} 3 \\ 7 \end{bmatrix} + (1 - t) \begin{bmatrix} 1 \\ 3 \end{bmatrix} \qquad (1.13.2)$$

This is sometimes referred to as taking a *convex combination* of the points $(1, 3)$ and $(3, 7)$, because t and $1 + t$ are both nonnegative and sum to 1.

Using the above style, write the parametric equations for the line segment that passes between the general points $P_0(x_0, y_0)$ and $P_1(x_1, y_1)$.

(c) An even more concise notation is to simply write $P(t) = (1 - t)P_0 + tP_1$. We will now use this notation to combine two or more of these parameterizations for line segments in a way that constructs curves that can be "controlled" in very interesting ways.

Consider three points, labeled P_0, P_1, and P_2. In the most recent form of $P(t)$ given above at (1.13.2), write parameterizations for the two line segments from P_0 to P_1 and from P_1 to P_2, as pictured below. Call the first parameterization $P^{(1)}(t)$ and the second parameterization $P^{(2)}(t)$.

In addition, determine the parameterizations $P^{(1)}(t)$ and $P^{(2)}(t)$ for the specific set of points $P_0(2, 3)$, $P_1(4, 7)$, and $P_2(7, 1)$. Show your work, and write each out in the expanded form where you have an expression for $x(t)$ and another for $y(t)$.

(d) From the two line-segment parameterizations in (c), we will now create a new parametric plot by taking similar combinations of $P^{(1)}(t)$ and $P^{(2)}(t)$.

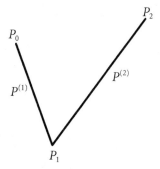

Figure 1.20 The line segments from P_0 to P_1 and P_1 to P_2.

Consider the function $Q(t)$ defined as follows:

$$Q(t) = (1-t) \cdot P^{(1)}(t) + t \cdot P^{(2)}(t) \qquad (1.13.3)$$

First, substitute in (1.13.3) your expressions for $P^{(1)}(t)$ and $P^{(2)}(t)$ from (c) that involve the general points P_0, P_1, and P_2. Simplify the result as much as possible in order to write the formula for Q in the following form:

$$Q(t) = a_0(t)P_0 + a_1(t)P_1 + a_2(t)P_2$$

where $a_0(t)$, $a_1(t)$, and $a_2(t)$ are polynomial functions of t.

Then, using the specific parameterizations for $P^{(1)}(t)$ and $P^{(2)}(t)$ for the points $P_0(2, 3)$, $P_1(4, 7)$, and $P_2(7, 1)$, determine the parametric equations for $x(t)$ and $y(t)$ that make up the function $Q(t)$. For each of these three parameterizations ($P^{(1)}$, $P^{(2)}$, and Q), use *Maple* to sketch a plot[8] and describe the results in detail. For example, how does $Q(t)$ look in comparison to the two line segments? What kind of functions make up the components $x(t)$ and $y(t)$ in Q?

What is true about $Q(0)$ relative to the points P_0, P_1, and P_2? $Q(1)$? What direction is a particle moving along $Q(t)$ headed as t starts out away from 0? As t gets near to 1?

(e) It turns out that we will have even more freedom and control in drawing curves if we start with four control points, P_0, P_1, P_2, and P_3. The development here is similar to what was done above, just using a greater number of points.

First, parameterize the segments from P_0 to P_1 (with $P^{(1)}(t)$), P_1 to P_2 (with $P^{(2)}(t)$), and from P_2 to P_3 (with $P^{(3)}(t)$). The usual formulas apply

---

[8] The *Maple* syntax to plot a parametric curve $(f(t), g(t))$ on the interval $[a, b]$ is
> `plot([f(t),g(t),t=a..b]);`.

here; write down the basic form of each $P^{(j)}(t)$, $j = 1, 2, 3$, in terms of the various points P_i.

Then combine, as in (d) above, the parameterizations for the first two segments to get a new function $Q^{(1)}$; also combine the parameterizations for the second two segments to get $Q^{(2)}$. These Q parameterizations are written as

$$Q^{(1)}(t) = (1 - t) \cdot P^{(1)}(t) + t \cdot P^{(2)}(t)$$

$$Q^{(2)}(t) = (1 - t) \cdot P^{(2)}(t) + t \cdot P^{(3)}(t)$$

Finally, combine $Q^{(1)}$ and $Q^{(2)}$ to get a new parametric function that we call $B(t)$ according to the natural formula

$$B(t) = (1 - t) \cdot Q^{(1)}(t) + t \cdot Q^{(2)}(t)$$

By substituting appropriately for $Q^{(1)}(t)$ and $Q^{(2)}(t)$ and then replacing these with the appropriate $P^{(j)}(t)$ functions, show that

$$B(t) = P_0(1 - t)^3 + 3P_1 t(1 - t)^2 + 3P_2 t^2(1 - t) + P_3 t^3.$$

$B(t)$ is called a *cubic Bézier curve*.

By finding and using appropriate t values, show that the points P_0 and P_3 both lie on the curve given by $B(t)$.

(f) Write the formulas for $x(t)$ and $y(t)$ that give the parameterizations for the cubic Bézier curve that has the four control points $P_0(2, 2)$, $P_1(5, 10)$, $P_2(40, 20)$, and $P_3(10, 5)$. Use *Maple* to plot each of the parametric curves given by $P^{(j)}(t)$, $j = 1 \ldots 3$, $Q^{(1)}(t)$, $Q^{(2)}(t)$, and $B(t)$ in the same window. Discuss how the various curves combine to form others.

(g) For the general Bézier curve with control points $P_0(x_0, y_0)$, $P_1(x_1, y_1)$, $P_2(x_2, y_2)$, and $P_3(x_3, y_3)$, derive the equation for the tangent line to the curve at the point (x_0, y_0), and prove that the point (x_1, y_1) lies on this tangent line. (Hint: to determine the slope of the tangent line, use the chain rule in the standard way for finding dy/dx for a parametric curve.)

(h) Laser printers and the program Postscript use Bézier curves to construct the fonts that we use to represent letters. For example, a picture of the letter g is shown below that reveals the control points and Bézier curves required to accomplish this.

In *Maple*, use two or more Bézier curves to sketch a reasonable representation of the letter S. (You need not try to emulate the thickness of the 'g' that is shown above.)

Then, use an appropriate number of Bézier curves to create an approximation of the lowercase letter 'a,' in the form shown here in quotes. State the control points required for the various curves.

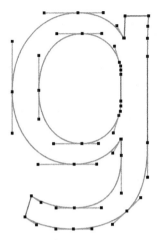

Figure 1.21   The letter g.

(i) Discuss the role that vectors and linear combinations play in the development of Bézier curves.

1.13.3 Discrete dynamical systems

A *linear discrete dynamical system* is a model that represents changes in a system from time k to time $k + 1$ by the rule

$$\mathbf{x}^{(k+1)} = \mathbf{A}\mathbf{x}^{(k)}$$

A discrete dynamical system is similar to a Markov chain, but we no longer require that the columns of the matrix $\mathbf{A}$ sum to 1. A key issue in either scenario is the long term behavior of the quantity $\mathbf{x}^{(k)}$ being modeled. In what follows, we explore the role of eigenvalues and eigenvectors in determining this long-term behavior and study an important application of these ideas.

(a) To begin investigating the long-term behavior of the system, we will assume that $\mathbf{A}$ is an $n \times n$ matrix with n real linearly independent eigenvectors $\mathbf{v}_1, \ldots, \mathbf{v}_n$. Furthermore, assume that the corresponding real eigenvalues of $\mathbf{A}$ satisfy the relationship

$$|\lambda_1| > |\lambda_2| \geq \cdots \geq |\lambda_n|$$

Consider an initial vector $\mathbf{x}^{(0)}$. Explain why there exist constants $c_1, \ldots, c_n$ such that

$$\mathbf{x}^{(0)} = c_1\mathbf{v}_1 + c_2\mathbf{v}_2 + \cdots + c_n\mathbf{v}_n$$

and show that

$$\mathbf{A}\mathbf{x}^{(0)} = c_1\lambda_1\mathbf{v}_1 + c_2\lambda_2\mathbf{v}_2 + \cdots + c_n\lambda_n\mathbf{v}_n$$

Furthermore, show that

$$\mathbf{x}^{(k)} = \mathbf{A}^k\mathbf{x}^{(0)} = c_1\lambda_1^k\mathbf{v}_1 + c_2\lambda_2^k\mathbf{v}_2 + \cdots + c_n\lambda_n^k\mathbf{v}_n \tag{1.13.4}$$

(b) In (1.13.4), divide both sides by λ_1^k. What can you conclude about $(\lambda_2/\lambda_1)^k$ as $k \to \infty$? Why can you make similar conclusions about $(\lambda_j/\lambda_1)^k$ for $j = 3 \ldots n$? Hence explain why for large k

$$\left(\frac{1}{\lambda_1}\right)^k \mathbf{A}^k\mathbf{x}^{(0)} \approx c_1\mathbf{v}_1$$

and thus why $\mathbf{A}^k\mathbf{x}^{(0)}$ is an approximate eigenvector of $\mathbf{A}$ corresponding to $\mathbf{v}_1$.

(c) In studying a population like spotted owls, mathematical ecologists often pay close attention to the various numbers of a species at different stages of life. For example, for spotted owls there are three pronounced groupings: juveniles (under 1 year), subadults (1 to 2 years old), and adults (2 years and older). The owls mate during the latter two stages, breed as adults, and can live for up to 20 years. A critical time in the life cycle and survival of these owls is when the juvenile leaves the nest to build a home of its own.[9] Let the number of spotted owls in year k be represented by the vector

$$\mathbf{x}^{(k)} = \begin{bmatrix} j_k \\ s_k \\ a_k \end{bmatrix}$$

where j_k is the number of juveniles, s_k the number of subadults, and a_k the number of adults. Using field data, mathematical ecologists have determined[10] that a particular spotted owl population is modeled by the discrete dynamical system

$$\mathbf{x}^{(k+1)} = \begin{bmatrix} 0 & 0 & 0.33 \\ 0.18 & 0 & 0 \\ 0 & 0.71 & 0.94 \end{bmatrix} \mathbf{x}^{(k)}$$

What does this model imply about the percent of juveniles that survive to become subadults? About the percent of subadults that survive to become adults? About the percent of adults that survive from one year to the next? What percent of adults produce juvenile offspring in a given year?

(d) Assume that in a given region, ecologists have measured the present populations as follows: $j_0 = 200$, $s_0 = 45$, and $a_0 = 725$. Use the model stated in (c) to determine the population $\mathbf{x}^{(k)} = [j_k \ s_k \ a_k]^T$ for $k = 1, \ldots, 20$. Do you think the spotted owl will become extinct? Give a

---

[9] To read more about the issue of spotted owl survival, see the introduction to chapter 5 of David C. Lay's *Linear Algebra and its Applications*.
[10] R. H. Lamberson et al., "A Dynamic Analysis of the Viability of the Northern Spotted Owl in a Fragmented Forest Environment," *Conservation Biology* 6 (1992), 505–512.

convincing argument using not only your computations of the population vectors but also the results of (b).

(e) Say that r is the fraction of juveniles that survive from one year to the next (that is, replace 0.18 in the matrix of the model with r) . By experimenting with different values of r, determine the minimum fraction of juveniles that must survive from one year to the next in order for the spotted owl population not to become extinct. How does your answer depend on the eigenvalues of the matrix?

(f) Let A be the $n \times n$ matrix of a discrete dynamical system and assume that A has n real linearly independent eigenvectors. Let $\mathbf{x}^{(0)}$ be an initial vector and let $\rho(A)$ denote the maximum absolute value of an eigenvalue of A. Show that the following are true:

(i) If $\rho(A) < 1$, then $\lim_{k \to \infty} A^k \mathbf{x}^{(0)} = \mathbf{0}$.

(ii) If $\rho(A) = 1$ and $\lambda = 1$ is the unique eigenvalue having this maximum absolute value, then $\lim_{k \to \infty} A^k \mathbf{x}^{(0)}$ is an eigenvector of A.

(iii) If $\rho(A) > 1$, then there exist choices of $\mathbf{x}^{(0)}$ for which $\|A^k \mathbf{x}^{(0)}\|$ grows without bound.

2

First-order differential equations

2.1 Motivating problems

Differential equations arise naturally in many problems encountered when modeling physical phenomena. To begin our study of this subject, we introduce two fundamental examples that demonstrate the central role that differential equations play in our world.

In section 1.1, we discussed how the amount of salt present in a system of two tanks can be modeled through a system of differential equations. Here, an even simpler situation is considered: our goal is to predict the amount of salt present in a city's water reservoir at time t, given a set of determining conditions.

Suppose that the reservoir is filled to its capacity of $10\,000\,\text{m}^3$, and that measurements indicate an initial concentration of salt of $C_0 = 0.02\,\text{g/m}^3$. Note that it follows there are $A_0 = 200$ g of salt initially present. As the city draws this solution from the reservoir for use, new solution (water with some salt concentration) from the local treatment facility flows into the reservoir so that the volume of water present in the tank stays constant. Let us assume that the concentration of salt in the inflowing solution is $0.01\,\text{g/m}^3$, and that the rate of this inflow is $1000\,\text{m}^3/\text{day}$. Since the city is also assumed to be drawing solution at an equal rate from the reservoir, the outflow also occurs at a rate of $1000\,\text{m}^3/\text{day}$.

We are interested in several key questions. How much salt is in the tank at time t? What is the concentration of salt in the water being used by the city at time t? What happens to these values over time?

We will let $A(t)$ denote the amount of salt in the tank at time t. The instantaneous rate of change dA/dt of $A(t)$ is given by the difference between the rate at which salt is entering the tank and the rate at which salt is leaving. Exploring the given information regarding inflow and outflow, we can determine these rates precisely.

Since solution is entering the reservoir at $1000 \, \text{m}^3/\text{day}$ containing a concentration of $0.01 \, \text{g/m}^3$, it follows that salt is entering the tank at a rate of

$$1000 \frac{\text{m}^3}{\text{day}} \cdot 0.01 \frac{\text{g}}{\text{m}^3} = 10 \frac{\text{g}}{\text{day}}$$

For salt leaving the reservoir, the situation is slightly more complicated. Since we do not know the exact amount of salt present in the reservoir at time t, we denote this by $A(t)$. Assuming that the solution in the reservoir is uniformly mixed, the concentration of salt in the outflowing solution is the ratio of the amount $A(t)$ of salt to the volume of the tank. That is, the outflowing concentration is

$$\frac{A(t)}{10\,000} \frac{\text{g}}{\text{m}^3}$$

Since this outflow is occurring at a rate of $1000 \, \text{m}^3/\text{day}$, it follows that salt is leaving the tank at a rate of

$$1000 \frac{\text{m}^3}{\text{day}} \cdot \frac{A(t)\text{g}}{10\,000 \, \text{m}^3} = \frac{A(t)}{10} \frac{\text{g}}{\text{day}}$$

It now follows that the instantaneous rate of change dA/dt of salt in the tank in grams per day is given by the difference of the rate of salt entering and the rate of salt leaving the tank. Specifically,

$$\frac{dA}{dt} = 10 - \frac{A(t)}{10} \tag{2.1.1}$$

Note carefully what this last equation is saying: $A(t)$ is an unknown function, but we have an equation that relates this unknown function to its derivative. Such an equation is called a *differential equation*. The *solution* to this equation is a function $A(t)$ that makes the equation true. If we can solve the equation for $A(t)$, we then will be able to predict the amount of salt in the tank at any time t. Determining such solutions and their long-term behaviors is the main focus of this chapter.

Another important application of differential equations involves population growth. Consider a population $P(t)$ of animals. As likelihood of reproduction depends on the number of animals present, it is natural to assume that the rate of change of $P(t)$ is directly proportional to $P(t)$. Phrased in terms of the derivative, this assumption means that

$$\frac{dP}{dt} = kP(t) \tag{2.1.2}$$

where k is some positive constant. Observe that (2.1.2) is a differential equation involving the function P. It is a standard exercise in calculus to show that functions of the form

$$P(t) = P_0 e^{kt}$$

are solutions to (2.1.2).

Because the function $P(t) = P_0 e^{kt}$ exhibits unbounded growth over time, it turns out that this exponential growth model is not realistic beyond a relatively short period of time. A related, but more sophisticated, model of population growth is the *logistic differential equation*

$$\frac{dP}{dt} = kP(t)\left(1 - \frac{P(t)}{A}\right)$$

where the constant k is considered the reproductive rate of the population and the constant A is the surrounding environment's *carrying capacity*. For example, if a population had a relative growth rate of $k = 0.02$ and a carrying capacity of $A = 100$, the population function would satisfy the differential equation

$$\frac{dP}{dt} = 0.02P(t)\left(1 - \frac{P(t)}{100}\right)$$

The logistic model, usually credited to the Dutch mathematician Pierre Verhulst, accounts not only for reproductive growth, but also for mortality by considering environmental limitations on maximum population. The logistic equation is more challenging to solve; we will do so in section 2.7.

In addition to mixing problems and models of population growth, differential equations enjoy widespread applications in other physical phenomena. Differential equations are also mathematically interesting in and of themselves, and in upcoming sections we will study not only their applications, but also their key properties and characteristics to better understand the subject as a whole.

2.2 Definitions, notation, and terminology

As we have seen with the examples

$$\frac{dA}{dt} = 10 - \frac{A}{10} \tag{2.2.1}$$

$$\frac{dP}{dt} = 0.02P\left(1 - \frac{P}{100}\right) \tag{2.2.2}$$

$$y'' + y = 0 \tag{2.2.3}$$

a *differential equation* is an equation relating an unknown function to one or more of its derivatives. Usually we will suppress the notation "$A(t)$" and instead simply write "A," as in (2.2.1). We will interchangeably use the notations y' and dy/dt to represent the first derivative; similarly, $y'' = d^2y/dt^2$. Other books sometimes employ the notations $y' = D(y) = \dot{y}$ and $y'' = D^2(y) = \ddot{y}$.

A *solution* of a differential equation is a differentiable function that satisfies the equation on some interval (a, b) of values for the independent variable. For example, the function $y = \sin t$ is a solution to (2.2.3) on $(-\infty, \infty)$ since $y'' = -\sin t$, and $-\sin t + \sin t = 0$ for all values of t.

Given any differential equation, we are interested in determining all of its solutions. But many, if not most, differential equations are difficult or impossible to solve. For example, the equation

$$y'' + ty = t$$

(which is only a slightly modified version of (2.2.3)) has no solution in terms of elementary functions.[1] In such situations, we may turn to qualitative or approximation methods that may enable us to analyze how a solution should behave, while perhaps not being able to determine an explicit formula for the function.

Equations (2.2.1), (2.2.2), and (2.2.3) are often called *ordinary* differential equations, in contrast to *partial* differential equations such as

$$\frac{\partial^2 u}{\partial x^2} + \frac{\partial^2 u}{\partial y^2} = 0$$

where the solution function $u(x, y)$ has two independent variables x and y. Our focus will be on ordinary differential equations, as partial differential equations are beyond the scope of this text. The *order* of a differential equation is the order of the highest derivative present. For example, (2.2.1) and (2.2.2) are first-order differential equations since they only involve first derivatives. Equation (2.2.3) is second-order. For now, we limit our attention to first-order equations; higher order equations will be discussed in detail in subsequent chapters.

It is important to note that every student of calculus learns to solve a certain class of differential equations through integration. For example, the problem, "find a function y whose derivative is te^t" can be restated as a differential equation. In particular, this problem can be stated as the differential equation

$$\frac{dy}{dt} = te^t \qquad\qquad (2.2.4)$$

Integrating both sides with respect to t and using integration by parts on the right, it follows that

$$y(t) = te^t - e^t + C$$

is a solution for any choice of the constant C. Here we see an important fact: differential equations typically have a family of infinitely many solutions. Determining all possible members of that family, like determining all solutions to systems of linear equations in linear algebra, will be a central component of our work.

Calculus students also know that if we are given one more piece of information about the function y along with (2.2.4), it is possible to uniquely determine the integration constant, C. For example, had the problem above read, "find a function y whose derivative is te^t such that $y(0) = 5$," we could integrate to find $y = te^t - e^t + C$, just as we did previously, and then use the *initial condition* $y(0) = 5$ to see that C must satisfy the equation

$$5 = 0 \cdot e^0 - e^0 + C$$

---

[1] This fact is not obvious.

and thus $C = 6$. When we are given a differential equation of order n along with n initial conditions, we say that we are solving an *initial-value problem*.[2] In the given example, $y = te^t - e^t + 6$ is the solution to the stated initial-value problem.

Based on the example above and our experience in calculus, it is clear that integration is an obvious (and often effective) approach to solving differential equations of the form

$$\frac{dy}{dt} = f(t)$$

where $f(t)$ is a given function. If we can integrate f symbolically, then the differential equation is solved. Even if $f(t)$ cannot by integrated symbolically with respect to t, we can still use techniques like numerical integration to successfully attack the problem. The situation grows more complicated when we want to solve differential equations that also involve the unknown function y, such as

$$\frac{dy}{dt} = te^y$$

In what follows in this chapter, we seek to classify first-order equations into types that can be solved in a straightforward way by symbolic means (often involving integration), as well as to develop methods that can be used to generate approximate solutions in situations where a symbolic solution is either difficult or impossible to attain. Throughout, the general form of the equations we are considering will be $y' = f(t, y)$, where the function $f(t, y)$ represents some combination of the independent variable t and the unknown function y.

It is also important to note that a wide range of first-order initial-value problems are guaranteed to have unique solutions. This is stated formally in the following theorem, whose proof may be found in more advanced texts.

Theorem 2.2.1   Consider the initial-value problem given by $y' = f(t, y)$, $y(t_0) = y_0$. If the function $f(t, y)$ is continuous on a rectangle that includes (t_0, y_0) in its interior and the partial derivative[3] $f_y(t, y)$ is continuous on that same rectangle, then there exists an interval containing t_0 on which the initial-value problem has a unique solution.

Often the dependent variable, or unknown function y, in a differential equation will model an important quantity in some physical problem: the amount of salt in a tank at time t, the number of members of a population at a given time, or the position of a mass attached to a spring. As such, we will place particular emphasis on the graph of the solution function in order to better understand what the differential equation is telling us about the physical situation it models.

---

[2] We often use the abbreviation IVP to stand for the phrase "initial-value problem."
[3] We typically use the notation $f_y(t, y) = \partial f / \partial y$.

Just as geometry and graphical interpretations shaped our understanding of linear algebra in chapter 1, these perspectives will prove extremely helpful in our study of differential equations. We begin our explorations of these graphical interpretations through the reservoir problem from section 2.1 and the earlier example $y' = te^t$.

So far in our references to derivatives in the reservoir and population models, we have viewed the derivative as measuring the instantaneous rate of change of a quantity that is varying. From a more geometric point of view, we also know that the derivative of a function measures the slope of the tangent line to the function's graph at a given point. For example, with the differential equation

$$\frac{dA}{dt} = 10 - \frac{A}{10} \tag{2.2.5}$$

we can say that if, at some time t, the amount of salt A is $A = 20$, then $dA/dt = 10 - 20/10 = 8$. Thus, if $A(t)$ is a solution to the differential equation, it follows that at any time where $A(t) = 20$, $A'(t) = 8$. Graphically, this means that at such a point, the slope of the tangent line to the curve must be 8.

Since we are interested in the function $A(t)$ over an interval of t-values, we also expect that $A(t)$ will take on a wide range of values. As such, it is natural to compute the slope of the tangent line determined by (2.2.5) for a large number of different values of A and t. Obviously computers are best suited to such a task, and, as we will see in the introduction to *Maple* commands at the end of this section, *Maple* and other computer algebra systems provide tools for doing so. Computing values of dA/dt over a grid of t and A values, we can plot a small portion of each corresponding tangent line at the point (t, A), and see the resulting *slope field* (or *direction field*). The slope field for (2.2.5) is shown in figure 2.1.

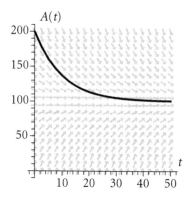

Figure 2.1 The slope field for (2.2.5); the graph of the solution corresponding to an initial condition $A(0) = 200$ is included.

Observe that a slope field provides an intuitive way to understand the information a first-order differential equation possesses: the slope at each point gives the direction of the solution at that point. Indeed, we use arrows instead of small lines in order to indicate the flow of the solution as time increases. In essence, the slope field is a map that the solution must navigate based on the initial point from which the function starts. For example, if we use the initial condition $A(0) = 200$ (as was given in the original example in section 2.1), we can start a graph at the point $(0, 200)$ and follow the map. Doing so yields the curve shown in figure 2.1.

Note particularly how we can clearly see the slope of the solution curve fitting with the slopes present in the direction field. Moreover, observe that the direction field provides an immediate overall sense of how every solution to the differential equation behaves: for any solution $A(t)$, $A(t) \to 100$ as $t \to \infty$. This makes sense physically, too, since the saltwater solution entering the reservoir has concentration 0.01 g/m^3. Over time, the concentration of solution in the reservoir should tend to that level, and with $10\,000$ m^3 of solution present in the reservoir, we expect the amount of salt to approach 100 g.

Another example of a differential equation's slope field provides further insights. For the differential equation

$$\frac{dy}{dt} = te^t \tag{2.2.6}$$

its slope field for the window $-2 \le t \le 1$ and $-2 \le y \le 2$ is given in figure 2.2.

We noted earlier that the general solution to this equation is $y = te^t - e^t + C$. Moreover, given any initial condition, we can determine C. For example, if $y(0) = 1/2$, $C = 3/2$. Likewise, if $y(0) = 0$, $C = 1$, and if $y(0) = -1$, $C = 0$. If we plot the corresponding three functions with the slope field, then (as shown in figure 2.2) the three members of the family of all solutions to the original differential equation appear as shown.

In integral calculus, students learn about *families of antiderivatives*[4] and how two members of such a family differ only by a constant. Here, we see this fact graphically in the slope field of figure 2.2, and can add the perspective that there exists a *family of solutions* to a certain differential equation. In upcoming sections, we will learn new techniques for how to determine solutions analytically in various circumstances, while not losing sight of the fact that every first-order differential equation can be interpreted graphically through a direction field.

Finally, there is an important type of first-order differential equation (DE) for which solutions can be determined algebraically. A first-order DE is said to be *autonomous* if it can be written in the form $y' = f(y)$. That is, the independent variable t is not involved explicitly in $f(y)$. For example, the equation

$$y' = 1 - y^2 \tag{2.2.7}$$

is autonomous.

---

[4] An antiderivative F of a function f is a function that satisfies $F' = f$.

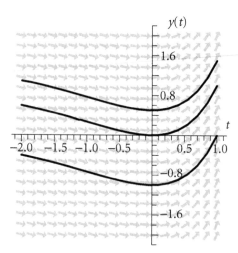

Figure 2.2 The slope field for (2.2.6) along with three solution functions for the initial conditions $y(0) = 1/2$, $y(0) = 0$, and $y(0) = -1$.

In addition, a solution y to a DE is called an *equilibrium* or *constant* solution if the function y is constant. In (2.2.7), both $y = 1$ and $y = -1$ are equilibrium solutions to the DE above. Such a solution is *stable* if all solutions with initial conditions $y(t_0) = y_0$ with y_0 close to the equilibrium solution result in the overall solution to the IVP tending toward the equilibrium solution. Otherwise, the equilibrium solution is called *unstable*.

We close this section with an example regarding an autonomous differential equation.

Example 2.2.1   Consider the differential equation $y' = (y^2 - 1)(y - 3)^2$. Determine all equilibrium solutions to the equation, as well as whether or not each is stable or unstable. Finally, plot the direction field for the equation and include plots of the equilibrium solutions.

Solution.   To find the equilibrium solutions, we assume that y is a constant function, and therefore $y' = 0$. Solving the algebraic equation

$$0 = (y^2 - 1)(y - 3)^2$$

we find that $y = -1$, $y = 1$, and $y = 3$ are the equilibrium solutions of the given DE.

We can decide the stability of each equilibrium solution by studying the sign of y' near the equilibrium value; note that $(y - 3)^2$ is always nonnegative. To consider the stability of $y = -1$, observe that when $y < -1$, $y' = (y + 1)(y - 1)(y - 3)^2 > 0$, since the first two terms are both negative and the third is positive. When $y > -1$ (and $y < 1$), it follows $y' = (y + 1)(y - 1)(y - 3)^2 < 0$

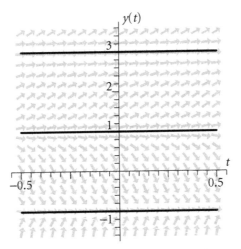

Figure 2.3 The slope field for $y' = (y^2 - 1)$ $(y - 3)^2$ along with its three equilibrium solutions.

since the middle term is negative while the other two are positive. Hence, if a solution starts just below $y = -1$, that solution will increase toward -1, whereas if a solution starts just above $y = -1$, it will decrease to -1. This makes the equilibrium $y = -1$ stable.

These observations are easiest to make visually in the direction field. As seen in figure 2.3, the constant solution $y = -1$ is stable, since any solution with an initial condition just above or just below $y = -1$ will tend to $y = -1$. However, the solution at $y = 1$ is unstable, since any solution with an initial value just above or just below $y = 1$ will tend away from 1 (and tend toward $y = 3$ or $y = -1$, respectively). Finally, although solutions just below $y = 3$ tend to 3, any solution that begins just above $y = 3$ will increase away from that constant solution, and hence $y = 3$ is also unstable.[5]

2.2.1 Plotting slope fields using *Maple*

Just as our work in linear algebra required the use of *Maple*'s Linear Algebra package, to take advantage of the software's support for the study of differential equations we use the DEtools package, loading it with the command

```
> with(DEtools):
```

---

[5] Some authors call a solution such as $y = 3$ in this example *semi-stable*, since there is stability on one side and instability on the other.

To plot the direction field associated with a given differential equation, it is convenient to first define the equation itself in *Maple*. This is accomplished (for the equation from the reservoir problem) through the following command:

```
> Eq1 := diff(A(t), t) = 10-1/10*A(t);
```

Note that the differential equation of interest is now stored in "Eq1". The slope field may now be generated by the command

```
> DEplot(Eq1, A(t), t = 0 .. 50, A(t) = 0 .. 200,
color = grey, arrows=large);
```

This command produces the slope field of figure 2.1, but without any particular solution satisfying an initial value included. It is important to note that the range of t and $A(t)$ values is extremely important. Without a well-chosen window selected by the user, the plot *Maple* generates may not be very insightful. For example, if the above command were changed so that the range of $A(t)$ values is 0 .. 10, almost no information can be gained from the slope field. As such, we will strive to learn to analyze the expected behavior of a differential equation from its form so that we can choose windows well in related plots; we may often have to experiment and explore to find graphs that are useful.

Finally, if we are interested in one or more related initial-value problems, a variation of the DEplot command enables us to sketch the graph of each corresponding solution. For example, the command

```
> DEplot(Eq1, A(t), t = 0 .. 50, A(t) = 0 .. 200,
color = grey, arrows=large, [[0,200]]);
```

will generate not only the slope field, but also the graph of the solution $A(t)$ that satisfies $A(0) = 200$, as shown in figure 2.1. Additional curves for different initial conditions may be plotted by listing the other conditions to be satisfied: for example, in the stated command above we could replace [[0,200]] with [[0,200], [0,100], [0,0]] to include the plots of the three solution curves that respectively satisfy $A(0) = 200$, $A(0) = 100$, and $A(0) = 0$.

Exercises 2.2

1. Consider the differential equation $y'' = 4y$.

   (a) What is the order of this equation?
   (b) Show via substitution that the function $y = e^{2t}$ is a solution to this equation.
   (c) Are there any other functions of the form $y = e^{rt}$ ($r \neq 2$) that are also solutions to the equation? If so, which? Justify your answer.

2. For a ball thrown straight up from an initial height $s(0) = 4$ meters at an initial velocity of $s'(0) = 10$ m/s, we know that after being thrown, the only force acting on the ball is gravity, provided we neglect air resistance. Knowing that acceleration due to gravity is constant at -9.81 m/s^2, it follows that $s''(t) = -9.81$. Use the given information to determine $s(t)$, the function that tells us the height of the ball at time t. Then determine the maximum height the ball reaches, as well as the time the ball lands.

3. In the differential equation $dA/dt = 10 - A/10$ from the reservoir problem, explain why the function $A(t) = 100$ is an equilibrium solution to the equation. Is it stable or unstable? Why?

4. Consider the logistic differential equation

$$\frac{dP}{dt} = 0.02P \left(1 - \frac{P}{100} \right)$$

Use *Maple* to plot the direction field for this equation. Print the output and, by hand, sketch the solutions that correspond to the initial conditions $P(0) = 10$, $P(0) = 75$, and $P(0) = 125$. What is the long-term behavior of every solution $P(t)$ for which $P(0) > 0$? Are there any constant (or equilibrium) solutions to the equation? Explain what these observations tell you about the behavior of the population being modeled.

5. For the logistic differential equation

$$\frac{dP}{dt} = 0.001P \left(1 - \frac{P}{25} \right)$$

how should the direction field appear? Use the constant/equilibrium solutions to the equation as well as the long-term behavior of the population to help you sketch, by hand, the direction field for this DE.

6. By constructing tangent lines over a grid with at least sixteen vertices, sketch a direction field by hand for each of the following differential equations.

(a) $y' = 1 - y$

(b) $y' = \frac{1}{2}(t - y)$

(c) $y' = \frac{1}{2}(t + y)$

(d) $y' = 1 - t$

7. Without using *Maple* to plot direction fields, match each of the following differential equations with its corresponding direction field. Write at least one sentence to explain the reasoning behind each of your choices.

(a) $\dfrac{dy}{dt} = y - t$ (b) $\dfrac{dy}{dt} = ty$ (c) $\dfrac{dy}{dt} = y$ (d) $\dfrac{dy}{dt} = t$

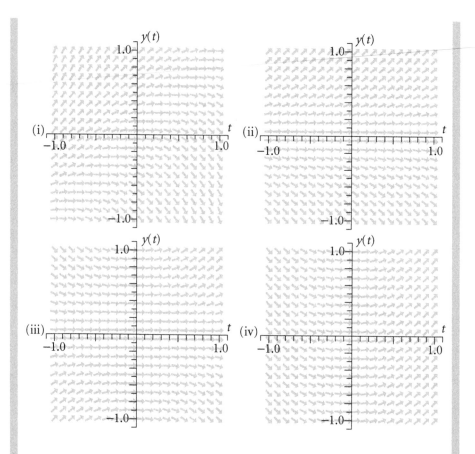

In exercises 8–15, use integration to find a family of solutions for the given differential equation.

8. $y' = t^2 + 2$

9. $y' = t + \cos t$

10. $y' = \dfrac{t}{t^2 + 1}$

11. $y'' = t^2 + 2$

12. $y''' = 5t$

13. $y' = t \sin t$

14. $y' = \dfrac{1}{t^2 + 5t + 6}$

15. $y' = te^{-t^2}$

In exercises 16–23, solve each of the following initial-value problems.

16. $y' = t^2 + 2, \quad y(1) = 4$

17. $y' = t + \cos t, \quad y(\pi/2) = 1$

18. $y' = \dfrac{t}{t^2 + 1}, \quad y(0) = 3$

19. $y'' = t^2 + 2, \quad y(1) = 4, y'(1) = -2$

20. $y''' = 5t, \quad y(-1) = 3, y'(-1) = -1, y''(-1) = 0$

21. $y' = t \sin t, \quad y(0) = 2$

22. $y' = \dfrac{1}{t^2 + 5t + 6}, \quad y(0) = 1$

23. $y' = te^{-t^2}, \quad y(0) = -1$

24. For an nth order IVP of the form $y^{(n)} = f(t)$, how many initial conditions are needed in order to uniquely determine the solution $y(t)$? Explain.

For each of the autonomous differential equations given in exercises 25–29, algebraically determine all equilibrium solutions to the DE. In addition, plot an appropriate direction field and use it to classify each equilibrium solution as stable or unstable.

25. $y' = 3 - 2y$

26. $y' = -y^2 - 5y - 6$

27. $y' = y - y^3$

28. $y' = e^{-y}(1 + y^2)$

29. $y' = (y - 1)(y - 3)^2$

2.3 Linear first-order differential equations

Some classes of differential equations can usually be solved by certain standard techniques. In this section, we consider the class of *linear* first-order differential equations and develop an approach for solving any such equation. Since any first-order DE is an equation that involves the functions y and y', it is natural for us to consider the different ways in which y and y' may be combined. For example, the equations

$$yy' = e^t \tag{2.3.1}$$

$$2ty + y' \sin t = \cos t \tag{2.3.2}$$

$$y' + \sin y = 2 \tag{2.3.3}$$

are all first-order DEs. Recall that in section 1.12 we discussed linear combinations of generalized vectors. Here we can view y and y' as functions that belong to a vector space, and thus think about whether a certain combination of y and y' is a linear combination or not. We say that any differential equation of the form

$$a_1(t)y' + a_0(t)y = b(t) \qquad (2.3.4)$$

is a *linear first-order differential equation*, since a linear combination of y and y' is being formed. Any other first-order differential equation is said to be *nonlinear*. If we stipulate that $a_1(t) \neq 0$, we can divide through by $a_1(t)$ and hence write

$$y' + p(t)y = f(t) \qquad (2.3.5)$$

as the standard form for a linear first-order equation. We call $f(t)$ the *forcing function*. Above, note that (2.3.1) and (2.3.3) are nonlinear equations, while (2.3.2) is linear.

The simplest linear first-order differential equations are those for which the forcing function is zero. We naturally call the equation

$$y' + p(t)y = 0 \qquad (2.3.6)$$

a *homogeneous* linear first-order DE. We consider a particular example that shows how every such homogeneous DE may be solved.

Example 2.3.1   Solve the differential equation $y' + (1 + 3t^2)y = 0$. In addition, solve the initial-value problem that is given by the same DE and the initial condition $y(0) = 4$.

Solution.   We will use integration to solve for y. Rearranging the given equation, we observe that $y' = -(1 + 3t^2)y$. Dividing both sides by y, we find that

$$\frac{y'}{y} = -(1 + 3t^2)$$

Keeping in mind the fact that y and y' are each unknown functions of t, we integrate both sides of the previous equation with respect to t:

$$\int \frac{y'}{y}\, dt = \int -(1 + 3t^2)\, dt$$

We recognize from the chain rule that the left-hand side is $\ln y$. Thus, integrating the polynomial in t on the right yields

$$\ln y = -t - t^3 + C$$

We note that while an arbitrary constant arises on each side of the equation when integrating, it suffices to simply include one constant on the right. Finally, we solve for y using properties of the natural logarithm and exponential functions to find that

$$y = e^{-t-t^3+C} = e^C e^{-t-t^3}$$

Since C is a constant, so is e^C, and thus we write

$$y = Ke^{-t-t^3}$$

Observe that we have found an entire family of functions that solve the original differential equation: regardless of the constant K, the above function y is a solution. If we consider the stated initial-value problem and apply the given initial condition $y(0) = 4$, we immediately see that $K = 4$, and the solution to the initial-value problem is

$$y = 4e^{-t-t^3}$$

The solution method in example 2.3.1 can be generalized to apply to any homogeneous linear first-order DE. Using the notation $p(t)$ to replace the function $1 + 3t^2$, which is the coefficient of y, the same steps above may be used to find the solution to the standard homogeneous linear first-order differential equation. We state this result in the following theorem.

Theorem 2.3.1   For any homogeneous linear first-order differential equation of the form

$$y' + p(t)y = 0,$$

the general solution is $y = Ke^{-P(t)}$, where P is any antiderivative of p. Moreover, for the initial condition $y(t_0) = y_0$, if $p(t)$ is continuous on an interval containing t_0, then the solution to the corresponding initial-value problem is unique.

The uniqueness of the solution to the initial-value problem follows from theorem 2.2.1. But perhaps the most important lesson to learn from this result is that a homogeneous linear first-order DE can always be solved. This is analogous to our experience with homogeneous linear systems of algebraic equations in chapter 1. In particular, note that by taking $K = 0$, the zero function ($y = 0$) is always a solution to $y' + p(t)y = 0$; in addition, the homogeneous linear first-order DE has infinitely many solutions. This is very similar to how, for a given matrix $\mathbf{A}$, the homogeneous equation $\mathbf{Ax} = \mathbf{0}$ always has the zero vector as a solution and, in the case where $\mathbf{A}$ is singular, $\mathbf{Ax} = \mathbf{0}$ has infinitely many solutions.

Having now completely addressed the case of a homogeneous linear first-order DE, we turn to the nonhomogeneous case. In particular, we are interested in solving the equation

$$y' + p(t)y = f(t) \tag{2.3.7}$$

where $f(t)$ is not identical to zero. Recalling the product rule from calculus,

$$\frac{d}{dt}[v(t) \cdot y] = v(t)y' + v'(t)y \tag{2.3.8}$$

we observe that the left-hand side of (2.3.7), $y' + p(t)y$, looks similar to the right-hand side of (2.3.8). If we multiply both sides of (2.3.7) by an unknown function $v(t)$, we have

$$v(t)y' + v(t)p(t)y = v(t)f(t) \tag{2.3.9}$$

Next, we observe that if $v(t)$ is a function such that $v'(t) = v(t)p(t)$, then it follows from the product rule that (2.3.9) has the form

$$\frac{d}{dt}[v(t)y] = v(t)f(t) \tag{2.3.10}$$

We assume temporarily that such a function $v(t)$ exists; we will proceed to discuss more about $v(t)$ shortly. Integrating both sides of (2.3.10), we now see that

$$v(t)y = \int v(t)f(t)\,dt \tag{2.3.11}$$

To solve for y, we divide both sides by $v(t)$, yielding

$$y(t) = \frac{1}{v(t)} \int v(t)f(t)\,dt \tag{2.3.12}$$

Prior to (2.3.10), we stipulated a condition on v that enabled us to proceed. In particular, we noted that "if $v(t)$ is a function such that $v'(t) = v(t)p(t)$," then we could find a solution in terms of v. Observe that the differential equation v satisfies is, in fact, a homogeneous linear first-order equation itself ($v' - p(t)v = 0$), and therefore its solution is

$$v(t) = Ke^{P(t)},$$

where $P(t) = \int p(t)\,dt$. Since we only need one such nonzero function v to proceed, we set $K = 1$. From this and our conclusion in (2.3.12), we have determined that

$$y(t) = e^{-P(t)} \int e^{P(t)}f(t)\,dt \tag{2.3.13}$$

where $P(t) = \int p(t)\,dt$. The function $v(t) = e^{P(t)}$ is usually called an *integrating factor*.

We next consider two examples of nonhomogeneous linear first-order differential equations and apply the method we just derived to solve them.

Example 2.3.2   Solve the differential equation $y' + 2y = 4$.

Solution.   In this equation, $p(t) = 2$, and therefore $P(t) = 2t$. From (2.3.13), it follows that

$$y(t) = e^{-P(t)} \int e^{P(t)}f(t)\,dt$$

$$= e^{-2t} \int e^{2t} \cdot 4\,dt$$

$$= e^{-2t}(2e^{2t} + C) \tag{2.3.14}$$

$$= 2 + Ce^{-2t} \tag{2.3.15}$$

There are several important observations to make from our work in example 2.3.2. First, the parentheses at (2.3.14) are essential. Without them, e^{-2t} is not multiplied by the entire antiderivative, and the function y would no longer be a solution to the given DE.

A second is that if we had instead solved the corresponding homogeneous differential equation $y' + 2y = 0$, we would have found the so-called *complementary* solution $y_h = Ce^{-2t}$. Moreover, by observing that $y' = 4 - 2y = 2(2 - y)$, if we consider the function $y_p = 2$, it is apparent that y_p is a solution to the nonhomogeneous equation $y' + 2y = 4$. In addition, if we omit the constant of integration C in (2.3.14), it follows that the method derived in (2.3.13) can be viewed as producing a so-called *particular solution* y_p that is a solution to the given nonhomogeneous linear first-order differential equation.

Thus we see that the method derived in (2.3.13) and implemented to find (2.3.15) ultimately expresses the solution to the original nonhomogeneous linear first-order DE in the form

$$y = y_p + y_h$$

where y_p is a particular solution to the nonhomogeneous equation, while y_h is the complementary solution, the solution to the corresponding homogeneous equation.

This situation reminds us of one way to view the general solution to a system of linear equations given by $\mathbf{A}\mathbf{x} = \mathbf{b}$, where in (1.5.1) in section 1.5 we found that $\mathbf{x} = \mathbf{x}_p + \mathbf{x}_h$. A further discussion of this property of linear first-order DEs will occur in theorem 2.3.2 to close the current section. Before doing so, we consider another example.

Example 2.3.3   Solve the nonhomogeneous first-order linear differential equation

$$y' + y \tan t = \cos t$$

In addition, solve the initial-value problem (IVP) that is given by the same DE and the initial condition $y(\pi/3) = 1$.

Solution.   We first determine the integrating factor $v(t)$. Since $p(t) = \tan t$, it follows that

$$P(t) = \int \tan t \, dt = -\ln(\cos t)$$

Thus, $v(t) = e^{-\ln(\cos t)}$. Applying the integrating factor and using properties of exponential and logarithmic functions, we now observe that

$$y = e^{\ln(\cos t)} \int \cos t \cdot e^{-\ln(\cos t)} \, dt$$

$$= \cos t \int \cos t \, \frac{1}{\cos t} \, dt$$

$$= \cos t \int 1 \, dt$$

$$= \cos t (t + C)$$

Thus, the general solution to the given differential equation is $y = t \cos t + C \cos t$.

To solve the corresponding IVP with the condition that $y(\pi/3) = 1$, it follows that $1 = \pi/3 \cdot 1/2 + C \cdot 1/2$, so that $C = 2 - \pi/3$. The solution is

$$y = t \cos t + (2 - \pi/3) \cos t$$

As in example 2.3.2, we note that the solution $y = t \cos t + C \cos t$ in example 2.3.3 is of the form $y = y_p + y_h$, where $y_h = C \cos t$ can easily be checked to be the solution to the corresponding homogeneous equation.

Two important results can now be stated in general. The first is a formal statement of our derivation in (2.3.12) that shows how we can use an integrating factor to solve any nonhomogeneous linear first-order DE. The second demonstrates that for any of these types of DEs, if y_p is a particular solution to the nonhomogeneous DE and y_h is a complementary solution to the corresponding homogeneous DE, then $y = y_p + y_h$ is also a solution to the nonhomogeneous DE.

Theorem 2.3.2 For any nonhomogeneous linear first-order differential equation of the form

$$y' + p(t)y = f(t),$$

the general solution is

$$y = e^{-P(t)} \int e^{P(t)} f(t) \, dt$$

where $P(t) = \int p(t) \, dt$. Moreover, for the initial condition $y(t_0) = y_0$, if $p(t)$ and $f(t)$ are continuous on an interval containing t_0, then the solution to the corresponding initial-value problem is unique.

The proof of the first part of theorem 2.3.2 is given above in the discussion of (2.3.7)–(2.3.12). The uniqueness of the solution to the IVP follows from theorem 2.2.1.

Finally, we observe that given a nonhomogeneous linear first-order differential equation $y' + p(t)y = f(t)$ and a particular solution y_p (so $y_p' + p(t)y_p = f(t)$) and complementary solution y_h to the corresponding homogeneous equation $(y_h' + p(t)y_h = 0)$, it follows that

$$(y_p + y_h)' + p(t)(y_p + y_h) = y_p' + y_h' + p(t)y_p + p(t)y_h$$
$$= (y_p' + p(t)y_p) + (y_h' + p(t)y_h)$$
$$= f(t) + 0$$
$$= f(t)$$

Therefore, $y_p + y_h$ is also a solution to the nonhomogeneous DE. Formally, we have the following result.

Theorem 2.3.3   For any nonhomogeneous linear first-order differential equation,

$$y' + p(t)y = f(t)$$

if y_p is a particular solution to the nonhomogeneous equation and y_h is a solution to the corresponding homogeneous equation, then $y = y_p + y_h$ is also a solution to the nonhomogeneous equation.

Exercises 2.3

In exercises 1–6, classify each equation as linear or nonlinear. Do not attempt to solve the equations.

1. $y' + 7y = e^t$

2. $\cos ty' + \sin ty = t^2$

3. $\cos y' + \sin y = t^2$

4. $ty' + t^2 y = t^3$

5. $y'y^2 = 3t$

6. $1 = y/y'$

In exercises 7–13, solve each of the given homogeneous linear first-order DEs.

7. $y' + y = 0$

8. $y' + 2y = 0$

9. $y' + ty = 0$

10. $y' + \dfrac{2}{t}y = 0$

11. $y' = -y \cot t$

12. $(1 + t^2)y' + 2ty = 0$

13. $y' = -\dfrac{2}{100 - t}y$

In exercises 14–20, solve each of the given nonhomogeneous linear first-order DEs.

14. $y' + y = 2$

15. $y' + 2y = 2t$

16. $y' + ty = 10t$

17. $y' + \dfrac{2}{t}y = e^t$

18. $y' = -(y-1)\cot t$

19. $(1+t^2)y' + 2ty = 2t$

20. $y' = 0.03 - \dfrac{2}{100-t}y$

In exercises 21–27, solve each of the given initial-value problems.

21. $y' + y = 2, \quad y(0) = 3$

22. $y' + 2y = 2t, \quad y(1) = 0$

23. $y' + ty = 10t, \quad y(0) = 5$

24. $y' + \dfrac{2}{t}y = e^t, \quad y(1) = 4, t > 0$

25. $y' = -(y-1)\cot t, \quad y(\pi/2) = 1, \ 0 < t < \pi$

26. $(1+t^2)y' + 2ty = 2t, \quad y(0) = 1$

27. $y' = 0.03 - \dfrac{2}{100-t}y, \quad y(0) = 1$

In exercises 28–33, plot a slope field in an appropriate window of t and y values for each of the given DEs. In addition, in the same window, plot the solution to each given IVP. Compare each graph to the solutions you found in the corresponding exercises 21–27.

28. $y' + y = 2, \quad y(0) = 3$

29. $y' + 2y = 2t, \quad y(1) = 0$

30. $y' + ty = 10t, \quad y(0) = 5$

31. $y' + \dfrac{2}{t}y = e^t, \quad y(1) = 4, t > 0$

32. $y' = -(y-1)\cot t, \quad y(\pi/2) = 1, 0 < t < \pi$

33. $(1+t^2)y' + 2ty = 2t, \quad y(0) = 1$

34. With matrix multiplication, we noted that for any matrix $\mathbf{A}$ and appropriately sized vectors $\mathbf{x}$ and $\mathbf{y}$, $\mathbf{A}(\mathbf{x} + \mathbf{y}) = \mathbf{A}\mathbf{x} + \mathbf{A}\mathbf{y}$. In addition, for any constant c, $\mathbf{A}(c\mathbf{x}) = c\mathbf{A}\mathbf{x}$. We called these properties "the linearity of matrix multiplication." In calculus, we learn that the derivative operator, D, satisfies similar properties of linearity. In particular, if f and g are differentiable functions and c is any constant, what can you say about $D(f+g)$ and $D(cf)$? (Recall that $D(f)$ is alternate notation for f'.)

2.4 Applications of linear first-order differential equations

A large number of important physical situations can be modeled by linear first-order differential equations. In this section we introduce several such applications through examples and explore further scenarios in the exercises.

2.4.1 Mixing problems

Recall that in section 2.1, we encountered a problem where a saltwater solution was entering and exiting a city's water reservoir. Specifically, in (2.1.1) we encountered the DE

$$\frac{dA}{dt} = 10 - \frac{A(t)}{10}$$

This equation, rewritten in the form

$$A' + \frac{1}{10}A = 10$$

is a linear first-order DE that we now can easily solve. With $p(t) = 1/10$, the integrating factor is $v(t) = e^{t/10}$, and therefore

$$A = e^{-t/10} \int e^{t/10} \cdot 10 \, dt \qquad (2.4.1)$$

$$= e^{-t/10}(100e^{t/10} + C) \qquad (2.4.2)$$

$$= 100 + Ce^{-t/10} \qquad (2.4.3)$$

From this result, we can also confirm our previous observation that as $t \to \infty$, $A(t) \to 100$, for any solution $A(t)$ to the differential equation. Moreover, if we consider the initial condition $A(0) = 200$ stated along with the original problem in section 2.1, it follows that

$$A(t) = 100 + 100e^{-t/10}$$

Certainly we can consider a wide range of variations on this mixing problem by changing concentrations, flow rates, and tank volumes. In every such scenario, the most important thing to keep in mind is that the rate of change of salt (or whatever quantity is under consideration) is the difference between the rate of salt entering and the rate exiting. Furthermore, an analysis of units is often very helpful. We consider one more example to demonstrate what can occur when the entering and exiting solutions are flowing at different rates.

Example 2.4.1   Consider a tank in which 1 g of chlorine is initially present in 100 m³ of a solution of water and chlorine. A chlorine solution concentrated at 0.03 g/m³ flows into the tank at a rate of 1 m³/min, while the uniformly mixed solution exits the tank at 2 m³/min. At what time is the maximum amount of chlorine present in the tank, and how much is present?

Solution.   To answer the questions posed, we set up and solve an IVP. We let $A(t)$ denote the amount of chlorine in the tank (in grams) at time t (in minutes). We note from the inflow that the rate at which chlorine is entering the tank is given by

$$\text{rate in} = 1 \, \frac{m^3}{\text{min}} \cdot 0.03 \, \frac{g}{m^3} \tag{2.4.4}$$

For the exiting flow, we must compute the concentration of chlorine present in the solution leaving the tank. This concentration is given by the ratio of amount present in grams to the total volume of solution in the tank at time t. In this problem, note that the volume is changing as a function of time. In particular, since solution enters at $1 \, m^3/\text{min}$ and exits at $2 \, m^3/\text{min}$, it follows that the volume $V(t)$ of solution present in the tank is decreasing at a rate of $1 \, m^3/\text{min}$. With $100 \, m^3$ initially present, we observe that $V(t) = 100 - t$ is the volume of solution in the tank at time t. Thus, the concentration of chlorine in the solution exiting the tank at time t is given by

$$\text{rate out} = 2 \, \frac{m^3}{\text{min}} \cdot \frac{A(t)}{V(t)} \, \frac{g}{m^3} = \frac{2 \cdot A(t)}{100 - t} \, \frac{g}{\text{min}} \tag{2.4.5}$$

It follows from (2.4.4) and (2.4.5) that the overall instantaneous rate of change of chlorine in the tank with respect to time is

$$\frac{dA}{dt} = \text{rate in} - \text{rate out} = 0.03 - \frac{2A}{100 - t}$$

Note that we also have the initial condition $A(0) = 1$. Rearranging the differential equation, we see that we must solve the nonhomogeneous linear first-order equation

$$A' + \frac{2}{100 - t} A = 0.03 \tag{2.4.6}$$

Applying the approach discussed in section 2.3, followed by the initial condition, it can be shown that the solution to (2.4.6) is

$$A(t) = 3 - 0.03t - 0.0002(100 - t)^2$$

From the quadratic nature of this solution, as well as from the direction field shown in figure 2.4, we can see that this function has a maximum value. It is a straightforward exercise to show that this maximum of $A(t)$ occurs when $t = 25 \, \text{min}$ and that the maximum is $A = 1.125 \, g$.

2.4.2 Exponential growth and decay

A radioactive substance emits particles; in doing so, the substance decreases its mass. This process is known as *radioactive decay*. For example, the radioactive isotope carbon-14 emits particles and loses half its mass over a period of 5730 years. For any such isotope, the instantaneous rate of decay is proportional to the mass of the substance present at that instant. Thus, assuming an initial

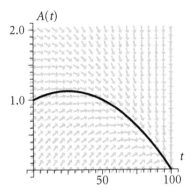

Figure 2.4 Direction field for (2.4.6) with solution corresponding to the initial condition $A(0) = 1$.

mass M_0 is present, it follows that the mass $M(t)$ of the substance at time t must satisfy the initial-value problem

$$M' = -kM, \quad M(0) = M_0 \tag{2.4.7}$$

for some positive constant k. Note that the minus sign is present in (2.4.7) since the mass $M(t)$ is decreasing. It follows from our work with homogeneous linear first-order DEs in section 2.3 that the solution to this equation is

$$M(t) = M_0 e^{-kt} \tag{2.4.8}$$

Similarly, experiments show that a population with zero death rate (e.g., a colony of bacteria with sufficient food and no predators) grows at a rate proportional to the size of the population at time t. In particular, if $P(t)$ is the population present at time t and P_0 is the initial population, then P satisfies the initial-value problem $P' = kP$, $P(0) = P_0$, for some positive constant k. Here, it follows that

$$P(t) = P_0 e^{kt} \tag{2.4.9}$$

Problems involving radioactive decay and exponential population growth are very similar and should be familiar to students from past courses in calculus and precalculus. We include one example here for review and several more in the exercises at the end of the section.

Example 2.4.2   A radioactive isotope initially has 40 g of mass. After 10 days of radioactive decay, its mass is 39.7 g. What is the isotope's half-life? At what time t will 1 g remain?

Solution.   Because the isotope decays radioactively, we know that its mass $M(t)$ must have the form $M(t) = M_0 e^{-kt}$. To answer the questions posed, we must first determine the constant k. In the given problem, we know that $M_0 = 40$

and that $M(10) = 39.7$. It follows that

$$39.7 = 40e^{-10k}$$

Dividing both sides of the equation by 40, taking natural logs, and solving for k, we find that

$$k = -\frac{1}{10}\ln\left(\frac{39.7}{40}\right)$$

To compute the half life, we now solve the equation

$$\frac{M_0}{2} = M_0 e^{-kt}$$

for t. In particular, we have

$$20 = 40e^{\frac{1}{10}\ln\left(\frac{39.7}{40}\right)t}$$

Dividing by 40 and taking natural logs,

$$\ln\left(\frac{1}{2}\right) = \frac{1}{10}\ln\left(\frac{39.7}{40}\right)t$$

so

$$t = \frac{\ln\left(\frac{1}{2}\right)}{\frac{1}{10}\ln\left(\frac{39.7}{40}\right)}$$

Thus the half-life of the isotope is approximately 921 days.

Finally, to determine when 1 g of the substance will remain, we simply solve the equation

$$1 = 40e^{\frac{1}{10}\ln\left(\frac{39.7}{40}\right)t}$$

Doing so shows that $t \approx 4900$ days.

2.4.3 Newton's law of Cooling

Suppose that $T(t)$ is the temperature of a body immersed in a cooler surrounding medium such as air or water. Sir Isaac Newton postulated (and experiments confirm) that the body will lose heat at a rate proportional to the difference between its present temperature and the temperature of its surroundings. If we assume that the temperature of the surrounding medium is constant, say T_m, and that the warmer body's initial temperature is $T(0) = T_0$, then *Newton's law of Cooling* can be expressed through the initial-value problem

$$T' = -k(T - T_m), \quad T(0) = T_0 \tag{2.4.10}$$

Written in the standard form of a nonhomogeneous linear first-order DE, we find that T satisfies the IVP

$$T' + kT = kT_m, \quad T(0) = T_0 \tag{2.4.11}$$

Solving this problem in the standard way reveals that the temperature of the cooling body must satisfy

$$T(t) = (T_0 - T_m)e^{-kt} + T_m \qquad (2.4.12)$$

We consider an example with some particular details given in order to analyze the behavior of the temperature function.

Example 2.4.3   A can of soda at room temperature $70°F$ is placed in a refrigerator that maintains a constant temperature of $40°F$. After 1 hour in the refrigerator, the temperature of the soda is $58°F$. At what time will the soda's temperature be $41°F$?

Solution.   Let $T(t)$ denote the temperature of the soda at time t in degrees F; note that $T_0 = 70$. Since the surrounding temperature is 40, T satisfies the initial-value problem

$$T' = -k(T - 40), \quad T(0) = 70$$

and therefore by (2.4.12) T has the form

$$T(t) = 30e^{-kt} + 40$$

In particular, note that the temperature is decreasing exponentially as time increases and tending towards $40°F$, the temperature of the refrigerator, as $t \to \infty$.

To determine the constant k, we use the additional given information that $T(1) = 58$, and therefore

$$58 = 30e^{-k} + 40$$

It follows that $e^{-k} = 3/5$, and thus $k = \ln(5/3)$. To now answer the original question, we solve the equation

$$41 = 30e^{-\ln(5/3)t} + 40$$

and find that $t = \ln(30)/\ln(5/3) \approx 6.658$ h.

Exercises 2.4

1. A population of bacteria is growing at a rate proportional to the number of cells present at time t. If initially 100 million cells are present and after 6 hours 300 million cells are present, what is the doubling time of the population? At what time will 100 billion cells be present?

2. The half-life of a radioactive element is 2000 years. What percentage of its original mass is left after 10 000 years? After 11 000 years?

3. The evaporation rate of moisture from a sheet hung on a clothesline is proportional to the sheet's moisture content. If one half of the moisture evaporates in the first 30 min, how long will it take for 95 percent of the moisture to evaporate?

4. A population of 200 million people is observed to grow at a rate proportional to the population present and to be increasing at a rate of 2 percent per year. How long will it take for the population to triple?

5. In a certain lake, wildlife biologists determine that the walleye population is growing very slowly. In particular, they conclude that the population growth is modeled by the differential equation $P' = 0.002P$, where P is measured in thousands of walleye, and time t is measured in years. The biologists estimate that the initial population of walleye in the lake is 100 000 fish. To enhance the fishery, the department of conservation begins planting walleye fingerlings in the lake at a rate of 5000 walleye per year.

   (a) Write an IVP that the population $P(t)$ of walleye in the lake in year t will satisfy under the assumption that walleye are being added to the lake at a rate of 5000 fish per year.
   (b) Solve the IVP stated in (a).
   (c) In 20 years, how many more walleye will be in the lake than if the biologists had not planted any fish?

6. Solve the IVP $A' = 0.03 - 2/(100 - t)A$, $A(0) = 1$, in order to verify the stated solution in example 2.4.1.

7. Brine (saltwater) is entering a 25 m^3 tank at flow rate of 0.25 m^3/min and at a concentration of 6 g/m^3. The uniformly mixed solution exits the tank at a rate of 0.25 m^3/min. Assume that initially there are 15 m^3 of solution in the tank at a concentration of 3 g/m^3.

   (a) State an IVP that is satisfied by $A(t)$, the amount of salt in grams in the tank at time t.
   (b) What will happen to the amount of salt in the tank as $t \to \infty$? Why?
   (c) Plot a direction field for the IVP stated in (a), including a plot of the solution.
   (d) At exactly what time will there be 75 g of salt present in the tank?

8. Brine is entering a 25-m^3 tank at flow rate of 0.5 m^3/min and at a concentration of 6 g/m^3. The uniformly mixed solution exits the tank at a rate of 0.25 m^3/min. Assume that initially there are 5 m^3 of solution in the tank at a concentration of 25 g/m^3.

   (a) State an IVP that is satisfied by the amount of salt $A(t)$ in grams in the tank at time t.
   (b) Solve the IVP stated in (a). For what values of t is this problem valid? Why?
   (c) At exactly what time will the least amount of salt be present in the tank? How much salt will there be at that time?
   (d) Plot a direction field for the IVP stated in (a), including a plot of the solution. Discuss why this direction field and the solution make sense in the physical context of the problem.

9. A body of water is polluted with mercury. The lake has a volume of 200 million cubic meters and mercury is present in a concentration of 5 grams per million cubic meters. Health officials state that any level above 1 g per million cubic meters is considered unsafe. If water unpolluted by mercury flows into the lake at a rate of 0.5 million cubic meters per day, and uniformly mixed lake water flows out of the lake at the same rate, how long will it take for the lake to reach a mercury concentration that is considered safe?

10. An average person takes eighteen breaths per minute and each breath exhales $0.0016 \, \text{m}^3$ of air that contains 4 percent more carbon dioxide (CO_2) than was inhaled. At the start of a seminar containing 300 participants, the room air contains 0.4 percent CO_2. The ventilation system delivers $10 \, \text{m}^3$ of fresh air per minute to the room whose volume is $1500 \, \text{m}^3$. Find an expression for the concentration level of CO_2 in the room as a function of time; assume that air is leaving the room at the same rate that it enters.

11. Solve the general Newton's law of Cooling IVP $T' = -k(T - T_m)$, $T(0) = T_0$ in order to verify the solution stated in (2.4.12).

12. A potato at room temperature of $72°F$ is placed in an oven set at $350°F$. After 30 min, the potato's temperature is $105°F$. At what time will the potato reach a temperature of $165°F$?

13. An object at a temperature of $80°C$ is placed in a refrigerator maintained at $5°C$. If the temperature of the object is $75°C$ at 20 min after it is placed in the refrigerator, determine the time (in hours) the object will reach $10°C$.

14. An object at a temperature of $9°C$ is placed in a refrigerator that is initially at $5°C$. At the same time the object is placed in the refrigerator, the refrigerator's thermostat is adjusted in order to raise the temperature inside from $5°C$ to $10°C$; the function that governs the temperature of the refrigerator is $R(t) = \dfrac{10}{1 + e^{-0.75t}}$.

   (a) Using the refrigerator's temperature constant k from exercise 13, modify Newton's law of Cooling appropriately to state an IVP whose solution is the temperature of the object.
   (b) Plot a direction field for the IVP from (a) and sketch an approximate solution to the IVP.
   (c) Discuss the qualitative behavior of the solution to the IVP. Estimate the minimum temperature the object achieves.

15. On a cold, winter evening with an outdoor temperature of $4°F$, a home's furnace fails at 10 pm. At the time of the furnace failure, the indoor temperature was $68°F$. At 2 am, the indoor temperature was $60°F$.

Assuming the outside temperature remains constant, at what time will the homeowner have to begin to worry about pipes freezing due to an indoor temperature below 32°F?

2.5 Nonlinear first-order differential equations

So far in our work with differential equations, we have seen that *linear* first-order differential equations have many interesting properties. One is that any IVP that corresponds to a linear first-order DE (with reasonably well-behaved functions $p(t)$ and $f(t)$) is guaranteed to have a unique solution. In addition, through our development of integrating factors, we have a method by which we can always (at least in theory) determine a solution for the differential equation.

Any differential equation that is not linear is called *nonlinear*. Thus, nonlinear differential equations constitute every other type of equation we can conceive. Unfortunately, nonlinear equations are (in general) far more difficult to solve than linear ones. We will limit ourselves in this section to considering a few relatively common special cases of nonlinear first-order differential equations that can be solved analytically. In section 2.6, we will consider qualitative and approximation techniques that enable us to gain valuable information from a nonlinear initial-value problem, even in the event that we cannot solve it explicitly.

2.5.1 Separable equations

In example 2.3.1 in section 2.3, we solved the differential equation $y' = -(1 + t^2)y$. While this equation is linear, our method provides insight into how to approach a class of nonlinear equations whose structure is similar. We begin by considering a slightly modified example.

Example 2.5.1   Solve the nonlinear first-order differential equation

$$y' = -(1 + t^2)y^2 \qquad (2.5.1)$$

Solution.   Following our approach in example 2.3.1, we can separate the variables y and t algebraically to arrive at the equation

$$y^{-2}\frac{dy}{dt} = -1 - 3t^2$$

Integrating both sides of this equation with respect to t,

$$\int (y(t))^{-2}\frac{dy}{dt}\,dt = \int (-1 - 3t^2)\,dt \qquad (2.5.2)$$

The left-hand side may be simplified to $\int y^{-2}\,dy$. Thus, evaluating each integral in (2.5.2), we find that

$$-y^{-1} = -t - t^3 + C \qquad (2.5.3)$$

We note again that since an arbitrary constant of integration arises on each side, it suffices to include just one. It is essential here to observe that by successfully integrating, we have removed the presence of y' in the equation, and now have only an algebraic, rather than differential, equation in t and y. Solving (2.5.3) algebraically for y, it follows

$$y = \frac{1}{t + \frac{1}{3}t^3 - C}$$

The strategy of example 2.5.1 may be applied to any differential equation of the form $y' = f(t, y)$ where $f(t, y)$ can be decomposed into a product of two functions of t and y only. That is, if we can write

$$f(t, y) = g(t) \cdot h(y)$$

then we are able to separate the variables in the equation, writing all of the y-terms on one side (multiplied by y'), and writing all of the t-terms on the other. Any differential equation of the form $y' = g(t) \cdot h(y)$ is said to be *separable*. We attempt to solve a separable differential equation by separating the variables and writing

$$\frac{1}{h(y)} y' = g(t) \tag{2.5.4}$$

Writing y' in the alternate notation dy/dt, we have

$$\frac{1}{h(y)} \frac{dy}{dt} = g(t) \tag{2.5.5}$$

Hence when we integrate both sides of (2.5.5) with respect to t, we find

$$\int \frac{1}{h(y)} \, dy = \int g(t) \, dt$$

Now, all of this work is only useful if we arrive at integrals we can actually evaluate. For example, if the left-hand side is $\int \sin \sqrt{y} \, dy$, we are really no closer to solving for y than we were when considering the initial differential equation.

In section 2.6, we will address ways to approximate the solution of such equations that we seem unable to solve analytically. For now, we consider a few examples of separable equations that we can solve, with more to follow in the exercises.

Example 2.5.2   Find a family of solutions to the differential equation

$$\frac{y'}{t} = e^{t+2y}$$

and a solution to the corresponding initial-value problem with the condition that $y(1) = 1$.

Solution.   First, we may write $e^{t+2y} = e^t e^{2y}$. Thus, we have

$$\frac{y'}{t} = e^t e^{2y}$$

Separating the variables, it follows that

$$e^{-2y} \frac{dy}{dt} = te^t$$

Integrating both sides with respect to t, we may now write

$$\int e^{-2y} \, dy = \int te^t \, dt$$

Using integration by parts on the right and evaluating both integrals, we have

$$-\frac{1}{2} e^{-2y} = (t-1)e^t + C$$

To now solve algebraically for y, we first multiply both sides by -2. Since C is an arbitrary constant, $-2C$ is just another constant, one that we will denote by C_1. Hence

$$e^{-2y} = -2(t-1)e^t + C_1$$

Taking logarithms and solving for y, we can conclude that

$$y = -\frac{1}{2} \ln(-2(t-1)e^t + C_1)$$

is the family of functions that provides the general solution to the original DE. To solve the corresponding IVP with $y(1) = 1$, we observe that

$$y(1) = -\frac{1}{2} \ln(-2(1-1)e^t + C_1) = -\frac{1}{2} \ln(C_1) = 1$$

so $\ln(C_1) = -2$, and therefore $C_1 = e^{-2}$. The solution to the IVP is

$$y = -\frac{1}{2} \ln(-2(t-1)e^1 + e^{-2})$$

Example 2.5.3   Is the following differential equation linear or nonlinear?

$$ty' + y^2 = 4$$

Classify the equation, and solve it to find a general family of solutions.

Solution.   We note that the given equation is nonlinear due to the presence of y^2 in the equation; said differently, the left-hand side is not a linear combination of y and y'. To separate the variables, we first write

$$ty' = 4 - y^2$$

Dividing both sides by $t(4 - y^2)$, it follows that

$$\frac{1}{4 - y^2} \frac{dy}{dt} = \frac{1}{t}$$

and therefore

$$\int \frac{dy}{4 - y^2} = \int \frac{dt}{t}$$

Evaluating both integrals, noting that the left-hand side requires integration by partial fractions or a table of integrals, we have

$$\frac{1}{4} \ln \left(\frac{y + 2}{y - 2} \right) = \ln t + C$$

It only remains to solve for y algebraically. Using rules of logarithms and letting $C = \ln K$, we can write

$$\ln \left(\frac{y + 2}{y - 2} \right)^{1/4} = \ln(Kt)$$

It now follows that

$$\left(\frac{y + 2}{y - 2} \right)^{1/4} = Kt$$

Raising both sides to the fourth power, multiplying by $(y - 2)$, and solving for y yields

$$y = 2 \frac{(Kt)^4 + 1}{(Kt)^4 - 1}$$

2.5.2 Exact equations

We will consider one other type of nonlinear differential equation that may be solved analytically. We explore this through an example. Let us solve the DE

$$(2 + t^2 y)y' + ty^2 = 0$$

We first observe that this equation is neither linear nor separable. The former is clear from the presence of y^2 and yy'; the latter is less obvious, but nonetheless true since the presence of the term $(2 + t^2 y)$ makes it impossible to separate the variables t and y. We therefore explore another algebraic approach. Considering the derivative in differential notation, we have

$$(2 + t^2 y) \frac{dy}{dt} + ty^2 = 0$$

and thus we may instead write

$$(ty^2)dt + (2 + t^2 y)dy = 0 \tag{2.5.6}$$

This form may remind us of the total differential $d\phi$ of a function $\phi(t, y)$, as studied in multivariable calculus. Recall that for a differentiable function $\phi(t, y)$, its total differential $d\phi$ is given by

$$d\phi = \phi_t \, dt + \phi_y \, dy$$

where $\phi_t = \partial \phi / \partial t$ and $\phi_y = \partial \phi / \partial y$. Note, therefore, from (2.5.6) that if there exists a function ϕ such that $\phi_t = ty^2$ and $\phi_y = 2 + t^2 y$, then (2.5.6) is actually of the form $d\phi = 0$, from which it follows that $\phi(t, y) = K$, for some constant K. Assuming that we can find the function $\phi(t, y)$, we have then transformed the original differential equation in t and y to an algebraic equation in t and y, one that we can hopefully solve for y.

In the current example, let us suppose that such a function $\phi(t, y)$ exists, and therefore that

$$\frac{\partial \phi}{\partial t} = ty^2 \tag{2.5.7}$$

and

$$\frac{\partial \phi}{\partial y} = 2 + t^2 y \tag{2.5.8}$$

Integrating both sides of (2.5.7) with respect to t, it follows that

$$\phi(t, y) = \frac{1}{2}t^2 y^2 + g(y)$$

The function $g(y)$ arises since the partial derivative with respect to t of any function of only y is zero. For ϕ to satisfy the condition in (2.5.8), we see that we must take the partial derivative with respect to y of our most recent result and set this equal to $2 + t^2 y$. Doing so, we find that

$$\frac{\partial \phi}{\partial y} = t^2 y + g'(y) = 2 + t^2 y$$

Therefore, $g'(y) = 2$, so $g(y) = 2y$, and we have found that

$$\phi(t, y) = \frac{1}{2}t^2 y^2 + 2y$$

Since it is the case that $d\phi = 0$, we know that $\phi(t, y) = K$, and therefore t and y are related by the algebraic equation

$$\frac{1}{2}t^2 y^2 + 2y = K$$

From the quadratic formula, it follows that

$$y = \frac{-2 \pm \sqrt{4 + 2Kt^2}}{t^2}$$

and we have solved the original equation. The choice of "+" or "−" in the solution would depend on the value given in an initial condition.

There are several important lessons to learn from this example. One is some terminology. If a differential equation can be written in the form

$$M(t, y)dt + N(t, y)dy = 0 \tag{2.5.9}$$

and there exists a function $\phi(t, y)$ such that $\phi_t(t, y) = M(t, y)$ and $\phi_y(t, y) = N(t, y)$, then since the differential equation is of the form $d\phi(t, y) = 0$, we say that the equation is *exact*.

So, certainly a first check of whether an equation might be exact consists in trying to write it in the form of (2.5.9). Still, there is the issue of whether or not ϕ exists. If ϕ does exist, and we further assume that $M(t, y)$ and $N(t, y)$ have continuous first-order partial derivatives, then it follows from Clairaut's Theorem in multivariable calculus that

$$M_y(t, y) = \phi_{ty} = \phi_{yt} = N_t(t, y)$$

Thus, if (2.5.9) is exact, then it must be the case that $M_y = N_t$. Said differently, if $M_y \neq N_t$, then the differential equation is not exact. In fact, it turns out that if $M_y = N_t$, then the equation is guaranteed to be exact, but this result is much more difficult to prove. As a consequence of this, it suffices for us to check if $M_y = N_t$ as a first step; if so, the equation is indeed exact and we then proceed to try to find the function ϕ in order to solve the differential equation. If not, another approach is needed.

An example is instructive.

Example 2.5.4   Solve the differential equation

$$\frac{t}{y} y' + \ln(ty) + 1 = 0$$

Solution.   We begin by observing that this equation is neither linear nor separable. Thus, writing the derivative in differential notation, we have

$$\frac{t}{y} \frac{dy}{dt} + \ln(ty) + 1 = 0$$

and then rearranging algebraically,

$$(\ln(ty) + 1)dt + \frac{t}{y} dy = 0 \qquad (2.5.10)$$

Letting $M(t, y) = \ln(ty) + 1$ and $N(t, y) = t/y$, we observe that

$$M_y = \frac{1}{ty} t = \frac{1}{y} \text{ and } N_t = \frac{1}{y}$$

and therefore, $M_y = N_t$. Hence the differential equation is exact and we can assume that a function ϕ exists such that $\phi_t = M(t, y)$ and $\phi_y = N(t, y)$.

Since the latter equation is more elementary, we consider $\phi_y = t/y$, and integrate both sides with respect to y. Doing so, we find that

$$\phi(t, y) = t \ln y + h(t) \qquad (2.5.11)$$

From (2.5.10), ϕ must also satisfy $\phi_t = \ln(ty) + 1$, so we take the partial derivative of both sides of (2.5.11) with respect to t to find that

$$\phi_t = \ln y + h'(t) = \ln(ty) + 1$$

From this and properties of the logarithm, we observe that

$$\ln y + h'(t) = \ln t + \ln y + 1$$

and thus $h'(t) = \ln t + 1$. It follows (integrating by parts and simplifying) that $h(t) = t \ln t$. Thus, we have demonstrated that the original equation is indeed exact by finding $\phi(t, y) = t \ln y + t \ln t = t \ln(ty)$. From here, we now know that $\phi(t, y) = K$, and so

$$t \ln(ty) = K$$

Solving for y, we have that

$$y = \frac{1}{t} e^{K/t}$$

Exercises 2.5

Classify each of the DEs in exercises 1–14 as linear, nonlinear, separable, or exact. Note that it is possible for an equation to satisfy more than one classification.

1. $y' = 10y$

2. $y' = 10y + 10$

3. $y' = 10y^2$

4. $y' = 10y^2 - 10$

5. $t^2 y' + y^2 = 1$

6. $e^{3t+y} \dfrac{dy}{dt} = 1$

7. $t\,dy - (y - 1)\,dt = 0$

8. $\dfrac{dy}{dt} = \dfrac{5ty - t}{4 + t^2}$

9. $y - t\dfrac{dy}{dt} = 6 - 3t^2\dfrac{dy}{dt}$

10. $\dfrac{dy}{dt} = \dfrac{-2ty}{t^2 + 1}$

11. $(2 + t^2)y' + 2ty = 0$

12. $3y^2 y' + t^2 = 0$

13. $(y + t)y' + y = t$

14. $y' \sin 2t + 2y \cos 2t = 0$

Solve each of the DEs in exercises 15–28.

15. $y' = 10y$

16. $y' = 10y + 10$

17. $y' = 10y^2$

18. $y' = 10y^2 - 10$

19. $t^2 y' + y^2 = 1$

20. $e^{3t+y} \dfrac{dy}{dt} = 1$

21. $t\, dy - (y-1)\, dt = 0$

22. $\dfrac{dy}{dt} = \dfrac{5ty - t}{4 + t^2}$

23. $y - t \dfrac{dy}{dt} = 6 - 3t^2 \dfrac{dy}{dt}$

24. $\dfrac{dy}{dt} = \dfrac{-2ty}{t^2 + 1}$

25. $(2 + t^2)y' + 2ty = 0$

26. $3y^2 y' + t^2 = 0$

27. $(y + t)y' + y = t$

28. $y' \sin 2t + 2y \cos 2t = 0$

Solve each of the IVPs stated in exercises 29–42. In addition, use a computer algebra system to plot an appropriate direction field for each, and sketch your solution within the plot.

29. $y' = 10y, \quad y(0) = 3$

30. $y' = 10y + 10, \quad y(0) = 2$

31. $y' = 10y^2, \quad y(1) = 4$

32. $y' = 10y^2 - 10, \quad y(1) = -1$

33. $t^2 y' + y^2 = 1, \quad y(2) = 0$

34. $e^{3t+y} \dfrac{dy}{dt} = 1, \quad y(0) = 0$

35. $t\, dy - (y-1)\, dt = 0, \quad y(1) = 3$

36. $\dfrac{dy}{dt} = \dfrac{5ty - t}{4 + t^2}, \quad y(1) = 1$

37. $y - t \dfrac{dy}{dt} = 6 - 3t^2 \dfrac{dy}{dt}, \quad y(1) = 5$

38. $\dfrac{dy}{dt} = \dfrac{-2ty}{t^2 + 1}, \quad y(0) = 4$

39. $(2 + t^2)y' + 2ty = 0, \quad y(1) = 1$

40. $3y^2 y' + t^2 = 0, \quad y(0) = 1$

41. $(y+t)y'+y=t, \quad y(0)=1$

42. $y'\sin 2t + 2y\cos 2t = 0, \quad y(\pi/4)=1/2$

43. Consider the IVP $y' = \sqrt{y}$, $y(0)=0$. Show that this IVP has more than one solution. Does this result contradict theorem 2.2.1?

2.6 Euler's method

While we have learned to solve certain classes of differential equations explicitly—including linear first-order, separable, and exact equations—we must also develop the ability to estimate solutions to initial-value problems that we cannot solve analytically. Direction fields will play a key role in motivating our work, as we see in the following introductory example.

Consider the initial-value problem

$$\frac{dy}{dt} + y^2 = t, \ \ y(0) = 1 \tag{2.6.1}$$

This DE is not linear due to the presence of y^2. In addition, since we can write $y' = t - y^2$, we see that the right-hand side may not be expressed as a product of two functions that each involve just one of the variables t and y. Thus, the equation is not separable. Finally, writing the equation in the form $dy + (y^2 - t)dt = 0$, it is straightforward to check that this equation is not exact.

While it may seem frustrating to not be able to use any of the solution methods we have discussed so far, it is important to realize that many differential equations cannot be solved explicitly by analytic techniques. As such, we must explore how we can use our understanding of derivatives to estimate certain values of the solution to an IVP.

For the given DE, writing $y' = t - y^2$, we can generate the direction field that is shown in figure 2.5. For the initial condition $y(0) = 1$, visually estimating how the solution $y(t)$ will flow through the direction field, we can roughly estimate that $y(1/2) \approx 0.75$. But if we think about the calculus underpinnings of slope fields, we can be much more precise in our estimate.

Recall that a direction field for a DE $y' = f(t, y)$ is created by observing that the slope of the tangent line to the solution curve $y(t)$ at the point (t_0, y_0) is $f(t_0, y_0)$. In the current example, we know that the solution to the IVP must pass through the point $(t_0, y_0) = (0, 1)$. At this point, the slope of the tangent line to the solution curve is $m = 0 - 1^2 = -1$; note also that $m \approx \Delta y/\Delta t$, where Δy is the exact change in y from $t = 0$ to $t = 1/2$, due to the fact that the tangent line approximates the solution curve for values near the point of tangency. Thus, as we step from $t_0 = 0$ to $t = 1/2$, a change of $1/2$ in the t-direction will generate an approximate change $\Delta y = \Delta t \cdot m = 1/2 \cdot (-1) = -1/2$ in y. Therefore, from our original y-value of 1, a change of $-1/2$ leads us to the approximation that $y(1/2) \approx 1/2$.

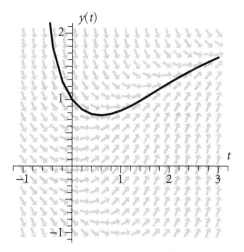

Figure 2.5 The direction field for (2.6.1).

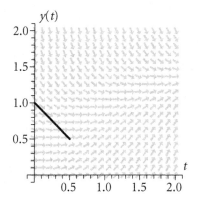

Figure 2.6 Taking one step to estimate $y(0.5)$ in (2.6.1).

Graphically, this estimation approach amounts to following the tangent
line to the solution curve for some prescribed change in t. We can see this in
figure 2.6, where it is immediately evident that our estimate is too small. In
calculus, we learn that while the tangent line approximation to a differentiable
function is good near the point of tangency, the approximation gets poorer
and poorer the further we move from the point of tangency. Thus, a natural
approach to the estimation problem at hand is to take a smaller step, then
search the direction field for a new direction to follow, and then take another
small step. In this situation, we are much like a hiker lost in the woods who is
attempting to navigate by compass: just as the hiker is best served by checking a
compass frequently, so are we best served by checking slopes frequently.

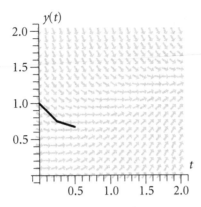

Figure 2.7 Two steps of size 0.25 to estimate $y(0.5)$ in (2.6.1).

So, rather than stepping the full distance of $1/2$ from $t = 0$ to $t = 1/2$, let us first step to $t = 1/4$, find an estimate to $y(1/4)$, and then proceed from there to estimate $y(1/2)$. Starting at $(0, 1)$, we know that the slope of the tangent line to the solution curve at this point is $m_0 = f(0, 1) = -1$. Stepping $\Delta t = 0.25$, it follows that we experience a change in y along the tangent line of $\Delta y = m_0 \Delta t = -1(0.25) = -0.25$. Thus, we have that $y(0.25) \approx y(0) + \Delta y = 1 - 0.25 = 0.75$.

Now we repeat this process from the point $(0.25, 0.75)$. At this point, the slope of the tangent line to the solution curve is $m_1 = f(0.25, 0.75) = 0.25 - (0.75)^2 = -0.3125$. Taking a step of $\Delta t = 0.25$, it follows that the change in y along the tangent line will be $\Delta y = m_1 \Delta t = -0.3125(0.25) = -0.078125$. Thus we have that $y(0.5) \approx 0.75 - 0.078125 = 0.671875$. We record our work graphically in figure 2.7, where our improved approximation is apparent, though the estimate is still too small.

It is evident from our work in this first example that we can significantly improve our ability to estimate an initial-value problem's solution at various t-values by developing an iterative process that uses reasonably small step sizes. In particular, we want to imitate the way in which we took two steps, but rather be able to take n steps using a step-size of $\Delta t = h$. Throughout, the key idea is always that we are estimating the solution function by determining its tangent line at a given point, and then following the tangent line for the determined step size. We observe that when moving along any line from a given point (t_{old}, y_{old}) to a new point (t_{new}, y_{new}), it follows that

$$y_{new} = y_{old} + \Delta y$$

$$= y_{old} + \frac{\Delta y}{\Delta t} \cdot \Delta t$$

$$= y_{old} + m \cdot \Delta t \tag{2.6.2}$$

Another essential observation to make is that the slope m at each step of our approximation is given by $m = y' = f(t, y)$ in the differential equation that we are attempting to solve. In particular, if we have some approximation at time t_k given by y_k, the slope of the tangent line to the solution curve at this point is given by $f(t_k, y_k)$. Therefore, using this value for m in (2.6.2) and letting $h = \Delta t$ be the step size, we now have

$$y_{new} = y_{old} + hf(t_{old}, y_{old}) \qquad (2.6.3)$$

Hence, starting from the initial condition (t_0, y_0), we are able to generate the sequence of points $(t_1, y_1), \ldots, (t_n, y_n)$, where for each $n \geq 0$,

$$t_{n+1} = t_n + h \text{ and } y_{n+1} = y_n + hf(t_n, y_n) \qquad (2.6.4)$$

The value y_n is an approximation of the exact solution value $y(t_n)$ at each step, so that $y_n \approx y(t_n)$ for each $n \geq 1$. This method of approximating the solution to an initial-value problem is known as *Euler's method*.

Example 2.6.1   For the initial-value problem

$$\frac{dy}{dt} + y^2 = t, \ \ y(0) = 1$$

that we have just considered, apply Euler's method to estimate the value of $y(1/2)$ using $h = 0.1$.

Solution.   At the end of this section, the implementation of Euler's method in a spreadsheet such as Excel will be discussed. Here, we simply report the results of such a computer implementation. If we use a step size of $h = 0.1$, we see that we will take five steps to move from $t_0 = 0$ to $t_5 = 0.5$, the point at which we seek to approximate y. Doing so yields the output shown in table 2.1.

With just five steps, we can see in the direction field in figure 2.8, together with a piecewise linear plot of the approximate solution, that we have an apparently good estimate in the above table for how the actual solution to this IVP behaves on this interval.

In the example we have been considering with various step sizes, one shortcoming is that we do not have a precise sense of how accurate our

Table 2.1
Euler's method applied to the IVP
$y' = t - y^2$, $y(0) = 1$, using $h = 0.1$

t_n	y_n
0	1
0.1	0.9
0.2	0.829
0.3	0.7802759
0.4	0.749392852
0.5	0.733233887

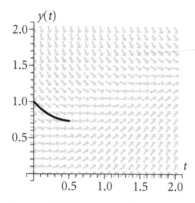

Figure 2.8 Five steps of size $h = 0.1$
to estimate $y(0.5)$.

approximations are. One way to explore this issue is to apply Euler's method to
an IVP that we can solve exactly, and then compare our estimates with actual
solution values. We do so in the following example.

Example 2.6.2   Solve the IVP $y' = y - t$, $y(0) = 0.5$ exactly, and use Euler's
method with the step sizes $h = 0.2$ and $h = 0.1$ to estimate the value of $y(1)$.
Hence analyze the effect that step size has on error in the method.

Solution.   We first observe that $y' = y - t$ is a linear first-order DE. Applying
our work from section 2.3, we can determine that the solution to this equation
is $y = 1 + t + Ce^t$. The initial condition $y(0) = 0.5$ then implies that $C = -1/2$,
so that the solution to the IVP is

$$y(t) = 1 + t - \frac{e^t}{2}$$

If we apply Euler's method with $h = 0.2$ and take 5 steps to determine y_n
at each, and also evaluate $y(t_n)$ at each stage, the resulting output is shown
in table 2.2.

Here, we observe the obvious pattern that the further we step away from
the initial condition, the greater the error we encounter. This is a natural
consequence of the use of linear approximations.

To get a further sense of how the error at a given step depends on step size,
we now apply the same method with $h = 0.1$. Doing so produces the results in
table 2.3. For ease of display and comparison to the case where $h = 0.2$, we only
report the results from every other step.

By comparing the approximations in the preceding two tables at the
common values of $t = 0.2, 0.4, 0.8, 1$ we can see that cutting the step size in
half appears to have reduced the error by a factor of approximately 2.

Table 2.2
Euler's method applied to the IVP $y' = y - t$, $y(0) = 0.5$, using $h = 0.2$

	Euler Est.	Solution	Error
t_n	y_n	$y(t_n)$	$\lvert y(t_n) - y_n \rvert$
0	0.5	0.5	0
0.2	0.6	0.5892986	0.0107014
0.4	0.68	0.6540877	0.0259123
0.6	0.736	0.6889406	0.0470594
0.8	0.7632	0.6872295	0.0759705
1.0	0.75584	0.6408591	0.1149809

Table 2.3
Euler's method applied to the IVP $y' = y - t$, $y(0) = 0.5$, using $h = 0.1$

	Euler Est.	Solution	Error
t_n	y_n	$y(t_n)$	$\lvert y(t_n) - y_n \rvert$
0	0.5	0.5	0
0.2	0.595	0.5892986	0.0057014
0.4	0.66795	0.6540877	0.0138623
0.6	0.7142195	0.6889406	0.0252789
0.8	0.728205595	0.6872295	0.0409761
1	0.70312877	0.6408591	0.0622697

In fact, there are sophisticated ways by which we can analyze the error of Euler's method in general; we explore these and related issues in depth in chapter 7 on numerical methods. And while Euler's method can give us an intuitive sense for how a solution is behaving locally, we must note here that its error grows too fast to make it reliable. More sophisticated algorithms for numerically estimating solutions to differential equations exist; several of these are developed in chapter 7.

2.6.1 Implementing Euler's method in Excel

Any spreadsheet program provides a straightforward way to implement Euler's method. In our calculations, we will use Microsoft Excel. Recall that in Euler's method, given an initial-value problem $y' = f(t, y)$, $y(t_0) = y_0$, we seek approximations $y_1, y_2, \ldots$ such that $y_n \approx y(t_n)$, where $t_n = t_0 + ht_n$ for some chosen step size h. In particular, we use the rule

$$y_{n+1} = y_n + hf(t_n, y_n)$$

In a given row of the spreadsheet, we will view the data (as labeled in the cells below) step number n, step size h, t-value t_n, approximate current y-value y_n, slope $f(t_n, y_n)$, and updated y-value y_{n+1}.

We will demonstrate the development of such an Excel spreadsheet for the particular example $y' = t - y^2$, $y(0) = 1$ using a step size of $h = 0.1$. To begin, we establish names for the various columns, say in cells A1, B1, C1, D1, E1, and F1, as shown below by entering the text "n", "h", etc., in the respective cells shown below.

	A	B	C	D	E	F
1	n	h	t_n	y_n	f(t_n,y_n)	y_n+1

In row 2, we now enter the given data at step zero. In particular, in cell A2 we enter the step number ("0"), in B2 the chosen step size ("0.1"), in C2 the starting t-value ("0"), in D2 the starting y-value ("1"), and in E2, we apply the function $f(t, y)$ to get the slope at the point at this step. That is, since in this IVP $f(t, y) = t - y^2$, we enter in E2 the command "=C2 - D2^2". We now also have enough information entered to compute y_1 in cell F2. Using the rule from Euler's method, we know $y_1 = y_0 + hf(t_0, y_0)$. In our spreadsheet, this implies we must enter "=D2 + B2*E2". Doing so, the result ($y_1 = 0.9$) appears in cell F2. Now our spreadsheet should appear as shown.

	A	B	C	D	E	F
1	n	h	t_n	y_n	f(t_n,y_n)	y_n+1
2	0	0.1	0	1	−1	0.9

In row 3, we may now build subsequent entries based on existing data. To increase the step number, in A3 we enter "=A2 + 1". Since the step-size stays constant throughout, in B3 we input "=B2". Because the next t-value will be the preceding t-value plus the step size ($t_1 = t_0 + h$), we enter in C3 the command "=C2 + B2". We also have the next y-value, so in D3 we enter "=F2" to have this data available in the given row. The slope at step 1 is computed according to the same rule (given by $f(t, y)$) as it was at step 0. Hence in cell E3 we simply paste a copy of cell E2, which ensures that Excel uses the same computations, but updates them for the current step. Equivalently, we can directly enter in E3 the text "=C3 - D3^2". Cell F3 computes the newest y-value: the same rule as in step 0 must be followed, so we can copy and paste cell F2 into F3, or equivalently enter in F3 "=D3 + B3*E3".

At this stage, we see on the screen the following.

	A	B	C	D	E	F
1	n	h	t_n	y_n	f(t_n,y_n)	y_n+1
2	0	0.1	0	1	−1	0.9
3	1	0.1	0.1	0.9	−0.71	0.829

Now we can harness the power of Excel to compute as many subsequent steps as we like. By using the mouse to highlight row 3 (cells A3 through F3), and then placing the cursor on the bottom right corner of cell F3, we can then click and drag downward to fill subsequent rows with similar calculations. For example, doing so through row 5 (i.e., down to F7) yields the following table.

	A	B	C	D	E	F
1	n	h	t_n	y_n	f(t_n,y_n)	y_n+1
2	0	0.1	0	1	-1	0.9
3	1	0.1	0.1	0.9	-0.71	0.829
4	2	0.1	0.2	0.829	-0.487241	0.7802759
5	3	0.1	0.3	0.7802759	-0.30883048	0.749392852
6	4	0.1	0.4	0.749392852	-0.161589647	0.733233887
7	5	0.1	0.5	0.733233887	-0.037631934	0.729470694

Besides the ease of iteration past the first two rows, there are further advantages Excel offers. One is that changing one appropriately-chosen cell will update all of our computations. For example, if we are interested in the change induced by a different step size, say $h = 0.05$, all we need to do is enter "0.05" in cell B2, and every other cell will update accordingly. In addition, if we desire to see the graphical results of our work, we can use Excel's Chart Wizard.

To plot our approximations, we can simultaneously highlight the t and y columns in our chart above (cells C2 through C7 and D2 through D7), and then go to Insert menu and select Chart (alternatively, we may click on the Chart Wizard icon on the toolbar). In the prompt window that arises, we choose "XY (Scatter)" and select one of the graph style options at the right by clicking on the desired one. By clicking "Next" in a few subsequent windows (in which advanced users can avail themselves of more options), we eventually get to a final window where our graph appears and the option to "Finish." Clicking on "Finish," the graph will appear in the spreadsheet and may be moved around

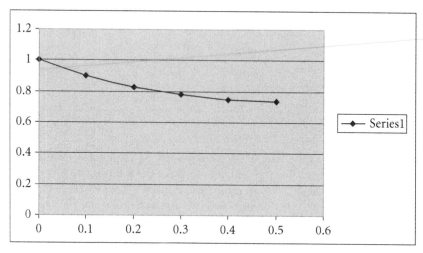

Figure 2.9 An Excel plot of an approximate solution to the IVP $y' = t - y^2$, $y(0) = 1$, for $0 \le t \le 0.5$.

by clicking and dragging it accordingly. We see the resulting plot displayed as in figure 2.9.

Exercises 2.6

1. Consider the IVP $y' = t/y$, $y(1) = 3$ (where we assume that y is always positive).

   (a) Program Excel to use Euler's method to determine an estimate of the value of $y(3)$. Do so using a step size of $h = 0.2$. Show the results in a table and create an appropriate plot of the approximate solution.

   (b) Use an established solution method to determine an algebraic formula for the unique solution $y(t)$ for the given IVP. Then determine $y(t_n)$ exactly and use Excel to determine the error in your approximation at each step n. Finally, compare a plot of $y(t)$ to your plot of the approximation above.

   (c) Use a computer algebra system appropriately to plot a direction field for the given differential equation. By hand, sketch a solution that satisfies the above IVP. Compare your work in (a) and (b) to the direction field.

2. Consider the IVP $y' = (1 - t)(1 + y)$, $y(0) = 2$.

   (a) Program Excel to use Euler's method to determine an estimate to the value of $y(1.6)$. Do so using step sizes of $h = 0.2$ and $h = 0.1$. Show the results in a table and create an appropriate plot of the approximate solution.

(b) Use an established solution method to determine an algebraic formula for the unique solution $y(t)$ for the given IVP. Then determine $y(t_n)$ exactly and use Excel to determine the error in your approximation at each step n. Finally, compare a plot of $y(t)$ to your plot of the approximation above.

(c) Use a computer algebra system appropriately to plot a direction field for the given differential equation. By hand, sketch a solution that satisfies the above IVP. Compare your work in (a) and (b) to the direction field.

3. Consider the IVP $y' = (t - y)^2/4, \ y(0) = 1/2$.

(a) Program Excel to use Euler's method to determine an estimate to the value of $y(1.5)$. Do so using step sizes of $h = 0.1$ and $h = 0.05$. Show the results in a table and create an appropriate plot of the approximate solution.

(b) Explain why you cannot solve the given IVP explicitly.

(c) Use a computer algebra system appropriately to plot a direction field for the given differential equation. By hand, sketch a solution that satisfies the above IVP. Compare your work in (a) to the direction field.

4. Consider the IVP $y' = e^t - \dfrac{2}{t} y, \ y(1) = 4, \ t > 0$.

(a) Program Excel to use Euler's method to determine an estimate to the value of $y(2.2)$. Do so using step sizes of $h = 0.1$ and $h = 0.05$. Show the results in a table and create an appropriate plot of the approximate solution.

(b) Use an established solution method to determine an algebraic formula for the unique solution $y(t)$ for the given IVP. Then determine $y(t_n)$ exactly and use Excel to determine the error in your approximation at each step n. Finally, compare a plot of $y(t)$ to your plot of the approximation above.

(c) Use a computer algebra system appropriately to plot a direction field for the given differential equation. By hand, sketch a solution that satisfies the above IVP. Compare your work in (a) and (b) to the direction field.

In each of exercises 5–10, find an approximate solution to the stated IVP by using Euler's method with $h = 0.1$ on the interval $[0, 1]$. In addition, find an exact solution and compare the values and plots of the approximate and exact solutions.

5. $y' + 2ty = 0, \quad y(0) = -2$

6. $y' = 2y - 1, \quad y(0) = 2$

7. $y' - y = 0, \quad y(0) = 2$

8. $(y')^2 + 2y = 0, \quad y(0) = 2$

9. $y'y^2 = 8, \quad y(0) = 1$

10. $(t+1)yy' = -1 - y^2, \quad y(0) = 2$

In each of exercises 11–14, find an approximation solution to the stated IVP by using Euler's method with $h = 0.1$ on the interval $[0, 1]$. In addition, explain why it is not possible to solve the IVP exactly by established methods.

11. $(y')^2 - 2y^2 = t, \quad y(0) = 2$

12. $y' - \sin y = 2e^t, \quad y(0) = 0$

13. $y' + y^3 = t^3, \quad y(0) = 2$

14. $(t+1)yy' = -1 - y^2 - t^2, \quad y(0) = 2$

2.7 Applications of nonlinear first-order differential equations

In this section, we explore two examples of nonlinear differential equations. It is important to recall that if an equation is nonlinear, it is possible that we may not be able to solve for the solution function explicitly. Regardless, we can use direction fields to qualitatively understand the behavior of solution curves; furthermore, if we are unable to find an exact solution function, we may employ Euler's method to generate approximate solutions.

2.7.1 The logistic equation

We have recently learned that if a population is assumed to grow at a constant relative growth rate (or in a way such that the rate of change of the population is proportional to the size of the population), then the population function satisfies the initial-value problem

$$P' = kP, \quad P(0) = P_0$$

This leads to the familiar population model $P(t) = P_0 e^{kt}$, which is also studied in algebra and calculus courses. While this model is a natural one, it is also unrealistic: over significant periods of time, the function P will grow to values that become unreasonable since the function exhibits unbounded growth.

Therefore, we now explore a more plausible population model. Let us assume we know that a given population P has the tendency over time to level off at a value A. The value A is often called the *carrying capacity* of the population; as the name indicates, it is the maximum population sustainable by the surrounding environment. It is natural to further assume that if P is close to, but less than A, then dP/dt will be small and positive, indicating that the population will be growing slowly. Similarly, if P is close to, but greater than A, we will want dP/dt to be negative and close to zero, so that the population will be decreasing slowly.

At the same time, we want to maintain the natural inherent exponential characteristic of growth, so when P is relatively small (in comparison to A), we would like for dP/dt to be approximately kP for some appropriate constant k. The combination of all these criteria led Dutch mathematician Pierre Verhulst (1804–1849) to propose the differential equation

$$\frac{dP}{dt} = kP\left(1 - \frac{P}{A}\right) \qquad\qquad (2.7.1)$$

as a more realistic model of population growth, where k and A are positive constants. Equation (2.7.1) is known as the *logistic differential equation.*

That the logistic equation may be solved in general (to determine an explicit solution P involving k and A) will be shown in the exercises. We consider here a specific example where k and A are given to provide further insight into the behavior of solutions to this equation.

Example 2.7.1   A population $P(t)$ exhibits logistic growth according to the model

$$\frac{dP}{dt} = 0.05P\left(1 - \frac{P}{75}\right), \quad P(0) = 10$$

(a) Determine the values of P for which P is an increasing function

(b) Plot the direction field for the differential equation

(c) Determine the value(s) of P for which P is increasing most rapidly

(d) Solve the IVP explicitly for P

Solution.

(a) To determine where P is increasing, we require that $dP/dt > 0$. If $P < 0$, note that $(1 - P/75) > 0$, which makes $dP/dt < 0$, so we need $P > 0$ and $(1 - P/75) > 0$ to make dP/dt positive. This occurs on the interval $0 < P < 75$, so for these P values, P is an increasing function of t. We note further that if $P > 75$ or $P < 0$, then $dP/dt < 0$ and P is a decreasing function. Finally, it is evident that both $P = 0$ and $P = 75$ are equilibrium solutions, which makes sense given the physical interpretation of the population model.

(b) Using familiar commands in *Maple*, we can plot the direction field for this differential equation. Note in advance the behavior we expect from our work above: two equilibrium solutions at 0 and 75, plus certain increasing and decreasing behavior. Finally, note that our analysis of the equation suggests a good range of values to select for P when plotting, say, $P = -10 \ldots 100$. As always, some experimentation with t may be necessary to get a useful plot. The plot is shown in figure 2.10.

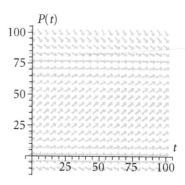

Figure 2.10 The slope field for $dP/dt = 0.05P(1 - P/75)$.

(c) To decide where P is increasing most rapidly, we seek the maximum value of P'. Graphically, we can observe in figure 2.10 that this appears to occur approximately halfway between $P = 0$ and $P = 75$. This is reasonable in light of the physical meaning of the logistic equation, since at this point the population has accumulated some substantial numbers to increase its growth rate, while not being close enough to the carrying capacity to have its growth slowed.

We can determine this point of greatest increase in P analytically as well. Note that $P' = 0.05P(1 - P/75) = 0.05P - 0.000\overline{6}P^2$, so that P' is determined by a quadratic function of P. We have already observed that this quadratic function has zeros at the equilibrium solutions ($P = 0$ and $P = 75$), and furthermore, we know that every quadratic function achieves is extremum (a maximum in this case, since the function $g(P) = 0.05P - 0.000\overline{6}P^2$ is concave down) at the midpoint of its zeros. Hence, P' is maximized precisely when $P = 75/2$.

(d) Our final task is to solve the given initial-value problem explicitly for P. We first solve the differential equation

$$\frac{dP}{dt} = 0.05P\left(1 - P/75\right)$$

for P. Note that this equation is separable and nonlinear. Separating variables, we first write

$$\frac{dP}{P(1 - P/75)} = 0.05dt \tag{2.7.2}$$

Because the left-hand side is a rational function of P, we may use the method of partial fractions to integrate the left-hand side of (2.7.2). Observe that

$$\frac{1}{P(1 - P/75)} = \frac{75}{P(75 - P)}$$

Now, letting
$$\frac{75}{P(75-P)} = \frac{A}{P} + \frac{B}{75-P}$$
it follows that $A = 1$ and $B = 1$, so that (2.7.2) may now be written as
$$\left(\frac{1}{P} - \frac{1}{P-75}\right) dP = 0.05 \, dt \tag{2.7.3}$$
Integrating both sides of (2.7.3), we find that P must satisfy the equation
$$\ln|P| - \ln|P-75| = 0.05t + C$$
Using a standard property of logarithms, the left-hand side may be expressed as $\ln|P|/|P-75|$, and hence using the definition of the natural logarithm, it follows that
$$\left|\frac{P}{P-75}\right| = e^{0.05t+C} = Ke^{0.05t}$$
where $K = e^C$. Since K is an arbitrary constant, the sign of K will absorb the $\pm$ that arises from the presence of the absolute value signs, and thus we may write
$$\frac{P}{P-75} = Ke^{0.05t}$$
Multiplying both sides by $P - 75$ and expanding, we see that
$$P = PKe^{0.05t} - 75Ke^{0.05t}$$
and gathering all terms involving P on the left,
$$P(1 - Ke^{0.05t}) = -75Ke^{0.05t}$$

Thus, it follows that
$$P = \frac{-75Ke^{0.05t}}{1 - Ke^{0.05t}}$$
Multiplying the top and bottom of the right-hand side by $-1/(Ke^{0.05t})$, it follows that
$$P = \frac{75}{1 - Me^{-0.05t}}$$
where $M = 1/K$. In this final form, it is evident that as $t \to \infty$, $P(t) \to 75$, which fits with the given carrying capacity in the original problem. At this point, we can use the initial condition $P(0) = 10$ to solve for M; doing so results in the equation $10 = 75/(1 - M)$, which yields that $M = -13/2$, and thus
$$P = \frac{75}{1 + \frac{13}{2}e^{-0.05t}}$$
A plot of this function (shown in figure 2.11), along with comparison to our work throughout this example, demonstrates that our solution is correct.

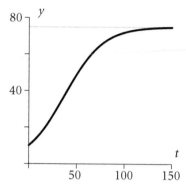

Figure 2.11 The solution $P = 75/$
$(1 + \frac{13}{2}e^{-0.05t})$ to the IVP $dP/dt =$
$0.05P(1 - P/75)$, $P(0) = 10$.

For the general logistic differential equation

$$\frac{dP}{dt} = kP\left(1 - \frac{P}{A}\right)$$

an argument similar to the one we just completed can be used to show that the solution to this equation is

$$P(t) = \frac{A}{1 + Me^{-kt}},$$

where M is a constant that may be determined by an initial condition. This fact will be shown in exercise 1 for this section.

2.7.2 Torricelli's law

Suppose that a water tank has a hole in its base with area a, through which water is flowing. Let $h(t)$ be the depth of the water and $V(t)$ be the volume of water in the tank at time t. At what rates are $h(t)$ and $V(t)$ changing?

Evangelista Torricelli (1608–1647) discovered what has come to be known as *Torricelli's law*, which describes the way water in an open tank will flow through a small hole in the bottom. To develop this law, let us consider[6] how water molecules will rearrange themselves as water exits the tank and the relationship between the potential and kinetic energy of a small mass m of water. The potential energy lost as a small mass m of water falls from a height $h > 0$ is mgh, where g is the gravitational constant; at the same time, the kinetic energy gained as an equal mass m exits the tank is $\frac{1}{2}mv^2$, where v is the velocity at which the water is flowing. Equating the potential and kinetic energy, we find

---

[6] Our approach follows that of R. D. Driver in "Torricelli's law: An Ideal Example of an Elementary ODE," *Amer. Math. Monthly*, **105**(5) (May 1998), pp. 453–455.

that $mgh = \frac{1}{2}mv^2$, so that

$$v = \sqrt{2gh}$$

This model assumes that no friction is present; a slightly more realistic model takes a fraction of this velocity, depending on the viscosity. For simplicity, we will consider the ideal case where friction is not considered.

If we now consider the water exiting the tank, it follows that the rate of change dV/dt of volume in the tank is determined by the product of the area a of the hole and the exiting water's velocity v. In other words,

$$\frac{dV}{dt} = -av = -a\sqrt{2gh} \tag{2.7.4}$$

At this point, observe that we have related the rate of change of volume to the height of the water in the tank at time t. Instead, we desire to either relate dV/dt and V or dh/dt and h. Of course, height and volume are related. If we assume that $A(y)$ denotes the tank's cross sectional area at height y, then integral calculus tells us that the volume of the tank up to height h is given by

$$V(h) = \int_0^h A(y)\,dy$$

Furthermore, by the Fundamental Theorem of Calculus, differentiating $V(h)$ implies $dV/dh = A(h)$, and thus by the chain rule,

$$\frac{dV}{dt} = \frac{dV}{dh}\frac{dh}{dt} = A(h)\frac{dh}{dt}$$

Using this new expression for dV/dt in (2.7.4), it follows that

$$A(h)\frac{dh}{dt} = -a\sqrt{2gh} \tag{2.7.5}$$

which is a differential equation in h. In particular, this nonlinear equation predicts, given a tank of a particular shape (as determined by $A(h)$) with a hole of area a, the behavior of the function $h(t)$ that describes the height of the water at time t. We explore this further in the following example.

Example 2.7.2   For a cylindrical tank of height 2 m and radius 0.3 m, filled to the top with water, how long does it take the tank to drain once a hole of diameter 4 cm is opened?

Solution.   In this situation, the cross sectional area $A(h)$ of the tank at height h is constant because each is a circle of radius 0.3, so that $A(h) = 0.09\pi$. In addition, the area of the hole in square meters is $a = \pi(0.02)^2 = 0.0004\pi$, and the gravitational constant is $g = 9.8$ m/s^2. Since we have already established that $A(h)dh/dt = -a\sqrt{2gh}$, we therefore conclude that h satisfies the equation

$$0.09\pi\frac{dh}{dt} = -0.0004\pi\sqrt{19.6h}$$

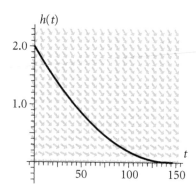

Figure 2.12 The slope field for $dh/dt = -0.019676\sqrt{h}$.

Simplifying, it follows that

$$\frac{dh}{dt} = -0.019676\sqrt{h}$$

Separating variables, we have

$$h^{-1/2}\,dh = -0.019676\,dt$$

and upon integrating, it follows that

$$2h^{1/2} = -0.019676t + C$$

Thus,

$$h(t) = (C_0 - 0.009838t)^2$$

Because $h(0) = 2$, $C_0 = \sqrt{2}$. Furthermore, with $h(t) = (\sqrt{2} - 0.009838t)^2$, we can see that $h(t) = 0$ when $t = 143.75$ sec, at which time the tank is empty. A plot of $h(t)$ confirms precisely the behavior observed in the direction field in figure 2.12.

Exercises 2.7

1. For a population $P(t)$ that exhibits logistic growth according to the general model

$$\frac{dP}{dt} = kP\left(1 - \frac{P}{A}\right), \quad P(0) = P_0$$

   (a) Determine the values of P (in terms of A and k) for which P is an increasing function.
   (b) Sketch by hand the direction field for the differential equation, clearly indicating the role of the constant A in your sketch.
   (c) Determine the value(s) of P (in terms of A and k) for which P is increasing most rapidly, and justify your answer.

(d) Solve the initial-value problem explicitly for P to show that

$$P(t) = \frac{A}{1 + Me^{-kt}}$$

and determine M in terms of A and P_0.

2. The growth of an animal population is governed by the equation

$$\frac{500}{P}\frac{dP}{dt} = 50 - P$$

where $P(t)$ is the number of animals in the herd at time t. The initial population is known to be 125. Determine the solution $P(t)$, sketch its graph, and decide whether there will ever be more than 125 or fewer than 50 animals present.

3. Consider the differential equation $dP/dt = -0.02P^2 + 0.08P$.

(a) What are the equilibrium solutions to this equation?
(b) Determine whether each equilibrium solution is stable or unstable.
(c) At what value of P is the function growing most rapidly?
(d) Under the initial condition $P(0) = 0.25$, determine the time at which $P(t) = 3$.

4. Consider a fish population that grows according to the model

$$\frac{dP}{dt} = 0.05P - 0.000005P^2$$

where t is measured in years, and P is measured in thousands.

(a) Determine the population of fish at time t if initially $P(0) = 1000$. What is the carrying capacity of the population?
(b) Suppose that the fish population is established as growing according to the above model in the absence of fish being removed from the lake. Suppose that harvesting begins at a rate of 20 000 fish per year. How does the differential equation governing the fish population change? Explain.
(c) Plot a direction field for the updated differential equation you found in part (b). Discuss the new equilibrium solutions for the fish population. Can you solve the IVP with $P(0) = 1000$?
(d) How would the DE change if wildlife biologists began planting 30 000 fish per year in the lake, and no harvesting occurred?

5. Solve the initial-value problem

$$\frac{dP}{dt} = 6 - 7P + P^2, \quad P(0) = 2$$

Sketch your solution curve $P(t)$ and explain why it makes sense in light of the equilibrium solutions to the given equation and your understanding of where dP/dt is positive and negative.

6. A cruise ship leaves port with 2500 vacationers aboard. At the time the boat leaves the dock, ten recent visitors of an amusement park are sick with the flu. Let $S(t)$ denote the number of people at time t who have had the flu at some time since leaving port.

   (a) Assuming that the rate at which the flu virus spreads is directly proportional to the product of the number of people who have had the flu times the number of people not yet infected, write a differential equation whose solution is the function $S(t)$. Explain why the differential equation is a logistic equation.

   (b) Solve the differential equation you found in (a). Assume that four days into the trip, 150 people have been sick with the flu. Clearly show how all constants are identified, and sketch a graph of your solution curve.

   (c) How many people have been sick seven days into the trip? How long would the boat have to stay at sea for half the vacationers to get ill?

7. A cylindrical tank of height 4 m and radius 1 m is full of water. A small hole of diameter 1 cm is opened in the bottom of the tank. Use Torricelli's law to determine how long it will take for all the water to drain from the tank.

8. A cylindrical tank of height 1.2 m and radius 30 cm is originally full of water. A small hole is opened in the bottom of the tank, and after 15 min, the water in the tank has dropped 10 cm. According to Torricelli's law, how large is the hole and how long will it take the tank to drain?

9. Consider a tank that is generated by taking the curve $x = \sqrt{y}$ and revolving it about the y-axis. Assume that the tank is full of water to a depth of 1.2 m and that a hole of diameter 1 cm is opened in the bottom. Use Torricelli's law to determine how long it will take for all the water to drain from the tank.

10. Suppose a hemispherical bowl has top radius of 30 cm and at time $t = 0$ is full of water. At that moment a circular hole with diameter 1.2 mm is opened in the bottom of the tank. Use Torricelli's law to determine how long it will take for all the water to drain from the tank.

11. For an open cylindrical tank, Torricelli's law tells us that if a small hole is opened, the height of the water at time t obeys the IVP

$$\frac{dh}{dt} = -k\sqrt{h}, \quad h(t_0) = h_0$$

where k is a constant that depends on the radius of the tank and the radius of the hole. In this exercise, we will take $k = 1$.

   (a) Explain why theorem 2.2.1 does not guarantee a unique solution to the IVP

$$\frac{dh}{dt} = -\sqrt{h}, \quad h(1) = 0$$

   (b) Explain why it is physically impossible to determine the height of the water at time $t < 1$ in a tank which satisfies $h(1) = 0$.

(c) Show that for any $c < 1$, the function

$$h(t) = \begin{cases} \left(\frac{1}{2}c - \frac{1}{2}t\right)^2 & \text{if } t < c \\ 0 & \text{if } t \geq c \end{cases}$$

is a solution to the IVP in (a).

(d) Explain how the result of (c) can be interpreted physically in light of the time when the tank becomes empty. Compare your findings to those in (a) and (b).

2.8 For further study

2.8.1 Converting certain second-order DEs to first-order DEs

Linear second-order differential equations such as

$$y'' + p(t)y' + q(t)y = f(t) \tag{2.8.1}$$

will be the focus of upcoming work in chapters 3 and 4. But there are some second-order equations we can solve at present. For example, if $q(t) = 0$ in (2.8.1), then we can perform a process called *reduction of order* to convert the equation to a first-order one.

(a) Consider the second-order equation $y'' + p(t)y' = f(t)$. Using the substitution $u = y'$, convert the equation to a new first-order DE involving the function u.

(b) Use a standard solution technique to state the solution u to the differential equation in (a) in terms of $p(t)$ and $f(t)$. (Your answer will involve integrals.)

(c) Explain how you would use your result in (b) to find the solution y to the original DE.

(d) Use reduction of order to solve each of the following second-order IVPs.

(i) $y'' + 2y' = 4$, $y(0) = 2$, $y'(0) = 1$

(ii) $y'' + \tan(t)y' = t$, $y(0) = 1$, $y'(0) = 0$

(iii) $y'' + \frac{2t}{1+t^2}y' = t^2$, $y(0) = 0$, $y'(0) = 1$

(iv) $y'' + \frac{1}{4-t}y' = 4 - t$, $y(0) = 1$, $y'(0) = 1$

(e) Reduction of order can be performed on certain nonlinear differential equations as well. For instance, suppose that we have an equation of form

$$y'' = g(y')h(t) \tag{2.8.2}$$

Show that the substitution $u = y'$ converts (2.8.2) to a first-order equation in u. Explain how you would approach solving the new equation in u.

(f) Solve each of the following second-order IVPs.

(i) $y'' = (y')^2 t^2$, $y(0) = 1$, $y'(0) = 0$

(ii) $y'' = \dfrac{t + t(y')^2}{y'}$, $y(0) = 2$, $y'(0) = 1$

(iii) $y'' = e^{2t+y'}$, $y(0) = 0$, $y'(0) = 0$

(iv) $y'' = \sqrt{y'}$, $y(0) = 3$, $y'(0) = 5$

2.8.2 How raindrops fall

The following questions and discussion are based on the article "Falling Raindrops" by Walter J. Meyer[7].

When a raindrop falls, various forces act upon it. We explore several different models that show the importance of adjusting assumptions appropriately to match physical conditions. Let us first assume that the only force acting upon the raindrop is the acceleration due to gravity. Under this assumption, Galileo (1564–1642) hypothesized that the falling raindrop would gain an extra 32 ft/s in velocity for every second for which it falls. In other words, the acceleration of the raindrop is constant and equal to 32 ft/sec^2.

(a) Let $y(t)$ denote the distance (in feet) traveled by the rain drop after it has been falling for t seconds. Write an initial-value problem involving $y(t)$ based on the above assumption. Solve this IVP; be sure to introduce appropriate initial conditions based on the context of the problem.

(b) Assuming that the raindrop starts from rest at an elevation of 3000 ft, how long does it take the raindrop to fall to earth? What is the raindrop's velocity when it hits the ground? Why is this model unrealistic?

(c) We next must attempt to account for the air resistance the raindrop encounters through a slightly more sophisticated model. For a raindrop having diameter $d \le 0.00025$ ft, this model, sometimes known as Stoke's law, states that the acceleration of the raindrop due to gravity is opposed by an acceleration directly proportional to the velocity of the raindrop at that instant. Suppose that the constant of proportionality is given by c/d^2, where $c \approx 3.29 \times 10^{-6}$ ft^2/s is an experimentally determined constant. Write a new IVP (again involving $y(t)$ and its relevant derivatives) for the raindrop having diameter d. Do not yet attempt to solve this equation. Leave d as an unknown constant.

(d) Letting $v = y'$ and using the fact that the raindrop starts from rest, convert the IVP in (c) to a first-order IVP involving v. Using $d = 0.00012$ ft (which can be considered a drizzle), produce a slope field corresponding to the

---

[7] See *Applications of Calculus*, MAA Notes Number 29, pp. 101–111.

differential equation in v. On this slope field, sketch a graphical approximation of the solution to the stated IVP. Describe the behavior of the raindrop's velocity based on the slope field you constructed in the problem above.

(e) In the model in (d), we will say that the long-term limiting velocity of the raindrop is its *terminal velocity*, denoted v_{term}. Calculate this terminal velocity by using the IVP to answer the following questions: What is the initial velocity of the raindrop? What is the equilibrium solution of the differential equation? What happens to the velocity of the raindrop if it ever reaches the equilibrium value? Why, in view of the differential equation, must the velocity of the raindrop increase from its initial value to the equilibrium value?

(f) Use your result from (e) to determine the terminal velocities for raindrops having diameters of 0.00009, 0.00012, and 0.00015 ft, respectively. Graph v_{term} as a function of d, and comment on the phenomena observed.

(g) Solve the IVP from (d) explicitly for v. Graph your solution, and then use your solution to calculate v_{term} as well.

(h) Assuming that a raindrop of diameter 0.00012 ft starts from rest at 3000 ft, how long does it take the raindrop to fall to the ground? What is its velocity at the instant it hits the ground? Do your answers surprise you? Is it raining hard or barely raining when raindrops are this size?

(i) When the diameter of the raindrop becomes too large, the force of air resistance on the raindrop becomes so appreciable that Stoke's model loses accuracy as well. This leads to a third model, known as the *velocity-squared model*. This model states that when a raindrop has diameter $d \geq 0.004$ ft, the acceleration due to gravity is opposed by an acceleration directly proportional to the square of the velocity of the raindrop at that instant. Here the constant of proportionality is given by k/d, where $k \approx 0.00046$.

(j) Repeat questions (c), (d), and (e) for the velocity-squared model. Compare your findings with those of Stoke's model. For example, how do the terminal velocities of small raindrops compare with those of large raindrops? For which type of raindrop, small or large, does the terminal velocity increase more rapidly as a function of diameter?

(k) Finally, explicitly solve the IVP arising from the velocity-squared model for the velocity function $v(t)$. Graph your solution $v(t)$ for an appropriate choice of d and compare the result to the results in (j).

2.8.3 Riccati's equation

The *Ricatti equation*

$$y' + p(t)y + q(t)y^2 = f(t) \tag{2.8.3}$$

and its study are attributed to the Italian mathematician Jacobo Riccati (1667–1748). Observe that this nonlinear equation is a modification of the standard linear first-order equation $y' + p(t)y = f(t)$. Through the following steps, we will use a change of variables to transform the Riccati equation into a linear, second-order differential equation.

(a) We consider a change of variables to convert (2.8.3) from being a differential equation in y to a new equation in v. Let v be a function that satisfies the relationship

$$v' = q(t)y(t)v(t)$$

(i) Differentiate $v' = qyv$ with respect to t to show that

$$v'' = (qyv)' = q'yv + qy'v + qyv' \tag{2.8.4}$$

(ii) Show that $q'yv = q'v'/q$.

(b) Multiply both sides of the Riccati equation (2.8.3) by qv and use (i) and (ii) to show that the left-hand side may be written

$$vqy' + vqpy + vq^2y^2 = v'' + \left(p - \frac{q}{q'}\right)v' \tag{2.8.5}$$

(c) Use your work in (b) to show that the Riccati equation may now be re-expressed as the second-order equation in v given by

$$v'' + \left(p - \frac{q}{q'}\right)v' - vqf = 0 \tag{2.8.6}$$

(d) Explain how you would solve the Riccati equation in the special case when $f(t) = 0$. Note particularly that to solve (2.8.6) with $f(t) = 0$, you must reduce the order of the equation through an appropriate substitution, say $u = v'$. See section 2.8.1 for further details on this technique. In addition, note that your goal is to find the solution y to the original equation (2.8.3). Be sure to explain how the functions v and u are used in this process.

(e) Solve the following differential equations, each of which is a Riccati equation.

(i) $y' + 2y + 4y^2 = 0$
(ii) $y' + \frac{1}{t}y + t^2y^2 = 0$
(iii) $y' + y\tan t + y^2\cos t = 0$

2.8.4 Bernoulli's equation

The Bernoulli brothers, James (1654–1705) and John (1667–1748), contributed to the solution of

$$y' + p(t)y = q(t)y^n, \quad n \neq 1 \tag{2.8.7}$$

the so-called *Bernoulli equation.* We will explore the approach credited to John through the following prompts. Similar to the Riccati equation, the Bernoulli equation may be transformed into a linear differential equation through a clever change of variables.

(a) First, multiply (2.8.7) by y^{-n} to obtain

$$y^{-n}y' + p(t)y^{1-n} = q(t) \tag{2.8.8}$$

Next, consider the change of variables $v = y^{1-n}$. Compute v' to show that

$$v' = (1-n)y^{-n}y' \tag{2.8.9}$$

Now use (2.8.8) and (2.8.9) to show that v satisfies the linear first-order equation

$$v' + (1-n)p(t)v = (1-n)q(t) \tag{2.8.10}$$

(b) Explain why in the cases when $n = 1$, $n = 2$, $q(t) = 0$, and $p(t) = 0$ the Bernoulli equation reduces to familiar equations whose solutions are known.

(c) Solve these differential equations, each of which is a Bernoulli equation.

(i) $y' + 2y = ty^3$
(ii) $y' + \frac{1}{t}y = 3y^3$
(iii) $y' + y\cot t = y^3 \sin t$

3

Linear systems of differential equations

3.1 Motivating problems

In section 1.1, we considered how the amount of salt present in a system of two tanks can be modeled through a system of differential equations. In that particular example, we assumed that the volume of solution in each tank (as seen in figure 3.1) remains constant and all inflows and outflows happen at the identical rate of 5 liter/min, and further that that the tanks are uniformly mixed so that the salt concentration in each is identical throughout each tank at a given time t.

With the additional premises that the volume of solution in tank A is 200 liters and the independent inflow entering A carries water contaminated with 4g/liter of salt, we can develop a differential equation that models $x_1(t)$, the amount of salt (in grams) in tank A at time t. Likewise, by presuming that tank B holds solution of volume 400 liters and the inflow entering B carries a concentration of salt of 7g/liter, a similar analysis produces a differential equation whose solution is $x_2(t)$, the amount of salt (in grams) in tank B at time t. In particular, we found in (1.1.6) that the following *system of differential equations* arose:

$$\frac{dx_1}{dt} = -\frac{x_1}{20} + \frac{x_2}{80} + 20$$
$$\frac{dx_2}{dt} = \frac{x_1}{40} - \frac{x_2}{40} + 35$$

(3.1.1)

With our experience in linear algebra, we can now represent this system in matrix notation. In particular, if we simultaneously consider the amounts of

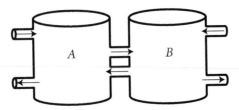

Figure 3.1 Two tanks with inflows, outflows, and connecting pipes.

salt $x_1(t)$ and $x_2(t)$ as entries in the vector function

$$\mathbf{x}(t) = \begin{bmatrix} x_1(t) \\ x_2(t) \end{bmatrix}$$

we know that

$$\mathbf{x}'(t) = \begin{bmatrix} dx_1/dt \\ dx_2/dt \end{bmatrix} \tag{3.1.2}$$

Moreover, in (3.1.1) we recognize the familiar form of a matrix product in the terms involving x_1 and x_2. Specifically,

$$\begin{matrix} -x_1/20 + x_2/80 \\ x_1/40 - x_2/40 \end{matrix} = \begin{bmatrix} -1/20 & 1/80 \\ 1/40 & -1/40 \end{bmatrix} \begin{bmatrix} x_1 \\ x_2 \end{bmatrix} \tag{3.1.3}$$

With the observations from (3.1.2) and (3.1.3) substituted into (3.1.1) and replacing the quantities 20 and 35 with the appropriate vector, we may now write the system of differential equations in the form

$$\mathbf{x}' = \begin{bmatrix} -1/20 & 1/80 \\ 1/40 & -1/40 \end{bmatrix} \mathbf{x} + \begin{bmatrix} 20 \\ 35 \end{bmatrix} \tag{3.1.4}$$

Letting $\mathbf{A}$ be the matrix of coefficients that multiplies the vector $\mathbf{x}$ and $\mathbf{b}$ the vector $[20 \ 35]^T$, we can also write the system in (3.1.4) in the simplified form

$$\mathbf{x}' = \mathbf{A}\mathbf{x} + \mathbf{b} \tag{3.1.5}$$

This form reminds us of the familiar nonhomogeneous linear first-order differential equation with constant coefficients, for instance, an equation such as

$$y' = 2y + 5 \tag{3.1.6}$$

In this chapter, we will study similarities between (3.1.5) and (3.1.6) with the specific goal of learning how to completely solve nonhomogeneous linear systems of differential equations with constant coefficients such as the system (3.1.4). We will be especially interested in the role that linear algebra plays in identifying certain characteristics of the coefficient matrix $\mathbf{A}$ that enable us to find all solutions to the system.

Before we proceed to an in-depth study of linear systems of differential equations, at least one more motivating example is appropriate. A *spring-mass system*

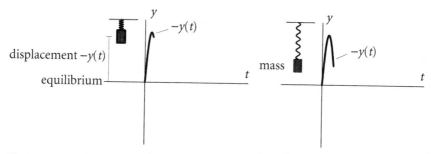

Figure 3.2 A spring-mass system shown at two different points in time; $-y(t)$ denotes the displacement of the mass from equilibrium (where displacements below the t-axis are considered positive).

is a physical situation that models vibrations; for example, such a system arises any time a mass attached to a spring is set in motion. We choose to envision this situation vertically, as seen in figure 3.2, though one can also imagine the mass resting on a table and moving horizontally.

We consider some of the physics of basic springs and motion under the influence of gravity in order to develop a differential equation that describes the spring-mass system. Initially, the mass will stretch the spring from its natural length. Hooke's law states that the force necessary to stretch a spring a distance x from its natural length is given by the equation

$$F(x) = kx$$

where k is the *spring constant*. Assume that the mass stretches the spring a distance L_0. Then from Hooke's law, when the system is in equilibrium, we see that the force F_s exerted by the spring must be

$$F_s = -kL_0$$

Here the minus sign indicates that the force is opposing the natural downward displacement of the spring. Note particularly that we view the downward direction as positive. We also know that gravity acts on the mass with force F_g given by

$$F_g = mg$$

If the system is in static equilibrium, we know that the sum of the two forces is zero. In other words,

$$F_g + F_s = 0$$

and therefore

$$mg = kL_0$$

Once the system is set in motion by some initial force or displacement, we track the location of the mass at time t with a function $y(t)$. In particular, $y(t)$ represents the displacement of the mass from the equilibrium position at time t; note that $y = 0$ is the equilibrium position of the system. We continue to designate the downward direction as positive, so $y(t) > 0$ means that the mass is below the equilibrium position, while $y(t) < 0$ means the mass is above the equilibrium position. We can see the role $y(t)$ plays in figure 3.2 as it tracks the displacement of the mass from equilibrium and thus traces out a curve with respect to time.

We can now use Newton's second law to obtain a differential equation that governs the system. The forces that act on the mass are:

- Gravity, with $F_g = mg$.

- The spring force F_s. Note now that at a given time t the displacement of the spring from its natural length is $L_0 + y(t)$, so that by Hooke's law we have $F_s = -k(L_0 + y)$.

- A possible damping force F_d. Motion may be damped due to air resistance, friction, or some sort of external damping system (usually called a dashpot). We assume that damping forces are directly proportional to the velocity of the mass. Under this assumption, it follows that $F_d = -cy'$. Again, the minus sign indicates that this force opposes the motion of the mass. The positive constant c is called the *damping constant*.

- Finally, there may be an external driving force present (such as the periodic force that drives a piston in an engine). We call this a *forcing function* $F(t)$; the role of forcing functions will be considered in detail later on in this chapter.

Newton's second law demands that the resultant force (that is, the sum of all the forces) on the mass must be equal to ma, where a is the body's acceleration (which is also y''). Summing all the aforementioned forces and equating the result with $ma = my''$, we find

$$my'' = F_g + F_s + F_d + F(t) \tag{3.1.7}$$

Using the formulas we developed earlier and substituting in (3.1.7) yields

$$my'' = mg - k(L_0 + y) - cy' + F(t) \tag{3.1.8}$$

Now recall that $mg - kL_0 = 0$, rearrange (3.1.8), and divide by m. This leads us to the standard form of the differential equation that governs a spring mass system,

$$y'' + \frac{c}{m}y' + \frac{k}{m}y = \frac{1}{m}F(t) \tag{3.1.9}$$

Note that (3.1.9) is a *nonhomogeneous linear second-order differential equation*.

To see how such a second-order linear differential equation is linked to a *system* of linear differential equations, let's consider the specific example where $c = 1$, $m = 1$, $k = 6$, and $F(t) = 0$, which results in the equation

$$y'' + y' + 6y = 0 \tag{3.1.10}$$

If we introduce the functions x_1 and x_2 through the substitutions $y = x_1$ and $y' = x_2$, then $x_1(t)$ represents the displacement of the mass at time t and $x_2(t)$ is the velocity of the mass at time t.

Observe first that

$$x_1' = x_2 \tag{3.1.11}$$

Moreover, since $x_2' = y''$, we can rewrite (3.1.10) as $x_2' + x_2 + 6x_1 = 0$. Equivalently,

$$x_2' = -6x_1 - x_2 \tag{3.1.12}$$

Thus (3.1.11) and (3.1.12) generate the system of differential equations

$$\begin{aligned} x_1' &= x_2 \\ x_2' &= -6x_1 - x_2 \end{aligned} \tag{3.1.13}$$

which may also be expressed in matrix form as

$$\mathbf{x}' = \begin{bmatrix} 0 & 1 \\ -6 & -1 \end{bmatrix} \mathbf{x} \tag{3.1.14}$$

We have therefore shown that the linear second-order differential equation (3.1.9) that describes a spring-mass system may be converted to the system of linear first-order equations (3.1.14) through the substitution $x_1 = y$, $x_2 = y'$.

In fact, any linear higher order differential equation may be converted through a similar substitution to a system of linear first-order equations. Therefore, by learning to understand and solve systems of linear equations, we will be able to determine the behavior of higher order linear equations as well. It is this fact that motivates us to study systems of linear equations prior to the study of higher order single equations.

3.2 The eigenvalue problem revisited

As we begin our study of linear systems of first-order differential equations, we are ultimately interested in two main questions: the first asks, for a linear system $\mathbf{x}' = \mathbf{A}\mathbf{x}$ such as

$$\mathbf{x}' = \begin{bmatrix} 2 & 3 \\ 2 & 1 \end{bmatrix} \mathbf{x}$$

how can we explicitly solve the system for $\mathbf{x}(t)$? In addition, what is the long-term behavior of the solution $\mathbf{x}(t)$ to such a system? How does its graph appear? We start our investigation by thinking carefully about the meaning of the matrix equation $\mathbf{x}' = \mathbf{A}\mathbf{x}$ and compare our experience with the single first-order differential equation $x' = ax$. Note that we naturally begin with the homogeneous system $\mathbf{x}' = \mathbf{A}\mathbf{x}$; later we will consider nonhomogeneous systems of the form $\mathbf{x}' = \mathbf{A}\mathbf{x} + \mathbf{b}$. In every case, we seek a vector function $\mathbf{x}(t)$ that solves the given system. An elementary example is instructive.

Example 3.2.1   Solve the linear system $\mathbf{x}' = \mathbf{A}\mathbf{x}$, where

$$\mathbf{A} = \begin{bmatrix} -3 & 0 \\ 0 & -1 \end{bmatrix}$$

Explain the role that the eigenvalues and eigenvectors of $\mathbf{A}$ play in the general solution, and graph and discuss the solution curves for different choices of initial conditions.

Solution.   First, we observe that the system

$$\begin{bmatrix} x_1' \\ x_2' \end{bmatrix} = \mathbf{x}' = \begin{bmatrix} -3 & 0 \\ 0 & -1 \end{bmatrix}\mathbf{x} = \begin{bmatrix} -3 & 0 \\ 0 & -1 \end{bmatrix}\begin{bmatrix} x_1 \\ x_2 \end{bmatrix} \qquad (3.2.1)$$

tells us that we seek two functions $x_1(t)$ and $x_2(t)$ such that $x_1' = -3x_1$ and $x_2' = -x_2$. Because the matrix of the system is diagonal, the problem is especially simple. In particular, the system is *uncoupled*, which means that the differential equation for x_1' does not involve x_2 and the equation for x_2' does not involve x_1.

From our experience with linear first-order equations, we know that the general solution to $x_1' = -3x_1$ is $x_1(t) = c_1 e^{-3t}$ and that the solution to $x_2' = -x_2$ is $x_2(t) = c_2 e^{-t}$. Writing the solution to the system as a single vector, we have

$$\mathbf{x} = \begin{bmatrix} x_1 \\ x_2 \end{bmatrix} = \begin{bmatrix} c_1 e^{-3t} \\ c_2 e^{-t} \end{bmatrix} \qquad (3.2.2)$$

Rewriting $\mathbf{x}$ in another form sheds further insight on the key components of this solution. Writing $\mathbf{x}$ as the sum of two vectors, we find

$$\mathbf{x} = \begin{bmatrix} c_1 e^{-3t} \\ 0 \end{bmatrix} + \begin{bmatrix} 0 \\ c_2 e^{-t} \end{bmatrix} = c_1 e^{-3t}\begin{bmatrix} 1 \\ 0 \end{bmatrix} + c_2 e^{-t}\begin{bmatrix} 0 \\ 1 \end{bmatrix} \qquad (3.2.3)$$

Here, we can make a key observation about the eigenvalues and eigenvectors of $\mathbf{A}$: because $\mathbf{A}$ is diagonal, its eigenvalues are its diagonal entries, $\lambda_1 = -3$ and $\lambda_2 = -1$. Moreover, its corresponding eigenvectors may be easily confirmed to be the vectors

$$\mathbf{v}_1 = \begin{bmatrix} 1 \\ 0 \end{bmatrix} \text{ and } \mathbf{v}_2 = \begin{bmatrix} 0 \\ 1 \end{bmatrix}$$

Thus, in (3.2.3), we see the interesting fact that the solution has the form $\mathbf{x} = c_1 e^{\lambda_1 t}\mathbf{v}_1 + c_2 e^{\lambda_2 t}\mathbf{v}_2$; the eigenvalues and eigenvectors therefore play a central role in the system's behavior.

Finally, we explore the solutions to several related initial-value problems for select initial conditions. If we have the initial condition $\mathbf{x}(0) = [4 \ 0]^{\mathrm{T}}$, we see in (3.2.3) that $c_1 = 4$ and $c_2 = 0$, so that the solution to the IVP is

$$\mathbf{x}(t) = 4e^{-3t}\begin{bmatrix} 1 \\ 0 \end{bmatrix}$$

Two key observations can be made about this solution curve: one is that its graph is a straight line, since for every value of t, $\mathbf{x}$ is a scalar multiple of the vector $[1 \ 0]^T$. Note particularly that the direction of this line is given by the eigenvector corresponding to $\lambda_1 = -3$. The other important fact is that $e^{-3t} \to 0$ as $t \to \infty$, and therefore $\mathbf{x}(t) \to \mathbf{0}$, so that the solution approaches the origin as time increases without bound.

For the initial condition $\mathbf{x}(0) = [0 \ 5]^T$, it follows from (3.2.3) that $c_1 = 0$ and $c_2 = 5$, and thus the solution to this IVP is

$$\mathbf{x}(t) = 5e^{-t} \begin{bmatrix} 0 \\ 1 \end{bmatrix}$$

Similar observations about the behavior of this solution may be made to those noted above for the first chosen initial condition: this solution curve is linear and approaches the origin as $t \to \infty$.

Finally, if we consider an initial condition that does not correspond to an eigenvector of the system, such as $\mathbf{x}(0) = [4 \ 5]^T$, (3.2.3) tells us that $c_1 = 4$ and $c_2 = 5$, and thus

$$\mathbf{x} = 4e^{-3t} \begin{bmatrix} 1 \\ 0 \end{bmatrix} + 5e^{-t} \begin{bmatrix} 0 \\ 1 \end{bmatrix}$$

This last solution's graph is not a straight line. As seen in figure 3.3, which shows the three different solutions based on the differing initial conditions, we see the consistent behavior that every solution tends to the origin as $t \to \infty$, as well as that the eigenvectors play a key role in how these graphs appear. We will discuss this graphical perspective further in sections 3.4 and 3.5.

The long-term behavior of the solutions to the system (3.2.1) in example 3.2.1 suggests that every solution tends to the zero vector. In fact, the origin itself is a solution, a so-called *constant* or *equilibrium* solution. That is, if

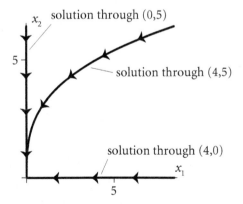

Figure 3.3 Plots of solutions to three IVPs for the system in example 3.2.1. Arrows indicate the direction of flow along the solution curve as time increases.

we consider whether there is any constant vector $\mathbf{x}$ that is a solution to $\mathbf{x}' = \mathbf{Ax}$, it follows that $\mathbf{x}' = \mathbf{0}$, and thus $\mathbf{x}$ must satisfy $\mathbf{Ax} = \mathbf{0}$. From our work with homogeneous linear equations, we know that $\mathbf{x} = \mathbf{0}$ is always a solution to this equation, and thus the zero vector is a constant solution to every homogeneous linear system of first-order differential equations. In sections 3.4 and 3.5 we will investigate the so-called *stability* of this equilibrium solution.

There is a second perspective from which we can see how eigenvectors and eigenvalues arise in the solution of linear systems of differential equations. After constant solutions, the next simplest type of solutions to such a system are *straight-line solutions*. In other words, solutions whose graph is a straight line in space form a particularly important type of solution to a system. In the preceding example, we saw two such straight-line solutions: each occurred in the direction of an eigenvector and passed through the origin.

In search of a general straight-line solution to $\mathbf{x}' = \mathbf{Ax}$, we know that any such solution must have the form $\mathbf{x}(t) = f(t)\mathbf{v}$, where $f(t)$ is a scalar function and $\mathbf{v}$ is a constant vector. This form guarantees that $\mathbf{x}(t)$ traces out a path that is a straight line through $\mathbf{0}$ in the direction of $\mathbf{v}$. In order for $\mathbf{x}(t)$ to satisfy the system, we observe that since $\mathbf{x}'(t) = f'(t)\mathbf{v}$, the equation

$$f'(t)\mathbf{v} = \mathbf{A}(f(t)\mathbf{v}) \tag{3.2.4}$$

must hold. Moreover, since $f(t)$ is a scalar, the linearity of matrix multiplication allows us to rewrite (3.2.4) as

$$f'(t)\mathbf{v} = f(t)\mathbf{Av} \tag{3.2.5}$$

Equation (3.2.5) is strongly reminiscent of the equation we use to define eigenvalues and eigenvectors: $\mathbf{Ax} = \lambda\mathbf{x}$. In fact, if $f'(t) = \lambda f(t)$, then (3.2.5) implies that

$$\lambda f(t)\mathbf{v} = f(t)\mathbf{Av}$$

Further, if $f(t) \neq 0$, then $\lambda\mathbf{v} = \mathbf{Av}$, and λ and $\mathbf{v}$ must be an eigenvalue-eigenvector pair of $\mathbf{A}$.

It is therefore natural for us to want f to satisfy the single differential equation $f'(t) = \lambda f(t)$. From our work in chapter 2, we know that $f(t) = Ce^{\lambda t}$ is the general solution to this equation. Substituting this form for f in (3.2.5), we now observe that

$$\lambda e^{\lambda t}\mathbf{v} = e^{\lambda t}\mathbf{Av} \tag{3.2.6}$$

and since $e^{\lambda t}$ is never zero, we can simplify (3.2.6) to

$$\lambda\mathbf{v} = \mathbf{Av} \tag{3.2.7}$$

which is satisfied precisely when $\mathbf{v}$ is an eigenvector of $\mathbf{A}$ with corresponding eigenvalue λ.

Our most recent work has demonstrated that if $\mathbf{x}(t)$ is a function of the form $\mathbf{x}(t) = e^{\lambda t}\mathbf{v}$ that is a solution to $\mathbf{x}' = \mathbf{Ax}$, then $(\lambda, \mathbf{v})$ is an eigenpair of the coefficient matrix $\mathbf{A}$. In fact, the converse also holds (as will be shown in the exercises), so that the following result is true for any $n \times n$ system of linear first-order differential equations.

Theorem 3.2.1   Let **A** be an $n \times n$ matrix. The vector function $\mathbf{x}(t) = e^{\lambda t}\mathbf{v}$ is a solution to the homogeneous linear system of first-order differential equations given by $\mathbf{x}' = \mathbf{A}\mathbf{x}$ if and only if $\mathbf{v}$ is an eigenvector of **A** with corresponding eigenvalue λ.

We close this section with one more example to demonstrate theorem 3.2.1 and one of its important consequences.

Example 3.2.2   Consider the system of differential equations given by

$$x_1' = -2x_1 - 2x_2$$
$$x_2' = -4x_1$$

Write the system in the form $\mathbf{x}' = \mathbf{A}\mathbf{x}$ and show that **A** has two real eigenvalues with corresponding linearly independent eigenvectors. Verify by substitution that for each eigenvalue-eigenvector pair, $\mathbf{x}(t) = e^{\lambda t}\mathbf{v}$ is a solution of the system. In addition, show that any linear combination of such solutions is also a solution to the system.

Solution.   First, we observe that the system can be expressed in the form $\mathbf{x}' = \mathbf{A}\mathbf{x}$ by using the matrix

$$\mathbf{A} = \begin{bmatrix} -2 & -2 \\ -4 & 0 \end{bmatrix}$$

We briefly review the process of determining the eigenvalues and eigenvectors of a matrix **A**; in most future occurrences, we will use *Maple* to determine this information using the commands introduced in section 1.10.2.

Since the eigenvalues are the roots of the characteristic equation, we solve $\det(\mathbf{A} - \lambda \mathbf{I}) = 0$. Doing so,

$$0 = \det(\mathbf{A} - \lambda \mathbf{I})$$

$$= \det \begin{bmatrix} -2 - \lambda & -2 \\ -4 & -\lambda \end{bmatrix} = -\lambda(-2 - \lambda) - 8$$

$$= \lambda^2 + 2\lambda - 8 = (\lambda + 4)(\lambda - 2)$$

so the eigenvalues of **A** are $\lambda = -4$ and $\lambda = 2$.

To find the eigenvector **v** that corresponds to $\lambda = -4$, we solve the equation $(\mathbf{A} - (-4\mathbf{I}))\mathbf{v} = \mathbf{0}$. Row-reducing the appropriate augmented matrix yields

$$\begin{bmatrix} 2 & -2 & 0 \\ -4 & 4 & 0 \end{bmatrix} \rightarrow \begin{bmatrix} 1 & -1 & 0 \\ 0 & 0 & 0 \end{bmatrix}$$

which shows that a corresponding eigenvector is any scalar multiple of the vector $\mathbf{v}_1 = [1 \ \ 1]^T$. Similar computations show that for $\lambda = 2$, a corresponding eigenvector is $\mathbf{v}_2 = [1 \ \ -2]^T$.

We now verify directly what theorem 3.2.1 guarantees: that $\mathbf{x}_1(t) = e^{-4t}[1 \ \ 1]^T$ and $\mathbf{x}_2(t) = e^{2t}[1 \ \ -2]^T$ are solutions to the given system of equations. Observe first that

$$\mathbf{x}_1'(t) = -4e^{-4t}\begin{bmatrix} 1 \\ 1 \end{bmatrix} \tag{3.2.8}$$

and that

$$\mathbf{A}\mathbf{x}_1(t) = \begin{bmatrix} -2 & -2 \\ -4 & 0 \end{bmatrix} e^{-4t}\begin{bmatrix} 1 \\ 1 \end{bmatrix} = e^{-4t}\begin{bmatrix} -2 & -2 \\ -4 & 0 \end{bmatrix}\begin{bmatrix} 1 \\ 1 \end{bmatrix}$$

$$= e^{-4t}\begin{bmatrix} -4 \\ -4 \end{bmatrix} = -4e^{-4t}\begin{bmatrix} 1 \\ 1 \end{bmatrix} \tag{3.2.9}$$

Equations (3.2.8) and (3.2.9) confirm that indeed $\mathbf{x}_1'(t) = \mathbf{A}\mathbf{x}_1(t)$ and demonstrate the role that eigenvalues and eigenvectors play in the solution.

Similarly, for the function $\mathbf{x}_2(t)$,

$$\mathbf{x}_2'(t) = 2e^{2t}\begin{bmatrix} 1 \\ -2 \end{bmatrix}$$

and

$$\mathbf{A}\mathbf{x}_2(t) = \begin{bmatrix} -2 & -2 \\ -4 & 0 \end{bmatrix} e^{2t}\begin{bmatrix} 1 \\ -2 \end{bmatrix} = e^{2t}\begin{bmatrix} -2 & -2 \\ -4 & 0 \end{bmatrix}\begin{bmatrix} 1 \\ -2 \end{bmatrix}$$

$$= e^{2t}\begin{bmatrix} 2 \\ -4 \end{bmatrix} = 2e^{2t}\begin{bmatrix} 1 \\ -2 \end{bmatrix} \tag{3.2.10}$$

This shows that $\mathbf{x}_2'(t) = \mathbf{A}\mathbf{x}_2(t)$.

Finally, we are asked to show that any linear combination of $\mathbf{x}_1(t)$ and $\mathbf{x}_2(t)$ is also a solution to the differential equation. While we could confirm this somewhat laboriously through direct computations, it is much easier to work more generally and consider known properties of differentiation and matrix multiplication.

In particular, differentiation is a linear operator and we know that if we let $\mathbf{y}(t) = c_1\mathbf{x}_1(t) + c_2\mathbf{x}_2(t)$ it follows that

$$\mathbf{y}'(t) = (c_1\mathbf{x}_1(t) + c_2\mathbf{x}_2(t))' = c_1\mathbf{x}_1'(t) + c_2\mathbf{x}_2'(t) \tag{3.2.11}$$

Similarly, matrix multiplication is a linear process, so

$$\mathbf{A}\mathbf{y}(t) = \mathbf{A}(c_1\mathbf{x}_1(t) + c_2\mathbf{x}_2(t)) = c_1\mathbf{A}\mathbf{x}_1(t) + c_2\mathbf{A}\mathbf{x}_2(t) \tag{3.2.12}$$

Since we have already established that $\mathbf{x}_1'(t) = \mathbf{A}\mathbf{x}_1(t)$ and $\mathbf{x}_2'(t) = \mathbf{A}\mathbf{x}_2(t)$, it follows that

$$c_1\mathbf{x}_1'(t) + c_2\mathbf{x}_2'(t) = c_1\mathbf{A}\mathbf{x}_1(t) + c_2\mathbf{A}\mathbf{x}_2(t)$$

so by (3.2.11) and (3.2.12) we have shown that $\mathbf{y}'(t) = \mathbf{A}\mathbf{y}(t)$ and thus indeed every linear combination of $\mathbf{x}_1(t)$ and $\mathbf{x}_2(t)$ is also a solution to $\mathbf{x}' = \mathbf{A}\mathbf{x}$.

Example 3.2.2 provides the foundation for much of our study of linear systems of differential equations. It shows that when we can find real eigenvalues and eigenvectors, these lead us directly to solutions of the system. In addition, any linear combination of such solutions is also a solution to the system; we state this formally in the next theorem.

Theorem 3.2.2 If $(\lambda_1, \mathbf{v}_1), (\lambda_2, \mathbf{v}_2), \ldots, (\lambda_k, \mathbf{v}_k)$ are eigenpairs of an $n \times n$ matrix $\mathbf{A}$ and $c_1, \ldots, c_k$ are any scalars, then

$$\mathbf{x}(t) = c_1 e^{\lambda_1 t} \mathbf{v}_1 + c_2 e^{\lambda_2 t} \mathbf{v}_2 + \cdots + c_k e^{\lambda_k t} \mathbf{v}_k$$

is a solution to $\mathbf{x}' = \mathbf{A}\mathbf{x}$.

In upcoming sections, we will determine whether we have found all of the solutions to a given system, address some subtle issues that arise when we cannot find enough real eigenvalues and eigenvectors, and better understand the graphical and long-term behavior of solutions. The exercises in this section will help further illuminate the roles of eigenvalues and eigenvectors as well as some of the issues that arise when there is an insufficient number of real eigenvectors for a given system's matrix.

Exercises 3.2
In exercises 1–7, compute by hand the eigenvalues and eigenvectors of the given matrix.

1. $\mathbf{A} = \begin{bmatrix} 1 & 4 \\ 2 & 3 \end{bmatrix}$

2. $\mathbf{A} = \begin{bmatrix} 0 & 4 \\ 1 & 0 \end{bmatrix}$

3. $\mathbf{A} = \begin{bmatrix} 0 & 3 \\ 3 & 8 \end{bmatrix}$

4. $\mathbf{A} = \begin{bmatrix} 2 & 2 \\ -1 & -1 \end{bmatrix}$

5. $\mathbf{A} = \begin{bmatrix} 2 & 2 & 0 \\ 1 & 2 & 1 \\ 1 & 2 & 1 \end{bmatrix}$

6. $\mathbf{A} = \begin{bmatrix} 3 & 0 & 1 \\ 0 & 2 & 0 \\ 5 & 0 & -1 \end{bmatrix}$

7. $\mathbf{A} = \begin{bmatrix} 2 & 1 & 0 \\ 0 & 2 & 1 \\ 0 & 0 & 2 \end{bmatrix}$

8. Consider the system of differential equations given by

$$x_1' = -2x_1 + 3x_2$$
$$x_2' = x_1 - 4x_2$$

(a) Determine a matrix $\mathbf{A}$ so that the system may be written in the form $\mathbf{x}' = \mathbf{A}\mathbf{x}$.
(b) Determine all constant (equilibrium) solutions to $\mathbf{x}' = \mathbf{A}\mathbf{x}$.
(c) Compute the eigenvalues and eigenvectors of $\mathbf{A}$.
(d) Determine all straight-line solutions to $\mathbf{x}' = \mathbf{A}\mathbf{x}$.
(e) Find a more general solution to $\mathbf{x}' = \mathbf{A}\mathbf{x}$ by taking all possible linear combinations of the straight-line solutions from (d).
(f) Solve the initial-value problem $\mathbf{x}' = \mathbf{A}\mathbf{x}$, $\mathbf{x}(0) = [1 \ 2]^{\mathrm{T}}$. Discuss the graphical behavior of this solution.

9. Consider the system of differential equations given by

$$x_1' = -x_1 + 2x_2$$
$$x_2' = -7x_1 + 8x_2$$

(a) Determine a matrix $\mathbf{A}$ so that the system may be written in the form $\mathbf{x}' = \mathbf{A}\mathbf{x}$.
(b) Determine all constant solutions to $\mathbf{x}' = \mathbf{A}\mathbf{x}$.
(c) Compute the eigenvalues and eigenvectors of $\mathbf{A}$.
(d) Determine all straight-line solutions to $\mathbf{x}' = \mathbf{A}\mathbf{x}$.
(e) Find a more general solution to $\mathbf{x}' = \mathbf{A}\mathbf{x}$ by taking all possible linear combinations of the straight-line solutions from (d).
(f) Solve the initial-value problem $\mathbf{x}' = \mathbf{A}\mathbf{x}$, $\mathbf{x}(0) = [-2 \ 0]^{\mathrm{T}}$. Discuss the graphical behavior of this solution.

10. Consider the system of differential equations given by

$$x_1' = 2x_1 + 3x_2$$
$$x_2' = -4x_2$$

(a) Determine a matrix $\mathbf{A}$ so that the system may be written in the form $\mathbf{x}' = \mathbf{A}\mathbf{x}$.
(b) Determine all constant solutions to $\mathbf{x}' = \mathbf{A}\mathbf{x}$.
(c) Compute the eigenvalues and eigenvectors of $\mathbf{A}$.
(d) Determine all straight-line solutions to $\mathbf{x}' = \mathbf{A}\mathbf{x}$.
(e) Find a more general solution to $\mathbf{x}' = \mathbf{A}\mathbf{x}$ by taking all possible linear combinations of the straight-line solutions from (d).
(f) Explain how you could find this same general solution *without* determining eigenvalues and eigenvectors. (Hint: focus on $x_2(t)$ first.)
(g) Solve the initial-value problem $\mathbf{x}' = \mathbf{A}\mathbf{x}$, $\mathbf{x}(0) = [0 \ 1]^{\mathrm{T}}$. Discuss the graphical behavior of this solution.

11. Consider the system of differential equations given by

$$x_1' = -2x_1 + x_2$$
$$x_2' = -2x_2$$

(a) Determine a matrix $\mathbf{A}$ so that the system may be written in the form $\mathbf{x}' = \mathbf{A}\mathbf{x}$.
(b) Determine all constant solutions to $\mathbf{x}' = \mathbf{A}\mathbf{x}$.
(c) Compute the eigenvalues and eigenvectors of $\mathbf{A}$.
(d) Determine all straight-line solutions to $\mathbf{x}' = \mathbf{A}\mathbf{x}$.
(e) Find a more general solution to $\mathbf{x}' = \mathbf{A}\mathbf{x}$ by taking all possible linear combinations of your straight-line solutions from (d).
(f) Attempt to solve the initial-value problem $\mathbf{x}' = \mathbf{A}\mathbf{x}$, $\mathbf{x}(0) = [1\ \ 1]^{\mathrm{T}}$. What does this tell you about the proposed general solution in (e)?

12. Consider the system of differential equations given by

$$x_1' = 2x_1 + 9x_2$$
$$x_2' = -x_1 - 2x_2$$

(a) Determine a matrix $\mathbf{A}$ so that the system may be written in the form $\mathbf{x}' = \mathbf{A}\mathbf{x}$.
(b) Determine all constant solutions to $\mathbf{x}' = \mathbf{A}\mathbf{x}$.
(c) Compute the eigenvalues and eigenvectors of $\mathbf{A}$.
(d) Are there any straight-line solutions to $\mathbf{x}' = \mathbf{A}\mathbf{x}$. Why or why not?

13. Consider the system of differential equations given by

$$x_1' = -3x_1 + x_2$$
$$x_2' = 3x_1 - x_2$$

(a) Determine a matrix $\mathbf{A}$ so that the system may be written in the form $\mathbf{x}' = \mathbf{A}\mathbf{x}$.
(b) Determine all constant solutions to $\mathbf{x}' = \mathbf{A}\mathbf{x}$. Compare and contrast your findings with preceding exercises.
(c) Compute the eigenvalues and eigenvectors of $\mathbf{A}$.
(d) Determine all straight-line solutions to $\mathbf{x}' = \mathbf{A}\mathbf{x}$. How many such solutions exist?
(e) Find a more general solution to $\mathbf{x}' = \mathbf{A}\mathbf{x}$ by taking all possible linear combinations of your straight-line solutions from (d).
(f) Solve the initial-value problem $\mathbf{x}' = \mathbf{A}\mathbf{x}$, $\mathbf{x}(0) = [3\ \ 0]^{\mathrm{T}}$. Discuss the graphical behavior of this solution.

14. Consider the system of differential equations given by

$$x_1' = 3x_1 + x_2 + x_3$$
$$x_2' = x_1 + 3x_2 + x_3$$
$$x_3' = x_1 + x_2 + 3x_3$$

(a) Determine a matrix $\mathbf{A}$ so that the system may be written in the form $\mathbf{x}' = \mathbf{A}\mathbf{x}$.

(b) Determine all constant solutions to $\mathbf{x}' = \mathbf{A}\mathbf{x}$.

(c) Compute the eigenvalues and eigenvectors of $\mathbf{A}$.

(d) Determine all straight-line solutions to $\mathbf{x}' = \mathbf{A}\mathbf{x}$.

(e) Find a more general solution to $\mathbf{x}' = \mathbf{A}\mathbf{x}$ by taking all possible linear combinations of your straight-line solutions from (d).

(f) Solve the initial-value problem $\mathbf{x}' = \mathbf{A}\mathbf{x}$, $\mathbf{x}(0) = [1\ \ 1\ \ 1]^{\mathrm{T}}$. Discuss the graphical behavior of this solution.

15. Consider the system of differential equations given by

$$x_1' = 8x_1 - x_2 - 11x_3$$
$$x_2' = 18x_1 - 3x_2 - 19x_3$$
$$x_3' = 2x_1 - x_2 - 5x_3$$

(a) Determine a matrix $\mathbf{A}$ so that the system may be written in the form $\mathbf{x}' = \mathbf{A}\mathbf{x}$.

(b) Determine all constant solutions to $\mathbf{x}' = \mathbf{A}\mathbf{x}$.

(c) Compute the eigenvalues and eigenvectors of $\mathbf{A}$.

(d) Determine all straight-line solutions to $\mathbf{x}' = \mathbf{A}\mathbf{x}$.

(e) Find a more general solution to $\mathbf{x}' = \mathbf{A}\mathbf{x}$ by taking all possible linear combinations of your straight-line solutions from (d).

(f) Solve the initial-value problem $\mathbf{x}' = \mathbf{A}\mathbf{x}$, $\mathbf{x}(0) = [1\ \ 1\ \ 1]^{\mathrm{T}}$. Discuss the graphical behavior of this solution.

Recall from section 3.1 that a second-order linear differential equation whose solution is $y(t)$ may be converted to a system of first-order linear equations whose solution is $\mathbf{x} = [x_1\ \ x_2]^{\mathrm{T}}$ through the substitution $x_1 = y$, $x_2 = y'$. See, for example, the discussion following (3.1.10). In exercises 16–22, convert each given higher order differential equation to a system of first-order equations through an appropriate substitution.

16. $y'' - 4y = 0$

17. $y'' + y' - 12y = 0$

18. $y'' + y' + y = 0$

19. $y'' - 2y' - 8y = e^t$

20. $y''' + 3y'' + 3y' + y = 0$

21. $y''' - 6y' + 5y = 0$

22. $y^{(4)} + 2y''' - 5y'' + y' - 9y = 0$

In sections 1.1 and 3.1, we showed how two connected tanks containing a solute lead to a system of linear first-order differential equations. In exercises 23–26, set up, but do not solve a system of differential equations or initial-value problem whose solution would give the amount of salt in each tank at time t. Write each system in matrix form.

23. A system of two tanks is connected in such a way that each of the tanks has an independent inflow that delivers salt solution to it, each has an independent outflow (drain), and each tank is connected to the other with an outflow and an inflow. The relevant information about each tank is given in the table below.

	Tank A	Tank B
Tank volume	100 liters	200 liters
Rate of inflow to the tank	5 liters/min	9 liters/min
Concentration of salt in inflow	7 g/liter	3 g/liter
Rate of drain outflow	4 liters/min	10 liters/min
Rates of outflows to other tank	to B: 3 liters/min	to A: 2 liters/min

24. Suppose that in exercise 23 all of the given information remains the same except for the fact that instead of saltwater flowing into each tank, pure water flows in; that is, the concentration of salt in the entering solution is 0 g/liter for each tank.

25. In a closed system of two tanks (i.e., one for which there are no input flows and no output flows), the following information is given. Tank A is filled with 100 liters of solution whose initial concentration is 0.25 g/liter. Tank B is filled with 50 liters of solution whose initial concentration is 3 g/liter. The two tanks are connected with two pipes having flows in opposite direction; mixed solution from Tank A flows to Tank B at a rate of 4 liters/min. Similarly, mixed solution flows from Tank B to Tank A at a rate of 4 liters/min.

26. In a closed system of three tanks (i.e., one for which there are no input flows and no output flows), the following information is given.

	Tank A	Tank B	Tank C
Tank Volume	100 liters	150 liters	125 liters
Rates of outflows to other tanks	to B: 3 liters/min	to C: 1 liter/min	to A: 4 liters/min
Rates of outflows to other tanks	to C: 4 liters/min	to A: 3 liters/min	to B: 1 liter/min

Tank A is filled with 100 liters of solution whose initial concentration is 8 g/liter. Tank B is filled with 150 liters of solution whose initial concentration is 3 g/liter. Tank C is initially filled with 125 liters of pure water. The three tanks are connected with pipes having flows in opposite directions; flow rates are given in the table above.

27. Show that if $(\lambda, \mathbf{v})$ is an eigenpair of the matrix $\mathbf{A}$, then $\mathbf{x}(t) = e^{\lambda t}\mathbf{v}$ is a solution to the homogeneous system of linear differential equations given by $\mathbf{x}' = \mathbf{A}\mathbf{x}$.

3.3 Homogeneous linear first-order systems

In preceding sections, we have encountered examples of systems of two (or three) linear differential equations in two (or three) unknown functions. More generally, a *linear system of n differential equations in n unknown functions* (or simply, a *linear system*) is a collection of differential equations for which we seek unknown functions $x_1(t), \ldots, x_n(t)$ when given n equations with coefficient functions $a_{ij}(t)$ and $b_i(t)$ in the form

$$\frac{dx_1}{dt} = a_{11}(t)x_1 + a_{12}(t)x_2 + \cdots + a_{1n}(t)x_n + b_1(t)$$

$$\frac{dx_2}{dt} = a_{21}(t)x_1 + a_{22}(t)x_2 + \cdots + a_{2n}(t)x_n + b_2(t)$$

$$\vdots \qquad\qquad\qquad \vdots$$

$$\frac{dx_n}{dt} = a_{n1}(t)x_1 + a_{n2}(t)x_2 + \cdots + a_{nn}(t)x_n + b_n(t)$$

It will be convenient to write the above system in matrix form. If we let $\mathbf{x}$ denote the vector function whose entries are $\mathbf{x}(t) = [x_i(t)]$, $\mathbf{A}(t)$ the $n \times n$ matrix of functions whose entries are $\mathbf{A} = [a_{ij}(t)]$, and $\mathbf{b}(t)$ the vector of functions whose entries are $\mathbf{b} = [b_i(t)]$, then the above system can be rewritten simply as

$$\mathbf{x}'(t) = \mathbf{A}(t)\mathbf{x}(t) + \mathbf{b}(t) \qquad\qquad (3.3.1)$$

In much of our work, we will suppress the independent variable t and write $\mathbf{x}' = \mathbf{A}\mathbf{x} + \mathbf{b}$. Moreover, it will most often be the case that, as in examples 3.2.1 and 3.2.2, the matrix $\mathbf{A}$ has all constant entries. Indeed, from this point on, unless otherwise noted, we will assume the matrix $\mathbf{A}$ has constant entries.

In the event that $\mathbf{b} = \mathbf{0}$, we say that the linear system is *homogeneous*. If $\mathbf{b}$ is nonzero, the system is *nonhomogeneous*. We have already encountered in theorems 3.2.1 and 3.2.2 the important facts that for any homogeneous first-order linear system $\mathbf{x}' = \mathbf{A}\mathbf{x}$, every solution of the form $\mathbf{x}(t) = e^{\lambda t}\mathbf{v}$ requires $(\lambda, \mathbf{v})$ to be an eigenpair of $\mathbf{A}$, and that any linear combination of such solutions is also a solution to the system.

Just as with individual differential equations, to each system of equations we can associate an *initial-value problem*. Using the matrix notation (3.3.1), if we assume that we also have the *initial condition* $x(t_0) = x_0$, then we have the standard initial-value problem

$$x'(t) = A(t)x(t), \quad x(t_0) = x_0 \tag{3.3.2}$$

We next consider a theoretical result (whose proof we omit) that will frame our overall work with systems. The following theorem is analogous to the earlier result we encountered in theorem 2.2.1 regarding the existence of a unique solution to the initial-value problem associated with a single first-order differential equation.

Theorem 3.3.1   In (3.3.2), let the entries of the matrix $A(t)$ be continuous functions on a common interval I that contains the value t_0. Then there exists a unique solution $x(t)$ to (3.3.2) on the interval I.

In particular, we note that in examples where the matrix A has constant coefficients, the entries are continuous functions, so that the IVP $x' = Ax$, $x(0) = x_0$ is guaranteed to have a unique solution. We now examine this result more closely through a particular example, revisiting a problem we considered in the preceding section.

Example 3.3.1   Determine the unique solution to the IVP given by

$$x' = \begin{bmatrix} -2 & -2 \\ -4 & 0 \end{bmatrix} x, \quad x(0) = \begin{bmatrix} -5 \\ 3 \end{bmatrix} \tag{3.3.3}$$

Solution.   We note, by theorem 3.3.1, that a unique solution exists. Moreover, from our work in example 3.2.2, every function of the form

$$x(t) = c_1 e^{-4t} \begin{bmatrix} 1 \\ 1 \end{bmatrix} + c_2 e^{2t} \begin{bmatrix} 1 \\ -2 \end{bmatrix} \tag{3.3.4}$$

is a solution to the system $x' = Ax$. We now explore whether we can find constants c_1 and c_2 in order that the function $x(t)$ will satisfy the given initial condition in (3.3.3).

The initial condition in (3.3.3) and (3.3.4) together imply

$$\begin{bmatrix} -5 \\ 3 \end{bmatrix} = x(0) = c_1 e^0 \begin{bmatrix} 1 \\ 1 \end{bmatrix} + c_2 e^0 \begin{bmatrix} 1 \\ -2 \end{bmatrix}$$

or equivalently

$$c_1 \begin{bmatrix} 1 \\ 1 \end{bmatrix} + c_2 \begin{bmatrix} 1 \\ -2 \end{bmatrix} = \begin{bmatrix} -5 \\ 3 \end{bmatrix} \tag{3.3.5}$$

We note that since the vectors $[1 \ 1]^T$ and $[1 \ -2]^T$ (which are eigenvectors of A) are linearly independent and span $\mathbb{R}^2$, we are guaranteed a unique solution to (3.3.5). Row-reducing the system (3.3.5), we find

$$\begin{bmatrix} 1 & 1 & -5 \\ 1 & -2 & 3 \end{bmatrix} \rightarrow \begin{bmatrix} 1 & 0 & -\frac{7}{3} \\ 0 & 1 & -\frac{8}{3} \end{bmatrix}$$

Thus, we have shown

$$\mathbf{x}(t) = -\frac{7}{3}e^{-4t}\begin{bmatrix} 1 \\ 1 \end{bmatrix} - \frac{8}{3}e^{2t}\begin{bmatrix} 1 \\ -2 \end{bmatrix}$$

is the unique solution to the given initial-value problem.

One especially important observation from example 3.3.1 can be made regarding the point at which we solved for the constants c_1 and c_2: we were guaranteed not only that a solution existed, but also that it was unique, due to the fact that two linearly independent eigenvectors of the 2×2 matrix $\mathbf{A}$ were present in the general solution (3.3.4). Indeed, if we imagine wanting to solve any similar IVP with the freedom to choose any initial vector $\mathbf{x}(0)$, it will be necessary that $\mathbf{x}(0)$ can be written as a linear combination of the vectors $\mathbf{v}_1$ and $\mathbf{v}_2$, whenever the general solution has form

$$\mathbf{x}(t) = c_1 e^{\lambda_1 t}\mathbf{v}_1 + c_2 e^{\lambda_2 t}\mathbf{v}_2$$

This situation is indicative of the general fact that for all 2×2 linear systems of DEs, we must have two parts to the general solution, in order to be able to uniquely determine the constants c_1 and c_2. Note further that for the solutions $\mathbf{x}_1(t) = e^{\lambda_1 t}\mathbf{v}_1$ and $\mathbf{x}_2(t) = e^{\lambda_2 t}\mathbf{v}_2$ we encountered above, $\mathbf{x}_1(0) = \mathbf{v}_1$ and $\mathbf{x}_2(0) = \mathbf{v}_2$ are linearly independent and form a basis for $\mathbb{R}^2$. This linear independence of the constant vectors $\mathbf{v}_1$ and $\mathbf{v}_2$ turns out to have an important analog in the linear independence of certain solutions to the system of differential equations.

More generally, we can consider these same issues for an $n \times n$ homogeneous system. Because theorem 3.3.1 guarantees the existence of a unique solution to the corresponding IVP for every initial condition $\mathbf{x}(0) \in \mathbb{R}^n$, when we think about the structure of the general solution, it is natural to think this solution will have form

$$\mathbf{x}(t) = c_1\mathbf{x}_1(t) + c_2\mathbf{x}_2(t) + \cdots + c_n\mathbf{x}_n(t)$$

where $\{\mathbf{x}_1(0), \mathbf{x}_2(0), \ldots, \mathbf{x}_n(0)\}$ form a basis for $\mathbb{R}^n$.

These observations, together with our earlier work in theorem 3.2.2 that showed that every linear combination of solutions to the general homogeneous linear system of DEs (3.3.1) is also a solution to (3.3.1), help explain why the set of all solutions to $\mathbf{x}' = \mathbf{Ax}$, where $\mathbf{A}$ is a matrix with constant coefficients, is a vector space of dimension n. We state this formally in the following result.

Theorem 3.3.2   The set of all solution vectors to the homogeneous linear system $\mathbf{x}' = \mathbf{Ax}$, where $\mathbf{A}$ is an $n \times n$ matrix with constant coefficients, forms a vector space of dimension n.

Theorem 3.3.2 shows us that in order to solve an $n \times n$ system of homogeneous first-order DEs, we must find n linearly independent solutions to the system. Said differently, the general solution to $\mathbf{x}' = \mathbf{Ax}$ will have form

$$\mathbf{x}(t) = c_1\mathbf{x}_1(t) + c_2\mathbf{x}_2(t) + \cdots + c_n\mathbf{x}_n(t) \tag{3.3.6}$$

where $\mathbf{x}_1(t), \ldots, \mathbf{x}_n(t)$ are linearly independent functions. Thus, our search for the general solution to the system requires us to find these n linearly independent functions $\mathbf{x}_1(t), \ldots, \mathbf{x}_n(t)$. While we need to discuss in more detail what it means for vector functions (rather than constant vectors) to be linearly independent, we can first note that we know by theorem 3.2.1 that when $(\lambda_i, \mathbf{v}_i)$ is an eigenpair of $\mathbf{A}$, the function

$$\mathbf{x}_i(t) = e^{\lambda_i t} \mathbf{v}_i$$

is a solution to $\mathbf{x}' = \mathbf{A}\mathbf{x}$. This fact, combined with theorem 3.3.2, implies the result depicted in theorem 3.3.3.

Theorem 3.3.3   If $\mathbf{A}$ is an $n \times n$ matrix with n linearly independent eigenvectors $\mathbf{v}_1, \mathbf{v}_2, \ldots, \mathbf{v}_n$, with corresponding eigenvalues $\lambda_1, \lambda_2, \ldots, \lambda_n$ (where the eigenvalues are not necessarily distinct), then the general solution to $\mathbf{x}' = \mathbf{A}\mathbf{x}$ is

$$\mathbf{x}(t) = c_1 e^{\lambda_1 t} \mathbf{v}_1 + c_2 e^{\lambda_2 t} \mathbf{v}_2 + \cdots + c_n e^{\lambda_n t} \mathbf{v}_n \tag{3.3.7}$$

The linear independence of $\mathbf{v}_1, \ldots, \mathbf{v}_n$ guarantees that we can solve the IVP $\mathbf{x}' = \mathbf{A}\mathbf{x}$, $\mathbf{x}(0) = \mathbf{x}_0$ for every possible choice of $\mathbf{x}_0 \in \mathbb{R}^n$, since we may write

$$\mathbf{x}_0 = c_1 \mathbf{v}_1 + c_2 \mathbf{v}_2 + \cdots + c_n \mathbf{v}_n$$

for a unique set of values $c_1, \ldots, c_n$. This shows that the general solution (3.3.7) indeed captures all possible solutions to the system.

In our original study of the eigenvalue problem in section 1.10, we observed (and proved in one of the exercises) that eigenvectors corresponding to distinct (real[1]) eigenvalues are linearly independent. This yields an important consequence of theorem 3.3.3: if $\mathbf{A}$ has n distinct real eigenvalues, then $\mathbf{A}$ has n linearly independent (real) eigenvectors. In particular, the following corollary is true.

Corollary 3.3.4   If $\mathbf{A}$ is an $n \times n$ matrix with n distinct real eigenvalues $\lambda_1, \lambda_2, \ldots, \lambda_n$, then the corresponding eigenvectors $\mathbf{v}_1, \mathbf{v}_2, \ldots, \mathbf{v}_n$ are linearly independent and the general solution to $\mathbf{x}' = \mathbf{A}\mathbf{x}$ is

$$\mathbf{x}(t) = c_1 e^{\lambda_1 t} \mathbf{v}_1 + c_2 e^{\lambda_2 t} \mathbf{v}_2 + \cdots + c_n e^{\lambda_n t} \mathbf{v}_n \tag{3.3.8}$$

We now consider a specific example in which we see corollary 3.3.4 at work.

Example 3.3.2   Determine the general solution to the homogeneous first-order system of DEs $\mathbf{x}' = \mathbf{A}\mathbf{x}$ and determine the unique solution to the initial-value problem

$$\mathbf{x}' = \mathbf{A}\mathbf{x} = \begin{bmatrix} -4 & 1 & -1 \\ -1 & -2 & 5 \\ -3 & 3 & 0 \end{bmatrix} \mathbf{x}, \quad \mathbf{x}(0) = \begin{bmatrix} 1 \\ -2 \\ 3 \end{bmatrix}$$

---

[1] We are interested in real solutions to the system $\mathbf{x}' = \mathbf{A}\mathbf{x}$; when eigenvalues and eigenvectors are complex, additional work is needed. See section 3.5.

Solution.   We begin by computing the eigenvalues and eigenvectors of **A**. Using the `Eigenvectors(A)` command in *Maple*, we find that the eigenvalues of **A** are $\lambda_1 = -6, \lambda_2 = -3, \lambda_3 = 3$, with corresponding eigenvectors

$$
\mathbf{v}_1 = \begin{bmatrix} 1 \\ -1 \\ 1 \end{bmatrix}, \mathbf{v}_2 = \begin{bmatrix} 1 \\ 1 \\ 0 \end{bmatrix}, \mathbf{v}_3 = \begin{bmatrix} 0 \\ 1 \\ 1 \end{bmatrix}
$$

Since the eigenvalues of **A** are distinct, we know immediately that the corresponding eigenvectors are linearly independent, and therefore by corollary 3.3.4 that the general solution to the given system is

$$
\mathbf{x}(t) = c_1 e^{-6t} \begin{bmatrix} 1 \\ -1 \\ 1 \end{bmatrix} + c_2 e^{-3t} \begin{bmatrix} 1 \\ 1 \\ 0 \end{bmatrix} + c_3 e^{3t} \begin{bmatrix} 0 \\ 1 \\ 1 \end{bmatrix} \qquad (3.3.9)
$$

To solve the IVP with

$$
\mathbf{x}(0) = \begin{bmatrix} 1 \\ -2 \\ 3 \end{bmatrix}
$$

we set $t = 0$ in (3.3.9) and apply the given condition, which leads to the vector equation

$$
c_1 \begin{bmatrix} 1 \\ -1 \\ 1 \end{bmatrix} + c_2 \begin{bmatrix} 1 \\ 1 \\ 0 \end{bmatrix} + c_3 \begin{bmatrix} 0 \\ 1 \\ 1 \end{bmatrix} = \begin{bmatrix} 1 \\ -2 \\ 3 \end{bmatrix}
$$

Writing this equation in augmented matrix form and row-reducing shows that

$$
\begin{bmatrix} 1 & 1 & 0 & 1 \\ -1 & 1 & 1 & -2 \\ 1 & 0 & 1 & 3 \end{bmatrix} \rightarrow \begin{bmatrix} 1 & 0 & 0 & 2 \\ 0 & 1 & 0 & -1 \\ 0 & 0 & 1 & 1 \end{bmatrix}
$$

and, therefore, the solution to the IVP is

$$
\mathbf{x}(t) = 2e^{-6t} \begin{bmatrix} 1 \\ -1 \\ 1 \end{bmatrix} - e^{-3t} \begin{bmatrix} 1 \\ 1 \\ 0 \end{bmatrix} + e^{3t} \begin{bmatrix} 0 \\ 1 \\ 1 \end{bmatrix}
$$

From corollary 3.3.4, we know that if we have an $n \times n$ matrix **A** with n linearly independent real eigenvectors, then we can completely solve the system $\mathbf{x}' = \mathbf{A}\mathbf{x}$. But what if **A** lacks n real linearly independent eigenvectors? While we will encounter this situation in more detail in section 3.5, here it is worthwhile to note that we will still be seeking n linearly independent solutions $\mathbf{x}_1(t), \ldots, \mathbf{x}_n(t)$ to the general system. For these vector functions, the fundamental meaning of linear independence remains the same as it does for constant vectors: the set of vector functions $\{\mathbf{x}_1(t), \dots, \mathbf{x}_n(t)\}$ is linearly independent if and only if the only values of $c_1, \dots, c_n$ that make

$$c_1\mathbf{x}_1(t) + \cdots + c_n\mathbf{x}_n(t) = \mathbf{0} \tag{3.3.10}$$

true for all values of t are $c_1 = \cdots = c_n = 0$. Testing the linear independence of vector functions is more involved; to do so, we introduce a new concept and a corresponding theorem.

Definition 3.3.1   Given vector functions $\mathbf{x}_1(t), \dots, \mathbf{x}_n(t)$ where each $\mathbf{x}_i(t) \in \mathbb{R}^n$ for all t, the *Wronskian* of these functions is

$$W[\mathbf{x}_1, \dots, \mathbf{x}_n] = \det[\mathbf{x}_1, \dots, \mathbf{x}_n] \tag{3.3.11}$$

That is, the Wronskian of a set of n vector functions, each of which lies in $\mathbb{R}^n$, is the determinant of the $n \times n$ matrix whose columns are $\mathbf{x}_1, \dots, \mathbf{x}_n$.

The Wronskian enables us to easily test whether or not vector functions are linearly independent through the following theorem, which will be stated without proof.

Theorem 3.3.5   Let $\mathbf{x}_1(t), \dots, \mathbf{x}_n(t)$ be vector functions continuous on an interval I, where $\mathbf{x}_i(t) \in \mathbb{R}^n$ for all $t \in I$. If at any point t_0 in I, $W[\mathbf{x}_1, \dots, \mathbf{x}_n]$ $(t_0) \neq 0$, then $\{\mathbf{x}_1(t), \dots, \mathbf{x}_n(t)\}$ is linearly independent on I.

We observe that this result appears reasonable since it is analogous to two statements that appear in the Invertible Matrix theorem: for a set of n constant vectors in $\mathbb{R}^n$, we know that the set is linearly independent if and only if the determinant of the matrix whose columns are these vectors is nonzero. Theorem 3.3.5 is a generalization of this result to the situation where the vectors are not constant.

An example will now demonstrate the use of the Wronskian in showing a set of vector functions is linearly independent.

Example 3.3.3   Consider the vector functions $\mathbf{x}_1 = [e^{-t} \quad -e^{-t} \quad e^{-t}]^{\mathrm{T}}$, $\mathbf{x}_2 = [3e^{2t} \quad e^{2t} \quad -2e^{2t}]^{\mathrm{T}}$, and $\mathbf{x}_3 = [e^{5t} \quad e^{5t} \quad e^{5t}]^{\mathrm{T}}$. Are $\mathbf{x}_1$, $\mathbf{x}_2$, and $\mathbf{x}_3$ linearly independent?

Solution.   We use the Wronskian of $\mathbf{x}_1$, $\mathbf{x}_2$, and $\mathbf{x}_3$ to determine their linearly independence. Observe that

$$\begin{aligned}
W[\mathbf{x}_1, \mathbf{x}_2, \mathbf{x}_3] &= \det \begin{bmatrix} e^{-t} & 3e^{2t} & e^{5t} \\ -e^{-t} & e^{2t} & e^{5t} \\ e^{-t} & -2e^{2t} & e^{5t} \end{bmatrix} \\
&= e^{-t}(e^{2t}e^{5t} + 2e^{5t}e^{2t}) - 3e^{2t}(-e^{-t}e^{5t} - e^{5t}e^{-t}) \\
&\quad + e^{5t}(2e^{2t}e^{-t} - e^{2t}e^{-t})
\end{aligned}$$

$$= e^{-t}(3e^{7t}) - 3e^{2t}(-2e^{4t}) + e^{5t}(e^t)$$

$$= 10e^{6t} \neq 0$$

Since $W[\mathbf{x}_1, \mathbf{x}_2, \mathbf{x}_3] \neq 0$ for at least one t-value (in fact, for all t), it follows by theorem 3.3.5 that the functions $\mathbf{x}_1$, $\mathbf{x}_2$, and $\mathbf{x}_3$ are linearly independent.

In conclusion, we now know that when we encounter a homogeneous system of n linear first-order differential equations in n unknown functions, the set of all solutions to the system forms an n-dimensional vector space. Hence, we seek n linearly independent solutions to the system $\mathbf{x}' = \mathbf{A}\mathbf{x}$. Such a set $\mathbf{x}_1, \ldots, \mathbf{x}_n$ of n linearly independent solution vectors to this system is called a *fundamental set*. Moreover, given a set of fundamental solutions $\mathbf{x}_1, \ldots, \mathbf{x}_n$ to $\mathbf{x}' = \mathbf{A}\mathbf{x}$, on some interval I, the general solution to the system is

$$\mathbf{x}(t) = c_1\mathbf{x}_1 + \cdots + c_n\mathbf{x}_n$$

We have also seen that if an $n \times n$ matrix $\mathbf{A}$ has n linearly independent real eigenvectors, then these eigenvectors and their corresponding eigenvalues generate a fundamental set for the system $\mathbf{x}' = \mathbf{A}\mathbf{x}$. In subsequent sections we will find that, even in the case when an insufficient number of real eigenvectors exists, the eigenvalue problem enables us to build a fundamental set. Moreover, we will investigate how fundamental solutions allow us to fully understand the graphical behavior of solutions and the stability of equilibrium solutions to the system.

Exercises 3.3

1. If $\mathbf{x}' = \mathbf{A}\mathbf{x}$ represents the system of differential equations given by a 4×4 matrix $\mathbf{A}$ with constant entries, how many linearly independent solutions to the system do we need to find in order to determine the general solution? What if $\mathbf{A}$ is 7×7?

2. Consider the second-order differential equation $y'' + y = 0$. Using the substitutions $y = x_1$ and $y' = x_2$, convert the given second-order differential equation to a system of first-order equations. What is the dimension of the solution space to the system? What does this tell you about the dimension of the solution space to the original second-order equation?

3. Consider the third-order differential equation $y''' + 3y'' + 3y' + y = 0$. Using the substitutions $y = x_1$, $y' = x_2$, and $y'' = x_3$, convert the given differential equation to a system of first-order equations. What is the dimension of the solution space to the system? What does this tell you about the dimension of the solution space to the original third-order equation?

In exercises 4–8, use the Wronskian to determine if the given set of vector functions is linearly independent.

4. $\mathbf{x}_1(t) = [e^{-t} \ -e^{-t}]^T, \mathbf{x}_2(t) = [e^{2t} \ 2e^{2t}]^T$

5. $\mathbf{x}_1(t) = [\cos t \ \sin t]^T, \mathbf{x}_2(t) = [\sin t \ \cos t]^T$

6. $\mathbf{x}_1(t) = [e^{-t} \ -e^{-t}]^T, \mathbf{x}_2(t) = [-3e^{-t} \ 3e^{-t}]^T$

7. $\mathbf{x}_1(t) = [e^t \ -e^t \ e^t]^T, \mathbf{x}_2(t) = [e^{7t} \ 2e^{7t} \ -3e^{7t}]^T, \mathbf{x}_3(t) = [4e^{-4t} \ e^{-4t} \ -e^{-4t}]^T$

8. $\mathbf{x}_1(t) = [\cos t \ -\sin t \ 0]^T, \mathbf{x}_2(t) = [\sin t \ \cos t \ 0]^T, \mathbf{x}_3(t) = [0 \ 0 \ e^t]^T$

9. Explain why for a set of two vector functions, the Wronskian is unneeded to check for linear independence. (Hint: what is the simple test for a pair of constant vectors to be linearly independent?)

10. Let $\mathbf{x}' = \mathbf{A}\mathbf{x}$ be given by the matrix

$$\mathbf{A} = \begin{bmatrix} -2 & 1 \\ 1 & -2 \end{bmatrix}$$

(a) Compute the eigenvalues and eigenvectors of $\mathbf{A}$. Explain why these enable you to find the general solution to $\mathbf{x}' = \mathbf{A}\mathbf{x}$.
(b) State the general solution to the system.
(c) Solve the IVP with the initial condition $\mathbf{x}(0) = [3 \ 2]^T$.

11. Let $\mathbf{x}' = \mathbf{A}\mathbf{x}$ be given by the matrix

$$\mathbf{A} = \begin{bmatrix} 3 & 1 \\ 0 & 3 \end{bmatrix}$$

(a) Compute the eigenvalues and eigenvectors of $\mathbf{A}$. Explain why you have found one linearly independent solution to the system, but still need to determine another.
(b) Verify through direct substitution that $\mathbf{x}_2(t) = te^{3t}[1 \ 0]^T + e^{3t}[0 \ 1]^T$ is a solution to the given system $\mathbf{x}' = \mathbf{A}\mathbf{x}$.
(c) Show that the solution you found in (a) above and the solution $\mathbf{x}_2(t)$ in (b) are linearly independent, and hence state the general solution to the system.
(d) Solve the IVP with the initial condition $\mathbf{x}(0) = [3 \ 2]^T$.

12. Let $\mathbf{x}' = \mathbf{A}\mathbf{x}$ be given by the matrix

$$\mathbf{A} = \begin{bmatrix} 3 & 0 \\ 0 & 3 \end{bmatrix}$$

(a) Compute the eigenvalues and eigenvectors of $\mathbf{A}$. Explain why, despite the repeated eigenvalue, you have found two linearly independent solutions to the system.
(b) State the general solution to the system.

(c) Solve the IVP with the initial condition $\mathbf{x}(0) = [3 \ 2]^T$.

(d) Explain how you could solve the original system given in this problem *without* using eigenvalues and eigenvectors.

13. Let $\mathbf{x}' = \mathbf{A}\mathbf{x}$ be given by the matrix

$$\mathbf{A} = \begin{bmatrix} 0 & -1 \\ 1 & 0 \end{bmatrix}$$

(a) Compute the eigenvalues and eigenvectors of $\mathbf{A}$. Explain why the eigenvalues and eigenvectors do not produce any real linearly independent solutions to the system.

(b) Verify through direct substitution that $\mathbf{x}_1(t) = [\cos t \ \sin t]^T$ and $\mathbf{x}_2(t) = [-\sin t \ \cos t]^T$ are solutions to the given system $\mathbf{x}' = \mathbf{A}\mathbf{x}$.

(c) Show that the solutions you verified in (b) are linearly independent, and hence state the general solution to the system.

(d) Solve the IVP with the initial condition $\mathbf{x}(0) = [3 \ 2]^T$.

14. Let $\mathbf{x}' = \mathbf{A}\mathbf{x}$ be given by the matrix

$$\mathbf{A} = \begin{bmatrix} 5 & 6 & 2 \\ 0 & -1 & -8 \\ 1 & 0 & -2 \end{bmatrix}$$

(a) Compute the eigenvalues and eigenvectors of $\mathbf{A}$. Explain why your work determines two linearly independent solutions to the system, but that one additional linearly independent solution remains to be found.

(b) Verify through direct substitution that
$\mathbf{x}_3(t) = te^{3t}[5 \ -2 \ 1]^T + e^{3t}[1 \ 1/2 \ 0]^T$ is a solution to the given system $\mathbf{x}' = \mathbf{A}\mathbf{x}$.

(c) Show that the set of three solutions from (a) and (b) is linearly independent, and hence state the general solution to the system.

(d) Solve the IVP with the initial condition $\mathbf{x}(0) = [3 \ 2 \ 1]^T$.

15. Consider the second-order differential equation $y'' + y = 0$. Convert this equation to a system of first-order equations and solve the system. Use your work to state the general solution y to the original equation. (Hint: See exercise 13.)

16. Convert the second-order differential equation $y'' + 3y' + 2y = 0$ to a system of first-order equations and solve the system. Use your work to state the general solution y to the original equation.

17. Convert the third-order differential equation $y''' - y' = 0$ to a system of first-order equations and solve the system. Use your work to state the general solution y to the original equation.

3.4 Systems with all real linearly independent eigenvectors

In this section, we closely examine the graphical and long-term behavior of solutions to 2×2 systems in the case where the coefficient matrix $\mathbf{A}$ has two real, linearly independent eigenvectors. We do so through a sequence of examples that demonstrate a variety of possibilities that naturally lead to discussion of the stability of equilibrium solutions.

We first review the graphical behavior of vector functions, a subject normally encountered in multivariable calculus. For the system $\mathbf{x}' = \mathbf{A}\mathbf{x}$ in the case where $\mathbf{A}$ is 2×2, every solution $\mathbf{x}(t)$ is a vector function whose output lies in $\mathbb{R}^2$. In particular, the graph of $\mathbf{x}(t)$ is the curve that is traced out by the vectors $\mathbf{x}(t)$ at various times t. For example, if

$$\mathbf{x}(t) = e^{-t}\begin{bmatrix} 1 \\ 0 \end{bmatrix} + e^{t}\begin{bmatrix} 0 \\ 1 \end{bmatrix} = \begin{bmatrix} e^{-t} \\ e^{t} \end{bmatrix} \tag{3.4.1}$$

is a function we have found by solving a system of differential equations, then evaluating $\mathbf{x}(t)$ at $t = -1, 0$, and 1 yields the vectors

$$\mathbf{x}(-1) \approx \begin{bmatrix} 2.719 \\ 0.368 \end{bmatrix}, \ \mathbf{x}(0) = \begin{bmatrix} 1 \\ 1 \end{bmatrix}, \ \text{and } \mathbf{x}(1) \approx \begin{bmatrix} 0.368 \\ 2.719 \end{bmatrix} \tag{3.4.2}$$

Plotting these vectors helps indicate how $\mathbf{x}(t)$ traces out the parametric curve given by $(x_1(t), x_2(t)) = (e^{-t}, e^{t})$, shown at left in figure 3.4.

In addition, it is important to recall the meaning of $\mathbf{x}'(t)$, the derivative of a vector function. The direction of the vector $\mathbf{x}'(t)$ indicates the instantaneous direction of motion of a particle traveling along the curve traced out by $\mathbf{x}(t)$, while the magnitude of $\mathbf{x}'(t)$ determines the instantaneous speed of the particle at time t. For our purposes, the direction of motion is most important because

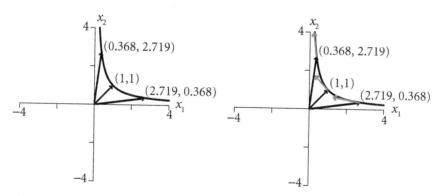

Figure 3.4 At left, the solution curve $\mathbf{x}(t)$ given in (3.4.1). At right, the solution curve $\mathbf{x}(t)$ given in (3.4.1), along with corresponding scaled derivative vectors at times $t = -1$, $t = 0$, and $t = 1$.

this indicates a *flow* along the solution curve as time increases. Thus, rather than plotting the vector $\mathbf{x}'(t)$ at various times, we plot scaled versions of it, each emanating from the tip of $\mathbf{x}(t)$. For example, since

$$\mathbf{x}'(t) = \begin{bmatrix} -e^{-t} \\ e^t \end{bmatrix} \tag{3.4.3}$$

it follows that

$$\mathbf{x}'(-1) \approx \begin{bmatrix} -2.719 \\ 0.368 \end{bmatrix}, \ \mathbf{x}'(0) = \begin{bmatrix} -1 \\ 1 \end{bmatrix}, \ \text{and} \ \mathbf{x}'(1) \approx \begin{bmatrix} -0.368 \\ 2.719 \end{bmatrix} \tag{3.4.4}$$

Plotting scaled versions of each of these vectors emanating from $\mathbf{x}(-1)$, $\mathbf{x}(0)$, and $\mathbf{x}(1)$, respectively, we see the updated image at the right in figure 3.4.

These plots of the derivative vectors and the flow of the solution curve remind us of our earlier work with slope fields for single differential equations. Indeed, since a solution curve such as $\mathbf{x}(t)$ will always be the result of solving some differential equation $\mathbf{x}' = \mathbf{A}\mathbf{x}$, we realize that we have a formula for $\mathbf{x}'$, just as we had a formula for y' in examples like $y' = -2y$. In the example discussed above, we can view $\mathbf{x}(t)$ as being the solution to the system $\mathbf{x}' = \mathbf{A}\mathbf{x}$ where $\mathbf{A}$ is the matrix

$$\mathbf{A} = \begin{bmatrix} -1 & 0 \\ 0 & 1 \end{bmatrix} \tag{3.4.5}$$

so that $\mathbf{x}'(t)$ satisfies the equation

$$\begin{bmatrix} x_1'(t) \\ x_2'(t) \end{bmatrix} = \mathbf{x}'(t) = \mathbf{A}\mathbf{x}(t) = \begin{bmatrix} -x_1(t) \\ x_2(t) \end{bmatrix} \tag{3.4.6}$$

In particular, (3.4.6) indicates how, for any point (x_1, x_2) in the plane, we can easily compute $\mathbf{x}'$ at that point, and hence know the direction of the flow of the solution curve that passes through that point. Using a computer to conduct such computations at points sampled throughout the plane (with each resulting vector scaled to be of equal length), we get a picture of the so-called *direction field* for the system, shown at left in figure 3.5, which is analogous to a direction field for a single differential equation.

If we now superimpose our plot of the solution curve in figure 3.4 in the direction field, now shown on the right in figure 3.5, we see clearly the role that the derivative $\mathbf{x}'$ and the direction field play in determining the graph of the solution $\mathbf{x}$, as well as the typical behavior of a solution as time increases.

The x_1–x_2 plane is usually called the *phase plane*; note that the independent variable t is implicit in the flow, while the behavior of the curve relative to the coordinate axes demonstrates the interrelationship between the components $x_1(t)$ and $x_2(t)$ of the solution $\mathbf{x}(t)$. Sample solution curves, such the one plotted in figure 3.5, are typically called *trajectories*. Each distinct trajectory is a solution to an initial-value problem; the one in figure 3.5 can be viewed as the solution to $\mathbf{x}' = \mathbf{A}\mathbf{x}, \mathbf{x}(0) = [1\ 1]^{\mathrm{T}}$.

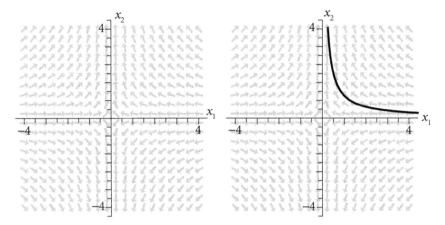

Figure 3.5 At left, the direction field for the system $\mathbf{x}' = \mathbf{A}\mathbf{x}$ given by (3.4.5). At right, the solution to (3.4.5) that is given by (3.4.1).

We will now explore the direction field, phase plane, and trajectories for several examples of 2×2 systems of linear differential equations for which the coefficient matrix has two real linearly independent eigenvectors. An important theme throughout will be the long-range behavior of solutions $\mathbf{x}(t)$ as $t \to \infty$. In addition, we will study the *equilibrium solutions* of each system; a solution $\mathbf{x}(t)$ is an *equilibrium* or *constant solution* if and only if $\mathbf{x}(t)$ is constant for all values of t.

Example 3.4.1   Consider the system of differential equations given by $\mathbf{x}' = \mathbf{A}\mathbf{x}$ where $\mathbf{A} = \begin{bmatrix} 3 & 2 \\ 2 & 3 \end{bmatrix}$. Compute the eigenvalues and eigenvectors of $\mathbf{A}$ and state the general solution to the system. In addition, determine all equilibrium solutions of the system. Finally, plot the direction field for the system, sketch several trajectories, and discuss the long-term behavior of solutions relative to the equilibrium solution(s).

Solution.   The *Maple* command $>$ Eigenvectors (A) produces the output

$$\begin{bmatrix} 5 \\ 1 \end{bmatrix} \begin{bmatrix} 1 & -1 \\ 1 & 1 \end{bmatrix}$$

so that $\mathbf{A}$ has eigenvalues $\lambda_1 = 5$ and $\lambda_2 = 1$, with corresponding eigenvectors $\mathbf{v}_1 = [1\ 1]^{\mathrm{T}}$ and $\mathbf{v}_2 = [-1\ 1]^{\mathrm{T}}$. We therefore know that the general solution to $\mathbf{x}' = \mathbf{A}\mathbf{x}$ is

$$\mathbf{x}(t) = c_1 e^{5t} \begin{bmatrix} 1 \\ 1 \end{bmatrix} + c_2 e^{t} \begin{bmatrix} -1 \\ 1 \end{bmatrix}$$

To find the equilibrium solution(s), we seek all constant vectors $\mathbf{x}$ that satisfy $\mathbf{x}' = \mathbf{A}\mathbf{x}$. In this situation, since $\mathbf{x}$ is constant with respect to t, we know that

$\mathbf{x}' = \mathbf{0}$, so therefore we must solve the system of linear equations given by $\mathbf{Ax} = \mathbf{0}$ where

$$\mathbf{A} = \begin{bmatrix} 3 & 2 \\ 2 & 3 \end{bmatrix}$$

Since $\det(\mathbf{A}) \neq 0$, it follows that $\mathbf{A}$ is an invertible matrix, so the only solution to $\mathbf{Ax} = \mathbf{0}$ is $\mathbf{x} = \mathbf{0}$. Thus the system has the origin as its only equilibrium solution.

At the end of this section, in subsection 3.4.1, we will show how to use *Maple* to plot direction fields for systems. In this and subsequent examples, well simply provide these plots for discussion. In figure 3.6, we see not only the direction field generated by the system, but also the plots of several trajectories, which are natural to sketch (even by hand, once the direction field is provided) by following the *map* that the direction field provides.

Note particularly the straight-line solutions that follow the eigenvectors $\mathbf{v}_1 = [1 \ 1]^T$ and $\mathbf{v}_2 = [-1 \ 1]^T$. Moreover, since both eigenvalues are positive, the respective scalar functions e^{5t} and e^t both increase without bound as $t \to \infty$. This explains why the flow along each straight-line solution is away from the origin. Indeed, every solution besides the zero solution flows away from the equilibrium solution at the origin.

In chapter 2, we considered single autonomous differential equations such as $y' = 2y - 4$. When we found equilibrium solutions to such equations, we also classified their stability based on the behavior exhibited in the direction field. We do likewise with equilibrium solutions for systems. In example 3.4.1,

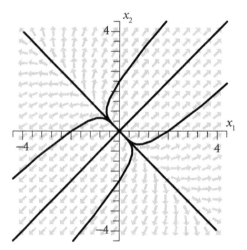

Figure 3.6 The direction field for the system $\mathbf{x}' = \mathbf{Ax}$ of example 3.4.1 along with several trajectories.

we found that $\mathbf{x} = \mathbf{0}$ is the only equilibrium solution of the system, and that every non-constant solution flows away from $\mathbf{0}$. This shows that $\mathbf{0}$ is an *unstable* equilibrium, and in this case we naturally call $\mathbf{0}$ a *repelling node*.

We next explore the behavior of a system where both eigenvalues are negative.

Example 3.4.2   Consider the system of differential equations given by $\mathbf{x}' = \mathbf{A}\mathbf{x}$ where $\mathbf{A} = \begin{bmatrix} -2 & 2 \\ 1 & -3 \end{bmatrix}$. Compute the eigenvalues and eigenvectors of $\mathbf{A}$, and state the general solution to the system. In addition, determine all equilibrium solutions to the system. Finally, plot the direction field for the system, sketch several trajectories, and discuss the long-term behavior of solutions relative to the equilibrium solution(s).

Solution.   Using *Maple*, we find that $\mathbf{A}$ has eigenvalues $\lambda_1 = -1$ and $\lambda_2 = -4$, with corresponding eigenvectors $\mathbf{v}_1 = [2 \ \ 1]^T$ and $\mathbf{v}_2 = [-1 \ \ 1]^T$. The general solution to $\mathbf{x}' = \mathbf{A}\mathbf{x}$ is therefore

$$\mathbf{x}(t) = c_1 e^{-t} \begin{bmatrix} 2 \\ 1 \end{bmatrix} + c_2 e^{-4t} \begin{bmatrix} -1 \\ 1 \end{bmatrix}$$

To find the equilibrium solution, we set $\mathbf{x}' = \mathbf{0}$. Solving the system of linear equations given by $\mathbf{A}\mathbf{x} = \mathbf{0}$, we see that since $\mathbf{A}$ is an invertible matrix, the only solution to $\mathbf{A}\mathbf{x} = \mathbf{0}$ is $\mathbf{x} = \mathbf{0}$, so the system has the origin as its only equilibrium solution.

Plotting the direction field and several trajectories, as shown in figure 3.7, we observe that all solutions flow towards the equilibrium solution at the origin. This makes sense due to the presence of the scalar functions e^{-4t} and e^{-t} in the general solution, as each approaches 0 as $t \to \infty$, and thus it follows that $\mathbf{x}(t) \to \mathbf{0}$ as $t \to \infty$. Moreover, note the two straight-line solutions that show flow along stretches of the two eigenvectors $\mathbf{v}_1 = [2 \ \ 1]^T$ and $\mathbf{v}_2 = [-1 \ \ 1]^T$.

Because every non-constant solution to the system in example 3.4.2 approaches the equilibrium solution at $\mathbf{0}$, we say that the origin is a *stable* equilibrium. Moreover, based on the patterns in the flow, we use the terminology that $\mathbf{0}$ is an *attracting node*.

We study the third case for a 2×2 linear system of differential equations with two real, nonzero eigenvalues in the next example: the eigenvalues have opposing signs.

Example 3.4.3   Let $\mathbf{A} = \begin{bmatrix} 3 & -2 \\ 2 & -2 \end{bmatrix}$ and consider the system of differential equations given by $\mathbf{x}' = \mathbf{A}\mathbf{x}$. Find the general solution of the system, determine all equilibrium solutions to the system, and plot the direction field for the system. Include sketches of several trajectories and discuss the long-term behavior of solutions relative to the equilibrium solution(s).

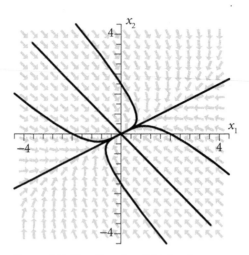

Figure 3.7 The direction field for the system $\mathbf{x}' = \mathbf{A}\mathbf{x}$ in example 3.4.2 along with several trajectories.

Solution. We find that $\mathbf{A}$ has eigenvalues $\lambda_1 = 2$ and $\lambda_2 = -1$, with corresponding eigenvectors $\mathbf{v}_1 = [2 \;\; 1]^{\mathrm{T}}$ and $\mathbf{v}_2 = [1 \;\; 2]^{\mathrm{T}}$. It follows that the general solution to $\mathbf{x}' = \mathbf{A}\mathbf{x}$ is

$$\mathbf{x}(t) = c_1 e^{2t} \begin{bmatrix} 2 \\ 1 \end{bmatrix} + c_2 e^{-t} \begin{bmatrix} 1 \\ 2 \end{bmatrix}$$

Since $\mathbf{A}$ is an invertible matrix, the only solution to $\mathbf{A}\mathbf{x} = \mathbf{0}$ is $\mathbf{x} = \mathbf{0}$, so the origin is only equilibrium solution of the system.

As figure 3.8 shows, the direction field and various trajectories exhibit a different type of behavior around the origin. In particular, solutions that do not lie on either eigenvector appear to initially flow toward the origin, and then turn away and tend toward the straight-line solution associated with the positive eigenvalue. More specifically, it appears that solutions that do not pass through a point on the line in the direction of the eigenvector $[1 \;\; 2]^{\mathrm{T}}$ are eventually attracted to stretches of the eigenvector $[2 \;\; 1]^{\mathrm{T}}$. This is reasonable since in the general solution, e^{-t} will tend to 0 as $t \to \infty$, leaving the function $c_1 e^{2t} [2 \;\; 1]^{\mathrm{T}}$ to dominate.

Since some solutions that pass through points near the origin tend away from the origin as $t \to \infty$, the origin is an unstable equilibrium in example 3.4.3. Moreover, as the trajectories remind us of the contour plot in multivariable calculus of a surface whose graph looks like a saddle, we say in this context as well that the origin is a *saddle point*.

The preceding examples demonstrate the three possible cases for a 2×2 system with real, nonzero eigenvalues: both positive, both negative, or opposites. Our next example investigates the situation when one eigenvalue is zero.

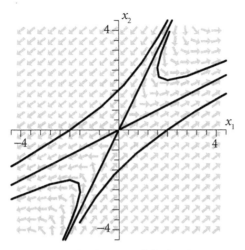

Figure 3.8 The direction field for the system $\mathbf{x}' = \mathbf{A}\mathbf{x}$ of example 3.4.3 along with several trajectories.

Example 3.4.4   For the matrix $\mathbf{A} = \begin{bmatrix} -3 & 1 \\ 3 & -1 \end{bmatrix}$ and the corresponding system of differential equations $\mathbf{x}' = \mathbf{A}\mathbf{x}$, find the general solution of the system and determine all equilibrium solutions. Furthermore, plot the direction field for the system along with sketches of several trajectories; discuss the long-term behavior of solutions relative to the equilibrium solution(s).

Solution.   We first do the standard computations to find that $\mathbf{A}$ has eigenvalues $\lambda_1 = -4$ and $\lambda_2 = 0$, with corresponding eigenvectors $\mathbf{v}_1 = [-1 \ \ 1]^{\mathrm{T}}$ and $\mathbf{v}_2 = [1 \ \ 3]^{\mathrm{T}}$. Thus, the general solution to $\mathbf{x}' = \mathbf{A}\mathbf{x}$ is

$$\mathbf{x}(t) = c_1 e^{-4t} \begin{bmatrix} -1 \\ 1 \end{bmatrix} + c_2 \begin{bmatrix} 1 \\ 3 \end{bmatrix}$$

We immediately notice something different about $\mathbf{x}(t)$. In particular, because the second eigenvalue is 0, the scalar function e^{0t} has no effect on the general solution. Furthermore, with e^{-4t} the only part of $\mathbf{x}(t)$ that changes with t, we can see that for any nonzero constant c_1 and any c_2, the graph of $\mathbf{x}(t)$ is always a straight line where the direction is given by the eigenvector corresponding to the nonzero eigenvalue.

In addition, the presence of a zero eigenvalue has a significant impact on the system's equilibrium solutions. The fact that the columns of $\mathbf{A}$ are scalar multiples of each other leads us to see immediately that $\mathbf{A}$ is not invertible; this can be equivalently deduced from the fact that $\mathbf{A}$ has a zero eigenvalue. The singularity of $\mathbf{A}$ further implies that the homogeneous equation $\mathbf{A}\mathbf{x} = \mathbf{0}$ has infinitely many solutions. In particular, row-reducing the appropriate augmented matrix, we find that

$$\begin{bmatrix} -3 & 1 & 0 \\ 3 & -1 & 0 \end{bmatrix} \rightarrow \begin{bmatrix} 1 & -1/3 & 0 \\ 0 & 0 & 0 \end{bmatrix}$$

This implies that any constant vector $\mathbf{x}$ of the form

$$\mathbf{x} = x_1 \begin{bmatrix} 1 \\ 3 \end{bmatrix}$$

satisfies the equation $\mathbf{x}' = \mathbf{A}\mathbf{x}$, and therefore is an equilibrium solution. Note especially that $\mathbf{x} = x_1 [1 \ 3]^{\mathrm{T}}$ is an eigenvector associated with $\lambda = 0$, and thus every eigenvector associated with the zero eigenvalue is an equilibrium solution to the system.

The interesting behaviors that we have discussed algebraically are seen in figure 3.9. Specifically, every non-constant solution is a straight line solution in the direction of the eigenvector $[-1 \ 1]^{\mathrm{T}}$ that is drawn toward an equilibrium point that lies on the eigenvector $[1 \ 3]^{\mathrm{T}}$ corresponding to the zero eigenvalue.

The flows in figure 3.9, as well as the long-term behavior of the function e^{-4t} in the general solution $\mathbf{x}(t)$, clearly demonstrate that every equilibrium solution to the system is stable. Moreover, we say that each such equilibrium point is an *attracting node.*

There are two important observations to make in closing. One is that we still must address the situations where $\mathbf{A}$ lacks two real linearly independent eigenvectors; we will do so in the next section. In addition, examples 3.4.1–3.4.4

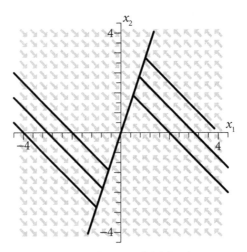

Figure 3.9 The direction field for the system $\mathbf{x}' = \mathbf{A}\mathbf{x}$ of example 3.4.4 along with several trajectories.

indicate that plotting a direction field is perhaps best left to a computer; however, in the case where **A** has two real, linearly independent eigenvectors, it is a straightforward exercise use the eigenvectors to plot these straight-line solutions by hand and to use the signs of the corresponding eigenvalues to understand the flows along the straight line solutions. Then, it is not difficult to imagine the overall appearance of the direction field and sketch several probable trajectories by hand, thus fully understanding the graphical behavior of all solutions to the system.

3.4.1 Plotting direction fields for systems using *Maple*

We again use the DEtools package, and load it with the command

```
> with(DEtools):
```

To plot the direction field associated with a given system of differential equations, we first define the system itself, similar to how we defined a single differential equation in order to plot its slope field. We do this through the following command for the system with coefficient matrix $\mathbf{A} = \begin{bmatrix} 3 & 2 \\ 2 & 3 \end{bmatrix}$ from example 3.4.1.

```
> sys := diff(x(t),t)= 3*x(t)+2*y(t),
diff(y(t),t)= 2*x(t)+3*y(t);
```

The system of differential equations of interest is now stored in "sys". While we typically use $x_1(t)$ and $x_2(t)$ to represent the component functions in our discussion of the theory and solution of systems, in working with *Maple* it is often simpler to use $x(t)$ and $y(t)$. The direction field may now be generated by the command

```
> DEplot([sys], [x(t),y(t)], t=-1..1, x=-4..4,
y=-4..4, arrows=large, color=gray);
```

This command produces the output shown at left in figure 3.10.

From here, it is a straightforward exercise to sketch trajectories by hand. Of course, *Maple* has the capacity to include trajectories that pass through any initial conditions we choose. For example, if we are interested in the various initial conditions $\mathbf{x}(0) = (2, 2), (0, 4), (4, 0)$, and $(-1, 1)$, we can modify the earlier DEplot command to

```
> DEplot([sys], [x(t),y(t)], t=-1.6..3.6, x=-4..4,
y=-4..4, arrows=large, color=gray, [[x(0)=-2,y(0)=0],
[x(0)=0,y(0)=-2], [x(0)=2,y(0)=0], [x(0)=0,y(0)=2],
```

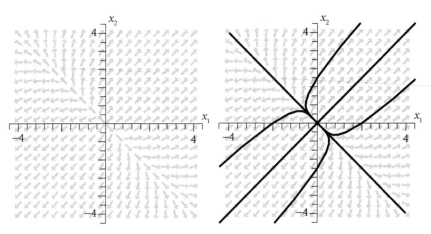

Figure 3.10 At left, the direction field for the system $\mathbf{x}' = \mathbf{A}\mathbf{x}$ of example 3.4.1. At right, the same direction field with several trajectories.

```
[x(0)=0.1,y(0)=0.1], [x(0)=-0.1,y(0)=-0.1],
[x(0)=0.1,y(0)=-0.1], [x(0)=-0.1,y(0)=0.1]]);
```

The results of this most recent DEplot command are shown at right in figure 3.10.

As always, the user can experiment some with the window in which the plot is displayed: the range of x- and y-values can affect how clearly the direction field is revealed, and the range of t-values determines how much of each trajectory is plotted.

Exercises 3.4

1. Consider the system of differential equations $\mathbf{x}' = \mathbf{A}\mathbf{x}$ given by

$$\mathbf{A} = \begin{bmatrix} 2 & -1 \\ 3 & -2 \end{bmatrix}$$

(a) Determine the general solution to the system $\mathbf{x}' = \mathbf{A}\mathbf{x}$.
(b) Classify the stability of all equilibrium solutions to the system.
(c) Sketch all straight-line solutions to the system and hence plot several nonlinear trajectories in the phase plane.

2. Consider the system of differential equations $\mathbf{x}' = \mathbf{A}\mathbf{x}$ given by

$$\mathbf{A} = \begin{bmatrix} 3 & 1 \\ 1 & 3 \end{bmatrix}$$

(a) Determine the general solution to the system $\mathbf{x}' = \mathbf{A}\mathbf{x}$.
(b) Classify the stability of all equilibrium solutions to the system.

(c) Sketch all straight-line solutions to the system and hence plot several nonlinear trajectories in the phase plane.

3. Consider the system of differential equations $\mathbf{x}' = \mathbf{A}\mathbf{x}$ given by

$$\mathbf{A} = \begin{bmatrix} -3 & 2 \\ 2 & -3 \end{bmatrix}$$

(a) Determine the general solution to the system $\mathbf{x}' = \mathbf{A}\mathbf{x}$.
(b) Classify the stability of all equilibrium solutions to the system.
(c) Sketch all straight-line solutions to the system and hence plot several nonlinear trajectories in the phase plane.

4. Consider the system of differential equations $\mathbf{x}' = \mathbf{A}\mathbf{x}$ given by

$$\mathbf{A} = \begin{bmatrix} -2 & 0 \\ 0 & -2 \end{bmatrix}$$

(a) Determine the general solution to the system $\mathbf{x}' = \mathbf{A}\mathbf{x}$.
(b) Classify the stability of all equilibrium solutions to the system.
(c) Sketch the straight-line solutions to the system that correspond to the two linearly independent eigenvectors. Why is every solution to this system also a straight-line solution?

5. Consider the system of differential equations $\mathbf{x}' = \mathbf{A}\mathbf{x}$ given by

$$\mathbf{A} = \begin{bmatrix} -2 & 2 \\ 1 & -1 \end{bmatrix}$$

(a) Determine the general solution to the system $\mathbf{x}' = \mathbf{A}\mathbf{x}$.
(b) Classify the stability of all equilibrium solutions to the system.
(c) Why is every non-constant solution to this system also a straight-line solution? How are these straight-line solutions related to the eigenvectors of the system?

In exercises 6–9, let $\mathbf{x}(t)$ be the stated general solution to some system $\mathbf{x}' = \mathbf{A}\mathbf{x}$. State the straight-line solutions to the system, classify the stability of the origin, and sketch some sample trajectories.

6. $\mathbf{x}(t) = c_1 e^{-2t} \begin{bmatrix} 1 \\ 3 \end{bmatrix} + c_2 e^{-5t} \begin{bmatrix} 3 \\ 1 \end{bmatrix}$

7. $\mathbf{x}(t) = c_1 e^{4t} \begin{bmatrix} -1 \\ 2 \end{bmatrix} + c_2 e^{-3t} \begin{bmatrix} 1 \\ 2 \end{bmatrix}$

8. $\mathbf{x}(t) = c_1 e^{2t} \begin{bmatrix} 2 \\ -1 \end{bmatrix} + c_2 \begin{bmatrix} 1 \\ 1 \end{bmatrix}$

9. $\mathbf{x}(t) = c_1 e^{0.1t} \begin{bmatrix} 1 \\ 1 \end{bmatrix} + c_2 e^{10t} \begin{bmatrix} -1 \\ 1 \end{bmatrix}$

10. For the system $\mathbf{x}' = \mathbf{Ax}$ whose general solution is given in exercise 6, determine a possible matrix $\mathbf{A}$ for the system. (Hint: If $\mathbf{A}$ is a matrix with all real linearly independent eigenvectors and those eigenvectors are the columns of a matrix $\mathbf{P}$, then $\mathbf{A}$ satisfies the equation $\mathbf{AP} = \mathbf{PD}$, where $\mathbf{D}$ is the diagonal matrix whose entries are the eigenvalues of $\mathbf{A}$ in order corresponding to the eigenvectors in the columns of $\mathbf{P}$.)

11. For the system $\mathbf{x}' = \mathbf{Ax}$ whose general solution is given in exercise 7, determine a possible matrix $\mathbf{A}$ for the system.

12. Consider the four systems of equations given by $\mathbf{x}' = \mathbf{Ax}$ where $\mathbf{A}$ is given by the matrices I, II, III, and IV below. Match each system with one of the four direction field plots (a), (b), (c), and (d) given below. Write one sentence for each to explain the reasoning behind your choice.

I. $\mathbf{A} = \begin{bmatrix} 5 & 3 \\ 3 & 5 \end{bmatrix}$ II. $\mathbf{A} = \begin{bmatrix} 2 & -4 \\ -1 & 2 \end{bmatrix}$ III. $\mathbf{A} = \begin{bmatrix} 2 & 7 \\ 7 & 2 \end{bmatrix}$ IV. $\mathbf{A} = \begin{bmatrix} 2 & 3 \\ 3 & -6 \end{bmatrix}$

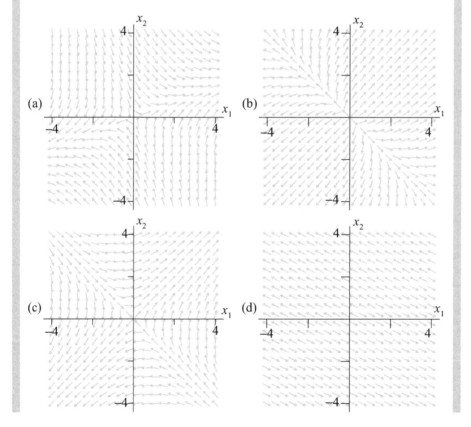

In exercises 13–17, solve the IVP $\mathbf{x}' = \mathbf{A}\mathbf{x}$ with the given matrix $\mathbf{A}$ and stated initial condition.

13. $\mathbf{A} = \begin{bmatrix} 2 & -1 \\ 3 & -2 \end{bmatrix}$, $\quad \mathbf{x}(0) = [1 \ \ 2]$

14. $\mathbf{A} = \begin{bmatrix} 3 & 1 \\ 1 & 3 \end{bmatrix}$, $\quad \mathbf{x}(0) = [-3 \ \ 1]^{\mathrm{T}}$

15. $\mathbf{A} = \begin{bmatrix} -3 & 2 \\ 2 & -3 \end{bmatrix}$, $\quad \mathbf{x}(0) = [1 \ \ -2]^{\mathrm{T}}$

16. $\mathbf{A} = \begin{bmatrix} -2 & 0 \\ 0 & -2 \end{bmatrix}$, $\quad \mathbf{x}(0) = [-2 \ \ -2]^{\mathrm{T}}$

17. $\mathbf{A} = \begin{bmatrix} -2 & 2 \\ 1 & -1 \end{bmatrix}$, $\quad \mathbf{x}(0) = [1 \ \ 4]^{\mathrm{T}}$

In exercises 18–22, use the standard substitution to convert the given second-order differential equation to a system of two linear first-order equations. Solve the system to hence determine the solution y to the second-order equation.

18. $y'' - y' - 6y = 0$

19. $y'' - 6y' + 5y = 0$

20. $y'' + 4y' = 0$

21. $y'' + 3y' + 2y = 0$

22. $y'' + y = 0$

3.5 When a matrix lacks two real linearly independent eigenvectors

We have seen repeatedly, both in theory and in specific examples, that when a 2×2 matrix $\mathbf{A}$ has two real linearly independent eigenvectors, we can determine the general solution to $\mathbf{x}' = \mathbf{A}\mathbf{x}$ and its graphical behavior. In this section, we address two remaining cases: when $\mathbf{A}$ has a repeated eigenvalue and only one associated real linearly independent eigenvector, and when $\mathbf{A}$ has complex eigenvalues and eigenvectors. In each case, we work through preliminary examples to discover general patterns and principles, expand these principles with appropriate theorems, and explore and discuss graphical behavior along the way.

Example 3.5.1   Consider the system of differential equations given by $\mathbf{x}' = \mathbf{A}\mathbf{x}$ where $\mathbf{A} = \begin{bmatrix} -2 & 1 \\ 0 & -2 \end{bmatrix}$. Compute the eigenvalues and eigenvectors of $\mathbf{A}$ and explain why this alone does not lead to the general solution of the system. By noting that the system is partially coupled, solve the system and determine a second real, linearly independent solution. Finally, state the general solution.

Solution.   By inspection, since $\mathbf{A}$ is a triangular matrix, we see that $\lambda = -2$ is a repeated eigenvalue of $\mathbf{A}$ with multiplicity 2. From this, we deduce that $\mathbf{v}_1 = [1 \quad 0]^{\mathrm{T}}$ is a corresponding eigenvector, and therefore one solution to $\mathbf{x}' = \mathbf{A}\mathbf{x}$ is $\mathbf{x}_1 = c_1 e^{-2t}[1 \quad 0]^{\mathrm{T}}$. However, $\mathbf{A}$ lacks a second linearly independent eigenvector associated with $\lambda = -2$; therefore, we need to find a second real linearly independent solution to the system in order to determine the general solution to $\mathbf{x}' = \mathbf{A}\mathbf{x}$. In this example, we are fortunate that the system is only partially coupled and that therefore we may solve the system directly by using techniques for single differential equations from chapter 2.

In particular, noting that the second equation in the system is $x_2' = -2x_2$, it follows immediately that the solution to this single differential equation is $x_2(t) = ce^{-2t}$. Substituting this result into the equation $x_1' = -2x_1 + x_2$, it remains for us to solve the single nonhomogeneous linear first-order differential equation

$$x_1' = -2x_1 + ce^{-2t}$$

Applying our understanding of such equations from section 2.3, via the integrating factor $v(t) = e^{2t}$ we know that

$$x_1(t) = \frac{1}{e^{2t}} \int e^{2t} \cdot ce^{-2t} \, dt = e^{-2t}(ct + k)$$

To summarize, with $x_1(t)$ and $x_2(t)$ as the components of $\mathbf{x}(t)$, we have found that a solution to the system is

$$\mathbf{x}(t) = \begin{bmatrix} x_1(t) \\ x_2(t) \end{bmatrix}$$

$$= \begin{bmatrix} e^{-2t}(ct + k) \\ ce^{-2t} \end{bmatrix} \tag{3.5.1}$$

If we factor this expression to write $\mathbf{x}(t)$ as a linear combination of two vectors in order to more clearly identify the role of the constants in (3.5.1), we see

$$\mathbf{x}(t) = k \begin{bmatrix} e^{-2t} \\ 0 \end{bmatrix} + c \begin{bmatrix} te^{-2t} \\ e^{-2t} \end{bmatrix} \tag{3.5.2}$$

In this form, two key observations can be made. First, each individual vector in (3.5.2) may be verified to be a solution to the given system. Moreover, these two vectors are linearly independent. Hence, (3.5.2) is the general solution to the given system.

While it is good that we were able to solve the system in example 3.5.1, it is still unclear how we will proceed in similar circumstances when neither equation in the system may be solved by techniques for single first-order equations. That is, if the equation for x_1' involves x_2 and the equation for x_2' involves x_1, but the system's matrix has only one linearly independent eigenvector, we cannot employ the approach used in example 3.5.1. However, the general form of the solution (3.5.2) can help us guess an appropriate form of the needed second linearly independent solution in the more general case.

Recall that we know that whenever $(\lambda, \mathbf{v})$ is a real eigenpair of $\mathbf{A}$, the function $\mathbf{x}(t) = e^{\lambda t}\mathbf{v}$ is a solution to $\mathbf{x}' = \mathbf{A}\mathbf{x}$, and moreover $\mathbf{x}(t)$ is a straight-line solution to the system. In example 3.5.1, we found that for the given matrix, which had a repeated eigenvalue and only one associated linearly independent eigenvector, the scalar function $te^{\lambda t}$ arose in the solution. If we recall that our original work with $e^{\lambda t}\mathbf{v}$ arose from guessing that a function of the form $f(t)\mathbf{v}$ was a solution to $\mathbf{x}' = \mathbf{A}\mathbf{x}$, example 3.5.1 now suggests that in the case where we are missing an eigenvector, we consider a vector function that somehow involves the scalar function $te^{\lambda t}$ as a second linearly independent solution to $\mathbf{x}' = \mathbf{A}\mathbf{x}$. A closer look at (3.5.2) suggests the form of this second solution we seek.

In particular, recalling that the matrix $\mathbf{A}$ in example 3.5.1 had $\mathbf{v}_1 = [1 \ 0]^{\mathrm{T}}$ as the eigenvector corresponding to $\lambda = -2$, rewriting (3.5.2) reveals the role $\mathbf{v}_1$ plays in the general solution. Specifically,

$$\mathbf{x}(t) = ke^{-2t}\begin{bmatrix} 1 \\ 0 \end{bmatrix} + cte^{-2t}\begin{bmatrix} 1 \\ 0 \end{bmatrix} + ce^{-2t}\begin{bmatrix} 0 \\ 1 \end{bmatrix} \tag{3.5.3}$$

and since $\mathbf{x}_1(t) = e^{-2t}[1 \ 0]^{\mathrm{T}}$ is the standard solution that arises through the eigenpair, we see from (3.5.3) that the second linearly independent solution

$$\mathbf{x}_2(t) = te^{-2t}\begin{bmatrix} 1 \\ 0 \end{bmatrix} + e^{-2t}\begin{bmatrix} 0 \\ 1 \end{bmatrix}$$

has the form $te^{-2t}\mathbf{v} + e^{-2t}\mathbf{u}$, where $\mathbf{u}$ is not an eigenvector of $\mathbf{A}$ corresponding to $\lambda = -2$. This suggests a form for the second solution when this case arises in general.

We now consider this situation for an arbitrary matrix with the appropriate properties. Let $\mathbf{A}$ be a 2×2 matrix with a single real, repeated eigenvalue λ with only one linearly independent eigenvector $\mathbf{v}$. Note specifically that we know $\mathbf{A}\mathbf{v} = \lambda\mathbf{v}$ and $\mathbf{x}_1(t) = e^{\lambda t}\mathbf{v}$ is a solution to $\mathbf{x}' = \mathbf{A}\mathbf{x}$. Now consider a second function

$$\mathbf{x}_2(t) = te^{\lambda t}\mathbf{v} + e^{\lambda t}\mathbf{u} \tag{3.5.4}$$

where $\mathbf{u}$ is an unknown constant vector and $(\lambda, \mathbf{v})$ remains an eigenpair of $\mathbf{A}$. We seek conditions on $\mathbf{u}$ that will make $\mathbf{x}_2(t)$ a solution to $\mathbf{x}' = \mathbf{A}\mathbf{x}$; as we have previously encountered in several instances, direct substitution into the differential equation reveals the constraints on $\mathbf{u}$.

First, differentiating (3.5.4) gives

$$\mathbf{x}_2'(t) = (\lambda te^{\lambda t} + e^{\lambda t})\mathbf{v} + \lambda e^{\lambda t}\mathbf{u} \tag{3.5.5}$$

Next, observe that multiplying $\mathbf{x}_2(t)$ by $\mathbf{A}$ yields

$$\mathbf{A}\mathbf{x}_2(t) = \mathbf{A}(te^{\lambda t}\mathbf{v} + e^{\lambda t}\mathbf{u}) = te^{\lambda t}(\mathbf{A}\mathbf{v}) + e^{\lambda t}(\mathbf{A}\mathbf{u}) \tag{3.5.6}$$

In order for $x_2(t)$ to be a solution to $x' = Ax$, it follows from (3.5.5) and (3.5.6) that we require the equality

$$(\lambda t e^{\lambda t} + e^{\lambda t})v + \lambda e^{\lambda t}u = t e^{\lambda t}(Av) + e^{\lambda t}(Au) \qquad (3.5.7)$$

to hold. Using the fact that $Av = \lambda v$ and expanding, we find

$$\lambda t e^{\lambda t}v + e^{\lambda t}v + \lambda e^{\lambda t}u = \lambda t e^{\lambda t}v + e^{\lambda t}(Au) \qquad (3.5.8)$$

With $\lambda t e^{\lambda t}v$ present on both sides of (3.5.8), we can simplify the equality to

$$e^{\lambda t}v + \lambda e^{\lambda t}u = e^{\lambda t}(Au) \qquad (3.5.9)$$

Since $e^{\lambda t}$ is never zero, we observe from (3.5.9) that u must satisfy the equation

$$v + \lambda u = Au \qquad (3.5.10)$$

In other words, $(A - \lambda I)u = v$, where (as we assumed earlier) v is an eigenvector of A that corresponds to the eigenvalue λ. In particular, note that v satisfies the equation $(A - \lambda I)v = 0$. We summarize our work above in the following theorem.

Theorem 3.5.1   If A is a 2×2 matrix with repeated eigenvalue λ and only one corresponding linearly independent eigenvector v, then the general solution to $x' = Ax$ is given by

$$x(t) = c_1 e^{\lambda t}v + c_2 e^{\lambda t}(tv + u)$$

where u satisfies the equation $(A - \lambda I)u = v$.

The vector u is often called a *generalized eigenvector* of A corresponding to λ. We now demonstrate the role of theorem 3.5.1 in the following example.

Example 3.5.2   Let $A = \begin{bmatrix} 1 & 4 \\ -1 & 5 \end{bmatrix}$ and consider the system of differential equations given by $x' = Ax$. Find the general solution of the system, determine all equilibrium solutions to the system, and plot the direction field for the system. Include sketches of several trajectories and discuss the long-term behavior of solutions relative to the equilibrium solution(s).

Solution.   We find that A has a single repeated eigenvalue $\lambda = 3$ with just one corresponding linearly independent eigenvector $v = [2\ 1]^T$. Thus, one linearly independent solution to $x' = Ax$ is $x_1(t) = e^{3t}v$. Applying theorem 3.5.1, we determine a second linearly independent solution to the system. Specifically, we first solve the vector equation $(A - 3I)u = v$. To do so, we row-reduce the appropriate augmented matrix and find

$$\begin{bmatrix} -2 & 4 & 2 \\ -1 & 2 & 1 \end{bmatrix} \rightarrow \begin{bmatrix} 1 & -2 & -1 \\ 0 & 0 & 0 \end{bmatrix}$$

It follows that the vector u must have components u_1 and u_2 that satisfy the equation $u_1 = 2u_2 - 1$, where u_2 is a free variable. Since we only need one

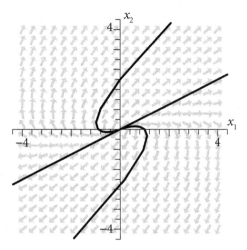

Figure 3.11 The direction field for the system $\mathbf{x}' = \mathbf{A}\mathbf{x}$ of example 3.5.2 along with several trajectories.

such vector $\mathbf{u}$, we choose $u_2 = 0$ and thus $u_1 = -1$. From theorem 3.5.1, it now follows that a second linearly independent solution to $\mathbf{x}' = \mathbf{A}\mathbf{x}$ is given by the function $\mathbf{x}_2(t) = e^{3t}(t\mathbf{v} + \mathbf{u})$. In particular, the general solution to $\mathbf{x}' = \mathbf{A}\mathbf{x}$ is

$$\mathbf{x}(t) = c_1 e^{3t} \begin{bmatrix} 2 \\ 1 \end{bmatrix} + c_2 e^{3t} \left(t \begin{bmatrix} 2 \\ 1 \end{bmatrix} + \begin{bmatrix} -1 \\ 0 \end{bmatrix} \right)$$

We note further that since $\mathbf{A}$ is an invertible matrix, the only solution to $\mathbf{A}\mathbf{x} = \mathbf{0}$ is $\mathbf{x} = \mathbf{0}$, so the origin is the only equilibrium solution of the system.

As figure 3.11 shows, the direction field and several trajectories exhibit behavior consistent with the fact that the system has just one straight-line solution, the one that corresponds to the single linearly independent eigenvector of $\mathbf{A}$. Note as well that since the system's only eigenvalue is positive, every non-constant solution flows away from the origin as $t \to \infty$.

In example 3.4.3, the origin is obviously an unstable equilibrium solution. Because there is only one linearly independent eigenvector for the system, we call the origin a *degenerate node*, and in this case where $\lambda = 3 > 0$ and all the trajectories flow away from the origin, this degenerate node is also called a *repelling node*.

We now consider an example that reveals the other possible situation that can arise when a matrix $\mathbf{A}$ lacks two real linearly independent eigenvectors: when $\mathbf{A}$ has no real eigenvalues and no real eigenvectors.

Example 3.5.3   Consider the system $\mathbf{x}' = \mathbf{A}\mathbf{x}$ given by the matrix

$$\mathbf{A} = \begin{bmatrix} 0 & -1 \\ 1 & 0 \end{bmatrix}$$

Compute the eigenvalues and eigenvectors of $\mathbf{A}$ and explain why this does not lead directly to the general solution of the system. In addition, plot the direction field for the system to confirm these observations from a graphical perspective. Using familiarity with solutions to single differential equations and the form of the equations for the given system, determine the general solution to the system.

Solution.   The eigenvalues of the matrix $\mathbf{A}$ are computed using the characteristic equation

$$\det(\mathbf{A} - \lambda\mathbf{I}) = \det\begin{bmatrix} -\lambda & -1 \\ 1 & -\lambda \end{bmatrix} = \lambda^2 + 1 = 0$$

We see that $\lambda^2 = -1$, so that $\lambda = \pm i$, where i is the complex number[2] $i = \sqrt{-1}$.

To determine the eigenvector associated with $\lambda = i$, we solve $(\mathbf{A} - i\mathbf{I})\mathbf{v} = \mathbf{0}$. Row-reducing the appropriate matrix with complex entries just as we would a matrix with real entries, we observe

$$\begin{bmatrix} -i & -1 & 0 \\ 1 & -i & 0 \end{bmatrix} \to \begin{bmatrix} 1 & -i & 0 \\ -i & -1 & 0 \end{bmatrix} \to \begin{bmatrix} 1 & -i & 0 \\ 0 & 0 & 0 \end{bmatrix}$$

where the first step was achieved by swapping the two rows, while the last step was achieved by computing the row replacement $iR_1 + R_2 \to R_2$. It follows that any eigenvector $\mathbf{v}$ associated with $\lambda = i$ must have components v_1 and v_2 that satisfy $v_1 = iv_2$. Choosing $v_2 = 1$, we see that an eigenvector $\mathbf{v}$ corresponding to $\lambda = i$ is $\mathbf{v} = [i\ 1]^{\mathrm{T}}$. Similar computations with $\lambda = -i$ show that a corresponding eigenvector is $\mathbf{v} = [-i\ 1]^{\mathrm{T}}$. While we might suggest at this point that

$$\mathbf{x}(t) = e^{it}\begin{bmatrix} i \\ 1 \end{bmatrix}$$

is a solution to $\mathbf{x}' = \mathbf{A}\mathbf{x}$, such a solution involves the complex number i, and is not a real solution to the system. A plot of the direction field for the system reveals further why no real solutions arise directly from the eigenvectors. In particular, if we examine figure 3.12, the direction field and various trajectories exhibit behavior consistent with the fact that the system has no straight-line solutions due to the fact that it has no real eigenpairs: every trajectory appears to be circular.

In this example, we will suspend our work with eigenvalues and eigenvectors and see whether we can determine a solution to the system more directly. If we examine the two equations given in the system $\mathbf{x}' = \mathbf{A}\mathbf{x}$, we observe that we

---

[2] A review of key concepts with complex numbers may be found in appendix B.

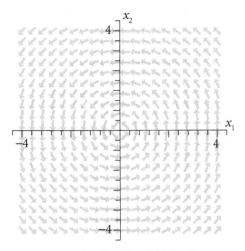

Figure 3.12 The direction field for the system
$\mathbf{x}' = \mathbf{A}\mathbf{x}$ of example 3.5.3.

are trying to solve the two equations $x_1' = -x_2$ and $x_2' = x_1$ simultaneously. In particular, we seek two functions $x_1(t)$ and $x_2(t)$ such that the derivative of the first is the opposite of the second and the derivative of the second is the first. This is a familiar scenario encountered in calculus and we recognize that $x_1(t) = \cos t$ and $x_2(t) = \sin t$ form a pair of such functions. Further consideration reveals that the choices $x_1(t) = -\sin t$ and $x_2(t) = \cos t$ also satisfy the system.

Our recent observations show that the vector functions

$$\mathbf{x}_1(t) = \begin{bmatrix} \cos t \\ \sin t \end{bmatrix} \text{ and } \mathbf{x}_2(t) = \begin{bmatrix} -\sin t \\ \cos t \end{bmatrix}$$

each form a real solution to $\mathbf{x}' = \mathbf{A}\mathbf{x}$; moreover, it is clear that $\mathbf{x}_1(t)$ and $\mathbf{x}_2(t)$ are not scalar multiples of one another, and thus these are two linearly independent solutions to the system. Therefore, theorem 3.3.2 implies that the general solution to the given system is

$$\mathbf{x}(t) = c_1 \begin{bmatrix} \cos t \\ \sin t \end{bmatrix} + c_2 \begin{bmatrix} -\sin t \\ \cos t \end{bmatrix} \tag{3.5.11}$$

The presence of the sine and cosine functions in the entries of $\mathbf{x}$ will also lead to the circular trajectories we expect from the direction field in figure 3.12.

Example 3.5.3 shows several new phenomena. In every preceding example we have considered for 2×2 systems $\mathbf{x}' = \mathbf{A}\mathbf{x}$, eigenpairs have directly provided at least one real solution to the system. But for the latest system we examined, the eigenpairs appeared to not produce any solutions to the system at all. Moreover, for the first time in our work with linear systems, the sine and cosine functions arose. An important question to consider at this point is whether the complex eigenpair

$$\lambda = i, \ \mathbf{v} = \begin{bmatrix} i \\ 1 \end{bmatrix} \tag{3.5.12}$$

can be linked to the general solution that we found in (3.5.11). It turns out that the key idea lies in understanding how the exponential function e^z behaves when the input z is a complex number.

The great Swiss mathematician Leonhard Euler (1707–1783) is credited with discovering *Euler's formula*, which states that for any real number t,

$$e^{it} = \cos t + i \sin t \tag{3.5.13}$$

In exercise 14 in this section, one way to derive Euler's formula through Taylor series for the exponential and trigonometric functions is explored. For now, we will simply accept (3.5.13) and put it to use.

Using the first complex eigenpair found in example 3.5.3, let us consider the standard form of a potential solution to $\mathbf{x}' = \mathbf{A}\mathbf{x}$, $\mathbf{x}(t) = e^{\lambda t}\mathbf{v}$, using the eigenpair identified in (3.5.12). Here, since the solution we are considering is in fact complex, we will use the notation $\mathbf{z}(t)$. Using Euler's formula and complex arithmetic, observe that

$$\mathbf{z}(t) = e^{it} \begin{bmatrix} i \\ 1 \end{bmatrix}$$

$$= (\cos t + i \sin t) \begin{bmatrix} i \\ 1 \end{bmatrix}$$

$$= \begin{bmatrix} i \cos t - \sin t \\ \cos t + i \sin t \end{bmatrix} \tag{3.5.14}$$

When working with complex numbers, it is often useful to identify the *real* and *imaginary* parts of the numbers. That is, for a complex number $z = a + ib$ where a and b are real, we call a the real part of z, and b the imaginary part of z. The same distinctions hold for vectors with complex entries. Considering (3.5.14), if we separate this vector into its real and imaginary parts, we may write

$$\mathbf{z}(t) = \begin{bmatrix} -\sin t \\ \cos t \end{bmatrix} + i \begin{bmatrix} \cos t \\ \sin t \end{bmatrix} \tag{3.5.15}$$

If we now compare the general solution to $\mathbf{x}' = \mathbf{A}\mathbf{x}$ that we found in (3.5.11) to (3.5.15) above, we can make a critical observation. The two linearly independent solutions to the system seen in (3.5.11) are in fact the real and complex parts of the vector $\mathbf{z}(t)$ which arose from considering $\mathbf{z}(t) = e^{\lambda t}\mathbf{v}$ where $(\lambda, \mathbf{v})$ was a complex eigenpair of $\mathbf{A}$. That this fact holds in general is our next stated theorem.

Theorem 3.5.2   If $\mathbf{A}$ is a real 2×2 matrix with a complex eigenvalue $\lambda = a + ib$ and corresponding eigenvector $\mathbf{v} = \mathbf{p} + i\mathbf{q}$, where a, b, $\mathbf{p}$, and $\mathbf{q}$ are real, then the real and imaginary parts of

$$\mathbf{z}(t) = e^{(a+bi)t}(\mathbf{p} + i\mathbf{q})$$

are real linearly independent solutions to $\mathbf{x}' = \mathbf{Ax}$.

We proceed to apply this result in another example involving complex eigenvalues and eigenvectors.

Example 3.5.4   Let $\mathbf{A} = \begin{bmatrix} -1 & -2 \\ 2 & -1 \end{bmatrix}$ and consider the system of differential equations given by $\mathbf{x}' = \mathbf{Ax}$. Find the general solution of the system, determine all equilibrium solutions to the system, and plot the direction field for the system. Include sketches of several trajectories and discuss the long-term behavior of solutions relative to the equilibrium solution(s).

Solution.   For matrices with complex eigenvalues, *Maple* provides an efficient and valuable approach: the program completes the necessary complex arithmetic automatically and produces the results we need. Doing so, we find that $\mathbf{A}$ has complex eigenvalues $\lambda = -1 \pm 2i$ with corresponding complex eigenvectors $\mathbf{v} = [\pm i \ \ 1]^{\mathrm{T}}$. We choose one of these complex eigenpairs and consider the complex function

$$\mathbf{z}(t) = e^{(-1+2i)t} \begin{bmatrix} i \\ 1 \end{bmatrix}$$

Observe that $e^{(-1+2i)t} = e^{-t}e^{2ti}$, so by Euler's formula

$$e^{(-1+2i)t} = e^{-t}(\cos 2t + i \sin 2t)$$

Substituting this fact into $\mathbf{z}(t)$, we observe that

$$\mathbf{z}(t) = e^{-t}(\cos 2t + i\sin 2t) \begin{bmatrix} i \\ 1 \end{bmatrix}$$

$$= e^{-t} \begin{bmatrix} -\sin 2t + i\cos 2t \\ \cos 2t + i\sin 2t \end{bmatrix}$$

$$= e^{-t} \begin{bmatrix} -\sin 2t \\ \cos 2t \end{bmatrix} + ie^{-t} \begin{bmatrix} \cos 2t \\ \sin 2t \end{bmatrix}$$

By theorem 3.5.2, it now follows that the real and imaginary parts of $\mathbf{z}(t)$ form two real linearly independent solutions to $\mathbf{x}' = \mathbf{Ax}$, and therefore the general solution to $\mathbf{x}' = \mathbf{Ax}$ is

$$\mathbf{x}(t) = c_1 e^{-t} \begin{bmatrix} -\sin 2t \\ \cos 2t \end{bmatrix} + c_2 e^{-t} \begin{bmatrix} \cos 2t \\ \sin 2t \end{bmatrix} \tag{3.5.16}$$

Since $\mathbf{A}$ is an invertible matrix, the origin is the only equilibrium solution of the system. Finally, as figure 3.13 shows, the direction field and plotted trajectories exhibit behavior consistent with the fact that the system has no

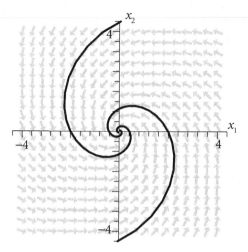

Figure 3.13 The direction field for the system $x' = Ax$ of example 3.5.4 along with several trajectories.

real eigenvectors and therefore no straight-line solutions. Moreover, since the real part of $\lambda = -1 + 2i$ is negative, the role of e^{-t} in the general solution (3.5.16) draws every solution to **0** and thus the origin is a stable equilibrium.

In cases such as the one in example 3.5.4 where there are no straight-line solutions and every nonconstant solution tends to **0** as $t \to \infty$, we naturally say that **0** is a *spiral sink*. Note that this case corresponds to the situation where the real part of a complex eigenvalue is negative. If the real part a of $\lambda = a + bi$ is positive, then we will have e^{at} present in the general solution, and this will drive every solution away from the origin. We therefore call **0** a *spiral source* and note that this equilibrium solution is unstable. Finally, in the event that $a = 0$ in the complex eigenvalue $\lambda = a + bi$, as it was in example 3.5.3, then all nonconstant solutions will orbit the origin while neither being drawn toward or repelled from the equilibrium solution. See, for example, figure 3.12. Such an equilibrium is called a *center* and is considered stable.

In our discussions in this section we have addressed the two possible cases for a 2×2 matrix **A** which lacks two linearly independent eigenvectors. Our work extends naturally to the case of more general $n \times n$ systems where the $n \times n$ matrix **A** may or may not have n real linearly independent eigenvectors. Of course, in the case where **A** has a full set of n real linearly independent eigenvectors, the eigenpairs allow the general solution to the system to be determined. In cases where some of the eigenvalues are complex, or repeated with missing eigenvectors, we can work with each individual eigenvalue to build real linearly independent solutions in ways similar to our preceding work. Some examples are explored in the exercises that follow.

Table 3.1
The stability of the origin as determined by the eigenvalues of a
2×2 matrix **A**

$0 < \lambda_1 \leq \lambda_2$	**0** is unstable and called a repelling node
$\lambda_1 < 0 < \lambda_2$	**0** is unstable and called a saddle
$\lambda_1 \leq \lambda_2 < 0$	**0** is stable and called an attracting node
$\lambda = a \pm bi$ and $a > 0$	**0** is unstable and called a spiral source
$\lambda = a \pm bi$ and $a = 0$	**0** is stable and called a center
$\lambda = a \pm bi$ and $a < 0$	**0** is stable and called a spiral sink

We close this section with a summary in table 3.1 of the stability of the origin as an equilibrium point of $\mathbf{x}' = \mathbf{A}\mathbf{x}$ in the cases where both eigenvalues are nonzero.

Exercises 3.5   For each of exercises 1–7, the general solution $\mathbf{x}(t)$ to a homogeneous linear 2×2 system of differential equations $\mathbf{x}' = \mathbf{A}\mathbf{x}$ is given. For each problem, sketch any straight-line solutions, classify the stability of the equilibrium solution $\mathbf{x} = \mathbf{0}$, and sketch a few trajectories that are not straight lines. Do not use a computer.

1. $\mathbf{x}(t) = c_1 e^{-2t} \begin{bmatrix} -1 \\ 2 \end{bmatrix} + c_2 e^{-3t} \begin{bmatrix} 1 \\ 2 \end{bmatrix}$

2. $\mathbf{x}(t) = c_1 e^{-2t} \begin{bmatrix} \cos t \\ \sin t \end{bmatrix} + c_2 e^{-2t} \begin{bmatrix} -\sin t \\ \cos t \end{bmatrix}$

3. $\mathbf{x}(t) = c_1 e^{2t} \begin{bmatrix} -1 \\ 1 \end{bmatrix} + c_2 e^{-t} \begin{bmatrix} 1 \\ 1 \end{bmatrix}$

4. $\mathbf{x}(t) = c_1 e^{-2t} \begin{bmatrix} -1 \\ 1 \end{bmatrix} + c_2 \begin{bmatrix} 1 \\ 1 \end{bmatrix}$

5. $\mathbf{x}(t) = c_1 \begin{bmatrix} 2\cos t \\ \sin t \end{bmatrix} + c_2 \begin{bmatrix} -\sin t \\ 2\cos t \end{bmatrix}$

6. $\mathbf{x}(t) = c_1 e^{t} \begin{bmatrix} 2\cos t \\ \sin t \end{bmatrix} + c_2 e^{3t} \begin{bmatrix} -\sin t \\ 2\cos t \end{bmatrix}$

7. $\mathbf{x}(t) = c_1 e^{2t} \begin{bmatrix} 4 \\ 1 \end{bmatrix} + c_2 e^{t} \begin{bmatrix} 1 \\ 4 \end{bmatrix}$

For each of exercises 8–13, the characteristic polynomial $p(\lambda)$ of a matrix **A** is given. That is, the zeros of the given polynomial are the eigenvalues of the matrix **A**. For each, classify the stability of the origin as an equilibrium point of the system given by $\mathbf{x}' = \mathbf{A}\mathbf{x}$.

8. $p(\lambda) = \lambda^2 - 4$

9. $p(\lambda) = \lambda^2 + 4$

10. $p(\lambda) = \lambda^2 + \lambda + 1$

11. $p(\lambda) = \lambda^2 - 10\lambda + 9$

12. $p(\lambda) = \lambda^2 - 2\lambda + 5$

13. $p(\lambda) = \lambda^2 + 3\lambda + 2$

14. Recall or look up the formulas for the Taylor series about $a = 0$ for each of the functions e^x, $\sin x$, and $\cos x$. Assuming that the Taylor series for e^x is valid for complex numbers x, compute e^{ib} and compare the result to the expansions for $\cos b$ and $i \sin b$ to show that

$$e^{ib} = \cos b + i \sin b$$

In addition, show that

$$e^{a+ib} = e^a(\cos b + i \sin b)$$

In exercises 15–19, a matrix **A** is given. For each, consider the system of differential equations $\mathbf{x}' = \mathbf{A}\mathbf{x}$ and respond to (a) - (d).

(a) Determine the general solution to the system $\mathbf{x}' = \mathbf{A}\mathbf{x}$.

(b) Classify the stability of all equilibrium solutions to the system.

(c) How many straight-line solutions does this system of equations have? Why?

(d) Use a computer algebra system to plot the direction field for this system and sketch several trajectories by hand.

15. $\mathbf{A} = \begin{bmatrix} 0 & -2 \\ 2 & 0 \end{bmatrix}$

16. $\mathbf{A} = \begin{bmatrix} 2 & -3 \\ 3 & 2 \end{bmatrix}$

17. $\mathbf{A} = \begin{bmatrix} -2 & 1 \\ 0 & -2 \end{bmatrix}$

18. $\mathbf{A} = \begin{bmatrix} -4 & 5 \\ -5 & 4 \end{bmatrix}$

19. $\mathbf{A} = \begin{bmatrix} 7 & -1 \\ 4 & 11 \end{bmatrix}$

In exercises 20–24, solve the IVP given by $\mathbf{x}' = \mathbf{A}\mathbf{x}$ and the stated initial condition.

20. $\mathbf{A} = \begin{bmatrix} 0 & -2 \\ 2 & 0 \end{bmatrix}$, $\mathbf{x}(0) = [1\ 3]^{\mathrm{T}}$

21. $\mathbf{A} = \begin{bmatrix} 2 & -3 \\ 3 & 2 \end{bmatrix}$, $\mathbf{x}(0) = [-3\ 1]^{\mathrm{T}}$

22. $\mathbf{A} = \begin{bmatrix} -2 & 1 \\ 0 & -2 \end{bmatrix}$, $\mathbf{x}(0) = [2\ -2]^{\mathrm{T}}$

23. $\mathbf{A} = \begin{bmatrix} -4 & 5 \\ -5 & 4 \end{bmatrix}$, $\mathbf{x}(0) = [-2\ -3]^{\mathrm{T}}$

24. $\mathbf{A} = \begin{bmatrix} 7 & -1 \\ 4 & 11 \end{bmatrix}$, $\mathbf{x}(0) = [0\ 5]^{\mathrm{T}}$

25. Consider the system of differential equations $\mathbf{x}' = \mathbf{A}\mathbf{x}$ given by

$$\mathbf{A} = \begin{bmatrix} 3 & 1 & -1 \\ 1 & 3 & 1 \\ -1 & 1 & 3 \end{bmatrix}$$

   (a) Determine the general solution to the system $\mathbf{x}' = \mathbf{A}\mathbf{x}$.
   (b) Classify the stability of all equilibrium solutions to the system.
   (c) How many straight-line solutions does this system of equations have? Why?

26. Repeat exercise 25 using the matrix

$$\mathbf{A} = \begin{bmatrix} 0 & 3/2 & -1/2 \\ -1 & -3/2 & 3/2 \\ -1 & 1/2 & -1/2 \end{bmatrix}$$

27. Explain why every 3×3 homogeneous linear system of differential equations of the form $\mathbf{x}' = \mathbf{A}\mathbf{x}$ must always have at least one straight-line solution. Must every 4×4 system have at least one straight-line solution? Explain. What can you say about any $n \times n$ homogeneous linear system?

In exercises 28–32, use the standard substitution to convert the given second-order differential equation to a system of two linear first-order equations. Solve the system to hence determine the solution y to the second-order equation.

28. $y'' + y' - 6y = 0$

29. $y'' + 2y' + 5y = 0$

30. $y'' + 4y = 0$

31. $y'' + 3y' - 28y = 0$

32. $y'' + y + 1 = 0$

3.6 Nonhomogeneous systems: undetermined coefficients

So far in our studies of systems of linear differential equations, we have focused almost exclusively on the case where the system is homogeneous and can be represented in the form $\mathbf{x}' = \mathbf{A}\mathbf{x}$. We now begin to investigate nonhomogeneous systems, which are systems of the form $\mathbf{x}' = \mathbf{A}\mathbf{x} + \mathbf{b}$ where $\mathbf{b} \neq \mathbf{0}$.

In section 3.1, we encountered a system of two tanks where we were interested in the amount of salt in each tank at time t. With the amount of salt in the two tanks represented respectively by $x_1(t)$ and $x_2(t)$, we saw that these component functions had to satisfy the system of differential equations given by

$$\mathbf{x}' = \begin{bmatrix} -1/20 & 1/80 \\ 1/40 & -1/40 \end{bmatrix} \begin{bmatrix} x_1 \\ x_2 \end{bmatrix} + \begin{bmatrix} 20 \\ 35 \end{bmatrix} \tag{3.6.1}$$

and that this system is naturally represented in the form

$$\mathbf{x}' = \mathbf{A}\mathbf{x} + \mathbf{b} \tag{3.6.2}$$

In our most recent work with the homogeneous equation $\mathbf{x}' = \mathbf{A}\mathbf{x}$, we noted several times the analogy to solving the single first-order differential equation $x' = ax$. In particular, we observed the key role that $e^{\lambda t}$ plays in the process of solving homogeneous systems of equations, much like e^{at} does in the solution of a single homogeneous linear first-order equation.

We next naturally consider the linear first-order analogy of (3.6.2), a nonhomogeneous equation such as

$$y' = 2y + 5 \tag{3.6.3}$$

In section 2.3, we made the observation in theorem 2.3.3 that for any linear first-order differential equation in the form

$$y' + p(t)y = f(t)$$

if y_p is any solution to the nonhomogeneous equation and y_h is a solution to the corresponding homogeneous equation, then $y = y_p + y_h$ is a solution to the nonhomogeneous equation.

In our studies of linear algebra in chapter 1, we made a similar observation in section 1.5: if we have a solution $\mathbf{x}_p$ to the nonhomogeneous equation $\mathbf{A}\mathbf{x} = \mathbf{b}$, and we add to $\mathbf{x}_p$ any solution $\mathbf{x}_h$ to the homogeneous equation $\mathbf{A}\mathbf{x} = \mathbf{0}$, the result $(\mathbf{x} = \mathbf{x}_p + \mathbf{x}_h)$ is also a solution to $\mathbf{A}\mathbf{x} = \mathbf{b}$. See (1.5.1) to revisit the details of this discussion. Note that in this purely linear algebra context, $\mathbf{x}$ is a vector whose entries are constant.

These two preceding observations for linear first-order differential equations and systems of linear algebraic equations are now applied to the nonhomogeneous system of linear first-order differential equations, $\mathbf{x}' = \mathbf{A}\mathbf{x} + \mathbf{b}$. We note specifically that in this context, $\mathbf{x}(t)$ is a function of t. Let's return to the known situation of the homogeneous system $\mathbf{x}' = \mathbf{Ax}$ and denote its solution by $\mathbf{x}_h(t)$. In addition, suppose we are able to determine a single solution $\mathbf{x}_p(t)$ to the nonhomogeneous equation $\mathbf{x}' = \mathbf{Ax} + \mathbf{b}$. We claim that the function $\mathbf{x}(t) = \mathbf{x}_h(t) + \mathbf{x}_p(t)$ is the general solution of the nonhomogeneous equation. To see this, we substitute directly into $\mathbf{x}' = \mathbf{Ax} + \mathbf{b}$ and verify that the equation is satisfied. By properties of linearity, observe that

$$\mathbf{x}'(t) = \mathbf{x}'_h(t) + \mathbf{x}'_p(t) \tag{3.6.4}$$

and furthermore

$$\mathbf{Ax} + \mathbf{b} = \mathbf{A}(\mathbf{x}_h + \mathbf{x}_p) + \mathbf{b} = \mathbf{Ax}_h + \mathbf{Ax}_p + \mathbf{b} \tag{3.6.5}$$

By how we defined $\mathbf{x}_h(t)$ and $\mathbf{x}_p(t)$, we know that $\mathbf{x}'_h(t) = \mathbf{Ax}_h(t)$ and $\mathbf{x}'_p(t) = \mathbf{Ax}_p(t) + \mathbf{b}$, and thus (3.6.5) implies

$$\mathbf{Ax} + \mathbf{b} = \mathbf{x}'_h(t) + \mathbf{x}'_p(t) \tag{3.6.6}$$

From (3.6.4) and (3.6.6), we see that $\mathbf{x} = \mathbf{x}_h + \mathbf{x}_p$ is indeed a solution to $\mathbf{x}' = \mathbf{Ax} + \mathbf{b}$. In fact, we have found the general solution to the nonhomogeneous system, as stated in the following theorem.

Theorem 3.6.1   Let $\mathbf{A}$ be an $n \times n$ matrix with constant coefficients. If $\mathbf{x}_h$ is the general solution to the homogeneous system $\mathbf{x}' = \mathbf{Ax}$ and $\mathbf{x}_p$ is any solution to the nonhomogeneous system $\mathbf{x}' = \mathbf{Ax} + \mathbf{b}$, then $\mathbf{x} = \mathbf{x}_h + \mathbf{x}_p$ is the general solution to $\mathbf{x}' = \mathbf{Ax} + \mathbf{b}$.

Theorem 3.6.1 provides an approach that will guide us throughout our efforts to solve nonhomogeneous systems of differential equations. First, we solve the associated homogeneous system to find $\mathbf{x}_h$, a process we are familiar with. We usually call $\mathbf{x}_h$ the *complementary solution* to the equation $\mathbf{x}' = \mathbf{Ax} + \mathbf{b}$. Next, we must find a so-called *particular solution* $\mathbf{x}_p$ to the nonhomogeneous system $\mathbf{x}' = \mathbf{Ax} + \mathbf{b}$. Although a more sophisticated approach will be introduced in the next section, for now we will investigate a few examples in which the process of finding such a particular solution $\mathbf{x}_p$ is relatively straightforward.

Example 3.6.1   From the system of two tanks discussed in sections 1.1 and 3.1, consider the nonhomogeneous system of linear differential equations given by

$$\mathbf{x}' = \begin{bmatrix} -1/20 & 1/80 \\ 1/40 & -1/40 \end{bmatrix} \mathbf{x} + \begin{bmatrix} 20 \\ 35 \end{bmatrix} \tag{3.6.7}$$

By solving the associated homogeneous system and determining a particular solution to the nonhomogeneous system, find the general solution to the given system. In addition, plot an appropriate direction field and discuss the long-term behavior of solutions and their meaning in the context of the salt in each tank. Determine and sketch the solution to the IVP with initial condition $\mathbf{x}(0) = [2000 \ \ 1000]^{\mathrm{T}}$.

Solution.   We begin by solving $\mathbf{x}' = \mathbf{A}\mathbf{x}$, where

$$\mathbf{A} = \begin{bmatrix} -1/20 & 1/80 \\ 1/40 & -1/40 \end{bmatrix}$$

The eigenvalues of $\mathbf{A}$ are approximately $\lambda_1 = -0.158$ and $\lambda_2 = -0.592$, with corresponding eigenvectors approximated by $\mathbf{v}_1 = [0.366 \ \ 1.000]^{\mathrm{T}}$ and $\mathbf{v}_2 = [-1.366 \ \ 1.000]^{\mathrm{T}}$. It follows that the general solution $\mathbf{x}_h$ is

$$\mathbf{x}_h(t) = c_1 e^{-0.158t} \begin{bmatrix} 0.366 \\ 1.000 \end{bmatrix} + c_2 e^{-0.592t} \begin{bmatrix} -1.366 \\ 1.000 \end{bmatrix}$$

Next, we must determine a particular solution $\mathbf{x}_p$ to the nonhomogeneous equation $\mathbf{x}' = \mathbf{A}\mathbf{x} + \mathbf{b}$. In this particular example, $\mathbf{b}$ is a constant vector. Therefore, it is natural to guess that a constant vector $\mathbf{x}_p$ will satisfy the nonhomogeneous equation. More than this, we should recall from earlier discussions of the problem leading to the given system that the vector $\mathbf{x}$ represents the amounts of salt in two connected tanks as streams of inflow deliver salt, each at a constant rate. Our intuition suggests that over time the two tanks should approach a stable equilibrium, and hence an equilibrium (and therefore constant) solution should be present.

Therefore, we assume that $\mathbf{x}_p$ is a constant vector and observe that this immediately implies that $\mathbf{x}_p' = \mathbf{0}$. Substituting into $\mathbf{x}' = \mathbf{A}\mathbf{x} + \mathbf{b}$, it follows that $\mathbf{x}_p$ must satisfy the system of linear equations $\mathbf{0} = \mathbf{A}\mathbf{x}_p + \mathbf{b}$ or $\mathbf{A}\mathbf{x}_p = -\mathbf{b}$. With the given entries of $\mathbf{A}$ and $\mathbf{b}$, this leads us to row reduce the appropriate augmented matrix and find that

$$\begin{bmatrix} -1/20 & 1/80 & -20 \\ 1/40 & -1/40 & -35 \end{bmatrix} \rightarrow \begin{bmatrix} 1 & 0 & 1000 \\ 0 & 1 & 2400 \end{bmatrix}$$

This shows $\mathbf{x}_p = [1000 \ \ 2400]^{\mathrm{T}}$ is a particular solution to $\mathbf{x}' = \mathbf{A}\mathbf{x} + \mathbf{b}$, and, more specifically, is an equilibrium solution of the system. Moreover, it now follows that the general solution to the system is given by

$$\mathbf{x}(t) = \mathbf{x}_h(t) + \mathbf{x}_p(t) = c_1 e^{-0.158t} \begin{bmatrix} 0.366 \\ 1.000 \end{bmatrix} + c_2 e^{-0.592t} \begin{bmatrix} -1.366 \\ 1.000 \end{bmatrix} + \begin{bmatrix} 1000 \\ 2400 \end{bmatrix}$$
$$(3.6.8)$$

If we add the initial condition that $\mathbf{x}(0) = [2000 \ \ 1000]^{\mathrm{T}}$, we can solve for the constants c_1 and c_2, and plot the appropriate corresponding trajectory, as shown in figure 3.14. In both (3.6.8) and figure 3.14 we can see how the long-term behavior of every solution tends to the equilibrium solution. Moreover, in the direction field we can also recognize the straight-line solutions that correspond to lines in the direction of each eigenvector but that now pass through the equilibrium solution $(1000, 2400)$.

From example 3.6.1, we observe that in cases where we want to solve $\mathbf{x}' = \mathbf{A}\mathbf{x} + \mathbf{b}$ and $\mathbf{b}$ is itself a constant vector, $\mathbf{x}_p$ may be determined by assuming that $\mathbf{x}_p$ is a constant vector and solving $\mathbf{0} = \mathbf{A}\mathbf{x}_p + \mathbf{b}$. If $\mathbf{x}_p$ is not constant, then the situation is more complicated, as we discover in the following example.

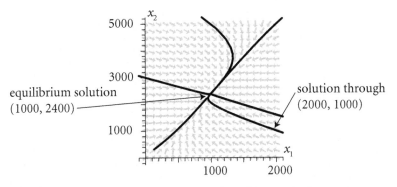

Figure 3.14   The direction field for the system $\mathbf{x}' = \mathbf{A}\mathbf{x} + \mathbf{b}$ of example 3.6.1.

Example 3.6.2   Find the general solution of the nonhomogeneous system given by

$$\mathbf{x}' = \begin{bmatrix} 2 & -1 \\ 3 & -2 \end{bmatrix} \mathbf{x} + \begin{bmatrix} \cos 2t \\ 0 \end{bmatrix} \tag{3.6.9}$$

Solution.   Since the eigenvalues of $\mathbf{A} = \begin{bmatrix} 2 & -1 \\ 3 & -2 \end{bmatrix}$ are $\lambda_1 = -1$ and $\lambda_2 = 1$ with corresponding eigenvectors $\mathbf{v}_1 = [1 \ 3]^{\mathrm{T}}$ and $\mathbf{v}_2 = [1 \ 1]^{\mathrm{T}}$, it follows that the complementary solution to the related homogeneous system is

$$\mathbf{x}_h = c_1 e^{-t} \begin{bmatrix} 1 \\ 3 \end{bmatrix} + c_2 e^{t} \begin{bmatrix} 1 \\ 1 \end{bmatrix}$$

To determine the particular solution $\mathbf{x}_p$ to the given nonhomogeneous system, we need to find a vector function $\mathbf{x}(t)$ that simultaneously satisfies the system (3.6.9). Due to the presence of $\cos 2t$ in the vector $\mathbf{b}$, it is natural to guess that the components of $\mathbf{x}_p$ will somehow involve $\cos 2t$. In addition, since $\mathbf{x}'_p$ plays a role in the system, we must account for the possibility that the derivative of $\cos 2t$ may also arise; moreover, since $\mathbf{A}\mathbf{x}$ will also be computed, linear combinations of vectors that involve the entries in $\mathbf{x}$ will be present. Therefore, we make the reasonable guess that $\mathbf{x}_p$ has the form

$$\mathbf{x}_p = \begin{bmatrix} a\cos 2t + b\sin 2t \\ c\cos 2t + d\sin 2t \end{bmatrix} \tag{3.6.10}$$

and attempt to determine values for the undetermined coefficients $a, b, c,$ and d that make $\mathbf{x}_p$ a solution to the system.

We accomplish this by direct substitution into (3.6.9). First, observe that

$$\mathbf{x}'_p = \begin{bmatrix} -2a\sin 2t + 2b\cos 2t \\ -2c\sin 2t + 2d\cos 2t \end{bmatrix} \tag{3.6.11}$$

Now substituting (3.6.10) and (3.6.11) into (3.6.9), it follows

$$\begin{bmatrix} -2a\sin 2t + 2b\cos 2t \\ -2c\sin 2t + 2d\cos 2t \end{bmatrix} = \begin{bmatrix} 2 & -1 \\ 3 & -2 \end{bmatrix} \begin{bmatrix} a\cos 2t + b\sin 2t \\ c\cos 2t + d\sin 2t \end{bmatrix} + \begin{bmatrix} \cos 2t \\ 0 \end{bmatrix}$$

If we now expand the matrix product and factor out the terms involving $\sin 2t$ and $\cos 2t$ on the right side,

$$-2a\sin 2t + 2b\cos 2t = (2b - d)\sin 2t + (2a - c + 1)\cos 2t \qquad (3.6.12)$$

$$-2c\sin 2t + 2d\cos 2t = (3b - 2d)\sin 2t + (3a - 2c)\cos 2t \qquad (3.6.13)$$

In (3.6.12), we can equate the coefficients of $\sin 2t$ to find that $-2a = 2b - d$. Doing likewise for the coefficients of $\cos 2t$, $2b = 2a - c + 1$. Similarly, (3.6.13) results in the two equations $-2c = 3b - 2d$ and $2d = 3a - 2c$. Reorganizing these four equations in four unknowns, we see that a, b, c, and d must satisfy the system

$$-2a - 2b + d = 0$$
$$-2a + 2b + c = 1$$
$$-3b - 2c + 2d = 0$$
$$-3a + 2c + 2d = 0$$

Row-reducing,

$$\begin{bmatrix} -2 & -2 & 0 & 1 & 0 \\ -2 & 2 & 1 & 0 & 1 \\ 0 & -3 & -2 & 2 & 0 \\ -3 & 0 & 2 & 2 & 0 \end{bmatrix} \rightarrow \begin{bmatrix} 1 & 0 & 0 & 0 & -2/5 \\ 0 & 1 & 0 & 0 & 2/5 \\ 0 & 0 & 1 & 0 & -3/5 \\ 0 & 0 & 0 & 1 & 0 \end{bmatrix}$$

which shows $a = -2/5$, $b = 2/5$, $c = -3/5$, and $d = 0$, so a particular solution to the nonhomogeneous system is

$$\mathbf{x}_p = \begin{bmatrix} -\frac{2}{5}\cos 2t + \frac{2}{5}\sin 2t \\ -\frac{3}{5}\cos 2t \end{bmatrix}$$

Finally, it follows that the general solution to the system is

$$\mathbf{x} = \mathbf{x}_h + \mathbf{x}_p = c_1 e^{-t}\begin{bmatrix} 1 \\ 3 \end{bmatrix} + c_2 e^{t}\begin{bmatrix} 1 \\ 1 \end{bmatrix} + \begin{bmatrix} -\frac{2}{5}\cos 2t + \frac{2}{5}\sin 2t \\ -\frac{3}{5}\cos 2t \end{bmatrix}$$

One lesson to take from example 3.6.2 is that while the process for trying to solve a nonhomogeneous system of differential equations is straightforward, the actual computation of a particular solution $\mathbf{x}_p$ can be quite cumbersome. Indeed, even in the case where the vector $\mathbf{b}$ is quite simple, as it is in the most recent example, tedious calculations can arise. Moreover, it is less clear how one might proceed in the situation where the vector $\mathbf{b}$ is particularly complicated. Specifically, making an appropriate guess for $\mathbf{x}_p$ may be difficult. We usually call the process of finding $\mathbf{x}_p$ through a guess involving unknown constants the *method of undetermined coefficients.*

To gain a better sense of the guesses that are involved in using undetermined coefficients, we turn to the following example.

Example 3.6.3   For nonhomogeneous linear systems of the form $\mathbf{x}' = \mathbf{A}\mathbf{x} + \mathbf{b}$ where $\mathbf{A}$ is a matrix with constant entries, state the natural guess to use for $\mathbf{x}_p$ when the vector $\mathbf{b}$ is

$$(a)\ \mathbf{b} = \begin{bmatrix} e^{-t} \\ 2e^{-t} \end{bmatrix} \qquad (b)\ \mathbf{b} = \begin{bmatrix} 1 \\ t \end{bmatrix} \qquad (c)\ \mathbf{b} = \begin{bmatrix} t^2 \\ 0 \end{bmatrix} \qquad (d)\ \mathbf{b} = \begin{bmatrix} e^{-3t} \\ -2 \end{bmatrix}$$

Solution.

(a) With $\mathbf{b} = [e^{-t}\ \ 2e^{-t}]^{\mathrm{T}}$, it is natural to expect that any particular solution must involve e^{-t} in its components. Specifically, we make the guess that

$$\mathbf{x}_p = \begin{bmatrix} Ae^{-t} \\ Be^{-t} \end{bmatrix}$$

and substitute directly into $\mathbf{x}' = \mathbf{A}\mathbf{x} + \mathbf{b}$ in order to attempt to find values of A and B for which $\mathbf{x}_p$ satisfies the given system.[3]

(b) Given $\mathbf{b} = [1\ \ t]^{\mathrm{T}}$, we must account for the fact that $\mathbf{x}_p$ and its derivative can involve constant and linear functions of t. In particular, we suppose that

$$\mathbf{x}_p = \begin{bmatrix} At + B \\ Ct + D \end{bmatrix}$$

and substitute appropriately in an effort to determine A, B, C, and D.

(c) For $\mathbf{b} = [t^2\ \ 0]^{\mathrm{T}}$, with one quadratic term present in $\mathbf{b}$, it is necessary to include quadratic terms in each entry of $\mathbf{x}_p$. But since the derivative of $\mathbf{x}_p$ will be taken, linear terms must be included as well. Finally, once linear terms are included, for the same reason we must permit the possibility that constant terms can be present in $\mathbf{x}_p$. Therefore, we guess the form

$$\mathbf{x}_p = \begin{bmatrix} At^2 + Bt + C \\ Dt^2 + Et + F \end{bmatrix}$$

(d) With $\mathbf{b} = [e^{-3t}\ \ -2]^{\mathrm{T}}$ having both an exponential and constant term present, we account for both of these scalar functions and their derivative by assuming that

$$\mathbf{x}_p = \begin{bmatrix} Ae^{-3t} + B \\ Ce^{-3t} + D \end{bmatrix}$$

---

[3] It is possible that the guess can fail to work, in which case a modified form for $\mathbf{x}_p$ is required. One setting where this may occur is when $\lambda = -1$ is an eigenvalue of $\mathbf{A}$, whereby a vector involving e^{-t} already appears in the complementary solution $\mathbf{x}_h$. See exercise 8 for further investigation of this issue.

The method of undetermined coefficients is not foolproof: it is certainly possible to guess incorrectly (as noted in the footnote related to part (a) of example 3.6.3). If our guess is incorrect, an inconsistent linear system of algebraic equations will arise, which tells us we need to modify our guess. Besides the possibility of guessing incorrectly, it can also be the case that the computations involved in determining $\mathbf{x}_p$ are very cumbersome. In the next section, we consider a different approach, one that parallels our solution of single linear first-order differential equations of the form $y' + p(t)y = f(t)$, that provides, at least in theory, an algorithmic approach to solving any nonhomogeneous system $\mathbf{x}' = \mathbf{Ax} + \mathbf{b}$ where the matrix $\mathbf{A}$ has real, constant entries.

Finally, we note that the presence of nonconstant entries in the vector $\mathbf{b}$ in a nonhomogeneous system $\mathbf{x}' = \mathbf{Ax} + \mathbf{b}$ makes it impossible to plot a direction field for the system. In particular, when we sketch direction fields, we rely on the fact that regardless of time, t, the direction vector $\mathbf{x}'$ to the solution curve $\mathbf{x}$ is dependent only on the location (x_1, x_2), and not on t. When $\mathbf{b}$ is nonconstant and a function of t, this is no longer the case and we therefore are left with only algebraic approaches to the problem. If $\mathbf{b}$ is constant, then we can generate the direction field for the system, such as the one shown in figure 3.14.

Exercises 3.6   In each of exercises 1–4, show by direct substitution that the given particular solution $\mathbf{x}_p$ is indeed a solution to the stated nonhomogeneous system of equations. Hence determine the general solution to the stated system.

1. $\mathbf{x}' = \begin{bmatrix} -1 & 3 \\ 2 & -3 \end{bmatrix} \mathbf{x} + \begin{bmatrix} 5 \\ -1 \end{bmatrix}$, $\mathbf{x}_p = \begin{bmatrix} -4 \\ -3 \end{bmatrix}$

2. $\mathbf{x}' = \begin{bmatrix} 1 & -2 \\ -2 & 1 \end{bmatrix} \mathbf{x} + \begin{bmatrix} e^{2t} \\ 0 \end{bmatrix}$, $\mathbf{x}_p = e^{2t} \begin{bmatrix} -1/3 \\ 2/3 \end{bmatrix}$

3. $\mathbf{x}' = \begin{bmatrix} 2 & 1 \\ 1 & 2 \end{bmatrix} \mathbf{x} + \begin{bmatrix} \sin t \\ 0 \end{bmatrix}$, $\mathbf{x}_p = \sin t \begin{bmatrix} -2/5 \\ 1/10 \end{bmatrix} + \cos t \begin{bmatrix} -3/10 \\ 1/5 \end{bmatrix}$

4. $\mathbf{x}' = \begin{bmatrix} -3 & 1 \\ 1 & -1 \end{bmatrix} \mathbf{x} + \begin{bmatrix} e^{2t} + 1 \\ 1 \end{bmatrix}$, $\mathbf{x}_p = \begin{bmatrix} 1 \\ 2 \end{bmatrix} + e^{2t} \begin{bmatrix} 3/14 \\ 1/14 \end{bmatrix}$

5. Consider the system of differential equations

$$\mathbf{x}' = \begin{bmatrix} 1 & 1 \\ 4 & 1 \end{bmatrix} \mathbf{x} + \begin{bmatrix} 1 \\ -3 \end{bmatrix}$$

   (a) Explain why it is reasonable to assume that $\mathbf{x}_p$ is a constant vector, and use this assumption to determine a particular solution to the given nonhomogeneous system.
   (b) Determine the complementary solution $\mathbf{x}_h$ to the associated homogeneous system, $\mathbf{x}' = \mathbf{Ax}$.
   (c) State the general solution to the system.
   (d) Is there an equilibrium solution to this system? If so, is it stable? Explain.

6. Consider the system of differential equations

$$\mathbf{x}' = \begin{bmatrix} 1 & 1 \\ 4 & 1 \end{bmatrix} \mathbf{x} + \begin{bmatrix} e^{4t} \\ 0 \end{bmatrix}$$

(a) Explain why it is reasonable to assume that $\mathbf{x}_p$ is a vector of the form

$$\mathbf{x}_p = \begin{bmatrix} ae^{4t} \\ be^{4t} \end{bmatrix}$$

Then use this assumption to determine a particular solution to the given nonhomogeneous system.

(b) Determine the complementary solution $\mathbf{x}_h$ to the associated homogeneus system, $\mathbf{x}' = \mathbf{A}\mathbf{x}$.

(c) State the general solution to the system.

7. Consider the system of differential equations

$$\mathbf{x}' = \begin{bmatrix} 1 & 1 \\ 4 & 1 \end{bmatrix} \mathbf{x} + \begin{bmatrix} e^{-2t} + 1 \\ 2e^{-2t} + 3 \end{bmatrix}$$

(a) Explain why it is reasonable to assume that $\mathbf{x}_p$ is a vector of the form

$$\mathbf{x}_p = \begin{bmatrix} ae^{-2t} + b \\ ce^{-2t} + d \end{bmatrix}$$

Use this assumption to determine a particular solution to the given nonhomogeneous system.

(b) Determine the complementary solution $\mathbf{x}_h$ to the associated homogeneus system, $\mathbf{x}' = \mathbf{A}\mathbf{x}$.

(c) State the general solution to the system.

8. Consider the system of differential equations

$$\mathbf{x}' = \begin{bmatrix} 1 & 1 \\ 4 & 1 \end{bmatrix} \mathbf{x} + \begin{bmatrix} e^{-t} \\ 0 \end{bmatrix}$$

(a) Explain why it is reasonable to assume that $\mathbf{x}_p$ is a vector of the form

$$\mathbf{x}_p = \begin{bmatrix} ae^{-t} \\ be^{-t} \end{bmatrix}$$

(b) Show that the form of $\mathbf{x}_p$ above does not result in a particular solution to the system.

(c) By assuming that $\mathbf{x}_p$ is a vector of the form

$$\mathbf{x}_p = \begin{bmatrix} ae^{-t} + bte^{-t} \\ ce^{-t} + dte^{-t} \end{bmatrix}$$

determine a particular solution to the given nonhomogeneous system

(d) Determine the complementary solution $\mathbf{x}_h$ to the associated homogeneus system, $\mathbf{x}' = \mathbf{A}\mathbf{x}$.

(e) State the general solution to the system.

For the nonhomogeneous linear systems of differential equations given in exercises 9–17, (a) determine a particular solution $\mathbf{x}_p$ by making an appropriate assumption about the form of $\mathbf{x}_p$, (b) determine the complementary solution $\mathbf{x}_h$ to $\mathbf{x}' = \mathbf{Ax}$, and (c) hence state the general solution to the system.

9. $\mathbf{x}' = \begin{bmatrix} -4 & 5 \\ 5 & -4 \end{bmatrix} \mathbf{x} + \begin{bmatrix} 1 \\ -1 \end{bmatrix}$

10. $\mathbf{x}' = \begin{bmatrix} -4 & 5 \\ 5 & -4 \end{bmatrix} \mathbf{x} + \begin{bmatrix} 3e^{-2t} \\ -e^{-2t} \end{bmatrix}$

11. $\mathbf{x}' = \begin{bmatrix} -1 & 1 \\ 0 & 1 \end{bmatrix} \mathbf{x} + \begin{bmatrix} 3e^{-2t} \\ -4e^{-2t} \end{bmatrix}$

12. $\mathbf{x}' = \begin{bmatrix} -1 & 1 \\ 0 & 1 \end{bmatrix} \mathbf{x} + \begin{bmatrix} 2 \\ -5 \end{bmatrix}$

13. $\mathbf{x}' = \begin{bmatrix} 0 & -1 \\ 1 & 0 \end{bmatrix} \mathbf{x} + \begin{bmatrix} 3 \\ -2 \end{bmatrix}$

14. $\mathbf{x}' = \begin{bmatrix} 0 & -1 \\ 1 & 0 \end{bmatrix} \mathbf{x} + \begin{bmatrix} e^t \\ -2e^t \end{bmatrix}$

15. $\mathbf{x}' = \begin{bmatrix} 0 & -1 \\ 1 & 0 \end{bmatrix} \mathbf{x} + \begin{bmatrix} 3 + e^t \\ -2 - 2e^t \end{bmatrix}$

16. $\mathbf{x}' = \begin{bmatrix} 2 & -1 \\ 3 & -2 \end{bmatrix} \mathbf{x} + \begin{bmatrix} t - 2 \\ 3t - 4 \end{bmatrix}$

17. $\mathbf{x}' = \begin{bmatrix} 2 & -1 \\ 3 & -2 \end{bmatrix} \mathbf{x} + \begin{bmatrix} \cos 3t \\ 4 \end{bmatrix}$

18. For the system of differential equations given in exercise 10, solve the IVP with initial condition $\mathbf{x}(0) = [1 \quad -2]^T$.

19. For the system of differential equations given in exercise 11, solve the IVP with initial condition $\mathbf{x}(0) = [-3 \quad -2]^T$.

20. For the system of differential equations given in exercise 14, solve the IVP with initial condition $\mathbf{x}(0) = [0 \quad 4]^T$.

21. For the system of differential equations given in exercise 15, solve the IVP with initial condition $\mathbf{x}(0) = [1 \quad -2]^T$.

22. Without actually computing $\mathbf{x}_p$, choose and justify the form you would guess for a particular solution to

$$\mathbf{x}' = \begin{bmatrix} -4 & 5 \\ 5 & -4 \end{bmatrix} \mathbf{x} + e^{-2t} \sin t \begin{bmatrix} 1 \\ -1 \end{bmatrix}$$

23. Without actually computing $\mathbf{x}_p$, choose and justify the form you would guess for a particular solution to

$$\mathbf{x}' = \begin{bmatrix} -4 & 5 \\ 5 & -4 \end{bmatrix} \mathbf{x} + \begin{bmatrix} \sin 3t \\ \cos 2t \end{bmatrix}$$

24. Suppose that $\mathbf{x}_1(t)$ and $\mathbf{x}_2(t)$ are solutions of

$$\mathbf{x}' = \mathbf{A}\mathbf{x} + \mathbf{f}_1(t) \text{ and } \mathbf{x}' = \mathbf{A}\mathbf{x} + \mathbf{f}_2(t)$$

respectively. Show that $\mathbf{x}(t) = \mathbf{x}_1(t) + \mathbf{x}_2(t)$ is a solution of

$$\mathbf{x}' = \mathbf{A}\mathbf{x} + \mathbf{f}_1(t) + \mathbf{f}_2(t)$$

3.7 Nonhomogeneous systems: variation of parameters

In section 3.6, we discovered that solving the nonhomogeneous linear system $\mathbf{x}' = \mathbf{A}\mathbf{x} + \mathbf{b}$ requires us to find one particular solution $\mathbf{x}_p$ to the nonhomogeneous system. We then combine this particular solution with the complementary solution $\mathbf{x}_h$—the general solution to the corresponding homogeneous system $\mathbf{x}' = \mathbf{A}\mathbf{x}$. While we were able to successfully solve a range of problems, the method of undetermined coefficients is somewhat dissatisfying: essentially we made an educated guess as to the form that $\mathbf{x}_p$ should take, and then substituted to see if our guess was appropriate and resulted in a particular solution. As was shown in exercise 8 in section 3.6, there are instances when the obvious guess fails to work and additional investigation of a possible solution $\mathbf{x}_p$ is needed. Moreover, with undetermined coefficients we only considered functions $\mathbf{b}(t)$ that had entries that were polynomial, sinusoidal, or exponential in nature. We desire a more systematic approach to finding $\mathbf{x}_p$; developing such a method is the purpose of this section.

In section 2.3, we learned that for any linear first-order differential equation of the form $y' + p(t)y = f(t)$, the solution y is given by

$$y = e^{-P(t)} \int e^{P(t)} f(t)\, dt \qquad (3.7.1)$$

where $P(t) = \int p(t)\, dt$. We now seek to establish a similar result for the case of systems of the form $\mathbf{x}' = \mathbf{A}\mathbf{x} + \mathbf{b}$, where $\mathbf{A}$ is an $n \times n$ matrix with constant entries and $\mathbf{b}$ is a vector function of t. Let us first consider the form of the general solution $\mathbf{x}_h$ to the corresponding homogeneous system. Recall that $\mathbf{x} = c_1\mathbf{x}_1 + \cdots + c_n\mathbf{x}_n$, where $\{\mathbf{x}_1, \ldots, \mathbf{x}_n\}$ is a set of n linearly independent solutions to $\mathbf{x}' = \mathbf{A}\mathbf{x}$.

Being more explicit about the vectors present, say with entries $x_{ij}(t)$, we can rewrite $\mathbf{x} = c_1\mathbf{x}_1 + \cdots + c_n\mathbf{x}_n$ as

$$\mathbf{x} = c_1 \begin{bmatrix} x_{11} \\ x_{21} \\ \vdots \\ x_{n1} \end{bmatrix} + c_2 \begin{bmatrix} x_{12} \\ x_{22} \\ \vdots \\ x_{n2} \end{bmatrix} + \cdots + c_n \begin{bmatrix} x_{1n} \\ x_{2n} \\ \vdots \\ x_{nn} \end{bmatrix} = \begin{bmatrix} c_1 x_{11} + c_2 x_{12} + \cdots + c_n x_{1n} \\ c_1 x_{21} + c_2 x_{22} + \cdots + c_n x_{2n} \\ \vdots \\ c_1 x_{n1} + c_2 x_{n2} + \cdots + c_n x_{nn} \end{bmatrix}$$

Now observe that the right side of the above equation—the overall vector formulation of **x**—can be expressed as a matrix product. In particular, we write

$$\mathbf{x} = \mathbf{\Phi C} \tag{3.7.2}$$

where **C** is the vector whose entries are the arbitrary constants $c_1, \ldots, c_n$ that arise in the formulation of the general solution **x**, and $\mathbf{\Phi}(t)$ is the matrix whose columns are the n linearly independent solutions to $\mathbf{x}' = \mathbf{Ax}$. We call $\mathbf{\Phi}(t)$ the *fundamental solution matrix* of the system.

At this point, it is essential to make two observations about $\mathbf{\Phi}(t)$. The first is that $\mathbf{\Phi}(t)$ is nonsingular for every relevant value of t. This holds because the columns of $\mathbf{\Phi}(t)$ are linearly independent since, by definition, they are linearly independent solutions of $\mathbf{x}' = \mathbf{Ax}$. Second, we note that $\mathbf{\Phi}'(t) = \mathbf{A\Phi}(t)$. Since the derivative of $\mathbf{\Phi}(t)$ is taken component-wise, this equation is simply the matrix way to say that each column of $\mathbf{\Phi}(t)$ satisfies the homogeneous system of equations $\mathbf{x}' = \mathbf{Ax}$.

Now, recall (3.7.2) where we expressed the complementary solution in the form $\mathbf{x}_h = \mathbf{\Phi}(t)\mathbf{C}$. As we now seek a particular solution $\mathbf{x}_p$ to the nonhomogeneous equation, it is natural to suppose that $\mathbf{x}_p$ has the form

$$\mathbf{x}_p(t) = \mathbf{\Phi}(t)\mathbf{u}(t) \tag{3.7.3}$$

where $\mathbf{u}(t)$ is a function yet to be determined. We now substitute this guess for $\mathbf{x}_p$ into $\mathbf{x}' = \mathbf{Ax} + \mathbf{b}(t)$ to see what conditions **u** must satisfy. For ease of display, in what follows we suppress the "(t)" notation in each of the functions $\mathbf{\Phi}, \mathbf{u}, \mathbf{u}'$, and **b**. By the product rule,

$$\mathbf{x}'_p = (\mathbf{\Phi u})' = \mathbf{\Phi u}' + \mathbf{\Phi}'\mathbf{u}$$

and so substituting into the system $\mathbf{x}' = \mathbf{Ax} + \mathbf{b}(t)$, we have

$$\mathbf{\Phi u}' + \mathbf{\Phi}'\mathbf{u} = \mathbf{A\Phi u} + \mathbf{b} \tag{3.7.4}$$

Recalling our observation above that $\mathbf{\Phi}' = \mathbf{A\Phi}$, we can substitute in (3.7.4) to find

$$\mathbf{\Phi u}' + \mathbf{A\Phi u} = \mathbf{A\Phi u} + \mathbf{b} \tag{3.7.5}$$

We next subtract $\mathbf{A\Phi u}$ from both sides of (3.7.5) to deduce that

$$\mathbf{\Phi u}' = \mathbf{b} \tag{3.7.6}$$

Since we are interested in determining the unknown function **u**, and we know that $\mathbf{\Phi}$ is nonsingular, we may now write

$$\mathbf{u}' = \mathbf{\Phi}^{-1}\mathbf{b} \tag{3.7.7}$$

and, therefore, **u** must have the form

$$\mathbf{u}(t) = \int \mathbf{\Phi}^{-1}(t)\mathbf{b}(t)\, dt \tag{3.7.8}$$

Finally, recalling the supposition we made in (3.7.3) that $\mathbf{x}_p = \Phi\mathbf{u}$, (3.7.8) now implies

$$\mathbf{x}_p(t) = \Phi(t) \int \Phi^{-1}(t)\mathbf{b}(t)\,dt \tag{3.7.9}$$

It is remarkable how this form of $\mathbf{x}_p$ aligns with our experience with a single linear first-order differential equation and the form of its solution given by (3.7.1). We summarize our above work in the following theorem.

Theorem 3.7.1   If $\mathbf{A}$ is an $n \times n$ matrix with constant entries, $\Phi(t)$ is the fundamental solution matrix of the homogeneous system of differential equations $\mathbf{x}' = \mathbf{A}\mathbf{x}$, and $\mathbf{b}(t)$ is a continuous vector function, then a particular solution $\mathbf{x}_p$ to the nonhomogeneous system $\mathbf{x}' = \mathbf{A}\mathbf{x} + \mathbf{b}(t)$ is given by

$$\mathbf{x}_p(t) = \Phi(t) \int \Phi^{-1}(t)\mathbf{b}(t)\,dt \tag{3.7.10}$$

The approach to finding a particular solution given in theorem 3.7.1 is often called *variation of parameters*. We next consider an example to see theorem 3.7.1 at work.

Example 3.7.1   Find the general solution of the nonhomogeneous system given by

$$\mathbf{x}' = \begin{bmatrix} 2 & -1 \\ 3 & -2 \end{bmatrix}\mathbf{x} + \begin{bmatrix} 0 \\ 4 \end{bmatrix}t$$

Solution.   From our determination of the eigenvalues and eigenvectors of the same coefficient matrix in example 3.6.2, the complementary solution is

$$\mathbf{x}_h = c_1 e^{-t} \begin{bmatrix} 1 \\ 3 \end{bmatrix} + c_2 e^t \begin{bmatrix} 1 \\ 1 \end{bmatrix}$$

Therefore, the fundamental matrix is

$$\Phi(t) = \begin{bmatrix} e^{-t} & e^t \\ 3e^{-t} & e^t \end{bmatrix}$$

According to (3.7.10), we next need to compute Φ^{-1}. While the inverse of this matrix of functions may be computed by row-reducing $[\Phi \mid \mathbf{I}]$ in the usual way, because of the function coefficients in Φ it is much easier to use a shortcut for computing the inverse of a 2×2 matrix that we established in exercise 19 of section 1.9. Specifically, if

$$\mathbf{A} = \begin{bmatrix} a & b \\ c & d \end{bmatrix}$$

is an invertible matrix, then

$$\mathbf{A}^{-1} = \frac{1}{\det(\mathbf{A})} \begin{bmatrix} d & -b \\ -c & a \end{bmatrix}$$

Here, since $\det(\Phi) = e^{-t}e^t - 3e^{-t}e^t = -2$, it follows

$$\Phi^{-1} = -\frac{1}{2}\begin{bmatrix} e^t & -e^t \\ -3e^{-t} & e^{-t} \end{bmatrix}$$

Thus, by (3.7.10), we now have

$$\mathbf{x}_p(t) = \Phi(t) \int \Phi^{-1}(t)\mathbf{b}(t)\,dt$$

$$= \begin{bmatrix} e^{-t} & e^t \\ 3e^{-t} & e^t \end{bmatrix} \int \begin{bmatrix} -\frac{1}{2}e^t & \frac{1}{2}e^t \\ \frac{3}{2}e^{-t} & -\frac{1}{2}e^{-t} \end{bmatrix} \begin{bmatrix} 0 \\ 4t \end{bmatrix} dt$$

$$= \begin{bmatrix} e^{-t} & e^t \\ 3e^{-t} & e^t \end{bmatrix} \int \begin{bmatrix} 2te^t \\ -2te^{-t} \end{bmatrix} dt$$

Integrating the vector function component-wise by parts and computing the subsequent matrix product,

$$\mathbf{x}_p(t) = \begin{bmatrix} e^{-t} & e^t \\ 3e^{-t} & e^t \end{bmatrix} \begin{bmatrix} 2(t-1)e^t \\ 2(t+1)e^{-t} \end{bmatrix}$$

$$= \begin{bmatrix} 2(t-1) + 2(t+1) \\ 6(t-1) + 2(t+1) \end{bmatrix}$$

$$= \begin{bmatrix} 4t \\ 8t - 4 \end{bmatrix}$$

Therefore, the general solution to the original nonhomogeneous system is

$$\mathbf{x} = \mathbf{x}_h + \mathbf{x}_p = c_1 e^{-t}\begin{bmatrix} 1 \\ 3 \end{bmatrix} + c_2 e^t\begin{bmatrix} 1 \\ 1 \end{bmatrix} + \begin{bmatrix} 4t \\ 8t - 4 \end{bmatrix}$$

Example 3.7.1 demonstrates that there are three key steps in the solution to systems of the form $\mathbf{x}' = \mathbf{A}\mathbf{x} + \mathbf{b}(t)$. The first is solving the related homogeneous system $\mathbf{x}' = \mathbf{A}\mathbf{x}$ to determine the fundamental solution matrix $\Phi(t)$. Next, we have to compute $\Phi^{-1}(t)$. And finally, we must integrate the vector function given by $\Phi^{-1}(t)\mathbf{b}(t)$. Since we are seeking just one particular solution $\mathbf{x}_p$, there is no need to include the arbitrary constants that arise in antidifferentiating $\Phi^{-1}(t)\mathbf{b}(t)$.

We close this section with a second example that shows the computations involved when more complicated functions are present in $\mathbf{b}(t)$.

Example 3.7.2 Find the general solution of the nonhomogeneous system given by

$$\mathbf{x}' = \begin{bmatrix} 2 & -1 \\ 3 & -2 \end{bmatrix}\mathbf{x} + \begin{bmatrix} 1/(e^t + 1) \\ 1 \end{bmatrix}$$

Solution.   We first find $\mathbf{x}_h$. By finding the eigenvalues and eigenvectors of the coefficient matrix $\mathbf{A}$, it is straightforward to show that

$$\mathbf{x}_h = c_1 e^{-t} \begin{bmatrix} 1 \\ 3 \end{bmatrix} + c_2 e^t \begin{bmatrix} 1 \\ 1 \end{bmatrix}$$

Therefore, the fundamental solution matrix is

$$\Phi(t) = \begin{bmatrix} e^{-t} & e^t \\ 3e^{-t} & e^t \end{bmatrix}$$

Moreover, we can show that

$$\Phi^{-1}(t) = -\frac{1}{2} \begin{bmatrix} e^t & -e^t \\ -3e^{-t} & e^{-t} \end{bmatrix}$$

We are now ready to compute $\mathbf{x}_p$ and write

$$\mathbf{x}_p(t) = \Phi(t) \int \Phi^{-1}(t)\mathbf{b}(t)\, dt$$

$$= \begin{bmatrix} e^{-t} & e^t \\ 3e^{-t} & e^t \end{bmatrix} \int -\frac{1}{2} \begin{bmatrix} e^t & -e^t \\ -3e^{-t} & e^{-t} \end{bmatrix} \begin{bmatrix} 1/(e^t + 1) \\ 1 \end{bmatrix} dt$$

$$= \begin{bmatrix} e^{-t} & e^t \\ 3e^{-t} & e^t \end{bmatrix} \int \begin{bmatrix} \frac{1}{2}\frac{e^{2t}}{e^t+1} \\ \frac{1}{2}\frac{2e^{-t}-1}{e^t+1} \end{bmatrix} dt$$

At this point, it is easiest to use a computer algebra system to integrate and complete our calculation of $\mathbf{x}_p$. Doing so, and then finding the required matrix product, we have

$$\mathbf{x}_p(t) = \begin{bmatrix} e^{-t} & e^t \\ 3e^{-t} & e^t \end{bmatrix} \begin{bmatrix} \frac{1}{2}e^t - \frac{1}{2}\ln(e^t + 1) \\ -e^{-t} - \frac{3}{2}t + \frac{3}{2}\ln(e^t + 1) \end{bmatrix}$$

$$= \begin{bmatrix} -\frac{1}{2} - \frac{1}{2}e^{-t}\ln(e^t + 1) - \frac{3}{2}te^t + \frac{3}{2}e^t\ln(e^t + 1) \\ \frac{1}{2} - \frac{3}{2}e^{-t}\ln(e^t + 1) - \frac{3}{2}te^t + \frac{3}{2}e^t\ln(e^t + 1) \end{bmatrix}$$

Hence, the general solution to the given nonhomogeneous system is

$$\mathbf{x} = \mathbf{x}_h + \mathbf{x}_p = c_1 e^{-t} \begin{bmatrix} 1 \\ 3 \end{bmatrix} + c_2 e^t \begin{bmatrix} 1 \\ 1 \end{bmatrix}$$

$$+ \begin{bmatrix} -\frac{1}{2} - \frac{1}{2}e^{-t}\ln(e^t + 1) - \frac{3}{2}te^t + \frac{3}{2}e^t\ln(e^t + 1) \\ \frac{1}{2} - \frac{3}{2}e^{-t}\ln(e^t + 1) - \frac{3}{2}te^t + \frac{3}{2}e^t\ln(e^t + 1) \end{bmatrix}$$

At each stage in applying variation of parameters it is essential to simplify. In particular, $\Phi^{-1}(t)$ should be simplified as much as possible before computing $\Phi^{-1}(t)\mathbf{b}(t)$, and similarly, $\int \Phi^{-1}(t)\mathbf{b}(t)\,dt$ should be simplified as much as possible before computing $\Phi(t)\int \Phi^{-1}(t)\mathbf{b}(t)\,dt$. One option, of course, is to use a computer algebra system to avoid the more tedious aspects of the computations. We offer some suggestions for how to use *Maple* to assist in the computations in the following subsection.

3.7.1 Applying variation of parameters using *Maple*

Here we address how *Maple* can be used to execute the computations in a problem such as the one posed in example 3.7.2, where we are interested in solving the nonhomogeneous linear system of equations given by

$$\mathbf{x}' = \begin{bmatrix} 2 & -1 \\ 3 & -2 \end{bmatrix}\mathbf{x} + \begin{bmatrix} 1/(e^t + 1) \\ 1 \end{bmatrix}$$

As usual, we load the `Linear Algebra` package.

```
> with(LinearAlgebra):
```

Because we already know how to find the complementary solution, we focus on determining $\mathbf{x}_p$ by variation of parameters. First, we use the complementary solution,

$$\mathbf{x}_h = c_1 e^{-t}\begin{bmatrix} 1 \\ 3 \end{bmatrix} + c_2 e^t \begin{bmatrix} 1 \\ 1 \end{bmatrix}$$

to define the fundamental matrix $\Phi(t)$:

```
> Phi := <<exp(-t),3*exp(-t)>|<exp(t),exp(t)>>;
```

We next use the `MatrixInverse` command to find Φ^{-1} by entering

```
> MatrixInverse(Phi);
```

The resulting output is

$$\begin{bmatrix} -\dfrac{1}{2}\dfrac{1}{e^{-t}} & \dfrac{1}{2}\dfrac{1}{e^{-t}} \\[2mm] \dfrac{3}{2}\dfrac{1}{e^t} & -\dfrac{1}{2}\dfrac{1}{e^t} \end{bmatrix}$$

We can simplify this result using negative exponents; *Maple* can do so through the following command, through which we also store Φ^{-1} in `PhiInv`:

```
> PhiInv := simplify(MatrixInverse(Phi));
```

Next, in order to compute $\Phi^{-1}(t)\mathbf{b}(t)$, we must enter the function $\mathbf{b}(t)$. We enter

```
> b := <<1/(exp(t)+1),1>>;
```

and then

```
> y := simplify(PhiInv.b);
```

At this point, y is a 2×1 array that holds the vector function $\Phi^{-1}(t)\mathbf{b}(t)$. Specifically, the output for y displayed by *Maple* is

$$y := \begin{bmatrix} \frac{1}{2}\frac{e^{2t}}{e^t+1} \\ -\frac{1}{2}\frac{e^{-t}(-2+e^t)}{e^t+1} \end{bmatrix}$$

To access the components in y, we reference them with the commands $y[1,1]$ and $y[2,1]$. In particular, since we have to integrate $\Phi^{-1}(t)\mathbf{b}(t)$ component-wise, we enter

```
> Y := <<int(y[1,1],t), int(y[2,1],t)>>;
```

This last command produces the output

$$Y := \begin{bmatrix} \frac{1}{2}e^t - \frac{1}{2}\ln(e^t+1) \\ -\frac{1}{e^t} - \frac{3}{2}\ln(e^t) + \frac{3}{2}\ln(e^t+1) \end{bmatrix}$$

and obviously stores $\Phi^{-1}(t)\mathbf{b}(t)$ in Y. Note that *Maple* has not made the obvious simplification $\ln(e^t) = t$. Finally, in order to compute $\Phi(t) \int \Phi^{-1}(t)\mathbf{b}(t)\, dt$, we need to enter Phi . Y. Of course, we again want to simplify, so we use

```
> simplify(Phi.Y);
```

which produces the output

$$\begin{bmatrix} -\frac{1}{2} - \frac{1}{2}e^{-t}\ln(e^t+1) - \frac{3}{2}e^t\ln(e^t) + \frac{3}{2}e^t\ln(e^t+1) \\ \frac{1}{2} - \frac{3}{2}e^{-t}\ln(e^t+1) - \frac{3}{2}e^t\ln(e^t) + \frac{3}{2}e^t\ln(e^t+1) \end{bmatrix}$$

This last result is the particular solution $\mathbf{x}_p$ to the original system of nonhomogeneous equations given in example 3.7.2. Note again that we can simplify $\ln(e^t)$ to t in each component.

Exercises 3.7

1. Consider the system of differential equations given by

$$\mathbf{x}' = \begin{bmatrix} 3 & 2 \\ 2 & 3 \end{bmatrix}\mathbf{x} + \begin{bmatrix} 5 \\ -1 \end{bmatrix}$$

   (a) Based on the form of $\mathbf{b}(t)$, make a guess and determine $\mathbf{x}_p$ by undetermined coefficients.
   (b) Use variation of parameters to determine $\mathbf{x}_p$.

2. Consider the system of differential equations given by

$$\mathbf{x}' = \begin{bmatrix} 3 & 2 \\ 2 & 3 \end{bmatrix} \mathbf{x} + \begin{bmatrix} e^{2t} \\ 0 \end{bmatrix}$$

(a) Based on the form of $\mathbf{b}(t)$, make a guess and determine $\mathbf{x}_p$ by undetermined coefficients.

(b) Use variation of parameters to determine $\mathbf{x}_p$.

3. Consider the system of differential equations given by

$$\mathbf{x}' = \begin{bmatrix} 3 & 2 \\ 2 & 3 \end{bmatrix} \mathbf{x} + \begin{bmatrix} 3e^t \\ e^t \end{bmatrix}$$

(a) Based on the form of $\mathbf{b}(t)$, what is the natural guess for $\mathbf{x}_p$? Show that this natural guess fails to work.

(b) Compute the complementary solution $\mathbf{x}_h$ to the stated system and use its form to explain why the natural guess in (a) is not a valid one.

(c) Use variation of parameters to determine $\mathbf{x}_p$.

4. Consider the system of differential equations given by

$$\mathbf{x}' = \begin{bmatrix} 0 & 2 \\ 1 & -1 \end{bmatrix} \mathbf{x} + \begin{bmatrix} 4\sin t \\ 2\sin t \end{bmatrix}$$

(a) Based on the form of $\mathbf{b}(t)$, what would be the natural guess to make for $\mathbf{x}_p$? How many undetermined coefficients would need to be computed?

(b) Use variation of parameters to determine $\mathbf{x}_p$.

In each of the exercises 5–12, determine the general solution to the given system by finding $\mathbf{x}_p$ using variation of parameters. Note that in each case, $\Phi(t)$ is given.

5. $\mathbf{x}' = \begin{bmatrix} 1 & 0 \\ -1 & 3 \end{bmatrix} \mathbf{x} + \begin{bmatrix} 2e^{-t} \\ 1 \end{bmatrix}$, $\Phi(t) = \begin{bmatrix} 2e^t & 0 \\ e^t & e^{3t} \end{bmatrix}$

6. $\mathbf{x}' = \begin{bmatrix} 1 & 0 \\ -1 & 3 \end{bmatrix} \mathbf{x} + \begin{bmatrix} e^{3t} \\ -e^{3t} \end{bmatrix}$, $\Phi(t) = \begin{bmatrix} 2e^t & 0 \\ e^t & e^{3t} \end{bmatrix}$

7. $\mathbf{x}' = \begin{bmatrix} 1 & 0 \\ -1 & 3 \end{bmatrix} \mathbf{x} + \begin{bmatrix} \cos 2t \\ 2\sin 2t \end{bmatrix}$, $\Phi(t) = \begin{bmatrix} 2e^t & 0 \\ e^t & e^{3t} \end{bmatrix}$

8. $\mathbf{x}' = \begin{bmatrix} 2 & 1 \\ 3 & 0 \end{bmatrix} \mathbf{x} + \begin{bmatrix} 10t \\ 10t \end{bmatrix}$, $\Phi(t) = \begin{bmatrix} e^{3t} & -e^{-t} \\ e^{3t} & 3e^{-t} \end{bmatrix}$

9. $\mathbf{x}' = \begin{bmatrix} 2 & 1 \\ 3 & 0 \end{bmatrix} \mathbf{x} + \begin{bmatrix} 2e^{-t} \\ 5e^{-t} \end{bmatrix}$, $\Phi(t) = \begin{bmatrix} e^{3t} & -e^{-t} \\ e^{3t} & 3e^{-t} \end{bmatrix}$

10. $\mathbf{x}' = \begin{bmatrix} 1 & 1 \\ -1 & 1 \end{bmatrix} \mathbf{x} + \begin{bmatrix} e^{2t} \\ 0 \end{bmatrix}$, $\Phi(t) = \begin{bmatrix} e^t \cos t & e^t \sin t \\ -e^t \sin t & e^t \cos t \end{bmatrix}$

11. $\mathbf{x}' = \begin{bmatrix} 1 & 1 \\ -1 & 1 \end{bmatrix} \mathbf{x} + \begin{bmatrix} 2t+2 \\ 0 \end{bmatrix}$, $\qquad \Phi(t) = \begin{bmatrix} e^t \cos t & e^t \sin t \\ -e^t \sin t & e^t \cos t \end{bmatrix}$

12. $\mathbf{x}' = \begin{bmatrix} 2 & 1 & 0 \\ 0 & 2 & 0 \\ 0 & 0 & 1 \end{bmatrix} \mathbf{x} + \begin{bmatrix} e^t \\ 1 \\ 0 \end{bmatrix}$, $\qquad \Phi(t) = \begin{bmatrix} e^{2t} & te^{2t} & 0 \\ 0 & e^{2t} & 0 \\ 0 & 0 & e^{-t} \end{bmatrix}$

3.8 Applications of linear systems

In this section, we consider three fundamental physical problems that may be modeled and studied using linear systems of differential equations.

3.8.1 Mixing problems

Through our study of the motivating example provided at the start of chapter 1 and reconsidered at the beginning of the current chapter, we have seen that mixing problems naturally lead to nonhomogeneous linear systems of differential equations. Below, we examine a slightly more complicated example.

Consider a system of three tanks connected in such a way that each of the tanks has an independent inflow that delivers salt solution to it, each has an independent outflow (drain), and each tank is connected to the other two with both outflow and inflow pipes. The relevant information about each tank is given in table 3.2.

We set up a system of differential equations whose solution represents the amount of salt in each tank at time t and state the system in matrix form. For tank A, we denote the amount of salt (in grams) in the tank at time t (in minutes) by $x_1(t)$. Similarly, we let $x_2(t)$ and $x_3(t)$ represent the amount of salt in tanks B and C. A careful check of the given data shows that for each tank the total rates

Table 3.2
Saltwater mixing in three tanks *A*, *B*, and *C*

	Tank *A*	Tank *B*	Tank *C*
Tank volume	50 liters	100 liters	200 liters
Rate of inflow to the tank	2 liters/min	4 liters/min	5 liters/min
Concentration of salt in inflow	0.25 g/liter	2 g/liter	0.9 g/liter
Rate of drain outflow	2 liters/min	4 liters/min	5 liters/min
Rates of outflows to other tanks	to *B*: 3 liters/min	to *C*: 1 liter/min	to *A*: 4 liters/min
Rates of outflows to other tanks	to *C*: 4 liters/min	to *A*: 3 liters/min	to *B*: 1 liter/min

of inflow and outflow of solution balance so that the volume of solution in each tank is constant.

From the given information on the independent inflow to the tank, we know that tank A gains salt at a rate of

$$0.25 \frac{\text{g}}{\text{liter}} \cdot 2 \frac{\text{liters}}{\text{min}} = 0.5 \frac{\text{g}}{\text{min}} \tag{3.8.1}$$

Furthermore, tank A also gains salt from the two inflows that come from tanks B and C. For tank B, which contains 100 liters of solution, solution flows to A at a rate of 3 liters/min with a concentration of $x_2(t)/100$ g/liter, so that salt is gained by tank A at a rate of

$$\frac{x_2}{100} \frac{\text{g}}{\text{liter}} \cdot 3 \frac{\text{liters}}{\text{min}} = \frac{3x_2}{100} \frac{\text{g}}{\text{min}} \tag{3.8.2}$$

Similarly, the flow from tank C to tank A results in A gaining salt at a rate of

$$\frac{x_3}{200} \frac{\text{g}}{\text{liter}} \cdot 4 \frac{\text{liters}}{\text{min}} = \frac{x_3}{50} \frac{\text{g}}{\text{min}} \tag{3.8.3}$$

Tank A is also losing salt through its three outflows: a drain, flow to tank B, and flow to tank C. Since the concentration of solution in tank A at time t is $x_1(t)/50$ g/liter, it follows that each outflow carries this concentration of salt, doing so at respective rates of 2 liters/min, 3 liters/min, and 4 liters/min. This shows that solution is leaving tank A at a cumulative rate of 9 liters/min, therefore causing the rate at which salt is lost from tank A to be

$$\frac{x_1}{50} \frac{\text{g}}{\text{liter}} \cdot 9 \frac{\text{liters}}{\text{min}} = \frac{9x_1}{50} \frac{\text{g}}{\text{min}} \tag{3.8.4}$$

Combining the rates of inflow and outflow in (3.8.1), (3.8.2), (3.8.3), and (3.8.4), it follows that $x_1(t)$ satisfies the differential equation

$$x_1' = 0.5 + \frac{3x_2}{100} + \frac{4x_3}{200} - \frac{9x_1}{50} \tag{3.8.5}$$

Similar reasoning shows that $x_2(t)$ and $x_3(t)$ satisfy the differential equations

$$x_2' = 8 + \frac{3x_1}{50} + \frac{x_3}{200} - \frac{8x_2}{100} \tag{3.8.6}$$

and

$$x_3' = 4.5 + \frac{4x_1}{50} + \frac{x_2}{100} - \frac{10x_3}{200} \tag{3.8.7}$$

Rearranging (3.8.5), (3.8.6), and (3.8.7) and writing the system they generate in matrix form, we see

$$\mathbf{x}' = \begin{bmatrix} -9/50 & 3/100 & 1/50 \\ 3/50 & -2/25 & 1/200 \\ 2/25 & 1/100 & -1/20 \end{bmatrix} \mathbf{x} + \begin{bmatrix} 0.5 \\ 8 \\ 4.5 \end{bmatrix} \tag{3.8.8}$$

We can easily determine the equilibrium solution to the system by setting $\mathbf{x}' = \mathbf{0}$ and row-reducing the resulting linear system of equations. Doing so results in

$$
\begin{bmatrix}
-9/50 & 3/100 & 1/50 & -0.5 \\
3/50 & -2/25 & 1/200 & -8 \\
2/25 & 1/100 & -1/20 & -4.5
\end{bmatrix}
\rightarrow
\begin{bmatrix}
1 & 0 & 0 & 50 \\
0 & 1 & 0 & 150 \\
0 & 0 & 1 & 200
\end{bmatrix}
$$

so that $x_1 = 50$, $x_2 = 150$, $x_3 = 200$ is the only equilibrium solution to the system. In addition, the eigenpairs of the coefficient matrix $\mathbf{A}$ are approximately $\lambda = -0.030, -0.204, -0.076$ and $\mathbf{v} = [0.203 \ 0.346 \ 1]^{\mathrm{T}}, [-2.041 \ 0.949 \ 1]^{\mathrm{T}},$ $[-0.168 \ -1.250 \ 1]^{\mathrm{T}}$. Since all three eigenvalues are real and negative, we can conclude that the above equilibrium is a stable attracting node. Moreover, we can determine the general solution to the system. The eigenvalues and eigenvectors provide us with $\mathbf{x}_h$, the complementary solution, while $\mathbf{x}_p$ is given by the equilibrium solution so that

$$
\mathbf{x}(t) = c_1 e^{-0.030t}
\begin{bmatrix}
0.203 \\
0.346 \\
1
\end{bmatrix}
+ c_2 e^{-0.204t}
\begin{bmatrix}
-2.041 \\
0.949 \\
1
\end{bmatrix}
$$

$$
+ c_3 e^{-0.076t}
\begin{bmatrix}
-0.168 \\
-1.250 \\
1
\end{bmatrix}
+
\begin{bmatrix}
50 \\
150 \\
200
\end{bmatrix}
$$

We conclude from this example that three connected tanks generate a natural example of a linear system of nonhomogeneous differential equations. Certainly, we can envision similar ideas being applied to more complicated scenarios, such as the spread of a pollutant through a connected chain of rivers and lakes.

3.8.2 Spring-mass systems

In section 3.1, we developed the linear second-order differential equation that governs the behavior of a spring-mass system and converted the equation to a system of two first-order equations. In particular, we learned that for a system with mass m, spring constant k, damping constant c, and driving force $F(t)$, the displacement $y(t)$ of the mass from its equilibrium position satisfies the DE

$$
y'' + \frac{c}{m} y' + \frac{k}{m} y = \frac{1}{m} F(t) \tag{3.8.9}
$$

Moreover, using the substitution $x_1 = y$ and $x_2 = x_1' = y'$, it follows that (3.8.9) can be represented by the system

$$
x_1' = x_2
$$

$$
x_2' = -\frac{k}{m} x_1 - \frac{c}{m} x_2 + \frac{1}{m} F(t) \tag{3.8.10}
$$

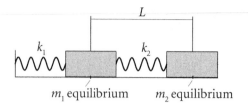

Figure 3.15 Two masses m_1 and m_2 joined
by two springs, at equilibrium.

Next, we consider the more complicated case of a system involving two
masses and two springs, but omit damping and driving forces. In particular,
suppose that a mass m_1 is attached to a spring with spring constant k_1 and that
from m_1 a second spring with constant k_2 and mass m_2 is attached, as shown
in figure 3.15. While we represent the masses with boxes, for our theoretical
work we assume we are working with point-masses, where all of the mass is
concentrated at a single point. We can envision these points as lying at the
centers of the respective boxes in figure 3.15.

To omit damping, we assume that the surface on which the masses rest is
frictionless. In addition, once the masses are set in motion by some collection
of initial displacements and velocities, we let $x_1(t)$ denote the displacement of
m_1 from its equilibrium position and $x_2(t)$ the displacement of m_2 from its
equilibrium position and set the system in motion, as shown in figure 3.16.

We seek a system of first-order differential equations that models this
situation. Note that m_1 has two springs attached to it, so each spring exerts
forces on m_1. One is $F_1 = -k_1 x_1$, which is the force the first spring exerts to
oppose the displacement of the first mass. Next, observe that when the system is
at equilibrium, the distance between the two masses is some constant L. Once
the system is set in motion, the distance between the two masses is $L + x_2 - x_1$.
As such, the second spring is being stretched a length of $x_2 - x_1$ beyond where
it is when the system is at equilibrium. On mass m_1 this exerts a force in the
opposite direction of F_1, specifically the force $F_2 = k_2(x_2 - x_1)$ on m_1. On the
second mass m_2 there is only this same force exerted by the second spring, but
in the opposite direction as on m_1. In particular, $F_3 = -k_2(x_2 - x_1)$ acts on m_2.

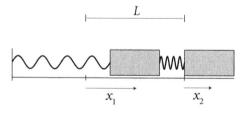

Figure 3.16 Two masses m_1 and m_2 and two
springs displaced from equilibrium.

Now, because we have omitted damping and forcing, these are the only forces acting on m_1 and m_2. Newton's second law tells us that the sum of all forces acting on an object must equal the object's mass times its acceleration. In particular, we have

$$m_1 x_1'' = -k_1 x_1 + k_2(x_2 - x_1)$$
$$m_2 x_2'' = -k_2(x_2 - x_1)$$

Dividing through by m_1 and m_2, respectively, these observations lead us to the system of linear second-order differential equations

$$x_1'' = -\frac{k_1}{m_1} x_1 + \frac{k_2}{m_1}(x_2 - x_1)$$
$$x_2'' = -\frac{k_2}{m_2}(x_2 - x_1) \tag{3.8.11}$$

To study the behavior of this system with the techniques that we have developed, we must convert each of the second-order equations to a system of two first-order equations. Before doing so, we introduce specific numerical values for the masses and spring constants to simplify our work. We let $k_1 = 2$ and $k_2 = 1$, and $m_1 = 2$ and $m_2 = 4$. This yields the system

$$x_1'' = -x_1 + 0.5(x_2 - x_1)$$
$$x_2'' = -0.25(x_2 - x_1) \tag{3.8.12}$$

Using the substitutions $y_1 = x_1$, $y_2 = y_1' = x_1'$, $y_3 = x_2$, $y_4 = y_3' = x_2'$, it follows that (3.8.12) results in the system of four first-order equations given by

$$y_1' = y_2$$
$$y_2' = -y_1 + 0.5(y_3 - y_1)$$
$$y_3' = y_4$$
$$y_4' = -0.25(y_3 - y_1) \tag{3.8.13}$$

Letting $\mathbf{y}$ be the vector $[y_1 \ y_2 \ y_3 \ y_4]^{\mathrm{T}}$, we can write (3.8.13) in matrix form,

$$\mathbf{y}' = \begin{bmatrix} 0 & 1 & 0 & 0 \\ -1.5 & 0 & 0.5 & 0 \\ 0 & 0 & 0 & 1 \\ 0.25 & 0 & -0.25 & 0 \end{bmatrix} \mathbf{y} \tag{3.8.14}$$

From this, we can now analyze the overall behavior of the coupled spring-mass system. In particular, the eigenvalues and eigenvectors of the coefficient matrix in (3.8.14) will enable us to find the general solution $\mathbf{y}$. Given initial conditions, we can fully describe the functions $y_i(t)$—particularly y_1 and y_3, which represent the respective displacements of the masses in the system—and understand the behavior of the system over time. This problem and others like it are explored further in the exercises at the end of this section.

3.8.3 RLC circuits

The flow of electricity through a circuit, much like the flow of water in a pipe, naturally involves relationships with rates of change. As such, the study of electrical current involves differential equations. Here, we explore some fundamental properties of electricity and how these lead to such equations.

Throughout what follows, we will make use of the analogy that the flow of charge carriers in an electrical circuit is like the flow of particles in a moving stream of water. Just as we consider flow of water in a pipe to be the number of water particles flowing past a given point during a certain time interval, the *current* $I(t)$ in a circuit at time t is proportional to the number of positive charge carriers that move past any given point per second in the conductor. Note particularly that current measures a rate of change of charge.

Current is measured in *amperes*(*amp*), the base unit through which all other units will be defined. One ampere corresponds to 6.2420×10^{18} charge carriers per second moving past a given point. The unit of charge is a *coulomb*, which is the amount of charge that flows through a cross section of a wire in one second when a one amp current is flowing. In other words,

$$1 \text{ amp} = 1 \text{ coulomb/s}$$

Here, we begin to see how derivatives and integrals are involved in the study of electricity. The current $I(t)$ at time t is by definition a rate of change of charge. Thus, by the Fundamental Theorem of Calculus, the total amount of charge that flows past a given point on a time interval $[t_0, t_1]$ is given by

$$\int_{t_0}^{t_1} I(s)\, ds \tag{3.8.15}$$

If we let $Q(t)$ measure the total accumulated charge at a given point in the circuit from time t_0 up to time t, then we have

$$Q(t) = Q(t_0) + \int_{t_0}^{t} I(s)\, ds \tag{3.8.16}$$

and therefore $Q'(t) = I(t)$.

As current flows through a circuit, the charge carriers and elements in the circuit exchange energy. We, therefore, define a *potential function* V throughout a circuit. The energy (per coulomb of charge) that has been exchanged by the charge carriers as they flow from point a to point b is computed as

$$V_{ab} = V_a - V_b$$

where V_a and V_b are the values of the potential function at points a and b in the circuit.

The difference V_{ab} is called the *voltage drop* from a to b and is measured in joules per coulomb, which are also known as *volts*. If we again think of the flow of water through a pipe, the concept of voltage drop is analogous to the change in water pressure between points a and b. Batteries, for example, maintain a voltage drop between two terminals; the energy provided by a battery's internal chemicals produces a constant amount of energy per coulomb as charge carriers move throughout the battery, which raises the function V by the voltage rating of the battery.

As current flows through a circuit, energy is lost. This makes the potential V at one end lower than the potential at the other. Over a portion of a circuit, say from a to b, where a substantial amount of energy is lost, we say that such a portion is called a *resistor*. Good examples of resistors are light bulbs and heating elements, because they show how electrical energy can be converted into light and heat.

The voltage drop across a resistor and the current flowing through it are modeled by *Ohm's law*, which says that the potential difference V_{ab} between the endpoints a and b of a resistor is proportional to the current flowing through the resistor. In other words,

$$V_{ab} = IR \tag{3.8.17}$$

where R is a constant called the *resistance*. The unit of resistance is the *ohm*, which is equal to one volt per ampere, or one volt-second per coulomb.

A changing electrical current $I(t)$ in a segment of a circuit will create a changing magnetic field that results in a voltage drop between the ends of a segment. When this effect is large, such as in a coil between points b and c (the effect can be magnified by different geometrical arrangements of the circuit), the device that induces the effect is called an *inductor*. *Faraday's law* tells us what happens with the voltage drops across inductors. In particular, the voltage drop across an inductor is proportional to the rate of change of the current, or, in other words

$$V_{bc} = L\frac{dI}{dt} \tag{3.8.18}$$

where L is a constant called the *inductance*. Note specifically that Faraday's law regards the rate of change of current. Inductance is measured in *henries*.

Finally, if a circuit is broken and we include two plates separated by an insulating material (such as air), and the terminals of the circuit are connected to a voltage source (such as a battery), then charges will build up on the plates. In the ongoing analogy to water, this is similar to a tank used to store water to provide a source of pressure. We call the set of plates a *capacitor*, and speak of the total charge $Q(t)$ on the capacitor.

From (3.8.16), since we know that current I is the rate of change of charge Q, if we know an initial charge $Q(t_0)$, then given a current $I(t)$ we can find the charge $Q(t)$ by the relationship

$$Q(t) = Q(t_0) + \int_{t_0}^{t} I(s)\,ds \tag{3.8.19}$$

Finally, *Coulomb's law* states that the voltage drop V_{cd} across a capacitor between points c and d is proportional to the charge on the capacitor, or

$$V_{cd} = \frac{1}{C}Q(t) = \frac{1}{C}\left(Q(t_0) + \int_{t_0}^{t} I(s)\,ds \right) \tag{3.8.20}$$

where C is called the *capacitance* of the capacitor and is measured in *farads*.

All three of the laws (3.8.17), (3.8.18), and (3.8.20) are based on experimental observations of circuits. Similarly, *Kirchoff's law* is a conservation law that tells us what we can expect for the voltage drops across various parts of a circuit. Simply stated, Kirchoff's law says that if we pick a sequence of points in a closed circuit, then the sum of the voltage drops across these segments is zero. Specifically, for points $a_1, a_2, \ldots, a_n$,

$$V_{a_1 a_2} + V_{a_2 a_3} + \cdots + V_{a_{n-1} a_n} + V_{a_n a_1} = 0 \qquad (3.8.21)$$

A final necessary law for us to consider is *Kirchoff's current law*, which tells us that at each point of a circuit, the sum of currents flowing into a point equals the sum of the currents flowing out. For a simple RLC circuit with one loop, Kirchoff's current law guarantees that we can use a single function $I(t)$ to model the current at any point at a given time t; for circuits with multiple loops, multiple functions $I(t)$ are needed.

Now we are prepared to see how these fundamental laws of electricity lead to a second-order differential equation, and hence a 2×2 system of first-order DEs. Let us consider an RLC circuit that consists of a resistor, inductor, and capacitor, along with some energy (voltage) source $E(t)$, arranged in series, as shown in figure 3.17. Kirchoff's law leads us directly to second-order differential equations that determine the behavior of the current $I(t)$ in the circuit and the charge $Q(t)$ on the capacitor.

By Ohm's law, we know that $V_{ab} = IR$. Similarly, Faraday's law implies that $V_{bc} = L\frac{dI}{dt}$ and Coulomb's law tells us that $V_{cd} = \frac{1}{C}Q(t) = \frac{1}{C}\left(Q(t_0) + \int_{t_0}^{t} I(s)\,ds\right)$. Finally, we know from the voltage source that $V_{da} = -E(t)$. Kirchoff's law now yields the equation $V_{ab} + V_{bc} + V_{cd} + V_{da} = 0$, or

$$RI(t) + LI'(t) + \frac{1}{C}Q(t) = E(t) \qquad (3.8.22)$$

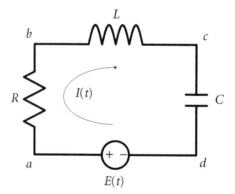

Figure 3.17 An RLC circuit with resistance R, inductance L, capacitance C, and energy source $E(t)$.

Recalling that $Q'(t) = I(t)$, we may rewrite (3.8.22) in two different ways. If we differentiate both sides of (3.8.22), and rearrange the terms in decreasing order of derivatives, it follows immediately that the current $I(t)$ must satisfy the linear second-order differential equation

$$LI''(t) + RI'(t) + \frac{1}{C}I(t) = E'(t) \tag{3.8.23}$$

If instead we substitute Q' for I in (3.8.22), then we see that Q is the solution to the linear second-order differential equation

$$LQ''(t) + RQ'(t) + \frac{1}{C}Q(t) = E(t) \tag{3.8.24}$$

We can therefore study the behaviors of different RLC circuits based on the given resistance, inductance, capacitance, and supplied voltage. Moreover, as we well know, any such linear second-order differential equation may be converted to a system of first-order equations. For example, letting $x_1 = I$ and $x_2 = I'$, we can convert (3.8.23) to the system of equations

$$x_1' = x_2$$

$$x_2' = -\frac{1}{CL}x_1 - \frac{R}{L}x_2 + \frac{1}{L}E'(t)$$

Example 3.8.1   Determine all solutions $I(t)$ for an RLC circuit when $L = 20$ H, $R = 80$ Ω, $C = 10^{-2}$ F, and the external voltage is given by the function $E(t) = 50\sin 2t$.

Solution.   From (3.8.23) and the given information, we can immediately determine the second-order differential equation that $I(t)$ satisfies. In particular, since $E(t) = 50\sin 2t$, we have $E'(t) = 100\cos 2t$, and using the values for L, C, and R, $I(t)$ is a solution to the equation

$$20I'' + 80I' + 100I = 100\cos 2t \tag{3.8.25}$$

Using the substitution $x_1 = I$ and $x_2 = I'$ and multiplying both sides of (3.8.25) by $1/20$, the system becomes

$$x_1' = x_2$$

$$x_2' = -5x_1 - 4x_2 + 5\cos 2t$$

From this, we can write the system in matrix form as

$$\mathbf{x}' = \begin{bmatrix} 0 & 1 \\ -5 & -4 \end{bmatrix} \mathbf{x} + \begin{bmatrix} 0 \\ 5\cos 2t \end{bmatrix} \tag{3.8.26}$$

For the coefficient matrix $\mathbf{A}$ in (3.8.26), we compute the eigenvalues and eigenvectors in order to find the complementary solution $\mathbf{x}_h$ of the system.

Doing so, we find that $\mathbf{A}$ has complex eigenvalues and eigenvectors; one eigenvalue-eigenvector pair is

$$\lambda = -2 + i, \ \mathbf{v} = \begin{bmatrix} -2 - i \\ 5 \end{bmatrix}$$

Writing

$$\mathbf{z}(t) = e^{(-2+i)t} \begin{bmatrix} -2 - i \\ 5 \end{bmatrix}$$

we know from theorem 3.5.2 that the real and imaginary parts of the vector function $\mathbf{z}(t)$ will form two real linearly independent solutions to the homogeneous system $\mathbf{x}' = \mathbf{A}\mathbf{x}$. Rewriting $\mathbf{z}$ using Euler's formula,

$$\mathbf{z}(t) = e^{-2t}(\cos t + i\sin t)\left(\begin{bmatrix} -2 \\ 5 \end{bmatrix} + i\begin{bmatrix} -1 \\ 0 \end{bmatrix}\right)$$

$$= e^{-2t} \begin{bmatrix} -2\cos t + \sin t \\ 5\cos t \end{bmatrix} + ie^{-2t}\begin{bmatrix} -\cos t - 2\sin t \\ 5\sin t \end{bmatrix}$$

The real and imaginary parts of $\mathbf{z}$ are real linearly independent solutions to $\mathbf{x}' = \mathbf{A}\mathbf{x}$, so we have determined that the complementary solution to the original system is

$$\mathbf{x}_h = c_1 e^{-2t} \begin{bmatrix} -2\cos t + \sin t \\ 5\cos t \end{bmatrix} + c_2 e^{-2t}\begin{bmatrix} -\cos t - 2\sin t \\ 5\sin t \end{bmatrix}$$

In theory, we are now ready to apply variation of parameters to find a particular solution $\mathbf{x}_p$. While we could do so here, the computations get remarkably cumbersome. In the next chapter on higher order differential equations, we will learn that for certain higher order equations, making a good guess at the form of a particular solution provides the simplest approach. In fact, we will even see that keeping certain second-order equations in that form, rather than converting them to systems of first-order equations, often is the best way to proceed.

For now, we will guess a form for $\mathbf{x}_p$. Since

$$\mathbf{b}(t) = \begin{bmatrix} 0 \\ 5\cos 2t \end{bmatrix}$$

we assume that a particular solution $\mathbf{x}_p$ has form

$$\mathbf{x}_p = \begin{bmatrix} a\cos 2t + b\sin 2t \\ c\cos 2t + d\sin 2t \end{bmatrix}$$

From this, it follows

$$\mathbf{x}'_p = \begin{bmatrix} -2a\sin 2t + 2b\cos 2t \\ -2c\sin 2t + 2d\cos 2t \end{bmatrix}$$

Substituting $\mathbf{x}_p$ and $\mathbf{x}'_p$ for $\mathbf{x}$ and $\mathbf{x}'$ in (3.8.26), we have

$$\begin{bmatrix} -2a\sin 2t + 2b\cos 2t \\ -2c\sin 2t + 2d\cos 2t \end{bmatrix} = \begin{bmatrix} c\cos 2t + d\sin 2t \\ -5a\cos 2t - 5b\sin 2t - 4c\cos 2t - 4d\sin 2t \end{bmatrix} + \begin{bmatrix} 0 \\ 5\cos 2t \end{bmatrix}$$

Equating the coefficients of $\sin 2t$ and $\cos 2t$ in the entries of the vectors in this most recent vector equation, the following system of four linear equations in a, b, c, and d arises:

$$-2a = d$$

$$2b = c$$

$$-2c = -5b - 4d$$

$$2d = -5a - 4c + 5$$

Rearranging this system to write it in matrix form and row-reducing, we observe

$$\begin{bmatrix} -2 & 0 & 0 & -1 & 0 \\ 0 & 2 & -1 & 0 & 0 \\ 0 & 5 & -2 & 4 & 0 \\ -5 & 0 & 4 & 2 & 5 \end{bmatrix} \rightarrow \begin{bmatrix} 1 & 0 & 0 & 0 & 1/13 \\ 0 & 1 & 0 & 0 & 8/13 \\ 0 & 0 & 1 & 0 & 16/13 \\ 0 & 0 & 0 & 1 & -2/13 \end{bmatrix}$$

Thus we conclude that a particular solution is

$$\mathbf{x}_p = \begin{bmatrix} 1/13\cos 2t + 8/13\sin 2t \\ 16/13\cos 2t - 2/13\sin 2t \end{bmatrix}$$

In conjunction with our earlier work to find $\mathbf{x}_h$, we have determined that the general solution to the system of first-order differential equations given by (3.8.25) is

$$\mathbf{x} = c_1 e^{-2t} \begin{bmatrix} -2\cos t + \sin t \\ 5\cos t \end{bmatrix} + c_2 e^{-2t} \begin{bmatrix} -\cos t - 2\sin t \\ 5\sin t \end{bmatrix} + \begin{bmatrix} 1/13\cos 2t + 8/13\sin 2t \\ 16/13\cos 2t - 2/13\sin 2t \end{bmatrix}$$

Recalling that $x_1 = I$ is the current in the given RLC circuit, we have shown that

$$I(t) = c_1 e^{-2t}(-2\cos t + \sin t) + c_2 e^{-2t}(-\cos t - 2\sin t) + \frac{1}{13}\cos 2t + \frac{8}{13}\sin 2t$$

Given initial conditions for $I(0)$ and $I'(0)$, we can find the values of the constants c_1 and c_2. Moreover, we note that as $t \to \infty$, the components of the solution that include e^{-2t} will die off, leaving us with long-term behavior of $I(t)$ modeled by $\frac{1}{13}\cos 2t + \frac{8}{13}\sin 2t$. We hence call $\frac{1}{13}\cos 2t + \frac{8}{13}\sin 2t$ the *steady-state solution* of the original equation (3.8.25) and $c_1 e^{-2t}(-2\cos t + \sin t) + c_2 e^{-2t}(-\cos t - 2\sin t)$ the *transient solution*.

Overall, we have now seen several examples of important phenomena governed by linear systems of differential equations. Further examples will be considered in the exercises.

Exercises 3.8

1. In a closed system of two tanks (i.e, one for which there are no input flows and no output flows), the following information is given. Tank A is filled with 100 liters of solution whose initial concentration is 0.25 g/liter. Tank B is filled with 50 liters of solution whose initial concentration is 1 g/liter. The two tanks are connected with two pipes having flows in opposite direction; mixed solution from Tank A flows to Tank B at a rate of 4 liters/min. Similarly, mixed solution flows from Tank B to Tank A at a rate of 4 liters/min.

   Set up and solve an initial-value problem whose solution will tell you the amount of salt in each tank at time t. Discuss the graphical behavior of the solution $\mathbf{x}(t)$ (whose components are the amount of salt in each tank at time t). Is there an equilibrium solution to the system? If so, what is it?

2. Consider a system of two tanks connected in such a way that each of the tanks has an independent inflow that delivers salt solution to it, each has an independent outflow (drain), and each tank is connected to the other with an outflow and an inflow. The relevant information about each tank is given in the table below.

	Tank A	Tank B
Tank volume	100 liters	200 liters
Rate of inflow to the tank	5 liters/min	9 liters/min
Concentration of salt in inflow	7 g/liter	3 g/liter
Rate of drain outflow	4 liters/min	10 liters/min
Rates of outflows to other tank	to B: 3 liters/min	to A: 2 liters/min

   Initially, Tank A has 20 g of salt present in its solution, and Tank B has 75 g of salt present in its solution.

   Set up and solve an initial-value problem whose solution will determine the amount of salt in each tank at time t. Discuss the graphical behavior of the solution $\mathbf{x}(t)$ (whose components are the amount of salt in each tank at time t). Is there an equilibrium solution to the system? If so, what is it?

3. Suppose that in exercise 2 all of the given information remains the same except for the fact that instead of saltwater flowing into each tank, pure water flows in. How do the results of your work in exercise 2 change?

4. In a closed system of three tanks (that is, one for which there are no input flows and no output flows), the following information is given.

	Tank A	Tank B	Tank C
Tank volume	100 liters	150 liters	125 liters
Rates of outflows to other tanks	to B: 3 liters/min	to C: 1 liters/min	to A: 4 liters/min
Rates of outflows to other tanks	to C: 4 liters/min	to A: 3 liters/min	to B: 1 liter/min

Tank A is filled with 100 liters of solution whose initial concentration is 8 g/liter. Tank B is filled with 150 liters of solution whose initial concentration is 3 g/liter. Tank C is initially filled with 125 liters of pure water. The three tanks are connected with pipes having flows in opposite directions; flow rates are given in the table above.

Set up and solve an initial-value problem whose solution will tell you the amount of salt in each tank at time t. Discuss the graphical behavior of the solution $\mathbf{x}(t)$ (whose components are the amount of salt in each tank at time t). Is there an equilibrium solution to the system? If so, what is it?

5. In a system of three tanks of saltwater, the following information is given.

	Tank A	Tank B	Tank C
Tank volume	400 liters	200 liters	300 liters
Rate of inflow to the tank	7 liters/min	0 liters/min	0 liters/min
Concentration of salt in inflow	10 g/liter	n/a	n/a
Rate of drain outflow	0 liters/min	0 liters/min	7 liters/min
Rates of outflows to other tanks	to B: 7 liters/min	to C: 7 liters/min	to A: 0 liters/min
Rates of outflows to other tanks	to C: 0 liters/min	to A: 0 liters/min	to B: 0 liters/min

Each tank is full; tank A contains solution whose initial concentration is 20 g/liter. Tank B contains solution whose initial concentration is 50 g/liter. Tank C contains pure water.

Without setting up a system of differential equations, first use your intuition to describe what you think will be the behavior of the functions $x_1(t)$, $x_2(t)$, and $x_3(t)$ that measure the amount of salt in each of the three respective tanks at time t.

Then, set up and solve an initial-value problem whose solution will tell you the amount of salt in each tank at time t. Discuss the graphical behavior of each component of the solution $\mathbf{x}(t)$ and compare it to your intuitive expectations. Is there an equilibrium solution to the system? If so, what is it?

6. In a system of three tanks of saltwater interconnected with pipes of inflow and outflow to and from each, the following information is given.

	Tank A	Tank B	Tank C
Tank volume	400 liters	800 liters	500 liters
Rate of inflow to the tank	5 liters/min	10 liters/min	5 liters/min
Concentration of salt in inflow	25 g/liter	15 g/liter	40 g/liter
Rate of drain outflow	4 liters/min	7 liters/min	9 liters/min
Rates of outflows to other tanks	to B: 6 liters/min	to C: 5 liters/min	to A: 4 liters/min
Rates of outflows to other tanks	to C: 4 liters/min	to A: 5 liters/min	to B: 1 liter/min

Assume that the system is such that initially there is a concentration of 10 g/liter of salt in each of the three tanks. Set up and solve an initial-value problem whose solution will tell you the amount of salt in each tank at time t. Discuss the graphical behavior of each component of the solution $\mathbf{x}(t)$. Is there an equilibrium solution to the system? If so, what is it?

7. Recall that for a spring-mass system of mass m, spring constant k, and damping constant c, the displacement $y(t)$ of the mass from equilibrium is governed by the linear second-order differential equation

$$y'' + \frac{c}{m}y' + \frac{k}{m}y = \frac{1}{m}F(t)$$

For a mass of 0.5 kg with spring constant $k = 2$ N/m in an undamped, unforced system, assume the mass is displaced 0.4 m from equilibrium and released (i.e., $y(0) = 0.4$ and $y'(0) = 0$).

(a) State the second-order IVP that models this situation.

(b) Convert the second-order equation to a system of first-order DEs using the standard substitution: $x_1 = y$, $x_2 = y'$.

(c) Solve the system in (b), and graph the component function $x_1(t)$. Discuss the long-term behavior of the spring-mass system.

8. For a mass of 0.5 kg with spring constant $k = 2$ N/m and damping constant $c = 0.5$ N·s/m in an unforced system, assume the mass is displaced 0.3 m from equilibrium and released.

(a) State the second-order IVP that models this situation.

(b) Convert the second-order equation to a system of first-order DEs using the standard substitution: $x_1 = y$, $x_2 = y'$.

(c) Solve the system in (b), and graph the component function $x_1(t)$. Discuss the long-term behavior of the spring-mass system.

9. For a mass of 0.5 kg with spring constant $k = 2$ N/m and damping constant $c = 0.5$ N·s/min a forced system with forcing function $F(t) = \cos 2t$ N, assume the mass is initially displaced 0.3 m from equilibrium and released.

(a) State the second-order IVP that models this situation.

(b) Convert the second-order equation to a system of first-order DEs using the standard substitution: $x_1 = y$, $x_2 = y'$.

(c) Use variation of parameters to solve the system in (b), and graph the component function $x_1(t)$. Discuss the long-term behavior of the spring-mass system.

10. In section 3.8.2, we considered a system of two masses attached to two springs in parallel, where a mass m_1 is attached to a spring with spring constant k_1 and from m_1 a second spring with constant k_2 and mass m_2 is attached. See figure 3.16.

If we assume that the surface on which the masses rest is frictionless and let let $x_1(t)$ denote the displacement of m_1 from its equilibrium position and $x_2(t)$ the displacement of m_2 from its equilibrium position and set the system in motion, then the system is governed by the system of second order differential equations

$$x_1'' = -\frac{k_1}{m_1}x_1 + \frac{k_2}{m_1}(x_2 - x_1)$$

$$x_2'' = -\frac{k_2}{m_2}(x_2 - x_1)$$

(a) Suppose that $k_1 = 2$, $m_1 = 1$, $k_2 = 4$ and $m_2 = 0.5$. Using the given constant values and the substitution $y_1 = x_1$, $y_2 = y_1' = x_1'$, $y_3 = x_2$, $y_4 = y_3' = x_2'$, convert the system of two second-order equations to a system of four first-order equations.

(b) Assume that the masses m_1 and m_2 are each displaced 1 unit from their natural equilibrium and released. That is, assume $x_1(0) = 1$, $x_1'(0) = 0$,

$x_2(0) = 1$, and $x_2'(0) = 0$. Solve this initial-value problem using the system in (a) and sketch the plots of y_1 and y_3 and discuss what they tell you about the system.

11. Recall that the current $I(t)$ in an RLC circuit is governed by the linear second-order differential equation

$$LI''(t) + RI'(t) + \frac{1}{C}I(t) = E'(t)$$

where L is the inductance, R the resistance, and C the capacitance of the circuit.

Suppose we have an RLC circuit for which an inductor of $L = 1$ henry and capacitor $C = 0.01$ farad are present. Assume further that $I(0) = 100$ and $I'(0) = 0$.

(a) State a second-order IVP whose solution is $I(t)$, the current at time t.
(b) Convert the IVP in (a) to a system of first-order IVPs using a standard substitution.
(c) Solve the system in (b) to determine the current $I(t)$ in the cases where the resistance is (i) $R = 0$ Ω, (ii) $R = 16$ Ω, (iii) $R = 20$ Ω, and (iv) $R = 25$ Ω, assuming consistent units. Sketch a plot of each solution $I(t)$ and discuss the impact that changing R has on the current.

12. Suppose we have an RLC circuit for which an inductor of $L = 1$ H, resistor $R = 16$ Ω, and capacitor $C = 0.01$ F are present. Assume further that $I(0) = 100$ A and $I'(0) = 0$. Finally, suppose that the system is provided a voltage source of $E(t) = 100 \sin 10t$

(a) State a second-order IVP whose solution is $I(t)$, the current in the circuit at time t.
(b) Convert the IVP in (a) to a system of first-order IVPs using a standard substitution.
(c) Solve the system in (b) to determine the current $I(t)$ at time t. Sketch a plot of the solution $I(t)$ and discuss the impact the forcing function has on the current.

3.9 For further study

3.9.1 Diagonalizable matrices and coupled systems

We have seen that in the case where a system of linear first-order differential equations is *uncoupled*, such as

$$\begin{bmatrix} x_1' \\ x_2' \end{bmatrix} = \begin{bmatrix} 3 & 0 \\ 0 & -2 \end{bmatrix} \begin{bmatrix} x_1 \\ x_2 \end{bmatrix} = \begin{bmatrix} 3x_1 \\ -2x_2 \end{bmatrix}$$

the system is particularly straightforward to solve. In addition, even when the coefficient matrix $\mathbf{A}$ of the system $\mathbf{x}' = \mathbf{A}\mathbf{x}$ is not a diagonal matrix, in the case where $\mathbf{A}$ is $n \times n$ and has n real, linearly independent eigenvectors, it is again a straightforward exercise to determine the general solution to $\mathbf{x}' = \mathbf{Ax}$. In what follows, we investigate the connections between $\mathbf{A}$ having n real linearly independent eigenvectors and the system being uncoupled.

(a) Solve the uncoupled system of linear first-order equations

$$\begin{bmatrix} x_1' \\ x_2' \end{bmatrix} = \begin{bmatrix} 3 & 0 \\ 0 & -2 \end{bmatrix} \begin{bmatrix} x_1 \\ x_2 \end{bmatrix} = \begin{bmatrix} 3x_1 \\ -2x_2 \end{bmatrix}$$

by directly solving the two individual equations $x_1' = 3x_1$ and $x_2' = -2x_2$.

(b) For the coefficient matrix

$$\mathbf{A} = \begin{bmatrix} 3 & 0 \\ 0 & -2 \end{bmatrix}$$

how are your solutions in (a) to the individual differential equations related to the eigenvalues and eigenvectors of $\mathbf{A}$?

(c) Determine the eigenvalues and eigenvectors of the matrix $\mathbf{A} = \begin{bmatrix} 1 & 6 \\ 5 & 2 \end{bmatrix}$ and show that $\mathbf{A}$ has two real, linearly independent eigenvectors.

(d) Let $\mathbf{D}$ be the diagonal 2×2 matrix whose diagonal entries are λ_1 and λ_2, the eigenvalues of $\mathbf{A}$ from (c), and let $\mathbf{P}$ be the 2×2 matrix whose columns are $\mathbf{x}_1$ and $\mathbf{x}_2$, the eigenvectors of $\mathbf{A}$ corresponding to λ_1 and λ_2. Show that $\mathbf{AP} = \mathbf{PD}$.

(e) More generally, let $\mathbf{A}$ be an $n \times n$ matrix with n linearly independent real eigenvectors $\mathbf{x}_1, \mathbf{x}_2, \ldots, \mathbf{x}_n$ that correspond to real eigenvalues $\lambda_1, \lambda_2, \ldots, \lambda_n$. As in (d), let $\mathbf{D}$ be the diagonal matrix whose diagonal entries are the eigenvalues of $\mathbf{A}$ and $\mathbf{P}$ be the matrix whose columns are the corresponding eigenvectors of $\mathbf{A}$. Explain why $\mathbf{AP} = \mathbf{PD}$ and thus why $\mathbf{A} = \mathbf{PDP}^{-1}$ and $\mathbf{D} = \mathbf{P}^{-1}\mathbf{DP}$.

A real $n \times n$ matrix $\mathbf{A}$ with the property that it has n real, linearly independent eigenvectors is called *diagonalizable*. When we factor $\mathbf{A}$ in the form $\mathbf{A} = \mathbf{PDP}^{-1}$, we say that we have *diagonalized* the matrix $\mathbf{A}$.

(f) For a 2×2 diagonalizable matrix $\mathbf{A}$, consider the system of differential equations given by $\mathbf{x}' = \mathbf{Ax}$. Let $\mathbf{D}$ and $\mathbf{P}$ be the matrices defined above in (d). Note that in this problem $\mathbf{A}$ is a arbitrary diagonalizable matrix: we are not specifying the values of λ_1 and λ_2, nor the values of the entries in the corresponding eigenvectors.

   (i) Let $\mathbf{y} = \mathbf{P}^{-1}\mathbf{x}$. Show that $\mathbf{x}' = \mathbf{Py}'$.

   (ii) Use the substitution $\mathbf{y} = \mathbf{P}^{-1}\mathbf{x}$ and the fact that $\mathbf{A} = \mathbf{PDP}^{-1}$ to show that the original system $\mathbf{x}' = \mathbf{Ax}$ may be equivalently represented by the system $\mathbf{y}' = \mathbf{Dy}$.

   (iii) Explain why the system $\mathbf{y}' = \mathbf{Dy}$ is preferable to the system $\mathbf{x}' = \mathbf{Ax}$.

(g) For the matrix $\mathbf{A} = \begin{bmatrix} 1 & 6 \\ 5 & 2 \end{bmatrix}$, solve the system $\mathbf{x}' = \mathbf{Ax}$ by executing the following steps.

   (i) Diagonalize $\mathbf{A}$ by determining matrices $\mathbf{D}$ and $\mathbf{P}$ such that $\mathbf{A} = \mathbf{PDP}^{-1}$. Recall that $\mathbf{D}$ is the diagonal matrix whose diagonal entries are the eigenvalues of $\mathbf{A}$ and $\mathbf{P}$ is the matrix whose columns are the corresponding eigenvectors of $\mathbf{A}$.
   (ii) Follow your work in (f) to introduce a substitution that converts the system $\mathbf{x}' = \mathbf{Ax}$ to a new system in the variable $\mathbf{y}$ that is uncoupled and of the form $\mathbf{y}' = \mathbf{Dy}$.
   (iii) Solve the uncoupled system in (ii) for $\mathbf{y}$.
   (iv) Determine the solution $\mathbf{x}$ to the original system by showing that $\mathbf{x} = \mathbf{Py}$ and using this substitution appropriately.

(h) Solve the system $\mathbf{x}' = \mathbf{Ax}$ given by

$$\mathbf{A} = \begin{bmatrix} 2 & 1 \\ 1 & 2 \end{bmatrix}$$

using the approach outlined in (g).

(i) Solve the system $\mathbf{x}' = \mathbf{Ax}$ given by

$$\mathbf{A} = \begin{bmatrix} 3 & -1 & 1 \\ -1 & 3 & -1 \\ 1 & -1 & 3 \end{bmatrix}$$

using the approach outlined in (g).

(j) Compare your work in (g)–(i) to how you learned to solve the system $\mathbf{x}' = \mathbf{Ax}$ in section 3.3. Is this new approach fundamentally the same or is it markedly different? Explain.

3.9.2 Matrix exponential

An important result in calculus is that e^x can be represented by its Taylor series expansion

$$e^x = 1 + x + \frac{x^2}{2!} + \frac{x^3}{3!} + \cdots + \frac{x^n}{n!} + \cdots \tag{3.9.1}$$

and that (3.9.1) holds for every real value of x. In what follows, we explore the notion of $e^{\mathbf{A}}$, where $\mathbf{A}$ is a matrix, through the use of an analogous expansion, as well as the role of $e^{\mathbf{A}}$ in the solution of systems of differential equations of the form $\mathbf{x}' = \mathbf{Ax}$.

(a) Let $\mathbf{A}$ be the diagonal matrix

$$\mathbf{A} = \begin{bmatrix} 3 & 0 \\ 0 & -2 \end{bmatrix}$$

Explain why

$$\mathbf{A}^n = \begin{bmatrix} 3^n & 0 \\ 0 & (-2)^n \end{bmatrix}$$

(b) For the matrix **A** in (a), show that

$$\mathbf{I} + \mathbf{A} + \frac{1}{2!}\mathbf{A}^2 + \cdots + \frac{1}{n!}\mathbf{A}^n = \begin{bmatrix} 1 + 3 + \frac{3^2}{2!} + \cdots + \frac{3^n}{n!} & 0 \\ 0 & 1 - 2 + \frac{(-2)^2}{2!} + \cdots + \frac{(-2)^n}{n!} \end{bmatrix}$$

$$(3.9.2)$$

Based on the entries in the right-hand matrix of (3.9.2), explain why it is reasonable to write that

$$e^{\mathbf{A}} = \mathbf{I} + \mathbf{A} + \frac{1}{2!}\mathbf{A}^2 + \frac{1}{3!}\mathbf{A}^3 + \cdots + \frac{1}{n!}\mathbf{A}^n + \cdots \qquad (3.9.3)$$

We use (3.9.3) as the definition of $e^{\mathbf{A}}$ for any diagonal matrix **A**.

(c) Now consider the matrix $\mathbf{B} = \begin{bmatrix} 2 & -2 \\ -2 & -1 \end{bmatrix}$. Find the eigenvalues and eigenvectors of **B** and diagonalize **B** by writing

$$\mathbf{B} = \mathbf{PDP}^{-1}$$

where **D** is the diagonal matrix whose diagonal entries are the eigenvalues of **B** and **P** is the matrix whose columns are the corresponding eigenvectors of **B**. For more on the notion of a matrix being 'diagonalizable', see subsection 3.9.1.

(d) For an arbitrary diagonalizable matrix **B** for which $\mathbf{B} = \mathbf{PDP}^{-1}$ (where **D** and **P** have the meaning ascribed in (c)), show that

$$\mathbf{B}^n = \mathbf{PD}^n\mathbf{P}^{-1}$$

(e) For an arbitrary diagonalizable matrix **B**, explain why

$$\mathbf{I} + \mathbf{B} + \frac{1}{2!}\mathbf{B}^2 + \frac{1}{3!}\mathbf{B}^3 + \cdots + \frac{1}{n!}\mathbf{B}^n + \cdots = \mathbf{P}\left(\mathbf{I} + \mathbf{D} + \frac{1}{2!}\mathbf{D}^2 + \frac{1}{3!}\mathbf{D}^3 + \cdots \right.$$

$$\left. + \frac{1}{n!}\mathbf{D}^n + \cdots \right)\mathbf{P}^{-1}$$

again where **D** and **P** have the meaning ascribed in (c). We thus define $e^{\mathbf{B}}$ for any diagonalizable matrix **B** by the equation

$$e^{\mathbf{B}} = \mathbf{I} + \mathbf{B} + \frac{1}{2!}\mathbf{B}^2 + \frac{1}{3!}\mathbf{B}^3 + \cdots + \frac{1}{n!}\mathbf{B}^n + \cdots \qquad (3.9.4)$$

(f) Show that if **B** is any diagonalizable matrix such that $\mathbf{B} = \mathbf{PDP}^{-1}$ (where **D** and **P** have the meaning ascribed in (c)), then

$$e^{\mathbf{B}} = \mathbf{P}e^{\mathbf{D}}\mathbf{P}^{-1}$$

(g) Use the result in (f) to compute $e^{\mathbf{B}}$ for the specific matrix $\mathbf{B}$ given in (c).

(h) Recall that when we solve a single homogeneous linear first-order DE such as

$$y' = 5y$$

one way to solve the equation is to guess that the solution is $y = e^{rt}$ and work to determine the value of r that satisfies the DE. Of course we find that $r = 5$ and $y = Ce^{5t}$ is the general solution. Indeed, for any constant a, the solution to $y' = ay$ is $y = Ce^{at}$.

Now let this consider solving the system of differential equations

$$\mathbf{x}' = \mathbf{Ax} = \begin{bmatrix} 3 & 0 \\ 0 & -2 \end{bmatrix} \tag{3.9.5}$$

noting that $\mathbf{A}$ is the diagonal matrix from (a) above.

(i) Viewing t as a scalar multiplier of $\mathbf{A}$, update your work from (3.9.3) to write a series expansion for $e^{\mathbf{A}t}$.

(ii) Noting that $e^{\mathbf{A}t}$ is a matrix, explain why it is reasonable to guess that $\Psi(t) = e^{\mathbf{A}t}$ is a solution matrix for the system $\mathbf{x}' = \mathbf{Ax}$.

(iii) Using your expression from (i) for $\Psi(t) = e^{\mathbf{A}t}$, compute both $\Psi'(t)$ and $\mathbf{A}\Psi(t)$ to verify that the matrix function $\Psi(t)$ satisfies the equation $\Psi'(t) = \mathbf{A}\Psi(t)$.

4

Higher order differential equations

4.1 Motivating equations

Through our study of linear systems of differential equations, we have already encountered higher order differential equations that arise naturally in physical applications. Two particularly important ones are those associated with spring-mass systems and RLC circuits. Here, we briefly revisit these equations.

In section 3.1, we considered a mass m suspended from a spring with spring constant k that is subject to damping with proportionality constant c. If $F(t)$ is an external forcing function on the system, then the displacement $y(t)$ of the mass from equilibrium satisfies

$$my'' + cy' + ky = F(t) \tag{4.1.1}$$

This is a nonhomogeneous linear second-order differential equation. While we have already studied this equation by using the substitution $x_1 = y$ and $x_2 = y'$ and considered the resulting linear system of first-order differential equations, there is further insight to be gained by examining (4.1.1) solely as a second-order equation. In fact, while it is theoretically possible to solve (4.1.1) using the corresponding linear system and ideas from chapter 3, doing so in the cases where $F(t) \neq 0$ is often cumbersome; we will see in section 4.4 that this equation may often be solved in a straightforward manner by leaving it in its original form as a second-order equation.

In section 3.8, we encountered another important nonhomogeneous linear second-order differential equation. By viewing the flow of electricity through a circuit as analogous to the flow of water in a pipe, we came to understand a differential equation that models the current $I(t)$. Using results from physics, including Ohm's law, Faraday's law, and Coulomb's law, we learned that the current $I(t)$ must satisfy the linear second-order differential equation

$$LI'' + RI' + \frac{1}{C}I = E'(t) \tag{4.1.2}$$

where L is the inductance, R is the resistance, C is the capacitance, and $E(t)$ represents an external voltage source.

We note specifically that the governing differential equations for spring-mass systems and RLC circuits are both linear nonhomogeneous second-order differential equations with constant coefficients. These differential equations therefore merit further study as we endeavor to more fully understand these physical systems. When the damping constant $c = 0$ and the resistance $R = 0$ in (4.1.1) and (4.1.2), these equations are often called *harmonic oscillator equations*. When small damping or resistance is present, we refer to them as *damped* harmonic oscillators.

4.2 Homogeneous equations: distinct real roots

If we consider our experience with single homogeneous linear first-order differential equations and systems thereof, we realize that the exponential function plays a central role in their solution. For example, if we solve the equation

$$y' - 5y = 0$$

the solution is $y = ce^{5t}$. Likewise, if we solve the system given by $\mathbf{x}' = \mathbf{Ax}$, where $\mathbf{A}$ is a matrix with eigenvalues $\lambda = 2$ and $\lambda = -3$, then the general solution is

$$\mathbf{x} = c_1 e^{2t} \mathbf{v}_1 + c_2 e^{-3t} \mathbf{v}_2$$

where $\mathbf{v}_1$ and $\mathbf{v}_2$ are eigenvectors that correspond to the eigenvalues $\lambda = 2$ and $\lambda = -3$.

Given this prominence of the exponential function, it is not surprising that functions of the form $y = e^{rt}$ play a central role in our study of higher order equations. For example, consider the second-order linear homogeneous differential equation with constant coefficients given by

$$y'' - y' - 6y = 0 \tag{4.2.1}$$

Even without our experience with first-order equations and systems, it is reasonable to think that one or more functions of the form $y = e^{rt}$ will be a solution to this equation because of the question the equation begs: "what function y is such that its second derivative minus its first derivative is equal to 6 times itself?" In essence, we are looking for a function y such that a certain linear combination of the function, its first derivative, and its second derivative, is the zero function. This makes it natural for us to expect that the solution is such that its derivatives are scalar multiples of itself, hence leading us to consider $y = e^{rt}$.

Letting $y = e^{rt}$, we observe that $y' = re^{rt}$ and $y'' = r^2 e^{rt}$. Substituting these functions into (4.2.1) requires r to satisfy the equation

$$r^2 e^{rt} - re^{rt} - 6e^{rt} = 0 \tag{4.2.2}$$

Factoring, we can rewrite (4.2.2) as

$$e^{rt}(r^2 - r - 6) = 0$$

and since e^{rt} is never zero, it follows that r must be such that $r^2 - r - 6 = (r-3)(r+2) = 0$. From this, $r = 3$ or $r = -2$, and therefore $y_1 = e^{3t}$ and $y_2 = e^{-2t}$ are both solutions to (4.2.1).

Since $y_1 = e^{3t}$ is not a scalar multiple of $y_2 = e^{-2t}$, it follows that y_1 and y_2 are linearly independent solutions to (4.2.1). Through our work with homogeneous linear systems, we are accustomed to taking linear combinations of linearly independent solutions in order to form a general solution; the same principle holds here, which we will verify directly. Letting $y = c_1 y_1 + c_2 y_2 = c_1 e^{3t} + c_2 e^{-2t}$, it follows that $y' = 3c_1 e^{3t} - 2c_2 e^{-2t}$ and $y'' = 9c_1 e^{3t} + 4c_2 e^{-2t}$. If we now consider $y'' - y' - 6y$, we have

$$y'' - y' - 6y = (9c_1 e^{3t} + 4c_2 e^{-2t}) - (3c_1 e^{3t} - 2c_2 e^{-2t}) - 6(c_1 e^{3t} + c_2 e^{-2t})$$

$$= (9c_1 e^{3t} - 3c_1 e^{3t} - 6c_1 e^{3t}) + (4c_2 e^{-2t} + 2c_2 e^{-2t} - 6c_2 e^{-2t})$$

$$= 0$$

Thus, we have shown that every function of the form $y = c_1 e^{3t} + c_2 e^{-2t}$ is a solution to (4.2.1). This shows that the solution space of (4.2.1) is at least two-dimensional; might there be any other linearly independent solutions to the equation? By our earlier work with systems, we know that the solution space of the equation $\mathbf{x}' = \mathbf{A}\mathbf{x}$, where $\mathbf{A}$ is $n \times n$, is n-dimensional. Since the second-order equation (4.2.1) can be converted to a 2×2 system of equations, it follows that its solution space has dimension exactly 2, and thus

$$y = c_1 e^{3t} + c_2 e^{-2t} \tag{4.2.3}$$

is the general solution to (4.2.1).

Our work to show that if y_1 and y_2 are solutions to (4.2.1), then $y = c_1 y_1 + c_2 y_2$ is also a solution may be generalized to any homogeneous linear second-order differential equation. We state this result in the following theorem.

Theorem 4.2.1 If y_1 and y_2 are solutions to the second-order linear homogeneous equation

$$y'' + a(t)y' + b(t)y = 0$$

then $y = c_1 y_1 + c_2 y_2$ is also a solution for any constants c_1 and c_2.

The important roles the constants c_1 and c_2 play are further exemplified by initial-value problems. For example, if we consider the initial-value problem

$$y'' - y' - 6y = 0, \quad y(0) = 2, \quad y'(0) = 1 \qquad (4.2.4)$$

we can show that this IVP has a unique solution. Using the general solution $y(t) = c_1 e^{3t} + c_2 e^{-2t}$, the condition $y(0) = 2$ implies that

$$2 = c_1 + c_2 \qquad (4.2.5)$$

Differentiating the general solution, we find that $y'(t) = 3c_1 e^{3t} - 2c_2 e^{-2t}$, and therefore $y'(0) = 1$ implies

$$1 = 3c_1 - 2c_2 \qquad (4.2.6)$$

Equations (4.2.5) and (4.2.6) form a linear system of two equations in two unknowns. Solving this system, $c_1 = 1$ and $c_2 = 1$, so that the function

$$y(t) = e^{3t} + e^{-2t}$$

is the unique solution to (4.2.4).

Our work with the example equation $y'' - y' - 6y = 0$ is indicative of many broader trends in the study of second-order linear differential equations. Because such equations can be converted to systems, we should not be at all surprised to learn that a broad class of initial-value problems associated with second-order equations have unique solutions, nor that the general solution to a second-order equation belongs to a two-dimensional solution space. We state two theorems in order to formalize these observations.

Theorem 4.2.2   Consider the second-order initial-value problem given by

$$y'' + p(t)y' + q(t)y = f(t) \quad y(t_0) = y_0, \quad y'(t_0) = y_1 \qquad (4.2.7)$$

where the coefficient functions $p(t)$ and $q(t)$ and the forcing function $f(t)$ are continuous on an open interval (a, b). Given any t_0 in (a, b), (4.2.7) has a unique solution in (a, b).

While the proof of theorem 4.2.2 is beyond the scope of this book, it is notable that in the case that $p(t)$ and $q(t)$ are constant functions, we can prove the theorem. Indeed, we will do so by actually constructing the solution in various cases in this section and those following.

Just as we almost exclusively considered matrices **A** with constant entries in our work with systems of linear first-order differential equations of the form $\mathbf{x}' = \mathbf{Ax}$, in our study of second-order linear differential equations, we will normally consider the situation where the coefficient functions $p(t)$ and $q(t)$ are constant. For this context, we can deduce the following result.

Theorem 4.2.3   The set of all solutions to the second-order homogeneous linear differential equation $y'' + a_1 y' + a_0 y = 0$, where a_0 and a_1 are constants, is a vector space of dimension 2.

This result can be viewed as a consequence of theorem 3.3.2 for linear systems of differential equations with constant coefficients. In particular, given

$$y'' + a_1 y' + a_0 y = 0 \qquad (4.2.8)$$

if we use the standard substitution $x_1 = y$, $x_2 = y'$, then it follows that (4.2.8) is equivalent to the system

$$\mathbf{x}' = \mathbf{A}\mathbf{x} = \begin{bmatrix} 0 & 1 \\ -a_0 & -a_1 \end{bmatrix} \mathbf{x}$$

which has a two-dimensional solution space.

Thus, in order to solve (4.2.8), we seek two linearly independent solutions that satisfy the equation. In particular, if we can find two functions $y_1 = e^{r_1 t}$ and $y_2 = e^{r_2 t}$ that are both solutions to (4.2.8), where $r_1 \neq r_2$, then the general solution must be

$$y = c_1 e^{r_1 t} + c_2 e^{r_2 t}$$

More specifically, if we recall our earlier approach following (4.2.1) in the first example in this section, we made the assumption that a solution y has form $y = e^{rt}$. Doing so and substituting in the general equation $y'' + a_1 y' + a_0 y = 0$, we see that r must satisfy

$$r^2 e^{rt} + a_1 r e^{rt} + a_0 e^{rt} = 0 \qquad (4.2.9)$$

Since e^{rt} is never zero, it follows that r must be a solution of the *characteristic equation* of the second-order homogeneous linear equation (4.2.8), which is

$$r^2 + a_1 r + a_0 = 0 \qquad (4.2.10)$$

If r_1 and r_2 are the roots of (4.2.10), then it follows that $y_1 = e^{r_1 t}$ and $y_2 = e^{r_2 t}$ are both solutions to the original equation (4.2.8). In particular, if $r_1 \neq r_2$, then y_1 and y_2 are linearly independent and we have found the general solution to (4.2.8), which is

$$y = c_1 e^{r_1 t} + c_2 e^{r_2 t}$$

We state this result formally in the following theorem.

Theorem 4.2.4   Given the second-order linear differential equation with constant coefficients

$$y'' + a_1 y' + a_0 y = 0$$

if the characteristic equation $r^2 + a_1 r + a_0 = 0$ has two distinct real roots r_1 and r_2, then the general solution to (4.2.4) is

$$y = c_1 e^{r_1 t} + c_2 e^{r_2 t}$$

We close this section with an example.

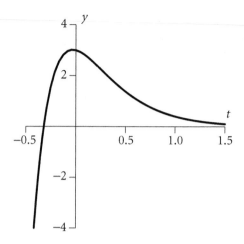

Figure 4.1 A plot of the solution $y(t)$ to the IVP given in (4.2.11).

Example 4.2.1 Solve the second-order initial-value problem given by

$$y'' + 7y' + 12y = 0, \quad y(0) = 3, \quad y'(0) = -1 \qquad (4.2.11)$$

Graph the solution and discuss its long-term behavior.

Solution. We begin by assuming that $y = e^{rt}$. Direct substitution into (4.2.11) and removing the factor e^{rt} results in the characteristic equation

$$r^2 + 7r + 12 = 0$$

Factoring, we find that $(r + 3)(r + 4) = 0$, and therefore, $r = -3$ or $r = -4$. Since the two r values are distinct, it follows that $y_1 = e^{-3t}$ and $y_2 = e^{-4t}$ are linearly independent solutions to (4.2.11) and the general solution is

$$y = c_1 e^{-3t} + c_2 e^{-4t} \qquad (4.2.12)$$

Applying the given initial conditions, we can solve for c_1 and c_2. Since $y(0) = 3$ and $y'(0) = -1$, (4.2.12) implies that

$$3 = c_1 + c_2$$

$$-1 = -3c_1 - 4c_2$$

It follows $c_1 = 11$ and $c_2 = -8$, and thus the unique solution to the given IVP (4.2.11) is $y = 11e^{-3t} - 8e^{-4t}$. Plotting $y(t)$ results in the graph shown in figure 4.1, where we clearly see the given initial behavior at $t = 0$ (the function value is 3 and the slope of the tangent line is -1) and that the solution's long-term behavior is that $y(t) \to 0$ as $t \to \infty$.

We can also observe from the negative constants present in the exponents of the general solution $y = c_1 e^{-3t} + c_2 e^{-4t}$, that every such solution must tend to zero as $t \to \infty$. We note that $y = 0$ is the only constant (equilibrium) solution to the original equation $y'' + 7y' + 12y = 0$, and that because every solution tends to $y = 0$, we say $y = 0$ is a stable equilibrium.

Exercises 4.2   In exercises 1–7, determine the general solution to the given second-order homogeneous linear DE.

1. $y'' - y' - 12y = 0$

2. $y'' + y' - 2y = 0$

3. $y'' - y = 0$

4. $y'' + 3y' = 0$

5. $y'' = 0$

6. $y'' + 4y' + 3y = 0$

7. $y'' + y' - y = 0$

In exercises 8–14, solve the stated IVP. In addition, graph your solution and discuss its long-term behavior. Note that the general solution to each equation has been found in exercises 1–7.

8. $y'' - y' - 12y = 0$, $\quad y(0) = -4$, $\quad y'(0) = 1$

9. $y'' + y' - 2y = 0$, $\quad y(0) = 2$, $\quad y'(0) = 2$

10. $y'' - y = 0$, $\quad y(0) = 1$, $\quad y'(0) = -1$

11. $y'' + 3y' = 0$, $\quad y(0) = 2$, $\quad y'(0) = 3$

12. $y'' = 0$, $\quad y(0) = -3$, $\quad y'(0) = 1$

13. $y'' + 4y' + 3y = 0$, $\quad y(0) = -2$, $\quad y'(0) = -6$

14. $y'' + y' - y = 0$, $\quad y(0) = 9$, $\quad y'(0) = -3$

In exercises 15–19, construct a second-order homogeneous linear DE having the given functions as solutions.

15. $y_1 = e^{-2t}$, $y_2 = e^{2t}$

16. $y_1 = e^{5t}$, $y_2 = e^{-3t}$

17. $y_1 = e^{4t}$, $y_2 = 1$

18. $y_1 = e^{2t}$, $y_2 = e^{3t}$

19. $y_1 = 1$, $y_2 = t$

20. Consider the second-order homogeneous linear equation
$y'' - 6y' + 9y = 0$.

(a) Use the substitution $y = e^{rt}$ to attempt to find two linearly independent solutions to the given equation.

(b) Explain why your work in (a) only results in one linearly independent solution, $y_1(t)$.

(c) Verify by direct substitution that $y_2 = te^{3t}$ is a solution to $y'' - 6y' + 9y = 0$. Explain why this function is linearly independent from y_1 found in (a).

(d) State the general solution to the given equation.

21. Consider the second-order homogeneous linear equation $y'' - 2y' + 5y = 0$.

(a) Use the substitution $y = e^{rt}$ to attempt to find two linearly independent solutions to the given equation.

(b) Explain why your work in (a) does not generate any real solutions to the given equation.

(c) Verify by direct substitution that $y_1 = e^t \cos 2t$ and $y_2 = e^t \sin 2t$ are solutions to $y'' - 2y' + 5y = 0$. Explain why these functions are linearly independent.

(d) State the general solution to the given equation.

22. Consider the second-order homogeneous linear equation $y'' + 4y = 0$.

(a) Use the substitution $y = e^{rt}$ to attempt to find two linearly independent solutions to the given equation.

(b) Explain why your work in (a) does not generate any real solutions to the given equation.

(c) Think about familiar functions that can satisfy the condition that "the second derivative equals -4 times the function itself." By making a natural guess and verifying by direct substitution, find two linearly independent functions y_1 and y_2 that satisfy the given differential equation.

(d) State the general solution to the given equation.

Recall that in a spring-mass system, the displacement $y(t)$ of the mass from its natural equilibrium is governed by the equation

$$y'' + \frac{c}{m}y' + \frac{k}{m}y = \frac{1}{m}F(t)$$

where c is the damping constant, k is the spring constant, m is the mass of the suspended object, and F is the forcing function.

23. For an unforced system with $c = 3$, $k = 2$, and $m = 1$, determine the displacement of the mass at time t if the system is set in motion via the initial conditions $y(0) = 2$, $y'(0) = 1$. Sketch a graph of the solution you determine and discuss the long-term behavior of the spring-mass system. Assume consistent units on all constants.

24. For an unforced spring-mass system with $k = 9$, $c = 12$, and $m = 3$, determine the displacement of the mass from equilibrium at time t if $y(0) = 0$ and $y'(0) = -1$. Assume consistent units on all constants.

Recall that in a standard RLC electrical circuit, the current $I(t)$ satisfies the equation

$$LI''(t) + RI'(t) + \frac{1}{C}I(t) = E'(t)$$

where L is the inductance, R is the resistance, C is the capacitance, and $E(t)$ represents an external voltage source.

25. For an RLC circuit with no external voltage source, $L = 20$, $R = 80$, and $C = 1/60$, determine the current at time t given the initial conditions $I(0) = 100$, $I'(0) = 25$. Graph the solution you determine and discuss the long-term behavior of the current. Assume consistent units on all constants.

26. For an RLC circuit with no external voltage source, $L = 20$, $R = 0$, and $C = 1/60$, determine the current at time t given the initial conditions $I(0) = 100$, $I'(0) = 25$. Graph the solution you determine and discuss the long-term behavior of the current. Assume consistent units on all constants.

4.3 Homogeneous equations: repeated and complex roots

In the preceding section, we observed that any time the characteristic equation of the second-order equation $y'' + a_1 y' + a_0 y$ has two real, distinct roots, the general solution of the differential equation is easily determined. However, in an equation such as

$$y'' - 6y' + 9y = 0 \qquad (4.3.1)$$

with characteristic equation $r^2 - 6r + 9 = 0$, the only root of this equation is $r = 3$. Although this leads us to the solution $y_1 = e^{3t}$, we do not immediately see how to find a second linearly independent solution. In a similar way, the equation

$$y'' - 2y' + 5y = 0 \qquad (4.3.2)$$

has characteristic equation is $r^2 - 2r + 5 = 0$ and its roots are

$$r = 1 \pm 2i$$

In this case, we see that no real solution to (4.3.2) results using our previous approach, so it remains for us to find two real linearly independent solutions. Now we will endeavor to understand how to address these two cases: when roots of the characteristic equation are repeated and when the roots of the characteristic equation are complex.

4.3.1 Repeated roots

Let us consider the second-order homogeneous linear DE given by

$$y'' + 4y' + 4y = 0 \qquad (4.3.3)$$

Its characteristic equation is $r^2 + 4r + 4 = (r+2)^2 = 0$, so that only the solution $y_1 = e^{-2t}$ results from the guess that $y = e^{rt}$. To find a second linearly independent solution, it is natural to think that we need to somehow complicate the function $y = e^{-2t}$, just as we did in section 3.5 when we encountered the similar case where the coefficient matrix of a 2×2 system of linear first-order DEs had a repeated eigenvalue.

Thus, we consider a second potential solution

$$y_2 = v(t)e^{-2t}$$

where $v(t)$ is a function yet to be determined. By using this function and substituting into the equation $y'' + 4y' + 4y = 0$, we find conditions that $v(t)$ must satisfy. First, observe by the product rule that

$$y_2' = -2ve^{-2t} + v'e^{-2t} \tag{4.3.4}$$

Similarly,

$$y_2'' = 4ve^{-2t} - 4v'e^{-2t} + v''e^{-2t} \tag{4.3.5}$$

Next, substituting into (4.3.3), we find

$$
\begin{aligned}
0 &= y_2'' + 4y_2' + 4y_2 \\
&= (4ve^{-2t} - 4v'e^{-2t} + v''e^{-2t}) + 4(-2ve^{-2t} + v'e^{-2t}) + 4(ve^{-2t}) \\
&= v''e^{-2t} \tag{4.3.6}
\end{aligned}
$$

Since e^{-2t} is never zero, it follows that $v''(t)$ must equal zero for all values of t. This implies that $v(t)$ can be any linear function. Because all we seek is one function $y_2 = v(t)e^{-2t}$ that is a solution to (4.3.3) and is linearly independent from $y_1 = e^{-2t}$, it suffices to choose $v(t) = t$. Specifically,

$$y_2 = te^{-2t}$$

is a second linearly independent solution to (4.3.3). The general solution is therefore

$$y(t) = c_1 e^{-2t} + c_2 te^{-2t}$$

The condition we derived at (4.3.6) for $v(t)$ will hold in any situation where the characteristic equation of a second-order linear homogeneous DE has a repeated root. This leads us to state the following theorem.

Theorem 4.3.1 For any second-order linear homogeneous differential equation of the form

$$y'' + 2ky' + k^2 y = 0$$

whose characteristic equation has repeated real root $r = -k$, the general solution to the differential equation is

$$y = c_1 e^{-kt} + c_2 te^{-kt}$$

Before proceeding to the case of complex roots, we consider one example to demonstrate theorem 4.3.1 at work.

Example 4.3.1   Determine the general solution to the equation

$$y'' - 10y' + 25y = 0 \qquad (4.3.7)$$

Solution.   The characteristic equation of the given DE is $r^2 - 10r + 25 = (r-5)^2 = 0$, which has the repeated root $r = 5$. By theorem 4.3.1, it follows that the general solution to (4.3.7) is

$$y = c_1 e^{5t} + c_2 t e^{5t}$$

4.3.2 Complex roots

We continue to be guided throughout our work with second-order linear homogeneous equations by the informed guess that the solution has form $y = e^{rt}$. When this guess and the corresponding characteristic equation result in two distinct, real values of r, we have found the general solution to the given differential equation. Likewise, we have just shown that when the characteristic equation has only one real root, we can still find the general solution to the DE. We next explore how, even in the complex case, we can find the general solution through our original guess, $y = e^{rt}$.

We return to the example

$$y'' - 2y' + 5y = 0 \qquad (4.3.8)$$

and recall that the roots of the characteristic equation are $r = 1 \pm 2i$. While this suggests that $z(t) = e^{(1+2i)t}$ should be a solution of the differential equation, the function $z(t)$ is complex-valued. When we encountered a similar situation in section 3.5 for a linear system whose coefficient matrix had complex eigenvalues and complex eigenvectors, we used Euler's formula to separate such a complex-valued function into real and imaginary parts in order to find real solutions. We proceed similarly here. Recall that Euler's formula states that $e^{i\theta} = \cos\theta + i\sin\theta$, so

$$e^{(a+bi)t} = e^{at} e^{ibt} = e^{at}(\cos bt + i\sin bt)$$

For the complex solution $z(t)$ to (4.3.8), we thus find that

$$z(t) = e^{(1+2i)t}$$

$$= e^t(\cos 2t + i\sin 2t)$$

$$= e^t \cos 2t + i e^t \sin 2t \qquad (4.3.9)$$

In (4.3.9), we see that $z(t)$ has been written in the form

$$z(t) = \text{Re}(z) + i\text{Im}(z)$$

where $\text{Re}(z)$ and $\text{Im}(z)$ are themselves real-valued functions of t. Based on our experience with systems of differential equations with complex-valued solutions, it is natural at this point to hope that both the real and imaginary parts of $z(t)$ will be linearly independent solutions to (4.3.8).

Indeed, if we let $y_1 = e^t \cos 2t$ and $y_2 = e^t \sin 2t$, then it can be shown by direct substitution that both y_1 and y_2 are solutions to (4.3.8). Because y_1 and y_2 are not scalar multiples of each other, these two functions are linearly independent, and therefore, by theorem 4.2.3, it follows that

$$y(t) = c_1 e^t \cos 2t + c_2 e^t \sin 2t$$

is the general solution to (4.3.8).

The direct substitution that is used to verify that the real and imaginary parts of $z(t)$ are solutions to the original equation is somewhat tedious, but not difficult. In fact, in the more general case where we have complex roots $a \pm bi$, it can be similarly verified by direct substitution into the corresponding second-order equation that $y_1 = e^{at} \cos bt$ and $y_2 = e^{at} \sin bt$ are each solutions to the equation. Note that this scenario implies that the characteristic equation has form $C(r) = 0$ where

$$C(r) = [r - (a + bi)][r - (a - bi)]$$

$$= r^2 - (a + bi)r - (a - bi)r + (a + bi)(a - bi)$$

$$= r^2 - 2ar + (a^2 + b^2) \tag{4.3.10}$$

This shows that, up to a scalar multiple of the equation, complex roots to the characteristic equation arise from second-order homogeneous linear differential equations of the form

$$y'' - 2ay' + (a^2 + b^2)y = 0 \tag{4.3.11}$$

Our work above now enables us to state a formal result on finding real, linearly independent solutions from complex-valued ones.

Theorem 4.3.2   Let a and b be real constants with $b \neq 0$. For the second-order homogeneous linear differential equation

$$y'' - 2ay' + (a^2 + b^2)y = 0$$

the roots of the corresponding characteristic equation are $r = a \pm bi$ and the general solution to the differential equation is given by

$$y = c_1 e^{at} \cos bt + c_2 e^{at} \sin bt$$

Note that it is precisely the presence of complex roots to the characteristic equation that produces the periodic functions $\cos bt$ and $\sin bt$ in the solution. In physical situations such as spring-mass systems and RLC circuits where we anticipate that solutions will have a sinusoidal component, we can expect that the characteristic equation will have complex roots.

We conclude this section by applying theorem 4.3.2 in the following example.

Example 4.3.2   Solve the initial-value problem given by

$$y'' + 2y' + 10y = 0, \quad y(0) = 1, \quad y'(0) = 1$$

Plot the solution and discuss its long-term behavior.

Solution.   We first find the general solution to the given differential equation. The corresponding characteristic equation is $r^2 + 2r + 10 = 0$, with roots

$$r = -1 \pm 3i$$

By theorem 4.3.2 it follows that the general solution is

$$y = c_1 e^{-t} \cos 3t + c_2 e^{-t} \sin 3t$$

To determine the solution to the stated IVP, first note that $y(0) = 1$ implies that

$$1 = c_1 e^0 \cos(0) + c_2 e^0 \sin(0)$$

so that $c_1 = 1$. In addition, since

$$y' = -c_1 e^{-t} \cos 3t - c_2 e^{-t} \sin 3t - 3c_1 e^{-t} \sin 3t + 3c_2 e^{-t} \cos 3t$$

it follows from the fact that $y'(0) = 1$ that

$$1 = -c_1 + 3c_2$$

Since $c_1 = 1$, we find that $c_2 = 2/3$ and hence the solution to the IVP is

$$y = e^{-t} \cos 3t + \frac{2}{3} e^{-t} \sin 3t$$

Plotting the function y in figure 4.2, we see that the function $y(t)$ oscillates due to the presence of the trigonometric functions, while $y(t) \to 0$ as $t \to \infty$ because of the damping effect of e^{-t}.

In fact, the graphical behavior demonstrated by $y(t)$ in figure 4.2 is precisely what we would expect if the given IVP was modeling a spring-mass system where relatively small damping is present: the mass will oscillate once sent in motion, but will eventually return to equilibrium.

Exercises 4.3   In exercises 1–9, use the characteristic equation to determine the general solution to the given second-order linear homogeneous differential equation.

1. $y'' - 8y' + 16y = 0$
2. $y'' + y' + y = 0$
3. $y'' + y' + \frac{1}{4}y = 0$
4. $y'' - 4y = 0$
5. $y'' + 4y = 0$
6. $y'' - 10y' + 50y = 0$

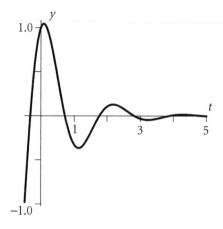

Figure 4.2 A plot of the solution $y(t)$ to the IVP given in example 4.3.2.

7. $y'' - 10y' + 25y = 0$

8. $y'' = 0$

9. $2y'' + 7y' + 5y = 0$

In exercises 10–18, solve the stated initial-value problem. In addition, graph your solution and discuss its long-term behavior. Note that the general solution to each equation has been found in corresponding problems in exercises 1–9.

10. $y'' - 8y' + 16y = 0$, $y(0) = -4$, $y'(0) = 1$

11. $y'' + y' + y = 0$, $y(0) = 2$, $y'(0) = 2$

12. $y'' + y' + \frac{1}{4}y = 0$, $y(0) = 0$, $y'(0) = -1$

13. $y'' - 4y = 0$, $y(0) = 7$, $y'(0) = -5$

14. $y'' + 4y = 0$, $y(0) = 2$, $y'(0) = 3$

15. $y'' - 10y' + 50y = 0$, $y(0) = -3$, $y'(0) = 1$

16. $y'' - 10y' + 25y = 0$, $y(0) = -2$, $y'(0) = -6$

17. $y'' = 0$, $y(0) = 0$, $y'(0) = 0$

18. $2y'' + 7y' + 5y = 0$, $y(0) = 9$, $y'(0) = -3$

19. Consider the second-order linear homogeneous equation
$y'' - 6y' + 9y = 0$.

(a) Find the general solution y of the given equation.
(b) Convert the given equation to a system $\mathbf{x}' = \mathbf{A}\mathbf{x}$ of two first-order equations using the substitution $x_1 = y$, $x_2 = y'$.

(c) Solve the system $\mathbf{x}' = \mathbf{A}\mathbf{x}$.

(d) Compare your results for y and x_1. What do you observe?

20. Consider the second-order linear homogeneous equation
$y'' + 6y' + 10y = 0$.

(a) Find the general solution y of the given equation.

(b) Convert the given equation to a system $\mathbf{x}' = \mathbf{A}\mathbf{x}$ of two first-order equations using the substitution $x_1 = y$, $x_2 = y'$.

(c) Solve the system $\mathbf{x}' = \mathbf{A}\mathbf{x}$.

(d) Compare your results for y and x_1. What do you observe?

21. Consider the general second-order linear homogeneous equation with constant coefficients given by
$$y'' + a_1 y' + a_0 y = 0$$
Under what conditions on a_1 and a_0 does the equation have two real distinct roots? one real repeated root? two distinct complex roots?

Recall that in a spring-mass system, the displacement $y(t)$ of the mass from its natural equilibrium is governed by the equation
$$y'' + \frac{c}{m}y' + \frac{k}{m}y = \frac{1}{m}F(t)$$
where c is the damping constant, k is the spring constant, m is the mass of the suspended object, and $F(t)$ is the forcing function. In the following exercises, we assume that units on all quantities and constants are consistent.

22. For an unforced spring-mass system with $c = 2$, $k = 1$, and $m = 1$, determine the displacement of the mass at time t if the system is set in motion with the initial conditions $y(0) = 2$, $y'(0) = 1$. Sketch the solution you determine and discuss the behavior of the spring-mass system.

23. For an unforced, undamped spring-mass system with $k = 9$ and $m = 3$, determine the displacement of the mass from equilibrium at time t if $y(0) = 2$ and $y'(0) = 1$. Sketch the solution you determine and discuss the behavior of the spring-mass system.

24. For an unforced spring-mass system with $c = 1$, $k = 2$, and $m = 1$, determine the displacement of the mass at time t if the system is set in motion with the initial conditions $y(0) = 2$, $y'(0) = 1$. Sketch the solution you determine and discuss the behavior of the spring-mass system.

Recall that in a standard RLC electrical circuit, the current $I(t)$ satisfies the equation
$$LI''(t) + RI'(t) + \frac{1}{C}I(t) = E'(t)$$
where L is the inductance, R is the resistance, C is the capacitance, and $E(t)$ represents an external voltage source. In the following exercises, we assume that units on all quantities and constants are consistent.

25. For an RLC circuit with no external voltage source, $L = 10$, $R = 40$, and $C = 1/40$, determine the current at time t given the initial conditions $I(0) = 100$, $I'(0) = 25$. Sketch the solution you determine and discuss the behavior of the current.

26. For an RLC circuit with no external voltage source, $L = 10$, $R = 40$, and $C = 1/50$, determine the current at time t given the initial conditions $I(0) = 100$, $I'(0) = 25$. Sketch the solution you determine and discuss the behavior of the current.

27. For an RLC circuit with no external voltage source, $L = 10$, $R = 0$, and $C = 1/90$, determine the current at time t given the initial conditions $I(0) = 100$, $I'(0) = 25$. Sketch the solution you determine and discuss the behavior of the current.

4.4 Nonhomogeneous equations

As motivated by a spring-mass system with a driving force or an RLC circuit with an external voltage source, we are now interested in solving second-order nonhomogeneous linear differential equations of the form

$$y'' + a_1 y' + a_0 y = f(t) \tag{4.4.1}$$

where $f(t)$ is not zero. We already know a theoretical way to solve such an equation: through the substitution $x_1 = y$ and $x_2 = y'$, we can convert (4.4.1) to a system of two first-order equations in the form $\mathbf{x}' = \mathbf{A}\mathbf{x} + \mathbf{b}$ and solve the two first-order DEs. While this approach works in theory, the actual execution of the process can be cumbersome. In fact, it is often much easier to solve (4.4.1) directly through the approaches we present in this section.

Analogous to several other types of linear algebraic and linear differential equations, a general principle from our work with nonhomogeneous equations guides us throughout: we first seek a *complementary solution* $y_h(t)$ to the corresponding homogeneous equation

$$y'' + a_1 y' + a_0 y = 0 \tag{4.4.2}$$

and then determine a *particular solution* $y_p(t)$ to the nonhomogeneous equation (4.4.1). It follows that $y = y_h + y_p$ will be the general solution to the nonhomogeneous equation. Indeed, we have the following theorem, a part of whose formal proof will be addressed in exercise 33 at the end of this section.

Theorem 4.4.1   Given the equation

$$y'' + a_1 y' + a_0 y = f(t) \tag{4.4.3}$$

where a_0 and a_1 are constants, if $y_h(t)$ is the general solution to the corresponding homogeneous equation $y'' + a_1 y' + a_0 y = 0$ and $y_p(t)$ is any solution to the nonhomogeneous equation (4.4.3) then $y = y_h + y_p$ is the general solution to (4.4.3).

We already understand how to find y_h, which depends entirely on the roots to the characteristic equation $r^2 + a_1 r + a_0 = 0$ as discussed in sections 4.2 and 4.3. It remains, however, to find y_p. To do so, we explore two methods: the guessing technique of undetermined coefficients, and the brute force technique of variation of parameters. Each of these methods is analogous to those that may be used to solve nonhomogeneous systems of the form $\mathbf{x}' = \mathbf{A}\mathbf{x} + \mathbf{b}$.

4.4.1 Undetermined coefficients

At this point in our discussion, examples are instructive. We consider several different nonhomogeneous linear second-order DEs to see how making reasonable guesses for the form of $y_p(t)$ can lead to the general solution in many elementary cases. Throughout, we use the following idea to guide our choice of the form of $y_p(t)$: since the first and second derivatives of many functions are similar to the original function (e.g., derivatives of sine and cosine functions are cosine and sine functions, derivatives of exponential functions are exponential functions, derivatives of polynomial functions are polynomials), and in equations of the form (4.4.3) we take linear combinations of y, y', and y'' to get $f(t)$, it is reasonable to guess that the form of $y_p(t)$ will be similar to the form of $f(t)$, the forcing function in the nonhomogeneous equation. We first see this for polynomial functions in the first example.

Example 4.4.1   Determine the general solution to

$$y'' - 3y' - 4y = 4t^2 + 2t - 9 \qquad (4.4.4)$$

Solution.   For the associated nonhomogeneous equation, $y'' - 3y' - 4y = 0$, by theorem 4.2.4 the complementary solution is $y_h = c_1 e^{-t} + c_2 e^{4t}$.

For a particular solution, we naturally guess that y_p has the form

$$y_p = at^2 + bt + c \qquad (4.4.5)$$

based on the form of the forcing function. The undetermined coefficients a, b, and c are found by direct substitution into (4.4.4). Note that $y_p' = 2at + b$ and $y_p'' = 2a$, so that from (4.4.4) we find

$$2a - 3(2at + b) - 4(at^2 + bt + c) = 4t^2 + 2t - 9$$

Rearranging the left-hand side of this equation, it follows

$$-4at^2 + (-6a - 4b)t + (2a - 3b - 4c) = 4t^2 + 2t - 9 \qquad (4.4.6)$$

Equating like coefficients of the power functions present in (4.4.6), the system of equations

$$-4a = 4$$

$$-6a - 4b = 2$$

$$2a - 3b - 4c = -9$$

must hold. We see that $a = -1$, from which it follows that $b = 1$ and $c = 1$ so that $y_p = -t^2 + t + 1$. Combining this with y_h, we have determined that the general solution to (4.4.4) is

$$y = c_1 e^{-t} + c_2 e^{4t} - t^2 + t + 1$$

We can imagine that if $f(t)$ was a polynomial other than $4t^2 + 2t - 9$, we would have guessed that y_p was a general polynomial of the same degree with unknown coefficients. This approach almost always works; we will discuss some exceptions that can arise after examples involving non-polynomial forcing functions.

Example 4.4.2   Determine the general solution to

$$y'' - y = 16e^{3t} \tag{4.4.7}$$

Solution.   Just as in example 4.4.1, we first solve the corresponding homogeneous equation and find y_h. Doing so, we observe that for $y'' - y = 0$, the solution y_h is

$$y_h = c_1 e^t + c_2 e^{-t}$$

For the particular solution, we use the natural guess that $y_p = Ae^{3t}$. From this, $y_p' = 3Ae^{3t}$ and $y_p'' = 9Ae^{3t}$, so substituting into (4.4.7), we find

$$9Ae^{3t} - Ae^{3t} = 16e^{3t}$$

Equating the coefficients of e^{3t}, it follows that $8A = 16$, so $A = 2$ and therefore $y_p = 2e^{3t}$.

Hence we have found the general solution of (4.4.7) to be

$$y = y_h + y_p = c_1 e^t + c_2 e^{-t} + 2e^{3t}$$

Here, we observe that if $f(t)$ in (4.4.7) were a different exponential function, say of the form $f(t) = Be^{kt}$, we would again guess that $y_p = Ae^{kt}$. This is based on the fact that our guess for y_p incorporates all the possible forms of the derivatives of $f(t)$. Just as with polynomial forcing functions, this approach almost always works. We will consider situations where these natural educated guesses can fail following one more example.

Example 4.4.3   Determine the general solution to
$$y'' - y' - 2y = 10\sin t \tag{4.4.8}$$

Solution.   First, we observe that the complementary solution can be shown to be

$$y_h = c_1 e^{2t} + c_2 e^{-t}$$

To find y_p, we guess that

$$y_p = A\sin t + B\cos t$$

Note that we must include the cosine function in y_p in order to account for the fact that the cosine function arises in the derivative of $f(t) = 10\sin t$.

From our guess for y_p, it follows that $y_p' = A\cos t - B\sin t$ and $y_p'' = -A\sin t - B\cos t$. Substituting in (4.4.8), we see that A and B must satisfy the equation

$$(-A\sin t - B\cos t) - (A\cos t - B\sin t) - 2(A\sin t + B\cos t) = 10\sin t \quad (4.4.9)$$

Rearranging (4.4.9) in order to compare coefficients of the sine and cosine functions, we have

$$(-A + B - 2A)\sin t + (-B - A - 2B)\cos t = 10\sin t$$

from which it follows that $-3A + B = 10$ and $-A - 3B = 0$. Consequently, $A = -3$ and $B = 1$, so that $y_p = -3\sin t + \cos t$. Therefore we have shown that the general solution of (4.4.8) is

$$y = y_h + y_p = c_1 e^{2t} + c_2 e^{-t} - 3\sin t + \cos t$$

In the more general setting where we imagine the forcing function $f(t)$ involving $\sin kt$ or $\cos kt$, it will be natural to make the guess that $y_p = A\sin kt + B\cos kt$, which again will work in most cases.

We have hinted that while the method of undetermined coefficients will usually work, it can occasionally fail. What can go wrong? First, if the forcing function $f(t)$ is particularly complicated, this can make determining a reasonable guess for y_p challenging. Moreover, even if $f(t)$ is a relatively simple function whose derivatives take on unusual forms—for example, $f(t) = \ln t$, where $f'(t)$ and $f''(t)$ are not logarithmic—we may find it difficult or impossible to find a form of y_p that works. These two situations will be addressed by the variation of parameters method that we introduce in the next subsection.

In addition, there is one more case in which undetermined coefficients can fail, yet the difficulty is straightforward to reconcile. An example is instructive.

Example 4.4.4   Find the general solution to the differential equation

$$y'' - y = 16e^{-t} \quad (4.4.10)$$

Solution.   Note that this differential equation is nearly identical to the one considered in example 4.4.2, but here the forcing function is $f(t) = 16e^{-t}$, rather than $f(t) = 16e^{3t}$.

As above, it still holds that $y_h = c_1 e^t + c_2 e^{-t}$. In addition, we naturally guess that $y_p = Ae^{-t}$, from which it follows that $y_p' = -Ae^{-t}$ and $y_p'' = Ae^{-t}$. Substituting in (4.4.10), we have

$$Ae^{-t} - Ae^{-t} = 16e^{-t}$$

But this last equality is clearly impossible, regardless of the value of A, since $0 = 16e^{-t}$ is never true.

We can determine where the method failed by observing that in this case, our guess for the particular solution y_p was actually part of the complementary solution. Note that $y_h = c_1 e^t + c_2 e^{-t}$, from which it follows that y_p cannot have the form Ae^{-t}, since this latter function belongs to y_h.

We therefore need a more complicated guess for y_p; a natural one to attempt is

$$y_p = Ate^{-t} \tag{4.4.11}$$

where we have introduced the additional multiplier t. From this, $y_p' = -Ate^{-t} + Ae^{-t}$ and $y_p'' = Ate^{-t} - Ae^{-t} - Ae^{-t}$. Substituting in (4.4.10), it now follows

$$(Ate^{-t} - 2Ae^{-t}) - (Ate^{-t}) = 16e^{-t}$$

Rearranging and simplifying this last equation in order to compare like coefficients of e^{-t} and te^{-t}, we see that the terms involving te^{-t} drop out and we are left with

$$-2Ae^{-t} = 16e^{-t}$$

so that $A = -8$ and $y_p = -8te^{-t}$.

We therefore have shown that the general solution is

$$y = y_h + y_p = c_1 e^t + c_2 e^{-t} - 8te^{-t}$$

The preceding example shows that if the form of the forcing function matches the form of one or more parts of the complementary solution y_h, then we have to use a different, more complicated guess for y_p than the most natural one. One more example will be helpful before we make some general conclusions.

Example 4.4.5 Find the general solution of

$$y'' - y' = 4t \tag{4.4.12}$$

Solution. From the characteristic equation $r^2 - r = 0$ for the corresponding homogeneous equation, we quickly deduce that

$$y_h = c_1 + c_2 e^t$$

Next, since $f(t) = 4t$, we naturally guess that y_p is a first order polynomial: $y_p = at + b$. From this, $y_p' = a$ and $y_p'' = 0$. Substituting in (4.4.12), we find

$$0 - a = 4t$$

Clearly, there is no value of a that makes $-a = 4t$ for all values of t, so there can be no particular solution y_p of the form $y_p = at + b$. From another perspective, we can see why this must be true by observing that the "b" in our guess for y_p is already part of the complementary solution since any constant function is a solution to $y'' - y' = 0$.

Therefore, we revise our guess for y_p and assume it has form $y_p = t(at + b) = at^2 + bt$. Doing so, we now have $y_p' = 2at + b$ and $y_p'' = 2a$, so substituting in (4.4.12) it follows

$$2a - (2at + b) = 4t$$

Rearranging so that we can equate like coefficients, we have

$$-2at + (2a - b) = 4t$$

so $-2a = 4$ and $2a - b = 0$. It follows that $a = -2$ and $b = -4$, and thus $y_p = -2t^2 - 4t$. Therefore, we have found the general solution of (4.4.12) to be

$$y = c_1 + c_2 e^t - 2t^2 - 4t$$

From our work with examples 4.4.1–4.4.5, we observe that the method of undetermined coefficients breaks down into two fundamental cases

Case 1. No functions in the assumed particular solution y_p are also solutions to the associated homogenous differential equation.

Case 2. A function in the assumed particular solution y_p is also a solution of the associated homogeneous differential equation.

Moreover, we can observe that when the forcing function $f(t)$ is a sum of polynomial, exponential, and sine and cosine functions, the linearity of the differential equation allows us to guess a form for y_p that is an appropriate sum of all the different types of functions represented. The following example shows some of the variety that arises in choosing the form of y_p.

Example 4.4.6   Write an appropriate guess for y_p for each of the following equations. Do not solve for the unknown coefficients.

(a) $y'' + y = 4e^{3t} + 5t^2$
(b) $y'' - 5y' - 6y = 3e^{-2t} + 4\cos 3t$
(c) $y'' - 2y' + 5y = 3te^t$
(d) $y'' - 4y' - 5y = 3e^{2t} \sin t$

Solution.

(a) The forcing function $f(t) = 4e^{3t} + 5t^2$ combines an exponential function and a second degree polynomial, so we would guess that
$y_p = Ae^{3t} + bt^2 + ct + d$.

(b) The natural guess is $y_p = Ae^{-2t} + B\cos 3t + C\sin 3t$ to account for the exponential and trigonometric functions present.

(c) $f(t) = 3te^t$ is a product of a linear function and an exponential one. Its derivatives will be sums of functions of the same form and constant multiples of exponential functions, so we assume that
$y_p = Ate^t + Be^t = e^t(At + B)$.

(d) We observe that every derivative of $f(t) = 3e^{2t} \sin t$ is the sum of functions of the form $Ae^{2t} \cos t + Be^{2t} \sin t$, so that we would guess that
$y_p = Ae^{2t} \cos t + Be^{2t} \sin t$.

Note the general rule we are using in case 1 and example 4.4.6: provided the terms of $f(t)$ do not belong to y_h, the form of y_p is a linear combination of all linearly independent functions that are generated by repeated differentiation of the forcing function $f(t)$.

For dealing with equations that fall into case 2, we make a guess y_p that is a sum of functions similar to those present in $f(t)$. We then have to tack on powers of t to modify any parts of y_p that already appear in y_h. In particular, we use the rule that if any part of y_p contains terms that duplicate terms in y_h, then we must multiply that part by t^n using the smallest possible value of n to eliminate the duplication.

For example, if we wanted to solve $y'' + 4y' + 4 = 3e^{-2t}$, which has characteristic equation $r^2 + 4r + 4 = (r+2)^2 = 0$, our work in section 4.3 implies that

$$y_h = c_1 e^{-2t} + c_2 t e^{-2t}$$

Therefore, for the form of y_p, which we initially might assume to be $y_p = Ae^{-2t}$, we see that we must in fact introduce a multiplier of t^2 in order to ensure that y_p does not appear in y_h. Thus, the appropriate form of y_p is $y_p = At^2 e^{-2t}$.

A few more examples of the possibilities that arise in case 2 are useful.

Example 4.4.7 Write an appropriate trial solution y_p for each of the following examples. Do not solve for the unknown coefficients.

(a) $y'' - y = 4e^t + 5e^{-t}$
(b) $y'' + 4y = 4\cos 2t$
(c) $y'' - 2y' + y = 3te^t$

Solution.

(a) Observe from the characteristic equation $r^2 - 1 = 0$ that $y_h = c_1 e^t + c_2 e^{-t}$, so both parts of the forcing function appear in y_h. We therefore assume that $y_p = Ate^t + Bte^{-t}$.

(b) The characteristic equation is $r^2 + 4 = 0$ with roots $r = \pm 2i$. It follows that $y_h = c_1 \sin 2t + c_2 \cos 2t$. Since $\cos 2t$ appears in the forcing function, and both $\sin 2t$ and $\cos 2t$ arise in y_h, the appropriate guess for y_p is $y_p = At \cos 2t + Bt \sin 2t$.

(c) Note that the characteristic equation is $r^2 - 2r + 1 = (r-1)^2 = 0$ so that $y_h = c_1 e^t + c_2 t e^t$. Since te^t is included in y_h, this implies that we must choose $y_p = At^2 e^t$.

Obviously the method of undetermined coefficients requires us to be experienced with a wide range of examples and to understand how the derivatives of the forcing function behave. The exercises at the end of this section will provide further practice in this regard.

4.4.2 Variation of parameters

Recall that we are focusing on solving the nonhomogeneous linear second-order equation

$$y'' + a_1 y' + a_0 y = f(t)$$

While the method of undetermined coefficients works well for a reasonable collection of forcing functions, it has some fairly strict limitations. In particular, it is unclear whether it is possible to make a reasonable guess for y_p in order to solve an equation such as $y'' + 4y' - 5y = \ln t$. In fact, we cannot: the derivative of the logarithm function is not a logarithm, and this is the main issue that prevents the use of this method.[1]

Here, we study a method that will enable us, in theory, to solve a much wider class of nonhomogeneous linear second-order equations; as always, the approach requires us to find the general solution to the related homogeneous equation first.

Let us again consider the equation

$$y'' + a_1 y' + a_0 y = f(t) \tag{4.4.13}$$

where a_0 and a_1 are constant and assume only that $f(t)$ is continuous. Suppose we know that $y_1(t)$ and $y_2(t)$ are linearly independent solutions of the associated homogeneous equation, so the complementary solution is $y_h = c_1 y_1(t) + c_2 y_2(t)$. In the method of undetermined coefficients, we made a guess of a particular solution y_p to (4.4.13) based on the form of $f(t)$. In the method of *variation of parameters*, we assume instead that the form of y_p is a more complicated version of y_h. In particular, we assume that y_p has the form

$$y_p = u_1(t)y_1(t) + u_2(t)y_2(t) \tag{4.4.14}$$

for unknown functions u_1 and u_2, where again y_1 and y_2 are the functions that arose in solving the related homogeneous equation.

The goal of variation of parameters is to find the functions $u_1(t)$ and $u_2(t)$ such that the function $y_p = u_1 y_1 + u_2 y_2$ is a particular solution to (4.4.13). Let us explore what conditions $u_1(t)$ and $u_2(t)$ must satisfy. Differentiating y_p yields

$$y_p' = u_1 y_1' + u_1' y_1 + u_2 y_2' + u_2' y_2 \tag{4.4.15}$$

While it seems natural at this point to differentiate again to find y_p'' and substitute into the differential equation, this becomes rather complicated.

Above we have seen that the two unknown functions must satisfy one condition (so far), that being the differential equation itself, as stated in (4.4.13). Because we have two functions, we have the freedom to set a second condition as well. In order to make the functions as simple as possible, and to eliminate

---

[1] If we tried the guess $y_p = A \ln t$, then $y_p' = A/t$, which introduces a function of an entirely new form. If we tried $y_p = A \ln t + B/t$, then the derivative leads us to a function involving $1/t^2$, again of a form not considered.

the second derivatives of u_1 and u_2 from arising in y_p'', we impose a second condition given by

$$u_1'y_1 + u_2'y_2 = 0 \qquad (4.4.16)$$

Observe now that by substituting the condition (4.4.16) in (4.4.15) we have

$$y_p' = u_1 y_1' + u_2 y_2'$$

so that

$$y_p'' = u_1 y_1'' + u_1' y_1' + u_2 y_2'' + u_2' y_2'$$

Substituting the above expressions for y_p'' and y_p' in (4.4.13) yields

$$(u_1 y_1'' + u_1' y_1' + u_2 y_2'' + u_2' y_2') + a_1(u_1 y_1' + u_2 y_2') + a_0(u_1 y_1 + u_2 y_2) = f(t) \qquad (4.4.17)$$

Reorganizing (4.4.17) according to the terms u_1, u_2, u_1', and u_2', we have

$$u_1(y_1'' + a_1 y_1' + a_0 y_1) + u_2(y_2'' + a_1 y_2' + a_0 y_2) + (u_1' y_1' + u_2' y_2') = f(t) \qquad (4.4.18)$$

Now, at this point we recall that y_1 and y_2 are fundamental solutions to the associated homogeneous equation $y'' + a_1 y' + a_0 = 0$, which shows that in (4.4.18) the coefficients of both u_1 and u_2 are zero. Therefore, (4.4.18) reduces to

$$u_1' y_1' + u_2' y_2' = f(t) \qquad (4.4.19)$$

Combining conditions (4.4.16) and (4.4.19) results in the system of linear equations in u_1' and u_2' given by

$$y_1 u_1' + y_2 u_2' = 0$$
$$y_1' u_1' + y_2' u_2' = f(t)$$

To solve for u_1' and u_2', we multiply the first equation by y_2' and the second equation by y_2, which gives

$$y_2' y_1 u_1' + y_2' y_2 u_2' = 0$$
$$y_2 y_1' u_1' + y_2 y_2' u_2' = y_2 f \qquad (4.4.20)$$

Subtracting the second equation from the first in (4.4.20), we have

$$y_2' y_1 u_1' - y_2 y_1' u_1' = -y_2 f$$

and therefore

$$u_1' = \frac{y_2 f}{y_2 y_1' - y_1 y_2'} \qquad (4.4.21)$$

Using similar algebra to solve for u_2', we may show that

$$u_2' = \frac{y_1 f}{y_1 y_2' - y_2 y_1'} \qquad (4.4.22)$$

Finally, to determine u_1 and u_2, we integrate to find

$$u_1 = \int \frac{y_2 f}{y_2 y_1' - y_1 y_2'} \, dt \quad \text{and} \quad u_2 = \int \frac{y_1 f(t)}{y_1 y_2' - y_2 y_1'} \, dt \qquad (4.4.23)$$

Once we integrate in (4.4.23) to solve for u_1 and u_2, we can conclude that a particular solution y_p to the original nonhomogeneous linear second-order differential equation is $y_p = u_1 y_1 + u_2 y_2$ where $y_h = c_1 y_1 + c_2 y_2$. Examples will be helpful to demonstrate the key steps of this method. First, we state the formal result proved by our discussion above.

Theorem 4.4.2 (Variation of Parameters Method) For the differential equation $y'' + a_1 y' + a_0 y = f(t)$, where f is continuous, assume that y_1 and y_2 are linearly independent solutions of the corresponding homogeneous equation $y'' + a_1 y' + a_2 y = 0$. Then, a particular solution to the non-homogeneous equation is $y_p = u_1 y_1 + u_2 y_2$, where u_1 and u_2 satisfy

$$u_1 = \int \frac{y_2 f}{y_2 y_1' - y_1 y_2'} \, dt \text{ and } u_2 = \int \frac{y_1 f}{y_1 y_2' - y_2 y_1'} \, dt \qquad (4.4.24)$$

Example 4.4.8   Solve the differential equation

$$y'' + y = \sec t \qquad (4.4.25)$$

where we assume that $-\frac{\pi}{2} < t < \frac{\pi}{2}$.

Solution.   We first observe that the corresponding characteristic equation is $r^2 + 1 = 0$ so that the complementary solution is $y_h = c_1 \cos t + c_2 \sin t$. In particular, $y_1 = \cos t$ and $y_2 = \sin t$.

We now seek two functions $u_1(t)$ and $u_2(t)$ that satisfy the equations (4.4.24). Since $y_1 = \cos t$ and $y_2 = \sin t$, it follows that $y_1' = -\sin t$ and $y_2' = \cos t$, and therefore, we have

$$u_1 = \int \frac{y_2 f}{y_2 y_1' - y_1 y_2'} \, dt = \int \frac{\sin t \sec t}{-\sin^2 t - \cos^2 t} \, dt$$

$$= -\int \sin t \sec t \, dt = -\int \frac{\sin t}{\cos t} \, dt = \ln(\cos t)$$

and

$$u_2 = \int \frac{y_1 f}{y_1 y_2' - y_2 y_1'} \, dt = \int \frac{\cos t \sec t}{\cos^2 t + \sin^2 t} \, dt$$

$$= \int 1 \, dt = t$$

Note that we have used the fundamental trigonometric identity $\sin^2 t + \cos^2 t = 1$ as well as other standard trigonometric relationships such as $\sec t = 1/\cos t$. Also, since we are seeking any two functions u_1 and u_2 that satisfy (4.4.24), it is not necessary to include the constants that can arise in integrating.

Hence we have found that $u_1 = \ln(\cos t)$ and $u_2 = t$. This enables us to conclude that a particular solution to the equation (4.4.25) is

$$y_p = u_1 y_1 + u_2 y_2 = \ln(\cos t) \cos t + t \sin t$$

and, therefore, the general solution is

$$y = y_h + y_p = c_1 \cos t + c_2 \sin t + \ln(\cos t) \cos t + t \sin t$$

Example 4.4.9 Solve the equation

$$y'' + 4y' + 4y = e^{-2t}\ln t \tag{4.4.26}$$

Solution. To begin, we solve the associated homogeneous equation and get

$$y_h = c_1 e^{-2t} + c_2 t e^{-2t}$$

Thus for variation of parameters, we assume that

$$y_p = u_1(t)e^{-2t} + u_2(t)te^{-2t}$$

and we seek u_1 and u_2. Since $y_1 = e^{-2t}$ and $y_2 = te^{-2t}$, it follows that $y_1' = -2e^{-2t}$ and $y_2' = -2te^{-2t} + e^{-2t}$, and therefore by (4.4.24)

$$u_1 = \int \frac{y_2 f}{y_2 y_1' - y_1 y_2'}\, dt = \int \frac{te^{-2t}(e^{-2t}\ln t)}{te^{-2t}(-2e^{-2t}) - e^{-2t}(-2te^{-2t} + e^{-2t})}\, dt$$

$$= \int \frac{te^{-4t}\ln(t)}{e^{-4t}(-2t + 2t - 1)}\, dt = -\int t\ln t\, dt = -\frac{1}{2}t^2\ln t + \frac{1}{4}t^2$$

and

$$u_2 = \int \frac{y_1 f}{y_1 y_2' - y_2 y_1'}\, dt = \int \frac{e^{-2t}(e^{-2t}\ln t)}{e^{-4t}(-2t + 1 + 2t)}\, dt$$

$$= \int \ln t\, dt = t\ln t - t$$

From these expressions for u_1 and u_2, we can conclude that the overall form of the solution y to (4.4.26) is

$$y = y_h + y_p$$

$$= c_1 e^{-2t} + c_2 t e^{-2t} + \left(-\frac{1}{2}t^2\ln t + \frac{1}{4}t^2\right)e^{-2t} + (t\ln t - t)te^{-2t}$$

$$= c_1 e^{-2t} + c_2 t e^{-2t} + \frac{1}{4}t^2 e^{-2t}(2\ln t - 3)$$

Exercises 4.4 In exercises 1–10, determine the complementary solution y_h and state the general form of y_p that you would guess in applying the method of undetermined coefficients.

1. $y'' - y' - 12y = 10e^{5t}$

2. $y'' + y' - 2y = 4t^2 - 1$

3. $y'' - y = 11e^t$

4. $y'' + 3y' = 3\sin 2t$

5. $y'' = t^2 + 3$

6. $y'' + 4y' + 3y = 2t + 4\cos t$

7. $y'' + 4y' + 4y = t^2$

8. $y'' + 4y = 2\sin 2t$

9. $y'' + 4y = 20e^t \cos t$

10. $y'' + y' - y = 3$

In exercises 11–20, solve the stated IVP using the method of undetermined coefficients. Note that the complementary solutions y_h and appropriate guesses for y_p were found in the corresponding exercises 1–10.

11. $y'' - y' - 12y = 10e^{5t}$, $y(0) = 2$, $y'(0) = -1$

12. $y'' + y' - 2y = 4t^2 - 1$, $y(0) = 1$, $y'(0) = 1$

13. $y'' - y = 11e^t$, $y(0) = -3$, $y'(0) = 2$

14. $y'' + 3y' = 3\sin 2t$, $y(0) = 0$, $y'(0) = 0$

15. $y'' = t^2 + 3$, $y(0) = -2$, $y'(0) = -2$

16. $y'' + 4y' + 3y = 2t + 4\cos t$, $y(0) = 2$, $y'(0) = 0$

17. $y'' + 4y' + 4y = t^2$, $y(0) = 5$, $y'(0) = 3$

18. $y'' + 4y = 2\sin 2t$, $y(0) = 1$, $y'(0) = -1$

19. $y'' + 4y = 20e^t \cos t$, $y(0) = 0$, $y'(0) = -1$

20. $y'' + y' - y = 3$, $y(0) = -1$, $y'(0) = -1$

In exercises 21–27, find the general solution of the given differential equation using variation of parameters.

21. $y'' + y = \tan t$, $-\frac{\pi}{2} < t < \frac{\pi}{2}$

22. $y'' + 5y' + 4y = te^t$

23. $y'' + 4y' + 4y = te^{-2t}$

24. $y'' + y = \csc t$, $0 < t < \pi$

25. $y'' - 2y' + y = \frac{e^t}{t}$, $t > 0$

26. $y'' - 4y' + 4y = e^t$

27. $y'' + y' - 6y = \frac{1}{e^t + 1}$, $t > 0$

28. For a forced spring-mass system with $c = 2$, $k = 1$, $m = 1$, and $F(t) = 20\cos 2t$, determine the displacement of the mass at time t if the system is set in motion by the initial conditions $y(0) = 2$, $y'(0) = 1$. Sketch the solution and discuss the long-term behavior of y_h and y_p separately and how these together influence the long-term behavior of the system.

29. For a forced undamped spring-mass system with $k = 8$, $m = 2$, and $F(t) = 5\cos 2.1t$, determine the displacement of the mass at time t if the system is set in motion by the initial conditions $y(0) = 2$, $y'(0) = 1$. Sketch the solution and discuss the long-term behavior of y_h and y_p separately and how these together influence the long-term behavior of the system.

30. For a forced undamped spring-mass system with $k = 8$, $m = 2$, and $F(t) = 5\cos 2t$, determine the displacement of the mass at time t if the system is set in motion by the initial conditions $y(0) = 2$, $y'(0) = 1$. Sketch the solution and discuss the long-term behavior of y_h and y_p separately and how these together influence the long-term behavior of the system.

31. For an RLC circuit with external voltage source $E(t) = 100\sin 20t$, $L = 10$, $R = 40$, and $C = 1/40$, determine the current at time t given the initial conditions $I(0) = 100$, $I'(0) = 25$. Sketch the solution and discuss the long-term behavior of the current.

32. For an RLC circuit with external voltage source $E(t) = 50\cos 40t$, $L = 10$, $R = 40$, and $C = 1/50$, determine the current at time t given the initial conditions $I(0) = 100$, $I'(0) = 25$. Sketch the solution and discuss the long-term behavior of the current.

33. Let

$$y'' + a_1 y' + a_0 y = f(t) \tag{4.4.27}$$

be a second-order nonhomogeneous linear differential equation with constant coefficients. If y_h is the general solution to the homogeneous equation $y'' + a_1 y' + a_0 y = 0$ and y_p is any solution to the nonhomogeneous equation (4.4.27), show that $y = y_h + y_p$ is a solution to (4.4.27).

4.5 Forced motion: beats and resonance

Based on our work with second-order differential equations, we are now able to completely solve the damped harmonic oscillator equation for a variety of forcing functions. In particular, we are able to determine the general solution of the spring-mass system equation

$$y'' + \frac{c}{m}y' + \frac{k}{m}y = \frac{1}{m}F(t) \tag{4.5.1}$$

by finding complementary and particular solutions. In this section, we explore some interesting phenomena related to periodic forcing functions $F(t)$. Our work will have important consequences for the study of other applications modeled by similar differential equations, including RLC circuits.

We begin by considering a sequence of related examples. As always, we assume that the units on all constants and related quantities are consistent.

Example 4.5.1   Determine the unique solution to the initial-value problem given by an undamped spring-mass system with $m = 1$ and $k = 4$ where $F(t) = \cos t$. Assume that the mass is initially released after being displaced 0.5 from equilibrium. Plot the solution and discuss its long-term behavior.

Solution.   Using the given information in (4.5.1), we see that the system is modeled by the initial-value problem

$$y'' + 4y = \cos t, \quad y(0) = 0.5, \quad y'(0) = 0 \qquad (4.5.2)$$

Solving the associated homogeneous equation $y'' + 4y = 0$ provides the complementary solution $y_h = c_1 \cos 2t + c_2 \sin 2t$. Applying the method of undetermined coefficients with the assumption that y_p has the form

$$y_p = A \cos t + B \sin t$$

we find upon substituting in (4.5.2) that A and B must satisfy the equation

$$(-A \cos t - B \sin t) + 4(A \cos t + B \sin t) = \cos t$$

Equating coefficients of $\cos t$ and $\sin t$, it follows

$$-A + 4A = 1$$

$$-B + 4B = 0$$

Therefore $A = 1/3$ and $B = 0$, so $y_p = \frac{1}{3} \cos t$ is a particular solution to (4.5.2). The general solution to the differential equation is

$$y = y_h + y_p = c_1 \cos 2t + c_2 \sin 2t + \frac{1}{3} \cos t$$

Finally, we use the stated initial conditions $y(0) = 1/2$ and $y'(0) = 0$ to determine the values of c_1 and c_2. The first condition implies that $1/2 = c_1 + 1/3$ and therefore $c_1 = 1/6$. Similarly, $y'(0) = 0$ implies that $0 = 2c_2$, and thus $c_2 = 0$. Hence the solution to the IVP is

$$y = \frac{1}{6} \cos 2t + \frac{1}{3} \cos t$$

In figure 4.3 we observe that the mass exhibits somewhat unusual behavior when negatively displaced due to the impact of the forcing function. With the undamped system and periodic forcing function, the observed behavior will repeat indefinitely.

We next explore how slight changes in the forcing function can result in substantially different behavior for the system.

Example 4.5.2   Determine the unique solution to the initial-value problem given by an undamped spring-mass system with $m = 1$ and $k = 4$ where $F(t) = \cos 1.75t$. Assume that the mass is initially released after being displaced 0.5 from equilibrium. Plot the solution and discuss its long-term behavior.

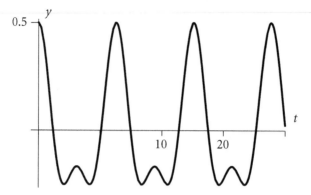

Figure 4.3   The solution y to the IVP in example 4.5.1.

Solution.   Similar to our work in example 4.5.1, we see that the system is modeled by the initial-value problem

$$y'' + 4y = \cos 1.75t, \quad y(0) = 0.5, \quad y'(0) = 0 \qquad (4.5.3)$$

Because only the forcing function has changed, the complementary solution is again $y_h = c_1 \cos 2t + c_2 \sin 2t$. Using the method of undetermined coefficients with

$$y_p = A \cos 1.75t + B \sin 1.75t$$

it follows that A and B must satisfy the equation

$$\left(-\frac{49}{16} A \cos 1.75t - \frac{49}{16} B \sin 1.75t \right) + 4(A \cos 1.75t + B \sin 1.75t) = \cos 1.75t$$

Equating like coefficients, we can deduce that $A = \frac{16}{15}$ and $B = 0$ so that

$$y_p = \frac{16}{15} \cos 1.75t$$

and the general solution is

$$y = c_1 \cos 2t + c_2 \sin 2t + \frac{16}{15} \cos 1.75t$$

Applying the initial condition $y(0) = 1/2$ shows that $c_1 = -17/30$. In addition, $y'(0) = 0$ implies that $c_2 = 0$. Hence the solution to the initial-value problem (4.5.3) is

$$y = -\frac{17}{30} \cos 2t + \frac{16}{15} \cos 1.75t$$

When we plot this solution, as shown in figure 4.4, we observe that while the solution is again periodic, in this instance there is an interesting pattern in which the amplitude of oscillation itself rises and falls. Because the system is undamped,

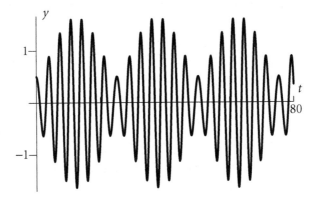

Figure 4.4 The solution to the IVP in example 4.5.2.

this behavior will repeat indefinitely. More importantly, observe how much the amplitude has increased in this example with $f(t) = \cos 1.75t$ as compared to what we saw in figure 4.3 with $f(t) = \cos t$: the amplitude of the solution in figure 4.4 is roughly 3 times that of the solution in figure 4.3, under identical other initial conditions.

In the solution

$$y = -\frac{17}{30}\cos 2t + \frac{16}{15}\cos 1.75t$$

to (4.5.3), we observe that we are adding two cosine functions of different frequencies to one another. In particular, these two frequencies are quite close to each other due to the coefficients "2" and "$\frac{7}{4}$." This results in the two functions' amplitudes sometimes reinforcing each other (such as when both amplitudes are large and positive), while at other times their amplitudes negate each other. The visual periodic phenomenon seen in figure 4.4 is known as *beats*. This is because the overall wave with the large wavelength appears as a beat and can often be heard when two sound waves have approximately the same frequency, such as when two instruments are out of tune. We will explore this phenomenon from a more rigorous, algebraic perspective shortly.

In the following example, we consider the case when the forcing function's frequency exactly matches that of the general solution to the corresponding homogeneous equation. Again, only a slight change to the forcing function will be made when compared to our work above.

Example 4.5.3   Determine the unique solution to the initial-value problem given by an undamped spring-mass system with $m = 1$ and $k = 4$ where $F(t) = \cos 2t$. Assume that the mass is initially released after being displaced 0.5 from equilibrium. Plot the solution and discuss its long-term behavior.

Solution.   As above, we see that the system is modeled by the initial-value problem

$$y'' + 4y = \cos 2t, \quad y(0) = 0.5, \quad y'(0) = 0 \qquad (4.5.4)$$

and the complementary solution is again $y_h = c_1 \cos 2t + c_2 \sin 2t$. Here, we observe that the forcing function is one of the linearly independent fundamental solutions present in y_h. Therefore we must make the modified guess that

$$y_p = At \cos 2t + Bt \sin 2t$$

in our attempt to find a particular solution. With

$$y_p' = A \cos 2t - 2At \sin 2t + B \sin 2t + 2Bt \cos 2t$$

and

$$y_p'' = -4A \sin 2t - 4At \cos 2t + 4B \cos 2t - 4Bt \sin 2t$$

we can substitute into (4.5.4) to see that A and B must satisfy the equation

$$(-4A\sin 2t - 4At\cos 2t + 4B\cos 2t - 4Bt\sin 2t) + 4(At\cos 2t + Bt\sin 2t) = \cos 2t$$

All terms involving $t \cos 2t$ and $t \sin 2t$ drop out, leaving us with

$$-4A \sin 2t + 4B \cos 2t = \cos 2t$$

from which it follows that $B = 1/4$ and $A = 0$. Hence $y_p = \frac{1}{4} t \sin 2t$ and thus

$$y = y_h + y_p = c_1 \cos 2t + c_2 \sin 2t + \frac{1}{4} t \sin 2t$$

Using the initial conditions $y(0) = 0.5$, $y'(0) = 0$, we can show that $c_1 = 1/2$ and $c_2 = 0$. Therefore, the solution to the IVP is

$$y = \frac{1}{2} \cos 2t + \frac{1}{4} t \sin 2t$$

A plot of this solution is shown in figure 4.5; observe the striking behavior that the solution not only oscillates periodically, but that its amplitude grows without bound as $t \to \infty$.

When we encounter the phenomenon in figure 4.5 where the solution to the harmonic oscillator initial-value problem grows without bound, we say that *resonance* occurs. This situation arises whenever the forcing function is a sine or cosine function whose frequency matches the natural frequency of the associated undamped homogeneous equation. In this case, the forcing function amplifies the natural oscillations of the system and causes them to grow without bound. In actual physical applications, such unbounded resonance is not realistic since either damping is present to limit the amplitude, the function is no longer a reasonable model for the phenomenon being modeled, or the structure simply fails. Large-amplitude oscillations do occur when forcing functions are close to or at the natural frequency of a structure or device, such as when the frequency of vortex shedding equals the natural frequency of bridge cables.

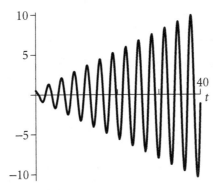

Figure 4.5 The solution to the IVP in example 4.5.3.

Our work from examples 4.5.1–4.5.3 can be generalized to the situation where the constants k and m present in (4.5.1) are arbitrary. In particular, for the undamped, undriven spring-mass system given by

$$y'' + \frac{k}{m}y = 0 \tag{4.5.5}$$

the general solution is

$$y_h = c_1 \cos\sqrt{\frac{k}{m}}t + c_2 \sin\sqrt{\frac{k}{m}}t \tag{4.5.6}$$

Since the mass will undergo one complete cycle as t goes from 0 to $2\pi\sqrt{m/k}$, the *period* of oscillation is $2\pi\sqrt{m/k}$. The number of cycles per second, or *frequency*, is the reciprocal of the period, or $\sqrt{k/m}/(2\pi)$. The *angular frequency* ω_0, which is measured in radians per second, is given by

$$\omega_0 = \sqrt{\frac{k}{m}}$$

This leads us to write the solution (4.5.6) to the equation (4.5.5) in the form

$$y_h = c_1 \cos\omega_0 t + c_2 \sin\omega_0 t$$

For the undamped spring-mass system driven by the periodic forcing function $F(t) = F_0 \cos\omega t$,

$$y'' + \frac{k}{m}y = \frac{1}{m}F_0 \cos\omega t \tag{4.5.7}$$

the method of undetermined coefficients can be used to show that

$$y_p = \frac{F_0}{m(\omega_0^2 - \omega^2)}\cos\omega t$$

provided that $\omega \neq \omega_0$. In this case, the general solution to (4.5.7) is

$$y = c_1 \cos\omega_0 t + c_2 \sin\omega_0 t + \frac{F_0}{m(\omega_0^2 - \omega^2)}\cos\omega t \tag{4.5.8}$$

When ω and ω_0 are nearly equal, the system demonstrates *near resonance* as the phenomenon of beats occurs.

In the case where $\omega = \omega_0$, the solution (4.5.8) obviously fails to hold; using work that generalizes example 4.5.3, it can be shown that a particular solution to (4.5.7) takes the form

$$y_p = \frac{F_0}{2m\omega_0} t \sin \omega_0 t$$

which produces the general solution

$$y = c_1 \cos \omega_0 t + c_2 \sin \omega_0 t + \frac{F_0}{2m\omega_0} t \sin \omega_0 t \qquad (4.5.9)$$

In the y_p term in (4.5.9), we see how the solution grows without bound as $t \to \infty$.

Having now discussed the phenomena of beats and resonance for undamped systems, we now briefly consider the situation where damping is present. Above we have observed that if $\omega \approx \omega_0$, then beats or resonance can occur. Regardless of the comparison of the frequencies of the system itself and the forcing function, a periodic forcing function will lead the system to oscillate indefinitely. The most important issue to understand is how large those oscillations can grow; this is especially critical for applications to vibrations and oscillations in physical structures such as bridges.

Two examples will be discussed to show the impact that different levels of damping can have on such a system.

Example 4.5.4 Determine the unique solution to the initial-value problem given by the damped spring-mass system with $m = 1$, $c = 0.1$, and $k = 4$, where $F(t) = \cos 2t$. Assume that the mass is initially released after being displaced 0.5 from equilibrium. Plot the solution and discuss its long-term behavior.

Solution. The system described above is modeled by the initial-value problem

$$y'' + 0.1y' + 4y = \cos 2t, \quad y(0) = 0.5, \quad y'(0) = 0 \qquad (4.5.10)$$

The characteristic equation is $r^2 + 0.1r + 4 = 0$, whose roots are approximately

$$r = -\frac{1}{20} \pm 1.999i$$

Thus, the complementary solution to (4.5.10) is

$$y_h = e^{-\frac{1}{20}t} (c_1 \cos 1.999t + c_2 \sin 1.999t)$$

Undetermined coefficients can be used in the usual way with the guess $y_p = A \cos 2t + B \sin 2t$ to find that $A = 0$ and $B = 5$ so that $y_p = 5 \sin 2t$. Thus, the general solution to the given differential equation is

$$y = e^{-\frac{1}{20}t} (c_1 \cos 1.999t + c_2 \sin 1.999t) + 5 \sin 2t$$

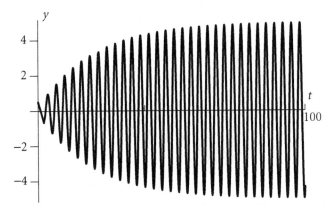

Figure 4.6   The solution to the IVP in example 4.5.4.

Applying the initial conditions, we can find the values of c_1 and c_2 to show that the solution to the stated IVP is

$$y = e^{-\frac{1}{20}t}\left(\frac{1}{2}\cos 1.999t - 4.989\sin 1.999t\right) + 5\sin 2t \qquad (4.5.11)$$

In figure 4.6, we see the plot of this solution over the interval $[0, 30\pi]$ and we observe that initially the amplitude of oscillations grows, much as it did in example 4.5.3 where resonance occurred. Here, however, we have a small amount of damping present in the system. Over time, this limits the amplitude of oscillations and keeps them from growing without bound, though such large-amplitude oscillation can result in damage to physical structures.

In the solution (4.5.11) to the IVP (4.5.10), we observe two very different behaviors in the complementary and particular solutions. Due to the presence of $e^{-t/20}$ in y_h, we see that as $t \to \infty$, $y_h(t) \to 0$. In contrast, $y_p = 5\sin 2t$ will oscillate continuously between -5 and 5. Because this is the behavior the system will tend to over time, we call y_p the *steady-state* solution. The solution y_h is called the *transient* solution, and is significant only for relatively small values of t.

Intuitively, increasing the damping that is present should decrease the amplitude of oscillations generated by a periodic forcing function. In example 4.5.4, a forcing function with amplitude 1 generated oscillations in the system that increased to an amplitude of nearly 5, in part due to the small damping constant, as well as the frequency of the forcing function which nearly matched the natural frequency of the system. In our next example, we increase the amount of damping present to see how this limits the size of the waves generated in the solution.

Example 4.5.5   Determine the unique solution to the initial-value problem given by the damped spring-mass system with $m = 1$, $c = 4$, and $k = 4$ where

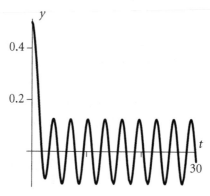

Figure 4.7 The solution to the IVP in example 4.5.5.

$F(t) = \cos 2t$. Assume that the mass is initially released after being displaced 0.5 from equilibrium. Plot the solution and discuss its long-term behavior.

Solution.   In this final modification of the spring-mass system we have studied in examples 4.5.1–4.5.4, here we have only increased the damping constant so that the system is modeled by the initial-value problem

$$y'' + 4y' + 4y = \cos 2t, \quad y(0) = 0.5, \quad y'(0) = 0 \qquad (4.5.12)$$

At this point in our work, we can show that $y_h = c_1 e^{-2t} + c_2 t e^{-2t}$ and $y_p = \frac{1}{8} \sin 2t$. Applying the initial conditions, the solution to the IVP (4.5.12) is

$$y = \frac{1}{2} e^{-2t} + \frac{3}{4} t e^{-2t} + \frac{1}{8} \sin 2t$$

Plotting this solution, as shown in figure 4.7, we observe that the amplitude decreases almost immediately because the complementary solution $y_h = \frac{1}{2} e^{-2t} + \frac{3}{4} t e^{-2t}$ vanishes quickly; moreover, only small steady-state oscillations persist due to $y_p = \frac{1}{8} \sin 2t$.

Exercises 4.5   In exercises 1–5, solve the given initial-value problem for $y(t)$ if $y(0) = y'(0) = 0$, given the stated parameters for an undamped spring-mass system. In addition, determine the maximum displacement of the mass, state if beats or resonance are present, and sketch the solution.

1. $m = 1$, $k = 25$, $f(t) = 0.01 \cos(5t)$

2. $m = 2$, $k = 32$, $f(t) = 2 \cos 4t$

3. $m = 1$, $k = 36$, $f(t) = 2e^{6t}$

4. $m = 3$, $k = 150$, $f(t) = 0.6 \cos 7t$

5. $m = 2$, $k = 100$, $f(t) = 4 \sin 7t$

6. A 2-kg mass is suspended from a spring with $k = 32$. A force $f(t) = 0.1 \sin 4t$ is applied to the mass. Calculate the time required for failure to occur if the spring breaks when the amplitude of oscillation exceeds 0.5. The motion starts at rest and there is no damping present. Assume consistent units.

7. A 20-N weight is suspended from a frictionless spring with $k = 98$. A force of $f(t) = 2 \cos 7t$ acts on the weight; the motion starts at rest. Does the system demonstrate resonance, beats, or neither? Explain, including a plot of the solution, assuming consistent units throughout.

In exercises 8–11, find the current $I(t)$ for each simple series circuit (with no resistor) if $I(0) = I'(0) = 0$, given the stated parameters for an undamped spring-mass system. In addition, determine the maximum current, state if beats or resonance are present, and sketch the solution. Assume consistent units.

8. $C = 10^{-3}$, $L = 0.1$, $E(t) = 120 \cos 101t$

9. $C = 0.02$, $L = 0.5$, $E(t) = 10 \sin 10t$

10. $C = 10^{-4}$, $L = 1.0$, $E(t) = 120 \sin 100t$

11. $C = 10^{-3}$, $L = 0.1$, $E(t) = 240 \cos 10t$

12. A forcing function $f(t) = 50 \cos 4t$ N is imposed on a spring-mass system for which $m = 2$, $k = 8$ N/m, and $c = 2$ kg/s. Determine the amplitude of the steady-state solution.

13. A forcing function $f(t) = 10 \sin(2t)$ N is imposed on a spring-mass system that starts from rest for which $m = 2$ kg and $k = 8$ N/m. Determine the damping coefficient necessary to limit the amplitude of the resulting motion to a maximum of 2 m.

14. A series circuit is composed of elements for which $R = 60\,\Omega$, $L = 10^{-4}$ H, and $C = 10^{-5}$ F. Find the steady-state current if a voltage of $E(t) = 120 \cos 120\pi t$ is applied.

4.6 Higher order linear differential equations

In the preceding sections of this chapter, we have focused on second-order linear differential equations. One reason we emphasize second-order equations is the importance of the (damped) harmonic oscillator equation. Moreover, second-order equations provide an appropriate setting in which to learn a variety of key ideas that may be generalized to linear equations of higher order. In this section, we consider several examples of higher order equations in order to gain exposure to important extensions of concepts we have already studied.

We first consider an example to see the natural approach to a third-order equation.

Example 4.6.1   Find the general solution to the differential equation

$$y''' - 2y'' - y' + 2 = 0 \qquad (4.6.1)$$

Solution.   For second-order linear homogeneous equations, we begin with the guess that $y = e^{rt}$ and determine the values of r for which e^{rt} is a solution to the equation. Doing likewise for this third-order equation, we note that $y' = re^{rt}$, $y'' = r^2 e^{rt}$, and $y''' = r^3 e^{rt}$. Substituting into (4.6.1), we find

$$r^3 e^{rt} - 2r^2 e^{rt} - re^{rt} + 2e^{rt} = 0$$

Factoring, it follows

$$e^{rt}[r^2(r-2) - 1(r-2)] = 0$$

or

$$e^{rt}(r-2)(r^2 - 1) = 0$$

We therefore see that the r-values for which $y = e^{rt}$ is a solution to (4.6.1) are $r = -1, 1$, and 2.

Using reasoning similar to our work with second-order equations, we now expect that the solutions $y_1 = e^{-t}$, $y_2 = e^t$, and $y_3 = e^{2t}$ are linearly independent and that the general solution is the linear combination

$$y = c_1 e^{-t} + c_2 e^t + c_3 e^{2t}$$

Just as with second-order equations, we call the equation $(r-2)(r^2 - 1) = 0$ that results from the guess $y = e^{rt}$ the *characteristic equation*. Roots of the characteristic equation play a central role in determining solutions to higher order equations. Furthermore, example 4.6.1 hints at the fact that we can expect several important theoretical results from second-order equations to hold for equations of order n. We state these results, which are analogous to theorems 4.2.1, 4.2.2, and 4.2.3, without proof. Observe that we will use the notation y''' to represent the third derivative of y, but for any derivative of order higher than 3, we use the notation $y^{(n)}$. For example, $y^{(5)}$ is the fifth derivative of y.

Theorem 4.6.1   If $y_1, y_2, \ldots, y_k$ are solutions to the nth-order linear homogeneous equation

$$y^{(n)} + a_{n-1}(t)y^{(n-1)} + \cdots + a_1(t)y' + a_0(t)y = 0$$

then $y = c_1 y_1 + c_2 y_2 + \cdots + c_k y_k$ is also a solution for any constants $c_1, \ldots, c_k$.

From theorem 4.6.1, we expect that linear combinations of fundamental solutions will play a key role in our solution to higher order equations. In addition, for corresponding initial-value problems, we are again guaranteed the existence of unique solutions under sufficiently nice conditions.

Theorem 4.6.2 Consider the nth-order initial-value problem given by

$$y^{(n)} + a_{n-1}(t)y^{(n-1)} + \cdots + a_1(t)y' + a_0(t)y = f(t)$$

$$y(t_0) = b_0, \quad y'(t_0) = b_1, \ldots, y^{(n-1)}(t_0) = b_{n-1}$$

(4.6.2)

where the coefficient functions $a_i(t)$ and the forcing function $f(t)$ are continuous on an open interval (a, b). Given any t_0 in (a, b), (4.6.2) has a unique solution in (a, b).

As in our earlier work with second-order linear DEs and systems of linear first-order DEs, in our current study of higher order differential equations, we usually consider the situation where the coefficient functions $a_i(t)$ are constant. In this setting, we can deduce the following result.

Theorem 4.6.3 The set of all solutions to the second-order homogeneous linear differential equation $y^{(n)} + a_{n-1}y^{(n-1)} + \cdots + a_1y' + a_0y = 0$, where $a_0, \ldots, a_{n-1}$ are constants, is a vector space of dimension n.

From these three results, we see that we can solve any homogeneous linear differential equation of order n provided that we can find n linearly independent solutions to the equation. Moreover, given such a general solution, we can determine the unique solution to any corresponding initial-value problem. With second-order equations, we normally verified the linear independence of two solutions by confirming that they were not scalar multiples of one another. For sets of more than two functions, a more sophisticated tool, the so-called *Wronskian*, is necessary to test for linear independence.

Definition 4.6.1 Suppose that $y_1, y_2, \ldots, y_n$ are each $(n-1)$-times differentiable functions on an interval $[a, b]$. The *Wronskian* W of these functions is given by

$$W(t) = \det \begin{pmatrix} y_1 & y_2 & \cdots & y_n \\ y_1' & y_2' & \cdots & y_n' \\ \vdots & \vdots & & \vdots \\ y_1^{(n-1)} & y_2^{(n-1)} & \cdots & y_n^{(n-1)} \end{pmatrix}$$

We emphasize that the Wronskian is itself a single scalar function of t. The most important feature of the Wronskian is that $W(t)$ is identically zero if and only if the functions $y_1, \ldots, y_n$ are linearly dependent. Hence, if $W(t)$ is not identically zero, then the functions are linearly independent. We consider an elementary example to demonstrate the use of the Wronskian.

Example 4.6.2 Use the Wronskian to show that the functions $y_1 = e^{-t}, y_2 = e^t$, and $y_3 = e^{2t}$ are linearly independent.

Solution. From the definition, observe that

$$W(t) = \det \begin{pmatrix} e^{-t} & e^t & e^{2t} \\ -e^{-t} & e^t & 2e^{2t} \\ e^{-t} & e^t & 4e^{2t} \end{pmatrix}$$

Computing the determinant, we find

$$W(t) = e^{-t}(4e^t e^{2t} - 2e^{2t} e^t) - e^t(-4e^{-t} e^{2t} - 2e^{2t} e^{-t}) + e^{2t}(-e^{-t} e^t - e^t e^{-t})$$

$$= 2e^{-t} e^{3t} + 6e^t e^t - 2e^{2t}$$

$$= 6e^{2t}$$

Since $W(t) \neq 0$, the functions $y_1 = e^{-t}$, $y_2 = e^t$, and $y_3 = e^{2t}$ are linearly independent.

Using the Wronskian, it can be shown that if the characteristic equation of a homogeneous linear differential equation of order n has n distinct, real solutions $r_1, \ldots, r_n$, then the corresponding functions $y_1 = e^{r_1 t}, \ldots, y_n = e^{r_n t}$ are linearly independent, and therefore can be used to form the general solution to the equation. In the cases where roots of the characteristic equation are repeated or complex, we use ideas similar to those encountered for second-order equations to find the required n real linearly independent solutions to the given differential equation. The next example examines this situation in the case of a repeated root.

Example 4.6.3   Determine three linearly independent solutions to the equation

$$y''' - 3y'' + 3y' - y = 0 \tag{4.6.3}$$

and hence state the general solution to the DE.

Solution.   The corresponding characteristic equation is

$$r^3 - 3r^2 + 3r - 1 = 0$$

Factoring, it follows that $(r - 1)^3 = 0$, so only one real, repeated root exists: $r = 1$. This shows that $y_1 = e^t$ is one solution to (4.6.3). Two more solutions remain to be found. Based on our experience with second-order equations and theorem 4.3.1, we naturally expect that $y_2 = te^t$ and $y_3 = t^2 e^t$ will be solutions to (4.6.3). It is a straightforward exercise to verify that each of these two functions is a solution to the given equation. Moreover, it may be shown that the Wronskian of these three functions is nonzero and, therefore, the functions are linearly independent, so the general solution to (4.6.3) is

$$y = c_1 e^t + c_2 te^t + c_3 t^2 e^t = (c_1 + c_2 t + c_3 t^2)e^t$$

The following result analogous to theorem 4.3.1 holds for repeated roots of multiplicity k in higher order equations.

Theorem 4.6.4 For any nth-order linear homogeneous differential equation of the form

$$y^{(n)} + a_{n-1}y^{(n-1)} + \cdots + a_1 y' + a_0 y = 0$$

whose characteristic equation has a repeated root r of multiplicity k, the k linearly independent solutions of the differential equation corresponding to r are

$$e^{rt}, te^{rt}, t^2 e^{rt}, \ldots, t^{k-1} e^{rt}$$

We deal with complex roots of the characteristic equation in exactly the same manner as in the case of second-order differential equations. In particular, if $r = a + ib$ is a complex root of the characteristic equation, we consider the complex-valued function

$$z(t) = e^{(a+ib)t} = e^{at} e^{ibt} = e^{at}(\cos bt + i \sin bt)$$

The real and imaginary parts of the complex solution then form linearly independent solutions to the differential equation. Our next example illustrates this in detail.

Example 4.6.4 Determine the general solution to the equation

$$y^{(4)} - 2y''' + 14y'' - 18y' + 45y = 0 \qquad (4.6.4)$$

Solution. If we consider the characteristic equation $r^4 - 2r^3 + 14r^2 - 18r + 45 = 0$ and factor, we find

$$r^4 - 2r^3 + 14r^2 - 18r + 45 = (r^2 + 9)(r^2 - 2r + 5) = 0$$

from which it follows that $r = \pm 3i$ and $r = 1 \pm 2i$. Thus one complex solution is

$$z_1(t) = e^{3it} = \cos 3t + i \sin 3t$$

so that $y_1 = \cos 3t$ and $y_2 = \sin 3t$ are solutions to (4.6.4). Similarly, another complex solution is

$$z_2(t) = e^{(1+2i)t} = e^t(\cos 2t + i \sin 2t)$$

so that $y_3 = e^t \cos 2t$ and $y_4 = e^t \sin 2t$ are also real solutions to (4.6.4). We can now conclude that the general solution to the given differential equation is

$$y = c_1 \cos 3t + c_2 \sin 3t + c_3 e^t \cos 2t + c_4 e^t \sin 2t$$

The only remaining case to consider for homogeneous equations is that of repeated complex roots. In this case, just as with that of repeated real roots, we multiply the basic solutions that arise by powers of t to build additional linearly independent solutions. For example, if $r = 1 \pm 2i$ is a repeated complex root of multiplicity two, the corresponding four real solutions would be $y_1 = e^t \cos 2t$, $y_2 = e^t \sin 2t$, $y_3 = te^t \cos 2t$, and $y_4 = te^t \sin 2t$.

We also observe at this point that we can solve corresponding initial-value problems for any given nth-order homogeneous linear equation. Since the general solution of an nth-order equation has n unknown constants $c_1, \ldots, c_n$, we will need n initial conditions to uniquely determine their values. The following example demonstrates the solution of a standard problem.

Example 4.6.5   Find the solution of the initial-value problem

$$y^{(4)} - y = 0, \quad y(0) = y'(0) = y''(0) = y'''(0) = 1 \qquad (4.6.5)$$

Solution.   With characteristic equation $r^4 - 1 = 0$, it is straightforward to verify that the roots of this equation are ± 1 and $\pm i$ so that the general solution to the DE in (4.6.5) is

$$y(t) = c_1 e^t + c_2 e^{-t} + c_3 \cos t + c_4 \sin t$$

The derivatives of y are

$$y'(t) = c_1 e^t - c_2 e^{-t} - c_3 \sin t + c_4 \cos t$$

$$y''(t) = c_1 e^t + c_2 e^{-t} - c_3 \cos t - c_4 \sin t$$

$$y'''(t) = c_1 e^t - c_2 e^{-t} + c_3 \sin t - c_4 \cos t$$

Using the stated initial conditions, observe that

$$y(0) = 1 = c_1 + c_2 + c_3$$

$$y'(0) = 1 = c_1 - c_2 + c_4$$

$$y''(0) = 1 = c_1 + c_2 - c_3$$

$$y'''(0) = 1 = c_1 - c_2 - c_4$$

Row-reducing this system of linear equations shows that the unique solution is given by $c_1 = 1$, $c_2 = c_3 = c_4 = 0$, and therefore, the solution to the IVP (4.6.5) is

$$y(t) = e^t$$

Finally, it remains for us to see how the previous methods of dealing with nonhomogeneous second-order equations extend to higher order equations. Just as with second-order equations, we first solve the corresponding homogeneous equation using the approach discussed above to find the *complementary* solution y_h. Then, in order to find a *particular* solution y_p to the nonhomogeneous equation, we can use extensions of the methods discussed in section 4.4.

For the method of undetermined coefficients, the approach is essentially identical: based on the form of the forcing function $f(t)$ and the presence of fundamental solutions within f, we make a reasonable guess of the form of a particular solution y_p involving unknown coefficients. By substituting into the given DE, we determine values for these coefficients and hence y_p. The general solution is then $y = y_h + y_p$. Variation of parameters may also be extended: given $y_h = c_1 y_1 + c_2 y_2 + \cdots + c_n y_n$, we seek functions $u_1, u_2, \ldots, u_n$ such that $y_p = u_1 y_1 + u_2 y_2 + \cdots + u_n y_n$ is a solution to the differential equation. This method is best understood through the theory developed for nonhomogeneous systems of first-order equations given in section 3.7 and reminding ourselves that any nth-order linear equation can be converted into a system of n first-order equations. While this approach provides a guaranteed particular solution in theory, the computational details are often very complicated. We therefore choose to focus on those higher order DEs that may be solved using undetermined coefficients. An example is instructive.

Example 4.6.6   Determine the general solution to the equation

$$y^{(5)} - y''' = 3e^t + t^2 - 4 \qquad (4.6.6)$$

Solution.   We first solve the corresponding homogeneous equation, $y^{(5)} - y''' = 0$ to determine y_h. Since the characteristic equation is $r^5 - r^3 = r^3(r^2 - 1) = 0$, we see that y_h is given by

$$y_h = c_1 + c_2 t + c_3 t^2 + c_4 e^t + c_5 e^{-t}$$

For the nonhomogeneous equation (4.6.6), based on the form of the forcing function $f(t)$, the natural form to assume for y_p is

$$y_p = Ae^t + B + Ct + Dt^2$$

However, since each part of our assumed form of y_p appears in y_h, we therefore modify our guess by multiplying by appropriate powers of t and assume instead that

$$y_p = Ate^t + Bt^3 + Ct^4 + Dt^5$$

From this, we observe that to substitute y_p into (4.6.6) we need to know y''' and $y^{(5)}$. By repeated differentiation,

$$y_p' = Ate^t + Ae^t + 3Bt^2 + 4Ct^3 + 5Dt^4$$

$$y_p'' = Ate^t + 2Ae^t + 6Bt + 12Ct^2 + 20Dt^3$$

$$y_p''' = Ate^t + 3Ae^t + 6B + 24Ct + 60Dt^2$$

$$y_p^{(4)} = Ate^t + 4Ae^t + 24C + 120Dt$$

$$y_p^{(5)} = Ate^t + 5Ae^t + 120D$$

Substituting in (4.6.6), it follows

$$Ate^t + 5Ae^t + 120D - (Ate^t + 3Ae^t + 6B + 24Ct + 60Dt^2) = 3e^t + t^2 - 4$$

so that

$$2Ae^t - 60Dt^2 - 24Ct + 120D - 6B = 3e^t + t^2 - 4$$

Equating like coefficients, we find that $A = 3/2$, $D = -1/60$, $C = 0$, and $B = 1/3$. Hence, we have found

$$y_p = \frac{3}{2}te^t + \frac{1}{3}t^3 - \frac{1}{60}t^5$$

Therefore, the general solution to (4.6.6) is

$$y = y_h + y_p = c_1 + c_2 t + c_3 t^2 + c_4 e^t + c_5 e^{-t} + \frac{3}{2}te^t + \frac{1}{3}t^3 - \frac{1}{60}t^5$$

Throughout this section we have seen that the approaches needed to solve nth-order linear equations are nearly identical to those we use for second-order equations. The main differences are that the characteristic equation is generally difficult, if not impossible, to factor, and we have to be especially cognizant of repeated roots in determining y_h and y_p.

4.6.1 Solving characteristic equations using *Maple*

While solving linear differential equations of order n requires nearly identical methods to DEs of order 2, there is one added challenge from the outset: solving the characteristic equation. The characteristic equation is a polynomial equation of degree n; while every such equation of degree 2 can be solved using the quadratic formula, equations of higher order can be much more difficult, and (for equations of degree 5 and higher) often impossible, to solve by algebraic means.

Computer algebra systems like *Maple* provide useful assistance in this matter with commands for solving equations exactly and approximately. For example, say we have the characteristic equation

$$r^4 - r^3 - 7r^2 + r + 6 = 0$$

To solve this exactly in *Maple*, we enter

```
> solve(r^4 - r^3 - 7*r^2 + r + 6 = 0, r);
```

Maple produces the output

$$-1, 1, -2, 3$$

showing that these are the four roots of the characteristic equation.

Of course, not all polynomial equations will have all integer solutions, much less all real solutions. For example, if we consider the equation

$$r^4 + r^3 + r^2 + r + 1 = 0$$

and use the `solve` command, we see that

```
> solve(r^4 + r^3 + r^2 + r + 1 = 0, r);
```

results in the output

$$-\frac{1}{4}+\frac{1}{4}\sqrt{5}+\frac{1}{4}I\sqrt{2}\sqrt{5+\sqrt{5}},\ -\frac{1}{4}\sqrt{5}-\frac{1}{4}+\frac{1}{4}I\sqrt{2}\sqrt{5-\sqrt{5}},$$

$$-\frac{1}{4}\sqrt{5}-\frac{1}{4}-\frac{1}{4}I\sqrt{2}\sqrt{5-\sqrt{5}},\ -\frac{1}{4}+\frac{1}{4}\sqrt{5}-\frac{1}{4}I\sqrt{2}\sqrt{5+\sqrt{5}}$$

In this case, we might prefer a decimal approximation to the roots rather than the exactness that *Maple* provides. One way to achieve this is to use the `fsolve` command:

```
> fsolve(r^4 + r^3 + r^2 + r + 1 = 0, r, complex);
```

which generates the result

$$-0.80902-0.58779I,\ -0.80902+0.58779I,\ 0.30902-0.95106I,$$
$$0.30902+0.95106I$$

Note that without the option "`complex`" in the `fsolve` command, the command will not generate any output. This is because the default setting for `fsolve` is to numerically approximate all of the real roots of the polynomial equation and to ignore complex ones. For polynomial equations of degree 5 or more, the `fsolve` command is the appropriate tool to use to determine accurate approximations of the equation's solutions.

Exercises 4.6   In exercises 1–12, use the characteristic equation to determine the general solution to the given higher order linear homogeneous DE.

1. $y''' - 2y'' - y' + 2y = 0$

2. $y''' - 2y'' - 3y' = 0$

3. $4y''' - 13y' - 6y = 0$

4. $y^{(4)} - 13y'' + 36y = 0$

5. $y''' + 3y'' + 3y' + y = 0$

6. $y^{(4)} - y''' - 7y'' + y' + 6y = 0$

7. $y''' - y'' + 4y' - 4y = 0$

8. $y^{(4)} - y = 0$

9. $y^{(5)} - 2y^{(4)} - y' + 2y = 0$

10. $y^{(6)} + 9y^{(4)} + 24y'' + 16y = 0$

11. $y^{(4)} + 4y''' + 6y'' + 4y' + y = 0$

12. $y^{(4)} + 3y''' + y'' - 5y' = 0$

In exercises 13–22, solve the given IVP.

13. $y''' - 4y' = 0$, $y(0) = 1$, $y'(0) = 0$, $y''(0) = 2$

14. $y''' - 3y'' + 2y' = 0$, $y(0) = 0$, $y'(0) = 2$, $y''(0) = 0$

15. $y''' - 6y'' + 11y' - 6y = 0$, $y(0) = 0$, $y'(0) = 2$, $y''(0) = 0$

16. $y^{(4)} - 2y''' - y'' + 2y' = 0$, $y(0) = 2$, $y'(0) = 0$, $y''(0) = 10$, $y'''(0) = 0$

17. $y''' + y'' + 4y' + 4y = 0$, $y(0) = 0$, $y'(0) = 10$, $y''(0) = 0$

18. $y^{(4)} + 5y'' + 4y = 0$, $y(0) = 4$, $y'(0) = 0$, $y''(0) = 10$, $y'''(0) = 0$

19. $y''' = 0$, $y(0) = 2$, $y'(0) = 0$, $y''(0) = 2$

20. $y^{(4)} - 16y = 0$, $y(0) = 4$, $y'(0) = 0$, $y''(0) = 0$, $y'''(0) = 0$

21. $y''' - 3y'' + 3y' - y = 0$, $y(0) = 1$, $y'(0) = 2$, $y''(0) = 1$

22. $y^{(5)} + y''' = 0$, $y(0) = 1$, $y'(0) = 0$, $y''(0) = 2$, $y'''(0) = 0$, $y^{(4)}(0) = 4$

In exercises 23–28, construct a homogeneous linear differential equation of the least possible order that has the given function(s) as solutions.

23. $y_1 = c$, $y_2 = e^t$

24. $y_1 = t^2 e^{2t}$

25. $y_1 = t$, $y_2 = \cos 3t$, $y_3 = e^{-t}$

26. $y_1 = te^{4t} \sin t$

27. $y_1 = e^{-t/2} \cos t$, $y_2 = \sin 5t$

28. $y_1 = \sin t$, $y_2(t) = t \sin t$

29. Find the general solution to $y^{(4)} + 2y'' + y = \cos t$.

30. Find a particular solution to $y^{(4)} + 2y'' + y = \sin t + 2\cos t$. How is your answer similar to the result in exercise 29?

In exercises 31–42, use undetermined coefficients to determine the general solution to the stated nonhomogeneous equation. Note that each of the corresponding homogeneous equations has been solved in exercises 1–12.

31. $y''' - 2y'' - y' + 2y = 2$

32. $y''' - 2y'' - 3y' = 2e^t$

33. $4y''' - 13y' - 6y = \cos t$

34. $y^{(4)} - 13y'' + 36y = t$

35. $y''' + 3y'' + 3y' + y = \sin t$

36. $y^{(4)} - y''' - 7y'' + y' + 6y = t^2 + 3$

37. $y''' - y'' + 4y' - 4y = e^{-t}$

38. $y^{(4)} - y = 3t$

39. $y^{(5)} - 2y^{(4)} - y' + 2y = 7$

40. $y^{(6)} + 9y^{(4)} + 24y'' + 16y = t^2$

41. $y^{(4)} + 4y''' + 6y'' + 4y' + y = t + \cos t$

42. $y^{(4)} + 3y''' + y'' - 5y' = 2t - \sin t + e^t$

4.7 For further study

4.7.1 Damped motion

Consider the general form of the spring-mass equation

$$my'' + cy' + ky = 0 \qquad (4.7.1)$$

where $c \neq 0$ so that viscous damping is present. In what follows, we explore how the values of the constants m, c, and k affect the behavior of the solution y. Note that in this context, m, c, and k are always positive.

(a) Show that the roots of the characteristic polynomial of (4.7.1) are

$$\lambda = \frac{-c \pm \sqrt{c^2 - 4mk}}{2m}$$

(b) We examine the three possible cases for the roots of the characteristic polynomial:

   (i) Suppose that $c^2 - 4km > 0$. Explain why $\sqrt{c^2 - 4mk} < c$ and thus why both roots of the characteristic equation must be negative. State the general solution to the equation (4.7.1) in terms of the constants c, m, and k.

   (ii) Suppose that $c^2 - 4km = 0$. Discuss the number of real roots of the characteristic polynomial and state the general solution to the equation (4.7.1) in terms of the constants c and m.

   (iii) Suppose that $c^2 - 4km < 0$. Explain why both roots of the characteristic polynomial are complex. Using $\Omega = \sqrt{4mk - c^2}/(2m)$, state the general solution to the equation (4.7.1) in terms of the constants c, m, and Ω.

(c) The respective cases (i), (ii), and (iii) in (b) are typically called *overdamping*, *critical damping*, and *underdamping*. How is the case of underdamping significantly different from overdamping and critical damping? Explain both in terms of the algebraic form of the solution as well as in terms of the solution's expected graph.

(d) A 4-kg mass is suspended from a spring with constant $k = 25$, and a dashpot with various levels of damping viscosity is present. The mass is displaced 0.5 m from its equilibrium and released. Determine the displacement $y(t)$ of the mass if

(i) $c = 15$, (ii) $c = 20$, (iii) $c = 25$, and (iv) $c = 30$

In each case, state whether the system is overdamped, critically damped, or underdamped, and sketch the solution curve.

(e) The case of underdamping is the most interesting of the three cases, for it is here that multiple oscillations through equilibrium occur. In (b)(iii), you should have shown that the general solution may be expressed in the form

$$y = e^{-\frac{c}{2m}t}(c_1 \cos \Omega t + c_2 \sin \Omega t)$$

Show that y may be alternatively expressed in the form

$$y = Ae^{-\frac{c}{2m}t}\cos(\Omega t - \theta) \qquad (4.7.2)$$

where $A = \sqrt{c_1^2 + c_2^2}$ and $\tan \theta = c_1/c_2$. (Hint: Set $A\cos(\Omega t - \theta) = c_1 \cos \Omega t + c_2 \sin \Omega t$ and equate like coefficients after using the trigonometric identity $\cos(\alpha - \beta) = \cos \alpha \cos \beta + \sin \alpha \sin \beta$.)

(f) In the underdamped case, we are interested in how fast the amplitude of the oscillations decays to zero. In what follows, we show how the ratio of consecutive local maxima (or minima) of $y(t)$ depends only on the constants c, m, and Ω.

(i) Using $y = Ae^{-\frac{c}{2m}t}\cos(\Omega t - \theta)$ from (e), determine y' and show that $y' = 0$ if and only if

$$\tan(\Omega t - \theta) = -\frac{c}{2m\Omega} \qquad (4.7.3)$$

(ii) If the solutions of (4.7.3) are denoted by t_n, then show that

$$t_n = \frac{\theta}{\Omega} + \frac{1}{\Omega}\arctan\left(-\frac{c}{2m\Omega}\right) + \frac{n\pi}{\Omega} \qquad (4.7.4)$$

Explain why we expect $y(t_n)$ and $y(t_{n+1})$ to be a local maximum and minimum (or local minimum and maximum), respectively, of $y(t)$, and hence why $y(t_n)$ and $y(t_{n+2})$ will be consecutive maxima or consecutive minima.

(iii) Let $y_n = y(t_n)$ and $y_{n+2} = y(t_{n+2})$. Using (4.7.2), evaluate $y(t_n)$ and $y(t_{n+2})$ and verify that

$$\frac{y_n}{y_{n+2}} = \frac{\cos(\Omega t_n - \theta)}{\cos(\Omega t_{n+2} - \theta)}e^{-\frac{c}{2m}(t_n - t_{n+2})} \qquad (4.7.5)$$

(iv) Show that (4.7.3) implies

$$(t_n - t_{n+2})\Omega = -2\pi$$

and thus

$$\frac{y_n}{y_{n+2}} = \frac{\cos(\Omega t_n - \theta)}{\cos(\Omega t_{n+2} - \theta)}e^{\pi c/m\Omega} \qquad (4.7.6)$$

(v) Show that

$$\Omega t_n - \theta = \Omega t_{n+2} - \theta - 2\pi$$

so that

$$\cos(\Omega t_n - \theta) = \cos(\Omega t_{n+2} - \theta)$$

Use this last result to prove that

$$\frac{y_n}{y_{n+2}} = e^{\pi c/m\Omega} \tag{4.7.7}$$

(g) The logarithm of (4.7.7),

$$D = \ln \frac{y_n}{y_{n+2}} = \ln e^{\pi c/m\Omega} = \frac{\pi c}{m\Omega} \tag{4.7.8}$$

is called the *logarithmic decrement*. Note that this quantity is independent of t as well as the initial conditions present in the underdamped case for the DE (4.7.1), and that the value of the logarithmic decrement tells us how rapidly consecutive oscillations diminish in the underdamped case.

For each of the following underdamped spring-mass systems, determine the solution function $y(t)$ and compute the logarithmic decrement. Explain how the value of the logarithmic decrement tells you whether oscillations will die out slowly or rapidly. Using a computer algebra system to execute the routine calculations is particularly appropriate here. In each case, assume the mass is displaced 1 m and released.

(i) $m = 4, c = 19, k = 25$
(ii) $m = 4, c = 10, k = 25$
(iii) $m = 4, c = 1, k = 25$
(iv) $m = 4, c = 0.1, k = 25$

4.7.2 Forced oscillations with damping

Consider the general form of the forced spring-mass equation

$$my'' + cy' + ky = f(t) \tag{4.7.9}$$

where $c > 0$ so that viscous damping is present. Again, we remark that in this context m and k are always positive.

(a) Show that if

$$\Omega = \frac{\sqrt{c^2 - 4km}}{2m}$$

then the complementary solution of (4.7.9) is

$$y_h(t) = e^{-\frac{c}{2m}t}\left(c_1 e^{\Omega t} + c_2 e^{-\Omega t}\right) \tag{4.7.10}$$

(b) Explain why

$$\lim_{t \to \infty} y_h(t) = 0$$

Recall that we call $y_h(t)$ the *transient* solution. What does this tell us about the role played by the particular solution $y_p(t)$ in the general solution $y = y_h + y_p$ as $t \to \infty$?

(c) We now consider the effects of the periodic forcing function $f(t) = F_0 \cos \omega t$. With this function, we have seen that resonance is only possible when no damping is present; here, we wish to explore the impact of the parameters in $f(t)$ on the steady-state solution y_p to (4.7.9).

(i) Use the method of undetermined coefficients to show that with $f(t) = F_0 \cos \omega t$, the particular solution y_p to (4.7.9) is

$$y_p = \frac{F_0(k - m\omega^2)}{(k - m\omega^2)^2 + \omega^2 c^2} \left(\cos \omega t + \frac{c\omega}{k - m\omega^2} \sin \omega t \right) \quad (4.7.11)$$

(ii) As in our study of undamped spring-mass systems and resonance, we let $\omega_0 = \sqrt{k/m}$. Show that $y_p(t)$ may be equivalently expressed in the form

$$y_p = \frac{F_0}{m^2(\omega_0^2 - \omega^2)^2 + \omega^2 c^2} \cos(\omega t - \theta) \quad (4.7.12)$$

Compare the result to (4.7.2).

(iii) Observe that the amplitude of the oscillation of y_p in (4.7.12) is

$$A(\omega) = \frac{F_0}{m^2(\omega_0^2 - \omega^2)^2 + \omega^2 c^2} \quad (4.7.13)$$

and that ω_0, m, and c are fixed constants determined by the given spring-mass system. We now examine how the size of these oscillations depends on ω.

First, compute

$$\frac{dA}{d\omega}$$

Then, set $dA/d\omega = 0$ to show that the maximum amplitude occurs when

$$\omega^2 = \omega_0^2 - \frac{c^2}{2m^2} \quad (4.7.14)$$

(iv) Explain why if c satisfies $c^2 > 2m^2\omega_0^2$, then there is no value of ω that produces a maximum amplitude of oscillation.

In addition, note that when a maximum amplitude exists (i.e., provided $c^2 < 2m^2\omega_0^2$), its value is given by $A(\omega)$ where ω satisfies (4.7.14). Use this condition to compute $A(\omega)$ and show that

$$A_{max} = \frac{2mF_0}{c\sqrt{4m^2\omega_0^2 - c^2}} \quad (4.7.15)$$

(v) Consider a particular spring-mass system for which $m = 1$ and $k = 4$ where we consider various damping constants c. In addition, assume we apply the forcing function $f(t) = \cos\omega t$, so that $F_0 = 1$. Recall that $\omega_0 = \sqrt{k/m}$, so $\omega_0 = 2$.
For each of the c-values $c = 0.1, 1, 2, 3, 4, 5, 6$, plot the function

$$\Delta(\omega) = \frac{F_0}{m^2(\omega_0^2 - \omega^2)^2 + \omega^2 c^2}$$

on the interval $\omega = 0 \ldots 10$. When a maximum oscillation exists, where does it occur? How is the size of the maximum oscillation correlated with c and ω? What should we ensure about the relationship between ω and ω_0 if we want to avoid large amplitude oscillations?

(d) Complete the following exercises which examine the magnitude of oscillations in damped, driven spring-mass systems.

   (i) A forcing function $f(t) = 10\sin 2t$ is imposed on a spring-mass system for which $m = 2$ kg and $k = 8$ N/m. Determine the damping constant necessary to limit the amplitude of the motion to a maximum of 2 m.
  (ii) A forcing function $f(t) = 50\cos\omega t$ is imposed on a spring-mass system for which $m = 4$ kg, $k = 100$ N/m, and $c = 2$ kg/s. Calculate the amplitude of the resulting motion for $\omega = 4$, $\omega = 4.5$, $\omega = 5$, and $\omega = 6$.
 (iii) Determine the input frequency ω that gives the maximum amplitude for the spring-mass system in (ii) above. For this frequency, what is the maximum amplitude?

4.7.3 The Cauchy–Euler equation

The vast majority of our efforts with higher order DEs have involved linear equations with constant coefficients. The Cauchy–Euler equation is an important example of a linear, second-order DE whose coefficients are not constant. In particular, the *Cauchy–Euler equation* is a differential equation of form

$$t^2 y'' + pty' + qy = 0 \qquad (4.7.16)$$

where p and q are real constants and $t > 0$.

(a) Explain why it is reasonable to guess that $y(t) = t^\lambda$ is a solution to (4.7.16). Show by direct substitution in (4.7.16) that the guess $y(t) = t^\lambda$ requires λ to be a solution to the *characteristic equation*

$$\lambda^2 + (p - 1)\lambda + q = 0 \qquad (4.7.17)$$

(b) In the case where (4.7.17) has two distinct real roots λ_1 and λ_2, then the general solution to the Cauchy–Euler equation is

$$y = c_1 t^{\lambda_1} + c_2 t^{\lambda_2}$$

Solve each of the following Cauchy–Euler initial-value problems:

(i) $t^2 y'' - 5ty' + 8y = 0, \quad y(1) = 1, y'(1) = 0$
(ii) $t^2 y'' + 9ty' + 12y = 0, \quad y(1) = 1, y'(1) = 0$

(c) When (4.7.17) has a repeated real root $\lambda_1 = \lambda_2 = \lambda$, then we have only determined one linearly independent solution $(y_1 = t^\lambda)$ of the Cauchy–Euler equation. Here we determine a second linearly independent solution.

(i) Assuming that λ is a repeated root of (4.7.17), show that $1 - p = 2\lambda$.
(ii) Letting $v(t)$ be an unknown function, consider the guess $y_2 = v \cdot t^\lambda$. By direct substitution in the Cauchy–Euler equation, show that v must satisfy the equation

$$t^\lambda [t^2 v'' + (2\lambda + p)tv' + (\lambda^2 + (p-1)\lambda + q)v] = 0 \qquad (4.7.18)$$

(iii) Use your work in (i) and (ii), as well as the fact that λ satisfies the equation

$$\lambda^2 + (p-1)\lambda + q = 0$$

to show that $y_2 = v \cdot t^\lambda$ is a solution of the Cauchy–Euler equation in the case of a repeated root provided that

$$tv'' + v' = 0 \qquad (4.7.19)$$

(iv) Show that $v(t) = \ln t$ is a solution of (4.7.19) and hence state the general solution of the Cauchy–Euler equation in the case where the characteristic equation has a single real repeated root.

(d) Solve each of the following Cauchy–Euler initial-value problems:

(i) $t^2 y'' + 7ty' + 9y = 0, \quad y(1) = 1, \quad y'(1) = 0$
(ii) $t^2 y'' - 9ty' + 25y = 0, \quad y(1) = 1, \quad y'(1) = 0$

(e) When (4.7.17) has complex roots, say $\lambda_1 = a + bi$ and $\lambda_2 = a - bi$, then we proceed with a corresponding complex solution to the Cauchy–Euler equation and verify that its real and imaginary parts are themselves real, linearly independent solutions to the equation. In particular, with $\lambda = a + bi$, observe that

$$z(t) = t^\lambda = t^{a+bi} = t^a t^{bi}$$

By writing

$$t^{bi} = e^{\ln(t^{bi})} = e^{bi \ln t}$$

and applying Euler's formula, show that

$$z(t) = t^a [\cos(b \ln t) + i \sin(b \ln t)] \qquad (4.7.20)$$

In addition, show by direct substitution that $y_1(t) = t^a \cos(b \ln t)$ is a solution to the Cauchy–Euler equation when $a + bi$ is a root of the characteristic polynomial. Likewise, show that $y_2(t) = t^a \sin(b \ln t)$ is a solution.

Hence, state the general solution to the Cauchy–Euler equation in the case where the characteristic polynomial has complex roots $\lambda = a \pm bi$.

(f) Solve each of the following Cauchy–Euler initial-value problems:

(i) $t^2 y'' + 3ty' + 5y = 0$, $y(1) = 1$, $y'(1) = 0$
(ii) $t^2 y'' - 3ty' + 13y = 0$, $y(1) = 1$, $y'(1) = 0$

4.7.4 Companion systems and companion matrices

Given a second-order linear differential equation with constant coefficients such as

$$y'' + by' + cy = 0 \qquad (4.7.21)$$

we know that through the substitution $x_1 = y$, $x_2 = y'$ we can convert (4.7.21) to the system of first-order equations given by

$$\begin{aligned} x_1' &= x_2 \\ x_2' &= -cx_1 - bx_2 \end{aligned} \qquad (4.7.22)$$

The system (4.7.22) is called the *companion system* of (4.7.21). In what follows, we explore the connections between the original equation and its companion system.

(a) Consider the homogeneous linear second-order DE

$$y'' + 3y' + 2y = 0 \qquad (4.7.23)$$

Using the guess $y = e^{rt}$, find the characteristic equation of (4.7.23) and the values of r that make $y = e^{rt}$ a solution of the given DE.

(b) Convert the DE (4.7.23) into a system of first-order equations in the form $\mathbf{x}' = \mathbf{Ax}$. In addition, determine the eigenvalues of the matrix $\mathbf{A}$.

(c) What do you observe about the roots of the characteristic equation in (a) and the eigenvalues of the matrix in (b)? Why is this result not surprising?

(d) Find the general solution of the second-order equation (4.7.23) using standard methods from chapter 4. Find the general solution of the first-order system you found in (b) using standard methods from chapter 3. Explain how your two results agree.

(e) Now consider the general equation (4.7.21) where b and c are arbitrary constants and its corresponding companion system.

(i) Show that the roots of the characteristic equation are

$$r = \frac{-b \pm \sqrt{b^2 - 4c}}{2}$$

and that the eigenvalues of the coefficient matrix of the companion system are

$$\lambda = \frac{-b \pm \sqrt{b^2 - 4c}}{2}$$

(ii) Assuming that $b^2 - 4c > 0$ so that the values of r in (i) are real and distinct, state the general solution of (4.7.21).

(iii) Show that the eigenvectors of the matrix of the companion system that correspond to λ_1 and λ_2 are given by

$$\mathbf{v}_1 = \begin{bmatrix} 1 \\ \lambda_1 \end{bmatrix} \text{ and } \mathbf{v}_1 = \begin{bmatrix} 1 \\ \lambda_1 \end{bmatrix}$$

where $\lambda_1 = (-b + \sqrt{b^2 - 4c})/2$ and $\lambda_2 = (-b - \sqrt{b^2 - 4c})/2$. State the general solution to the companion system.

(iv) Compare your result from (ii) to the result for x_2 in (iii). Do your solutions agree?

(f) Our work above shows that for any second-order differential equation, there exists a companion system of two first-order equations whose vector solution contains the solution of the second-order equation.
For the third-order equation

$$y''' + 2y'' - y' - 2y = 0$$

find the solution of the system directly by using standard methods from chapter 4. Then, find the general solution of the first-order companion system constructed from the substitution $x_1 = y$, $x_2 = y'$, $x_3 = y''$ using standard methods from chapter 3. Compare your results.

(g) In both the direct solution of higher order linear differential equations and in the solution of systems of linear first-order equations, the solution methods require us to find roots of polynomials. Our work above enables us to see the fact that any polynomial has an associated matrix, a so-called *companion matrix*, whose eigenvalues are the same as the zeros of the polynomial. In general, given a polynomial function

$$p(t) = t^n + a_{n-1}t^{n-1} + a_{n-2}t^{n-2} + \cdots + a_1 t + a_0$$

the *companion matrix* of $p(t)$ is given by

$$\mathbf{C} = \begin{bmatrix} 0 & 1 & 0 & 0 & \cdots & 0 \\ 0 & 0 & 1 & 0 & \cdots & 0 \\ \vdots & \vdots & & & & \vdots \\ 0 & 0 & \cdots & 0 & 0 & 1 \\ -a_0 & -a_1 & -a_2 & \cdots & -a_{n-2} & -a_{n-1} \end{bmatrix} \qquad (4.7.24)$$

That is, $\mathbf{C}$ is an $n \times n$ matrix whose first $n - 1$ rows are all zero except for the entry just above the diagonal, whose value is 1. The final row consists of the opposites of the coefficients of the constant, linear, etc., terms of the polynomial p.

It can be proved that, in general, the eigenvalues of C are the same as the zeros of $p(t)$. We verify this fact through a few examples.

(i) For the polynomial $p(t) = t^2 + 3t + 2$, determine the companion matrix C. Compute the eigenvalues of C directly and compare the result to the zeros of $p(t)$.

(ii) For the polynomial $p(t) = t^3 + 3t^2 + 3t + 1$, determine the companion matrix C. Compute the eigenvalues of C directly and compare the result to the zeros of $p(t)$.

(iii) For the polynomial $p(t) = t^4 - 1$, determine the companion matrix C. Compute the eigenvalues of C directly and compare the result to the zeros of $p(t)$.

(h) For the nth-order linear homogeneous equation

$$y^{(n)} + a_{n-1}y^{(n-1)} + \cdots a_1 y' + a_0 y = 0 \qquad (4.7.25)$$

show that the coefficient matrix of the corresponding companion system is in fact that companion matrix of the characteristic polynomial of (4.7.25).

5

Laplace transforms

5.1 Motivating problems

In this chapter, we again consider solving nonhomogeneous linear differential equations such as

$$y'' + a_1 y' + a_2 y = f(t)$$

but in contexts where the forcing function is different from those we have previously encountered. While we have developed the methods of undetermined coefficients and variation of parameters to approach this problem, there are several reasons to consider a different means of solution. Perhaps, most prominent is that in every example to date, we have assumed that the function $f(t)$ is continuous. Indeed, it has also typically been the case that $f(t)$ is a standard function, one belonging to the library of basic functions like $\sin 2t$ and $\ln t$ that we encounter in calculus. In many applications, however, it is possible for $f(t)$ to be piecewise defined, discontinuous, or worse. We consider two examples that demonstrate these possibilities.

Electrical circuits with a voltage source provide a common situation where the forcing function $f(t)$ is not continuous. If we flip a switch to turn the voltage on, then the forcing function is actually a *step* function that leaps from zero to a constant value. Recall that the charge $Q(t)$ in an RLC circuit is modeled by the second-order equation

$$LQ'' + RQ' + \frac{1}{C}Q = E(t) \tag{5.1.1}$$

where $E(t)$ is an external voltage source. Suppose that we are given an RLC circuit with an initial charge $Q(0)$ and initial current $Q'(0)$, and that the voltage

$E(t) = 1000$ is turned on at $t = 4$. The voltage function $E(t)$ is, therefore, defined piecewise by the formula

$$E(t) = \begin{cases} 0, & \text{if } 0 \le t < 4 \\ 1000, & \text{if } t \ge 4 \end{cases}$$

Let us further assume that $L = 20$ H, $R = 40\ \Omega$, $C = 10^{-2}$ F, and that $Q(0) = 25$ and $Q'(0) = 0$. From the given information and (5.1.1), we know that $Q(t)$ is modeled by the initial-value problem

$$20Q'' + 40Q' + 100Q = E(t), \quad Q(0) = 25, \ Q'(0) = 0 \tag{5.1.2}$$

We have not yet encountered means to deal with a step function as the forcing function in an initial-value problem. In section 5.4, we will discuss step functions in detail, learning how they may be used to turn other functions on and off; in addition, we will show how the Laplace transform provides an ideal tool for dealing with piecewise-defined functions in initial-value problems. With these tools, we will be able to determine the solution $Q(t)$ for (5.1.2) whose graph is shown in figure 5.1. Observe that we see the expected damped oscillation in $Q(t)$ up until time $t = 4$ when the forcing function $E(t)$ is turned on, at which point we see the solution driven vertically away from zero so that as t increases, $Q(t) \to 10$. That $Q(t)$ approaches 10 should not surprise us since $Q(t) = 10$ is a constant solution to the equation

$$20Q'' + 40Q' + 100Q = 1000$$

In fact, $Q(t) = 10$ is a stable equilibrium solution of the equation.

In addition to functions that get turned on or off at a certain time, another important forcing function to consider is a so-called *impulse function*. These functions are ones where a force is imparted over an extremely short time interval such as a hammer striking a mass. In section 5.4, we introduce the Dirac delta function, $\delta(t)$, study its properties, and see how it may be used in settings such as the following.

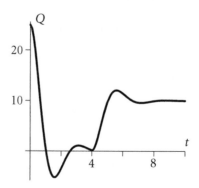

Figure 5.1 The solution $Q(t)$ to (5.1.2).

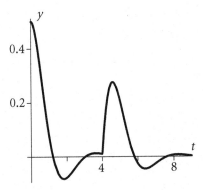

Figure 5.2 The solution curve $y(t)$ to (5.1.3).

Suppose that a mass of 1 kg is attached to a spring with constant $k = 4$ and the system's damping constant is $c = 2$. In addition, assume that the mass is initially displaced 0.5 m from equilibrium and released. At time $t = 4$, the mass is struck with a hammer imparting a unit impulse in the positive direction. The combination of all of these conditions leads to the initial-value problem

$$y'' + 2y' + 4y = \delta(t - 4), \quad y(0) = 0.5, \quad y'(0) = 0 \qquad (5.1.3)$$

where the function $\delta(t - 4)$ represents the hammer imparting the unit force of impulse.

Just as with piecewise-defined functions, we will learn that the Laplace transform provides an ideal tool for dealing with impulses. Once we develop the appropriate theory, we will be able to solve initial-value problems such as (5.1.3) and see that the solution behaves as shown in figure 5.2. In the solution, we see the noticeable impact of the impulse as the problem appears to restart, almost as if new initial conditions have been given at time $t = 4$.

In addition to being able to address discontinuous and impulse forcing functions, the Laplace transform is a powerful tool because it handles all allowable forcing functions in the same manner. Moreover, in each case it proceeds directly to the solution of initial-value problems without first finding the general solution to the differential equation. These ideas and more will be studied in subsequent sections.

5.2 Laplace transforms: getting started

The motivating idea behind the Laplace transform is natural: to solve a differential equation, our desire is to integrate. For the simplest examples, such as $y' = y$, we know that we can separate variables and integrate in order to determine y. However, if we approach the problem

$$y' + a_0 y = f(t) \qquad (5.2.1)$$

by attempting to integrate both sides from 0 to s with respect to t in order to eliminate y', doing so leads to the equation

$$\int_0^s y'(t)\,dt + a_0 \int_0^s y(t)\,dt = \int_0^s f(t)\,dt \tag{5.2.2}$$

While $\int_0^s y'(t)\,dt = y(s) - y(0)$ eliminates the derivative y' from the equation, and $\int_0^s f(t)\,dt$ can usually be computed for a given f, in (5.2.2) we are left with the expression $\int_0^s y(t)\,dt$, where y is an unknown function. Essentially this step of integrating has replaced the derivative of the unknown function y with its integral in the equation we are endeavoring to solve. This leaves us no closer to finding the solution function $y(t)$.

Rather than simply trying to integrate, the Laplace transform uses a modified approach in which every function in (5.2.1) is multiplied by another function before integrating; this approach will enable us to convert differential equations in $y(t)$ and $y'(t)$ to algebraic equations in a new unknown function $Y(s)$ that we can solve for $Y(s)$. This method is similar to the use of integrating factors when solving linear first-order equations.

Before we formally define the Laplace transform, we discuss a few preliminary ideas, some of which are familiar concepts from calculus. First, we assume throughout this chapter that all forcing functions are piecewise continuous functions defined for $t > 0$ and that

$$f(0) = f(0^+) = \lim_{t \to 0+} f(t) \tag{5.2.3}$$

That is, f cannot be discontinuous at the origin itself, though it is allowed to have finitely many discontinuities for $t > 0$.

Furthermore, we assume that the forcing function does not grow more rapidly than an exponential function. Formally, we will assume that $f(t)$ is of *exponential order*, which means that for sufficiently large t,

$$|f(t)| \le Me^{bt} \tag{5.2.4}$$

for positive constants M and b. Functions that are piecewise continuous and meet conditions (5.2.3) and (5.2.4) are called *acceptable*. For example, polynomial functions, $\sin kt$, e^{kt}, and sums and products of these functions are acceptable, as are piecewise-defined functions with finitely many discontinuities whose pieces consist of these basic functions. In particular, linear combinations of acceptable functions are acceptable. Functions such as

$$e^{t^2}, \quad t^{-1/2}, \quad (t-1)^{-1}$$

are not acceptable. The first grows too rapidly to be of exponential order, the second fails to meet the condition (5.2.3) that a limit exists from the right at the origin, and the third is not piecewise continuous on any interval containing $t = 1$.

In addition, from calculus we recall the following important concepts:

- If $y' = f(t)$ and $y(0) = 0$, then $y = \int_0^t f(s)\, ds$.
- The improper integral $\int_0^\infty f(t)\, dt$ is said to *converge* whenever

$$\lim_{r \to \infty} \int_0^r f(t)\, dt$$

exists. If this limit fails to exist, we say the improper integral *diverges.*

- Given a function of two variables $K(s, t)$, if we integrate this function with respect to t from $t = a$ to $t = b$, the result is a function of s. That is,

$$\int_a^b K(s, t)\, dt$$

is a function of s.

Recall our earlier note regarding the overall approach with Laplace transforms: in order to solve an initial-value problem, we integrate both sides of the differential equation after both sides have been multiplied by a more complicated function. The main idea is that we use the *transformation* given by

$$\int_0^\infty K(s, t) f(t)\, dt$$

Knowing the prominent role that the exponential function has played throughout our work with differential equations to date, it is not surprising that we choose to use $K(s, t) = e^{-st}$. Specifically, we make the following definition.

Definition 5.2.1   Let $f(t)$ be an acceptable function defined on the interval $[0, \infty)$. The *Laplace transform* of $f(t)$, denoted $\mathcal{L}[f]$, is the function defined by

$$\mathcal{L}[f] = \int_0^\infty e^{-st} f(t)\, dt \tag{5.2.5}$$

We note that because $\mathcal{L}[f]$ is a function of s, we often write $F(s)$ rather than the more explicit $\mathcal{L}[f(t)]$. We consider an example to see the Laplace transform at work.

Example 5.2.1   Compute the Laplace transform of $f(t) = t$.

Solution.   By definition,

$$\mathcal{L}[t] = \int_0^\infty t e^{-st}\, dt \tag{5.2.6}$$

Replacing the improper integral with a limit and integrating by parts, we observe that

$$\mathcal{L}[t] = \lim_{r \to \infty} \int_0^r te^{-st}\, dt$$

$$= \lim_{r \to \infty} \left[-\frac{1}{s}\left(t + \frac{1}{s}\right)e^{-st}\Big|_0^r \right]$$

$$= \lim_{r \to \infty} \left[-\frac{1}{s}\left(r + \frac{1}{s}\right)e^{-sr} + \frac{1}{s}\left(0 + \frac{1}{s}\right)e^0 \right]$$

$$= \lim_{r \to \infty} \left[-\frac{r}{s}e^{-sr} - \frac{1}{s^2}e^{-sr} + \frac{1}{s^2} \right] \tag{5.2.7}$$

By L'Hopital's Rule,[1] we know that $re^{-sr} \to 0$ as $r \to \infty$ for each $s > 0$. Combined with the fact that $e^{-sr} \to 0$ as $r \to \infty$, it follows from (5.2.7) that

$$\mathcal{L}[t] = F(s) = \frac{1}{s^2} \tag{5.2.8}$$

Soon we will apply the Laplace transform in order to solve initial-value problems. This process will require us to also use the *inverse Laplace transform* which asks, "given a function $F(s)$, what function $f(t)$ is such that $\mathcal{L}[f(t)] = F(s)$?" For instance, (5.2.8) tells us we may write

$$\mathcal{L}^{-1}\left[\frac{1}{s^2}\right] = t \tag{5.2.9}$$

Much more on inverse transforms will follow as we progress in our study.

It is not obvious that the Laplace transform of every acceptable function exists. While we omit the proof, it is possible to prove the following theorem by showing that not only does $f(t)$ being acceptable guarantee that $\mathcal{L}[f(t)] = F(s)$ exists, but that $F(s)$ is a function that must tend to 0 as $s \to \infty$.

Theorem 5.2.1 If $f(t)$ is acceptable, then the Laplace transform $F(s)$ of $f(t)$ exists. Moreover,

1. $sF(s)$ is bounded as $s \to \infty$, from which it follows that

2. $\lim_{s \to \infty} F(s) = 0$.

Although it is not necessary for a function to be acceptable in order to have a Laplace transform, our focus will be almost exclusively on acceptable functions. In addition, we note that not all elementary functions can be generated by taking the Laplace transform of an acceptable function. For instance, $F(s) = 1$ cannot be the Laplace transform of an acceptable function since both parts of theorem 5.2.1 are contradicted.

---

[1] $\lim_{r \to \infty} \dfrac{r}{e^{sr}} = \lim_{r \to \infty} \dfrac{1}{se^{sr}} = 0$.

The next three examples further illustrate the definition and notational conventions we use with Laplace transforms.

Example 5.2.2   Compute the Laplace transform of $f(t) = 1$.

Solution.   From the definition, we observe that

$$\mathcal{L}[1] = \int_0^\infty e^{-st}\, dt = \lim_{r \to \infty} -\frac{1}{s} e^{-st}\Big|_0^r = \lim_{r \to \infty}\left[-\frac{1}{s} e^{-sr} + \frac{1}{s}\right] = \frac{1}{s}$$

since $e^{-sr} \to 0$ as $r \to \infty$.

Example 5.2.3   Find the Laplace transform of $f(t) = e^{at}$.

Solution.   We compute

$$\mathcal{L}[e^{at}] = \int_0^\infty e^{at} e^{-st}\, dt = \int_0^\infty e^{(a-s)t}\, dt = \lim_{r \to \infty}\int_0^r e^{(a-s)t}\, dt$$

$$= \lim_{r \to \infty}\frac{1}{a-s} e^{(a-s)t}\Big|_0^r = \lim_{r \to \infty}\left[\frac{1}{a-s} e^{(a-s)r} - \frac{1}{a-s}\right] = \frac{1}{s-a}$$

provided that $s > a$, for then $e^{(a-s)r} \to 0$ as $r \to \infty$.

At times, we will need to restrict the values of s in order for the Laplace transform to exist. Above, we observed that $\mathcal{L}[e^{at}] = 1/(s-a)$, provided that $s > a$. Usually, we will suppress the discussion of the restriction on s-values and simply assume that the domain of the Laplace transform is as large as possible.

Example 5.2.4   Find $\mathcal{L}[\cos kt]$ and $\mathcal{L}[\sin kt]$.

Solution.   By definition,

$$\mathcal{L}[\cos kt] = \int_0^\infty \cos kt\, e^{-st}\, dt$$

Integrating by parts twice or using a table of integrals,

$$\mathcal{L}[\cos kt] = \lim_{r \to \infty}\frac{1}{s^2 + k^2}\left(k^2 \sin kt - s\cos kt\right) e^{-st}\Big|_0^r$$

$$= \lim_{r \to \infty}\left[\frac{1}{s^2 + k^2}\left(k^2 \sin kr - s\cos kr\right) e^{-sr} - \frac{1}{s^2 + k^2}(0 - s)\right]$$

$$= \lim_{r \to \infty}\left[\frac{e^{-sr}k^2 \sin kr}{s^2 + k^2} - \frac{e^{-sr}s\cos kr}{s^2 + k^2} + \frac{s}{s^2 + k^2}\right] \qquad (5.2.10)$$

Since $e^{-sr} \to 0$ as $r \to \infty$ and $|\sin kr|$ and $|\cos kr|$ are bounded by 1 as $r \to \infty$, it follows from (5.2.10) that

$$\mathcal{L}[\cos kt] = \frac{s}{s^2 + k^2}$$

Similar computations show

$$\mathcal{L}[\sin kt] = \frac{k}{s^2 + k^2}$$

Table 5.1
Laplace transforms of some basic functions

$f(t)$	$F(s) = \mathcal{L}[f(t)] = \int_0^\infty f(t)e^{-st}\,dt$
1	$1/s$
t	$1/s^2$
t^2	$2/s^3$
e^{at}	$1/(s - a)$
$\cos kt$	$s/(s^2 + k^2)$
$\sin kt$	$k/(s^2 + k^2)$

We close this section with table 5.1, which summarizes the Laplace transforms we have computed so far.

Observe that each line in the table may also be written in inverse form. For example, $\mathcal{L}^{-1}[1/(s - a)] = e^{at}$. This will be particularly useful in the next section as we see the first example of how the transform and its inverse can be used to solve an initial-value problem. In order to apply the Laplace transform successfully, we need to develop a deeper understanding of its properties and explore the impact of the transform on a wide range of functions. The following exercises and our investigations in the next section continue our work to this end.

Exercises 5.2   In exercises 1–4, explain why the limit of each function $g(r)$ is 0 as $r \to \infty$. In each, assume $s > 0$.

1. $g(r) = re^{-sr}$

2. $g(r) = r^2 e^{-sr}$

3. $g(r) = r^n e^{-sr}$

4. $g(r) = e^{-sr} \sin kr$

In exercises 5–16, use the definition of the Laplace transform to compute $\mathcal{L}[f(t)]$. For each, state the domain of s-values on which $\mathcal{L}[f(t)] = F(s)$ is defined.

5. $f(t) = 2t$

6. $f(t) = t - 3$

7. $f(t) = 2 - t$

8. $f(t) = t^2$

9. $f(t) = t^2 - 3$

10. $f(t) = (t - 2)^2$

11. $f(t) = e^{3t}$

12. $f(t) = e^{2t-3}$

13. $f(t) = e^{3t+5}$

14. $f(t) = \cos 4t$

15. $f(t) = te^{at}$

16. $f(t) = t \sin 2t$

From examples 5.2.2 and 5.2.1, we know that

$$\mathcal{L}[1] = \frac{1}{s} \text{ and } \mathcal{L}[t] = \frac{1}{s^2}$$

Use these facts to compute the Laplace transform of each of the functions in exercises 17–19 with as little computation as possible. What properties of integrals and limits are being used?

17. $f(t) = 1 + t$

18. $f(t) = 3t - 2$

19. $f(t) = c + kt$

20. Explain why the Laplace transform is a linear operator on the vector space of acceptable functions.[2] That is, explain why for any real numbers a and b and any acceptable functions f and g,

$$\mathcal{L}[af(t) + bg(t)] = a\mathcal{L}[f(t)] + b\mathcal{L}[g(t)]$$

5.3 General properties of the Laplace transform

In many ways, the Laplace transform resembles the differentiation and integration operators from calculus. For example, given a function $f(t) = 3t^4 + 5t + 1$, taking the derivative results in a new function $f'(t)$. Using the alternate notation $D[f]$ for the derivative of f with respect to t, we see that

$$D[3t^4 + 5t + 1] = 12t^3 + 5$$

---

[2] See appendix D for further discussion on linear transformations of vector spaces.

In particular, the "D" operator transforms one function into another. Likewise, if we consider the definite integral of $f(t) = t - 1$ from $t = 0$ to $t = x$, we find that

$$\int_0^x (t - 1)\, dt = \frac{1}{2}x^2 - x$$

Letting $I(f) = \int_0^x f(t)\, dt$, we see that I transforms one function $f(t)$ into another function $F(x)$ by the process of integration. In the same way, as we have seen in examples 5.2.1–5.2.4, the Laplace transform takes an acceptable function $f(t)$ and transforms it into a new function $F(s)$ by a process slightly more complicated than standard integration.

From calculus and our preceding work with differential equations, we know that taking the derivative of a function is a linear process, as is calculating the definite integral. More specifically, for any constants a and b and functions $f(t)$ and $g(t)$ that are differentiable and integrable, we know that

$$D[af(t) + bg(t)] = aD[f(t)] + bD[g(t)]$$

and

$$\int_0^x [af(t) + bg(t)]\, dt = a \int_0^x f(t)\, dt + b \int_0^x g(t)\, dt$$

Similarly, because the Laplace transform's definition involves limits and integrals, it has the same properties of linearity as the derivative and integral operators. In particular, as was shown in exercise 20 of section 5.2, the following theorem holds.

Theorem 5.3.1   For every pair of scalars a and b and acceptable functions $f(t)$ and $g(t)$,
$$\mathcal{L}[af(t) + bg(t)] = a\mathcal{L}[f(t)] + b\mathcal{L}[g(t)] \tag{5.3.1}$$

Theorem 5.3.1 shows that the Laplace transform, like the differential and integral operators, is a *linear transformation* or *linear operator*. Formally, a linear transformation is a function T that maps one vector space V to another vector space W where T satisfies the property that for all constants a and b and all elements $\mathbf{u}$ and $\mathbf{v}$ in V, $T(a\mathbf{u} + b\mathbf{v}) = aT(\mathbf{u}) + bT(\mathbf{v})$. Appendix D provides further discussion on linear transformations of vector spaces.

In calculus, following the definitions of the derivative and the definite integral, we quickly discover more general properties that enable us to compute derivatives and integrals without using the definition directly. In the same way, while we have seen a few examples of how to use the definition to compute the Laplace transform of certain functions $f(t)$, we can use results such as theorem 5.3.1 to more easily determine the Laplace transform of more complicated functions. Two examples follow.

Example 5.3.1   Find the Laplace transform of $f(t) = 7 - 3e^{2t}$.

Solution.   We know from examples 5.2.2 and 5.2.3 that $\mathcal{L}[1] = 1/s$ and $\mathcal{L}[e^{2t}] = 1/(s-2)$. By theorem 5.3.1 it now follows that

$$\mathcal{L}[7 - 3e^{2t}] = 7\mathcal{L}[1] - 3\mathcal{L}[e^{2t}] = \frac{7}{s} - \frac{3}{s-2}$$

We note that the individual Laplace transforms are defined on different domains: $7/s$ is valid for $s > 0$ while $3/(s-2)$ is defined if $s > 2$. We usually suppress discussion of this issue and assume that $\mathcal{L}[f(t)]$ is defined on the largest interval possible. In example 5.3.1, this domain is $\{s|s > 2\}$.

Example 5.3.2   Find the Laplace transform of $\cosh kt$ and $\sinh kt$.

Solution.   By definition, the hyperbolic cosine function is given by $\cosh kt = \frac{1}{2}e^{kt} + \frac{1}{2}e^{-kt}$. By the linearity of the Laplace transform, it follows that

$$\mathcal{L}[\cosh kt] = \frac{1}{2}\mathcal{L}[e^{kt}] + \frac{1}{2}\mathcal{L}[e^{-kt}]$$

$$= \frac{1}{2}\left(\frac{1}{s-k} + \frac{1}{s+k}\right) = \frac{s}{s^2 - k^2}$$

Similarly,

$$\mathcal{L}[\sinh kt] = \mathcal{L}\left[\frac{1}{2}(e^{kt} - e^{-kt})\right] = \frac{1}{2}\left(\frac{1}{s-k} - \frac{1}{s+k}\right) = \frac{k}{s^2 - k^2}$$

In addition to taking linear combinations of functions, we often want to multiply a given function by t or some power of t. For example, it is natural to wonder if we can use our work in preceding examples to compute $\mathcal{L}[te^{at}]$. If we first consider the Laplace transforms of the simple power functions 1, t, t^2, and so on, we find evidence for a conjecture on how we might approach $\mathcal{L}[te^{at}]$. In particular, note that

$$\mathcal{L}[1] = \frac{1}{s} \quad \mathcal{L}[t] = \frac{1}{s^2} \quad \mathcal{L}[t^2] = \frac{2}{s^3} \tag{5.3.2}$$

The last result was shown in exercise 8 of section 5.2. In fact, we could go on to show that $\mathcal{L}[t^3] = 6/s^4$. This sequence of results reminds us of derivatives: in particular,

$$\frac{d}{ds}\left[\frac{1}{s}\right] = -\frac{1}{s^2} \quad \frac{d}{ds}\left[\frac{1}{s^2}\right] = -\frac{2}{s^3} \quad \frac{d}{ds}\left[\frac{2}{s^3}\right] = -\frac{6}{s^4} \tag{5.3.3}$$

From this sequence of examples, it appears that each time we take a given function $f(t) = t^n$ and multiply it by t, the impact on its Laplace transform is that the transform of the new function is the opposite of the derivative of the transform of the original. Using a result from multivariable calculus known as Leibniz's rule, a formal proof of this fact may be established, not only for power functions, but also for all functions having Laplace transforms. We defer this work to exercise 25 and state the following theorem.

Theorem 5.3.2   If $\mathcal{L}[f(t)] = F(s)$, then

$$\mathcal{L}[tf(t)] = -F'(s) = -\frac{d}{ds}F(s) \qquad (5.3.4)$$

Theorem 5.3.2 enables us to expand on our observations above regarding the Laplace transforms of the power functions t, t^2, t^3, and so on. In particular, replacing $F(s)$ with $\mathcal{L}[t]$, we can take the perspective that (5.3.4) implies

$$\mathcal{L}[tf(t)] = -\frac{d}{ds}\mathcal{L}[f(t)] \qquad (5.3.5)$$

This shows that, for example,

$$\mathcal{L}[t^4] = \mathcal{L}[t \cdot t^3] = -\frac{d}{ds}\mathcal{L}[t^3] = -\frac{d}{ds}\left[\frac{6}{s^4}\right] = \frac{24}{s^5}$$

In addition, a generalization of this reasoning can be used to show the following corollary to theorem 5.3.2. See exercise 26.

Corollary 5.3.3   For each positive integer n,

$$\mathcal{L}[t^n f(t)] = (-1)^n F^{(n)}(s) \qquad (5.3.6)$$

We next consider two examples that show how we can use recent results to compute the Laplace transform of familiar functions multiplied by t.

Example 5.3.3   Find $\mathcal{L}[te^{at}]$ and $\mathcal{L}[t^2 e^{at}]$.

Solution.   We know from earlier work that $\mathcal{L}[e^{at}] = 1/(s-a)$. It follows from theorem 5.3.2 that

$$\mathcal{L}[te^{at}] = -\frac{d}{ds}\mathcal{L}[e^{at}] = -\frac{d}{ds}\left[\frac{1}{s-a}\right] = \frac{1}{(s-a)^2}$$

Similarly,

$$\mathcal{L}[t^2 e^{at}] = -\frac{d}{ds}\mathcal{L}[te^{at}] = -\frac{d}{ds}\left[\frac{1}{(s-a)^2}\right] = \frac{2}{(s-a)^3}$$

In fact, as we will see in exercise 27, we can show in general that

$$\mathcal{L}[t^n e^{at}] = \frac{n!}{(s-a)^{n+1}} \qquad (5.3.7)$$

Example 5.3.4   Find $\mathcal{L}[t \sin kt]$.

Solution. In example 5.2.4, we showed that

$$\mathcal{L}[\sin kt] = \frac{k}{s^2 + k^2}$$

Applying theorem 5.3.2, we know that

$$\mathcal{L}[t \sin kt] = -\frac{d}{ds}\left[\frac{k}{s^2 + k^2}\right] = \frac{2ks}{(s^2 + k^2)^2}$$

As we have noted, we are motivated to develop the Laplace transform by the need to solve initial-value problems that involve unusual forcing functions. For example, we will soon work to solve equations of the form

$$y' + a_0 y = f(t) \tag{5.3.8}$$

where $f(t)$ is a step function or other piecewise defined function. We will use our understanding of the Laplace transform to solve these equations by taking the Laplace transform of each side of (5.3.8) to transform the differential equation (in t) into an algebraic equation (in s). Our hope is that upon doing so, we can solve the new algebraic equation in order to ultimately solve the differential one.

To see how this process begins, we take the Laplace transform of both sides of (5.3.8) and apply the linearity property. Doing so results in the equation

$$\mathcal{L}[y'] + a_0\mathcal{L}[y] = \mathcal{L}[f(t)] \tag{5.3.9}$$

Here, we realize that while we can compute $\mathcal{L}[f(t)]$ using the definition or established results, it is unclear how to work with $\mathcal{L}[y']$ and $\mathcal{L}[y]$. Ideally, if we could understand how the Laplace transform $\mathcal{L}[y']$ of the derivative of an unknown function is related to the Laplace transform $\mathcal{L}[y]$ of the function itself, that would enable us to work with one unknown quantity. To this end, we return to the definition and show how $\mathcal{L}[y']$ depends on $\mathcal{L}[y]$.

Let us suppose that y and y' are acceptable functions and that y is continuous. By definition,

$$\mathcal{L}[y'(t)] = \int_0^\infty y'(t)e^{-st}\,dt = \lim_{r\to\infty}\int_0^r y'(t)e^{-st}\,dt \tag{5.3.10}$$

To evaluate $\int_0^r y'(t)e^{-st}\,dt$, we use integration by parts with $u = e^{-st}$ and $dv = y'(t)dt$. It follows that $du = -se^{-st}\,dt$ and $v = y(t)$. Integrating[3] (5.3.10),

$$\mathcal{L}[y'(t)] = \lim_{r\to\infty} y(t)e^{-st}\Big|_0^r + s\int_0^r y(t)e^{-st}\,dt$$

$$= \lim_{r\to\infty} y(r)e^{-sr} - y(0) + s\int_0^r y(t)e^{-st}\,dt \tag{5.3.11}$$

---

[3] The integration by parts formula holds since y is continuous. If y has a jump discontinuity, then this part of the argument is more complicated.

Since y is an acceptable function, it is of exponential order and $|y(t)| \leq Me^{bt}$ for some positive constants M and b. Assuming that $s > b$, it follows $y(r)e^{-sr} \to 0$ as $r \to \infty$. In addition, in (5.3.11) we observe

$$\lim_{r \to \infty} s \int_0^r y(t)e^{-st} \, dt = s \int_0^\infty y(t)e^{-st} \, dt = s\mathcal{L}[y(t)]$$

by the definition of the Laplace transform. Hence, (5.3.11) implies

$$\mathcal{L}[y'(t)] = s\mathcal{L}[y(t)] - y(0) \tag{5.3.12}$$

Our work has proved the following theorem.

Theorem 5.3.4 Suppose $y(t)$ is continuous and $y(t)$ and $y'(t)$ are acceptable. Then

$$\mathcal{L}[y'(t)] = s\mathcal{L}[y(t)] - y(0) \tag{5.3.13}$$

Note particularly the appearance of $y(0)$ in the conclusion of theorem 5.3.4. This foreshadows how we will use the Laplace transform to solve an initial-value problem directly without resorting to a general solution of the associated differential equation. To see further how we will use the Laplace transform, we consider the following example.

Example 5.3.5 Use the Laplace transform to solve the initial-value problem

$$y' + y = e^{-t}, \quad y(0) = 0 \tag{5.3.14}$$

Solution. We begin by taking the Laplace transform of both sides of (5.3.14) to achieve

$$\mathcal{L}[y'] + \mathcal{L}[y] = \mathcal{L}[e^{-t}] \tag{5.3.15}$$

From example 5.2.3, we know that $\mathcal{L}[e^{-t}] = 1/(s+1)$. Furthermore, we just established

$$\mathcal{L}[y'] = s\mathcal{L}[y] - y(0) \tag{5.3.16}$$

Combining (5.3.15), (5.3.16), and the given fact that $y(0) = 0$, we have

$$s\mathcal{L}[y] + \mathcal{L}[y] = \frac{1}{s+1} \tag{5.3.17}$$

Letting $Y(s) = \mathcal{L}[y]$, factoring, and solving for $Y(s)$,

$$Y(s) = \frac{1}{(s+1)^2} \tag{5.3.18}$$

To solve the initial-value problem, it remains for us to determine the function $y(t)$ whose Laplace transform is $Y(s) = 1/(s+1)^2$. That is, we must find $\mathcal{L}^{-1}[Y(s)] = \mathcal{L}^{-1}[1/(s+1)^2]$. In example 5.3.3, we saw that $\mathcal{L}[te^{at}] = 1/(s-a)^2$. In particular,

$$\mathcal{L}[te^{-t}] = \frac{1}{(s+1)^2} \quad \text{or} \quad \mathcal{L}^{-1}\left[\frac{1}{(s+1)^2}\right] = te^{-t}$$

From (5.3.18), it now follows that

$$y(t) = te^{-t}$$

This is precisely the solution we expect had we applied another method (such as using an integrating factor) to solve (5.3.14).

Note particularly that our work in (5.3.14)–(5.3.17) converted the given initial-value problem (5.3.14) involving y' to an algebraic equation (5.3.17) involving $\mathcal{L}[y] = Y(s)$. We then had to use the inverse Laplace transform in order to determine $y(t)$. This process is typical for how the transform is used to solve IVPs; at this point, we largely need to gain experience with more complicated functions and situations in order to solve more advanced problems.

We make note of one more result that relates the Laplace transform of a higher order derivative to the transform of the original function in order to help us solve higher order IVPs before proceeding to establish additional results on products of familiar functions and piecewise-defined functions in order to more fully understand the workings of the Laplace transform.

Corollary 5.3.5 Suppose $y(t)$ and $y'(t)$ are continuous and $y(t)$, $y'(t)$, and $y''(t)$ are acceptable. Then

$$\mathcal{L}[y''(t)] = s^2\mathcal{L}[y(t)] - sy(0) - y'(0) \tag{5.3.19}$$

The proof of corollary 5.3.5 is straightforward by two applications of theorem 5.3.4; see exercise 28.

In theorem 5.3.2, we computed the Laplace transform of $tf(t)$ in terms of the Laplace transform of $f(t)$. In addition to multiplying by t (or powers of t), another function that arises frequently in the study of differential equations is e^{at}. Hence we are naturally interested in how $\mathcal{L}[e^{at}f(t)]$ is related to $\mathcal{L}[f(t)]$.

Letting $f(t)$ be an acceptable function and $\mathcal{L}[f(t)] = F(s)$, we have by definition that

$$F(s) = \int_0^\infty f(t)e^{-st}\, dt \tag{5.3.20}$$

For the Laplace transform of $e^{at}f(t)$, we note that $e^{at}f(t)$ is an acceptable function and, by definition,

$$\mathcal{L}[e^{at}f(t)] = \int_0^\infty e^{at}f(t)e^{-st}\, dt = \int_0^\infty f(t)e^{-(s-a)t}\, dt \tag{5.3.21}$$

From the right-hand sides of (5.3.20) and (5.3.21), we observe that the only difference is that s has been replaced by $s - a$. In particular, $\mathcal{L}[e^{at}f(t)] = F(s-a)$, where $\mathcal{L}[f(t)] = F(s)$. We say that $F(s)$ has been *shifted* by multiplying $f(t)$ by e^{at} and call the theorem we have just proved the *first shifting property*, which is stated as follows.

Theorem 5.3.6 (First Shifting Property).   Let $f(t)$ be acceptable and $\mathcal{L}[f(t)] = F(s)$. For any real value of a,

$$\mathcal{L}[e^{at}f(t)] = F(s-a)$$

In the next example, we compute three Laplace transforms to show the straightforward application of theorem 5.3.6.

Example 5.3.6   Find $\mathcal{L}[e^{at}\cos kt]$, $\mathcal{L}[e^{at}\sin kt]$, and $\mathcal{L}[e^{at}t^2]$.

Solution.   We have already established that

$$\mathcal{L}[\cos kt] = \frac{s}{s^2 + k^2}$$

so by the first shifting property,

$$\mathcal{L}[e^{at}\cos kt] = \frac{s-a}{(s-a)^2 + k^2}$$

Similarly, from the fact that

$$\mathcal{L}[\sin kt] = \frac{k}{s^2 + k^2}$$

we observe

$$\mathcal{L}[e^{at}\sin kt] = \frac{k}{(s-a)^2 + k^2}$$

Finally,

$$\mathcal{L}[t^2] = \frac{2}{s^3}$$

and theorem 5.3.6 together imply

$$\mathcal{L}[e^{at}t^2] = \frac{2}{(s-a)^3}$$

A summary of the results we established in this section follows in table 5.2.

Exercises 5.3   In exercises 1–5, use the linearity property and the transforms derived in the examples to find the Laplace transform of the given function.

1. $f(t) = 3 - e^t$

2. $f(t) = 4\cos t + 2\sin t$

3. $f(t) = 3e^{2t} - 3\sin 2t$

4. $f(t) = 2 + 5\sin 3t$

5. $f(t) = 4\cos 5t - 6e^{-2t}$

Table 5.2
Summary of results on the Laplace Transform from section 5.3

$f(t)$	$F(s) = \mathcal{L}[f(t)] = \int_0^\infty f(t)e^{-st}\,dt$
$af(t) + bg(t)$	$a\mathcal{L}[f(t)] + b\mathcal{L}[g(t)]$
$tf(t)$	$-F'(s) = -\frac{d}{ds}\mathcal{L}[f(t)]$
$t^n f(t)$	$(-1)^n F^{(n)}(s)$
$f'(t)$	$s\mathcal{L}[f(t)] - f(0) = sF(s) - f(0)$
$f''(t)$	$s^2\mathcal{L}[f(t)] - sf(0) - f'(0) = s^2 F(s) - sf(0) - f'(0)$
$e^{at} f(t)$	$F(s - a)$

In exercises 6–11, use theorem 5.3.2 or corollary 5.3.3 and the transforms derived in the examples to find the Laplace transform of the given function.

6. $f(t) = 3te^{3t}$

7. $f(t) = t^2 e^{-t}$

8. $f(t) = 3t\cos 4t$

9. $f(t) = t^3 \sin t$

10. $f(t) = t^2 \cos t$

11. $f(t) = 4\cos 5t - 6e^{-2t}$

In exercises 12–17, use the first shifting property and the transforms derived in the examples to find the Laplace transform of the given function.

12. $f(t) = 3te^{3t}$

13. $f(t) = t^2 e^{-t}$

14. $f(t) = e^{-2t}\cos 4t$

15. $f(t) = e^{-t}\sin 2t$

16. $f(t) = e^{4t}\sinh 2t$

17. $f(t) = \cosh 2t \sin 3t$

In exercises 18–24, use established general properties and the transforms derived in the examples to find the Laplace transform of the given function.

18. $f(t) = 3te^{3t} - e^{2t}\cos t$

19. $f(t) = 4t^2 e^{-t} + 7e^{-3t}\sin t$

20. $f(t) = e^{-2t}(t^2 + 4t + 5)$

21. $f(t) = (t^2 - t)\sin t$

22. $f(t) = t(\cos 4t - 2\sin 4t)$

23. $f(t) = te^{-t}\sin 2t$

24. $f(t) = t^2 e^{-t}\sin 2t$

25. In multivariable calculus, students may have encountered Leibniz's rule, which allows differentiation across the integral sign. In particular, the rule states that under reasonable hypotheses on a function $K(s, t)$,

$$\frac{d}{ds}\int_{t=a}^{t=b} K(s, t)\, dt = \int_{t=a}^{t=b} \frac{\partial}{\partial s}[K(s, t)]\, dt$$

Use Leibniz's rule to explain why theorem 5.3.2 is true. In particular, show that if $F(s) = \mathcal{L}[f(t)]$, then $-F'(s) = \mathcal{L}[tf(t)]$

26. Using the rule established in theorem 5.3.2, show why corollary 5.3.3 is true. Specifically, show that if n is a positive integer, then

$$\mathcal{L}[t^n f(t)] = (-1)^n F^{(n)}(s)$$

(**Hint:** Apply the theorem to $\mathcal{L}[t \cdot t^{n-1} f(t)]$ to show that

$$\mathcal{L}[t^n f(t)] = -\frac{d}{ds}\mathcal{L}[t^{n-1} f(t)]$$

and then repeat this line of reasoning on the expression $\mathcal{L}[t^{n-1} f(t)]$.)

27. Use corollary 5.3.3 to show that

$$\mathcal{L}[t^n e^{at}] = \frac{n!}{(s - a)^{n+1}}$$

28. Apply theorem 5.3.4 twice to prove corollary 5.3.5.

29. Express $\mathcal{L}\left[f^{(4)}(t)\right]$ in terms of $\mathcal{L}[f(t)]$ and the first three derivatives of $f(t)$ at $t = 0$ by using theorem 5.3.4.

30. We have established that $\mathcal{L}[e^{at}] = 1/(s - a)$ for any real number a. Assume now that this formula holds for any complex number $a = \alpha + \beta i$, and hence compute the Laplace transform

$$\mathcal{L}[e^{(\alpha + \beta i)t}]$$

Use Euler's formula and properties of complex numbers to show that

$$\mathcal{L}[e^{\alpha t}(\cos \beta t + i \sin \beta t)] = \frac{s - \alpha}{(s - \alpha)^2 + \beta^2} + i\frac{\beta}{(s - \alpha)^2 + \beta^2}$$

Explain how equating real and imaginary parts produces an alternate derivation for the Laplace transforms of $e^{\alpha t}\cos \beta t$ and $e^{\alpha t}\sin \beta t$.

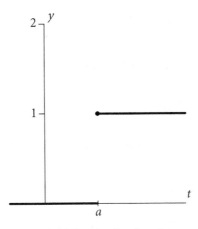

Figure 5.3 The translated unit step function $u(t-a)$.

5.4 Piecewise continuous functions

In physical applications, we sometimes encounter *step functions* that represent some quantity being turned on or off, such as an electric switch. If a mass in a spring-mass system is struck with a hammer or a drug is delivered by muscle injection, *impulse functions* that involve forces acting over very short time periods play a key role. To help us address these and related situations, we study the application of the Laplace transform to two important functions—the *Heaviside function* and the *Dirac delta function*.

5.4.1 The Heaviside function

We define the *Heaviside function*, or *unit step function*, denoted $u(t)$, to be the function that is 0 for all $t < 0$ and 1 for all $t \geq 0$. That is,

$$u(t) = \begin{cases} 0, & \text{if } t < 0 \\ 1, & \text{if } t \geq 0 \end{cases} \tag{5.4.1}$$

Often, we will make use of a step function that turns on at $t = a$, rather than $t = 0$. Thus we employ the *translated unit step function*, $u(t-a)$, which by (5.4.1) is given by

$$u(t-a) = \begin{cases} 0, & \text{if } t < a \\ 1, & \text{if } t \geq a \end{cases} \tag{5.4.2}$$

A plot of the translated unit step function is given in figure 5.3.

Step functions may be used to turn other functions on or off. For example, if we consider the function $f(t) = (4-t)u(t-4)$, we observe that since $u(t-4) = 0$

for $t < 4$ and $u(t - 4) = 1$ for $t \geq 4$, it follows

$$f(t) = \begin{cases} 0, & \text{if } t < 4 \\ 4 - t, & \text{if } t \geq 4 \end{cases} \tag{5.4.3}$$

From this perspective, we see that the function $(4 - t)$ is off until $t = 4$, at which time it is turned on.

To see how we can use step functions to turn another function both on and off at various times, we consider the function $g(t) = u(t - a) - u(t - b)$, where $a < b$. This difference of translated unit step functions turns on for $a \leq t < b$ and turns off when $t \geq b$. More specifically, for $t < a$, both $u(t - a)$ and $u(t - b)$ are zero, so $g(t) = 0$. For $a \leq t < b$, $u(t - a) = 1$ and $u(t - b) = 0$, thus $g(t) = 1$. And finally, once $t \geq b$, both $u(t - a) = 1$ and $u(t - b) = 1$, so that $g(t) = 0$. This can be written equivalently as

$$g(t) = \begin{cases} 0, & \text{if } t < a \\ 1, & \text{if } a \leq t < b \\ 0, & \text{if } t \geq b \end{cases} \tag{5.4.4}$$

This property of the function $u(t - a) - u(t - b)$ enables us to write a single formula for any piecewise-defined function that arises, rather than the traditional cases format where we stipulate the different formulas on different intervals, as in (5.4.4). The next example demonstrates the role of $u(t - a) - u(t - b)$.

Example 5.4.1 Define the following piecewise function using unit step functions.

$$f(t) = \begin{cases} t, & \text{if } 0 \leq t < 2 \\ 2, & \text{if } 2 \leq t < 4 \\ 0, & \text{otherwise} \end{cases}$$

Solution.   We use the fact that the function $u(t) - u(t - 2)$ is 1 in the interval $0 \leq t < 2$ and 0 otherwise, and $u(t - 2) - u(t - 4)$ is 1 on $2 \leq t < 4$ and 0 otherwise. Thus, we turn on t for $0 \leq t < 2$ and turn on 2 for $2 \leq t < 4$ by writing

$$f(t) = t[u(t) - u(t - 2)] + 2[u(t - 2) - u(t - 4)]$$
$$= tu(t) + (2 - t)u(t - 2) - 2u(t - 4)$$

A plot of $f(t)$ is shown in figure 5.4

At this point, we should again not lose sight of our goal: we are interested in using Laplace transforms to solve initial-value problems such as

$$y'' + 2y' + 5y = u(t - 2), \quad y(0) = 1, \ y'(0) = 0$$

where the forcing function is turned on at time $t = 2$. Since we will solve such equations by taking the Laplace transform of both sides, we must understand the

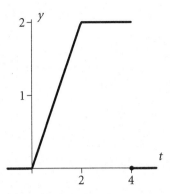

Figure 5.4 The function $f(t)$ in example 5.4.1.

transform of basic step functions. In fact, since step functions will be used to turn other functions on and off, we are more generally interested in $\mathcal{L}[u(t-a)f(t)]$. We return to the definition to explore this situation further.

Because we will employ a change of variables in our work, we begin by using z as a different variable of integration than the usual t in the definition. Specifically, from the definition of the Laplace transform we have

$$\mathcal{L}[u(t-a)f(t)] = \int_0^\infty u(z-a)f(z)e^{-sz}\,dz = \int_a^\infty f(z)e^{-sz}\,dz$$

The second equality follows from the fact that $u(z-a) = 0$ for all $z < a$ and $u(z-a) = 1$ for all $z \geq a$, which allows us to eliminate the presence of the unit step function.

We now employ the substitution $z = t + a$ and note that $t = z - a$ and $dz = dt$. From this and our work above, we see

$$\mathcal{L}[u(t-a)f(t)] = \int_a^\infty f(z)e^{-sz}\,dz$$

$$= \lim_{r \to \infty} \int_{z=a}^{z=r} f(z)e^{-sz}\,dz$$

$$= \lim_{r \to \infty} \int_{t=0}^{t=r-a} f(t+a)e^{-s(t+a)}\,dt$$

$$= \lim_{r \to \infty} \int_{t=0}^{t=r-a} f(t+a)e^{-st}e^{-as}\,dt \qquad (5.4.5)$$

In (5.4.5), since e^{-as} is constant with respect to t, we can remove it from the integral. Moreover, we can take the limit as $r \to \infty$ and note that $(r - a) \to \infty$ as well. From this, we now have

$$\mathcal{L}[u(t-a)f(t)] = e^{-as} \int_0^\infty f(t+a)e^{-st}\,dt$$

On the right, we observe that the Laplace transform of $f(t+a)$ has arisen, and therefore

$$\mathcal{L}[u(t-a)f(t)] = e^{-as}\mathcal{L}[f(t+a)]$$

We call this result the *second shifting property* and state it formally in the next theorem.

Theorem 5.4.1 (Second Shifting Property) If $f(t)$ has a Laplace transform, then

$$\mathcal{L}[u(t-a)f(t)] = e^{-as}\mathcal{L}[f(t+a)] \tag{5.4.6}$$

When working with inverse transforms, we'll often use the equivalent formulations of this result that

$$\mathcal{L}[u(t-a)f(t-a)] = e^{-as}\mathcal{L}[f(t)] \quad \text{or} \quad \mathcal{L}^{-1}[e^{-as}F(s)] = u(t-a)f(t-a) \tag{5.4.7}$$

which come from replacing t with $t-a$ in the argument of f. To see how the second shifting property works and gain more experience with the roles played by unit step functions, we consider several examples.

Example 5.4.2   Determine the Laplace transform of the step function, $u(t-3)$.

Solution.   We can view $u(t-3)$ as the function $u(t-3) \cdot 1$. Since we know that $\mathcal{L}[1] = 1/s$, by the second shifting property it follows that

$$\mathcal{L}[u(t-3)] = \mathcal{L}[u(t-3) \cdot 1] = e^{-3s}\mathcal{L}[1] = \frac{e^{-3s}}{s}$$

More generally, we can show that for any $a \geq 0$,

$$\mathcal{L}[u(t-a)] = \frac{e^{-as}}{s} \tag{5.4.8}$$

Example 5.4.3   Determine the Laplace transform of $f(t) = u(t-3)t^2$.

Solution.   With $f(t) = t^2$, by the second shifting property we have

$$\mathcal{L}[u(t-3)t^2] = e^{-3s}\mathcal{L}[(t+3)^2]$$
$$= e^{-3s}\mathcal{L}[t^2 + 6t + 9]$$
$$= e^{-3s}\left(\frac{2}{s^3} + \frac{6}{s^2} + \frac{9}{s}\right)$$

Example 5.4.4   Determine the Laplace transform of $f(t) = u(t-a) - u(t-b)$.

Solution.   Because we know $\mathcal{L}[u(t-a)] = e^{-as}/s$, we can use the linearity of the Laplace transform to find

$$\mathcal{L}[u(t-a) - u(t-b)] = \frac{1}{s}e^{-as} - \frac{1}{s}e^{-bs} = \frac{1}{s}(e^{-as} - e^{-bs})$$

With our understanding of the Laplace transform of step functions and the second shifting property, we are now prepared to compute transforms of a wide range of step functions.

Example 5.4.5   Find the Laplace transform of

$$f(t) = \begin{cases} 1, & \text{if } 0 \le t < 1 \\ t, & \text{if } 1 \le t < 2 \\ 2, & \text{if } 2 \le t \end{cases}$$

Solution.   We first use step functions to write $f(t)$ with a single formula. Using $u(t) - u(t-1)$ to turn 1 on and off, and similar ideas for t and 2, we have

$$f(t) = 1[u(t) - u(t-1)] + t[u(t-1) - u(t-2)] + 2u(t-2)$$
$$= u(t) + (t-1)u(t-1) + (2-t)u(t-2)$$

Using the linearity of the Laplace transform, the second shifting property, and familiar transforms,

$$\mathcal{L}[f(t)] = \mathcal{L}[u(t)] + \mathcal{L}[(t-1)u(t-1)] + \mathcal{L}[(2-t)u(t-2)]$$

$$= \frac{1}{s} + e^{-s}\mathcal{L}[(t+1) - 1] + e^{-2s}\mathcal{L}[2 - (t+2)]$$

$$= \frac{1}{s} + e^{-s}\mathcal{L}[t] + e^{-2s}\mathcal{L}[-t]$$

$$= \frac{1}{s} + \frac{1}{s^2}e^{-s} - \frac{1}{s^2}e^{-2s}$$

Example 5.4.6   Find the Laplace transform of $f(t)$, where $f(t)$ is the piecewise linear function shown in the following graph.

Solution.   From the graph, we see that f has slope 1 on $[0, 2)$ and slope -2 on $[2, 3)$. Therefore, f can be defined piecewise by the rule

$$f(t) = \begin{cases} t, & \text{if } 0 \le t < 2 \\ 6 - 2t, & \text{if } 2 \le t < 3 \\ 0, & \text{if } 3 \le t \end{cases}$$

Using step functions, we can write f according to the formula
$$f(t) = t[u(t) - u(t-2)] + (6 - 2t)[u(t-2) - u(t-3)]$$
$$= tu(t) + (6 - 3t)u(t-2) - (6 - 2t)u(t-3)$$

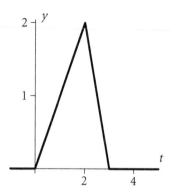

Applying the second shifting property, linearity, and familiar transforms, we see that

$$\mathcal{L}[f(t)] = \mathcal{L}[tu(t)] + \mathcal{L}[(6 - 3t)u(t - 2)] - \mathcal{L}[(6 - 2t)u(t - 3)]$$

$$= \mathcal{L}[t] + e^{-2s}\mathcal{L}[6 - 3(t + 2)] - e^{-3s}\mathcal{L}[6 - 2(t + 3)]$$

$$= \mathcal{L}[t] + e^{-2s}\mathcal{L}[-3t] - e^{-3s}\mathcal{L}[-2t]$$

$$= \frac{1}{s^2} - \frac{3}{s^2}e^{-2s} + \frac{2}{s^2}e^{-3s}$$

At this point, we have become familiar with piecewise-defined functions and how the Laplace transform may be applied to them. In the near future, we will be solving initial-value problems of the form

$$y' + 2y = 6 \cdot u(t - 4), \quad y(0) = 1$$

through the use of Laplace transforms. In order to assess our progress to date, we explore this approach briefly here. Taking the transform of both sides of the differential equation,

$$s\mathcal{L}[y] - 1 + 2\mathcal{L}[y] = \frac{6e^{-4s}}{s}$$

Letting $Y(s) = \mathcal{L}[y]$ and solving for $Y(s)$, it follows that

$$Y(s)(s + 2) = 1 + \frac{6e^{-4s}}{s}$$

so

$$Y(s) = \frac{1}{s + 2} + 6e^{-4s}\frac{1}{s(s + 2)} \tag{5.4.9}$$

Here, it remains to determine the function $y(t)$ whose Laplace transform is $Y(s)$. That is, we must compute the inverse Laplace transform of the righthand side of (5.4.9). Doing so involves using the inverse perspective on the second shifting property, as well as some algebraic work with the quantity $1/s(s + 2)$. We will pursue these and related ideas further in subsequent sections.

Next, however, we turn our attention to the study of impulse functions that can model phenomena such as the striking of a hammer.

5.4.2 The Dirac delta function

In physical situations where a large force is delivered over a very short time interval, unit step functions are no longer sufficient to model the forcing function. For example, if a hammer is used to strike a mass attached to a spring at a given time, it is not immediately clear how we should represent this forcing function. To address this situation, physicist Paul Dirac proposed what is today called the *Dirac delta function*, denoted $\delta(t)$. We seek to understand this function by first examining what happens when a force of constant magnitude acts over a smaller and smaller time interval.

Suppose that a force F_h of constant magnitude acts on an object over the time interval $[a - h, a + h]$, where $a > 0$. Assume that the force is zero otherwise. The *impulse* (or amount of *push*) of the force is defined by

$$I = \int_{a-h}^{a+h} F_h \, dt \tag{5.4.10}$$

If we want this constant force F_h to deliver a one-unit impulse, it follows that

$$F_h = \frac{1}{2h}$$

More specifically, if we wish to view the delivered force F_h as being generated by a forcing function $F_h(t)$, we can use the unit step function to express $F_h(t)$ through the formula

$$F_h(t) = \frac{1}{2h}[u(t - (a - h)) - u(t - (a + h))] \tag{5.4.11}$$

A plot of $F_h(t)$ for several different values of h is shown in figure 5.5; the vertical lines in each are technically not a part of the graph of $F_h(t)$, but are

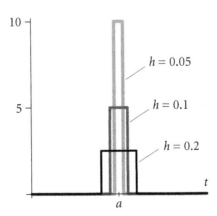

Figure 5.5 The forcing function $F_h(t)$ for $h = 0.2$, $h = 0.1$, and $h = 0.05$.

included to help contrast the different values of h. Note particularly that $F_h(t)$ satisfies the property that

$$\int_{-\infty}^{\infty} F_h(t)\, dt = 1 \tag{5.4.12}$$

and that as $h \to 0$, the magnitude of the force grows without bound in order to maintain the same total amount of push being delivered.

For an actual impulse, such as when a hammer strikes a mass, we want the force to act instantaneously at time $t = a$, where $a > 0$. This instantaneous impulse function is known as the *Dirac delta function*,[4] denoted $\delta(t - a)$, and is determined by letting $h \to 0$ in $F_h(t)$. In particular, we note two key properties of $\delta(t - a)$:

I. $\delta(t - a) = \lim\limits_{h \to 0} F_h(t) = \lim\limits_{h \to 0} \dfrac{1}{2h}[u(t - (a - h)) - u(t - (a + h))]$

II. $\int_{-\infty}^{\infty} \delta(t - a)\, dt = 1$

Property I is the definition of the Dirac delta function; Property II is a consequence of (5.4.12) and taking the limit as $h \to 0$.

A good way to think of $\delta(t - a)$ is as a function that is zero everywhere except at a, but infinite right at a. Actually, $\delta(t - a)$ is a limit of step functions that are nonzero over shorter and shorter intervals, but that always enclose an area of one unit, thus having spikes that grow in magnitude as the interval width shrinks. In situations such as a mass being struck with a hammer, we can now use the delta function to model the forcing function. For instance, if a hammer strikes the mass at $t = 3$, we can model the forcing function by $f(t) = \delta(t - 3)$.

In order to solve initial-value problems that involve the delta function, it will be essential to know the Laplace transform of $\mathcal{L}[\delta(t - a)]$. To do so, we first apply the definition of the transform to the step function $F_h(t)$. In particular, by familiar properties of the Laplace transform,

$$\mathcal{L}[F_h(t)] = \mathcal{L}\left[\frac{1}{2h}[u(t - (a - h)) - u(t - (a + h))] \right]$$

$$= \frac{1}{2h}\mathcal{L}[u(t - (a - h))] - \frac{1}{2h}\mathcal{L}[u(t - (a + h))]$$

$$= \frac{1}{2h}\left(\frac{1}{s}e^{-(a-h)s} - \frac{1}{s}e^{-(a+h)s} \right)$$

$$= \frac{e^{-as}}{2hs}\left(e^{hs} - e^{-hs} \right) \tag{5.4.13}$$

---

[4] Technically, the Dirac delta function is not a function, because it has the unusual property that it is zero everywhere but a, and infinite at $t = a$. Ultimately, the Laplace transform is what enables us to make sense of this function.

Since $\delta(t-a)$ is defined as the limit of $F_h(t)$ as $h \to 0$, we naturally define the Laplace transform of $\delta(t-a)$ to be the limit of the Laplace transform of $F_h(t)$ as $h \to 0$. In particular, from (5.4.13), some algebraic rearrangement, and an application of L'Hopital's Rule, we can state that

$$\lim_{h\to 0} \mathcal{L}[F_h(t)] = \lim_{h\to 0} \frac{e^{-as}}{2hs}\left(e^{hs} - e^{-hs}\right)$$

$$= \frac{e^{-as}}{s}\lim_{h\to 0} \frac{e^{hs} - e^{-hs}}{2h}$$

$$= \frac{e^{-as}}{s}\lim_{h\to 0} \frac{se^{hs} + se^{-hs}}{2}$$

$$= \frac{e^{-as}}{s} \cdot s = e^{-as}$$

We therefore define $\mathcal{L}[\delta(t-a)] = e^{-as}$. We close this section with an example that foreshadows the use of the delta function in a spring-mass system and the role of Laplace transforms in solving the corresponding IVP.

Example 5.4.7   Consider a spring mass system where $m = 1$, $k = 13$, and $c = 4$. Assume that the mass is initially displaced 1 m and released. Finally, assume that at $t = 3$, the mass is struck with a hammer in the positive direction. Set up and solve an initial-value problem that describes this situation.

Solution.   Using the delta function, the given problem is a standard damped harmonic oscillator equation with an impulse forcing function. In particular, the displacement y of the mass satisfies the initial-value problem

$$y'' + 4y' + 13y = \delta(t-3), \quad y(0) = 1, \; y'(0) = 0 \qquad (5.4.14)$$

Before we solve the IVP, we can use our intuition as a guide: we expect the size of the oscillations of the mass to decrease in magnitude until $t = 3$, at which time we expect the problem to restart as the blow from the hammer will increase the displacement of the mass, from which oscillations should eventually decrease to zero. We begin to solve (5.4.14) by using the Laplace transform in order to see how far our method enables us to progress.

Taking the Laplace transform of both sides of (5.4.14),

$$\mathcal{L}[y''] + 4\mathcal{L}[y'] + 13\mathcal{L}[y] = \mathcal{L}[\delta(t-3)]$$

From corollary 5.3.5, it follows that

$$s^2\mathcal{L}[y] - sy(0) - y'(0) + 4s\mathcal{L}[y] - 4y(0) + 13\mathcal{L}[y] = \mathcal{L}[\delta(t-3)]$$

Using the conditions $y(0) = 1$ and $y'(0) = 0$, as well as the fact that $\mathcal{L}[\delta(t-3)] = e^{-3s}$, we now have

$$s^2\mathcal{L}[y] - s + 4s\mathcal{L}[y] - 4 + 13\mathcal{L}[y] = e^{-3s}$$

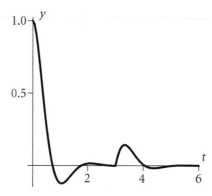

Figure 5.6 The solution to the IVP (5.4.14).

Solving for $\mathcal{L}[y] = Y(s)$, we see that

$$Y(s)(s^2 + 4s + 13) = s + 4 + e^{-3s}$$

or

$$Y(s) = \frac{s+4}{s^2 + 4s + 13} + \frac{e^{-3s}}{s^2 + 4s + 13} \qquad (5.4.15)$$

It remains for us to learn how to compute the inverse Laplace transform of (5.4.15) in order to find the solution y to the IVP. The following sections are devoted to these ideas. Upon further study, we will be able to show that the function $y(t)$ that satisfies (5.4.15) is

$$y = \frac{1}{3}e^{-2t}(3\cos 3t + 2\sin 3t) + \frac{1}{3}u(t-3)e^{-2(t-3)}\sin 3(t-3)$$

A plot of this solution is shown in figure 5.6, where $y(t)$ demonstrates precisely the type of behavior we expect.

The Laplace transform helps us make sense of the Dirac delta function in several ways. One is that we can imagine wanting to say that a hammer strikes a mass with different intensities. If, say, we want to compare the results of the initial-value problems where a hammer strikes a mass to deliver a given impulse versus what happens when the hammer strikes the mass three times as hard, this at first seems to be nonsense: $\delta(t-3)$ and $3\delta(t-3)$ are both zero everywhere and infinite at $t = 3$. But the power of the Laplace transform rescues us again. Since by linearity, $\mathcal{L}[3\delta(t-3)] = 3\mathcal{L}[\delta(t-3)] = 3e^{-3s}$, the transform detects the difference in the amount of push delivered by the hammer strike, and the results are shown accordingly in the solution to the initial-value problem. In addition, since $\mathcal{L}[\delta(t-a)] = e^{-as}$, we know that the presence of e^{-as} in $Y(s)$ will lead to the presence of $u(t-a)$ in $y(t)$: here we see how the delta function leads to a restart at $t = a$ as the function $u(t-a)$ turns on at this time in the function $y(t)$.

5.4.3 The Heaviside and Dirac functions in *Maple*

Both the Heaviside and Dirac functions belong to *Maple*'s library of basic functions. The syntax for the Heaviside function is simply `> Heaviside(t);`. Similarly, the Dirac function is given by `> Dirac(t);`.

For work with the Heaviside function, we often denote the function by $u(t)$. In *Maple*, this can be accomplished with the command

```
> u := t -> Heaviside(t);
```

Then, to enter and plot a piecewise-defined function such as

$$f(t) = t(u(t) - u(t-2)) + (6 - 2t)(u(t-2) - u(t-3))$$

we may use the syntax

```
> f := t -> t*(u(t)-u(t-2)) + (6-2*t)*(u(t-2)-u(t-3));
> plot(f(t), t=-1..5, color=black, thickness=2);
```

to generate the plot shown in figure 5.7.

More on both the Heaviside function and the Dirac function in *Maple*, particularly related to their roles in solving initial-value problems with Laplace transforms, can be found in section 5.6.1.

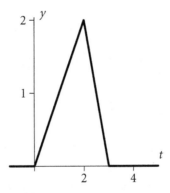

Figure 5.7 The function $f(t) = t(u(t) - u(t-2)) + (6 - 2t)(u(t-2) - u(t-3))$.

Exercises 5.4   In exercises 1–7, sketch a graph of each of the following functions and write each in terms of unit step functions.

1. $f(t) = \begin{cases} 0, & \text{if } 0 \leq t < 1 \\ 1, & \text{if } 1 \leq t < 2 \\ 0, & \text{if } 2 \leq t \end{cases}$

2. $f(t) = \begin{cases} 1, & \text{if } 0 \le t < 4 \\ 2, & \text{if } 4 \le t \end{cases}$

3. $f(t) = \begin{cases} 0, & \text{if } 0 \le t < 1 \\ t, & \text{if } 1 \le t < 2 \\ t^2, & \text{if } 2 \le t \end{cases}$

4. $f(t) = \begin{cases} t, & \text{if } 0 \le t < 2 \\ 0, & \text{if } 2 \le t \end{cases}$

5. $f(t) = \begin{cases} \sin t, & \text{if } 0 \le t < 2\pi \\ 0, & \text{if } 2\pi \le t \end{cases}$

6. $f(t) = \begin{cases} \sin t, & \text{if } 0 \le t < 2\pi \\ \sin 2t, & \text{if } 2\pi \le t \end{cases}$

7. $f(t) = \begin{cases} t, & \text{if } 0 \le t < 2 \\ 2, & \text{if } 2 \le t < 4 \\ 4 - t, & \text{if } 4 \le t \end{cases}$

8. Determine the Laplace Transform of the function $f(t)$ given in

   (a) Exercise 1
   (b) Exercise 2
   (c) Exercise 3
   (d) Exercise 4
   (e) Exercise 5
   (f) Exercise 6
   (g) Exercise 7

In exercises 9–11, compute the Laplace transform of $f(t)$.

9. $f(t) = 2[u(t - 1) - u(t - 3)] + \delta(t - 5)$

10. $f(t) = 2\sin 5t + \delta(t - 3)$

11. $f(t) = 2e^{-3t}\sin 2t + \delta(t - 8)$

12. Set up, but do not solve, an initial-value problem that represents a spring-mass system with $m = 4$ kg, spring constant $k = 10$, and damping constant $c = 2$, where a unit impulse is delivered by a hammer at $t = 6$. Assume the units on all quantities are consistent and that the mass is initially displaced 0.25 m and released.

13. Set up, but do not solve, an initial-value problem that represents a spring-mass system with $m = 4$ kg, spring constant $k = 10$, and damping constant $c = 2$, where a forcing function $f(t) = 3\sin 2t$ is turned on at $t = 4$ and an impulse of magnitude 4 is delivered by a hammer at $t = 10$.

Assume the units on all quantities are consistent and that the mass is initially displaced 0.25 m and released.

5.5 Solving IVPs with the Laplace transform

As we have seen in examples 5.3.5 and 5.4.7, in order to solve initial-value problems using the Laplace transform, the final step in the process is to answer the question "what function $y(t)$ has Laplace transform $Y(s)$?" In this section, we will further study the *inverse Laplace transform*, the process that takes the Laplace transform of an unknown function back to the function itself. Throughout, we motivate our work through examples of solving initial-value problems to see some of the typical functions $Y(s)$ that arise in this approach and the steps necessary to determine $y(t) = \mathcal{L}^{-1}[Y(s)]$.

Example 5.5.1   Use Laplace transforms to solve the initial-value problem

$$y' - 2y = 5, \quad y(0) = 4$$

Solution.   We begin by taking the Laplace transform of both sides of the differential equation. Using the linearity of the transform,

$$\mathcal{L}[y'] - 2\mathcal{L}[y] = 5\mathcal{L}[1]$$

By theorem 5.3.4 and the familiar transform of the function $f(t) = 1$, it follows that

$$s\mathcal{L}[y] - y(0) - 2\mathcal{L}[y] = \frac{5}{s}$$

Using the given fact that $y(0) = 4$ and denoting $\mathcal{L}[y] = Y(s)$,

$$sY(s) - 2Y(s) = 4 + \frac{5}{s} \tag{5.5.1}$$

Note particularly that (5.5.1) is now an algebraic equation in the unknown function $Y(s)$. Solving for $Y(s)$, we find

$$Y(s) = \frac{4s + 5}{s(s - 2)}$$

At this point, we recall that $Y(s) = \mathcal{L}[y]$, where $y(t)$ is the original unknown function we seek as the solution to the stated IVP. Solving the IVP has now been reduced to finding the function $y(t)$ that has Laplace transform $Y(s)$. That is, we seek $y(t) = \mathcal{L}^{-1}[Y(s)]$.

With a bit of algebraic rearrangement and insight, we can find the function $y(t)$. In particular, using a partial fraction decomposition, we can show that

$$Y(s) = \frac{4s + 5}{s(s - 2)} = -\frac{5/2}{s} + \frac{13/2}{s - 2} \tag{5.5.2}$$

Recalling that $\mathcal{L}[1] = 1/s$ and $\mathcal{L}[e^{2t}] = 1/(s-2)$, (5.5.2) implies

$$y(t) = -\frac{5}{2} + \frac{13}{2}e^{2t}$$

This is precisely the solution we would find to the IVP were we to use an integrating factor or separation of variables to solve the differential equation.

Whenever we use the Laplace transform to solve an IVP, we will employ a process similar to our work in example 5.5.1:

(1) Take the transform of both sides of the stated differential equation to transform the differential equation in $y(t)$ into an algebraic equation in $Y(s) = \mathcal{L}[y]$;

(2) Use algebra to solve for $Y(s)$;

(3) Determine which function $y(t)$ has the Laplace transform $Y(s)$.

As we have noted previously, given a function $F(s)$, a function $f(t)$ such that $\mathcal{L}[f(t)] = F(s)$ is called the *inverse Laplace transform* of F. We use the notation $\mathcal{L}^{-1}[F(s)] = f(t)$. For our purposes, a good way to view the operator $\mathcal{L}^{-1}$ is as one that reverses the work of the Laplace transform.

A key step in working backward will be to decompose the function $F(s)$ into more manageable pieces, often through a partial fraction decomposition. A review of partial fractions can be found in appendix A; partial fractions are an algebraic technique that proves useful for more than just integration, as we will see throughout this section. Once the pieces of $F(s)$ are in a recognizable form, we use standard rules we have developed for Laplace transforms to compute the inverse transform. For example, after using partial fractions to decompose $Y(s)$ in example 5.5.1, we showed that since $\mathcal{L}[e^{2t}] = 1/(s-2)$, it follows that

$$\mathcal{L}^{-1}\left[\frac{1}{s-2}\right] = e^{2t}$$

More generally, we can state that

$$\mathcal{L}^{-1}\left[\frac{1}{s-a}\right] = e^{at} \qquad (5.5.3)$$

Indeed, we realize that we can turn around any known relationship generated by the Laplace transform in order to make a statement about the inverse transform. For example, the inverse transform satisfies the linearity property stated in the following theorem.

Theorem 5.5.1 For every pair of constants a and b,

$$\mathcal{L}^{-1}[aF(s) + bG(s)] = a\mathcal{L}^{-1}[F(s)] + b\mathcal{L}^{-1}[G(s)]$$

Both shifting properties we have developed are regularly used in their inverse form. For the first shifting property, given $\mathcal{L}[f(t)] = F(s)$, we know that for any real value of a,

$$\mathcal{L}[e^{at}f(t)] = F(s-a)$$

Stated differently, this first shifting property implies

$$\mathcal{L}^{-1}[F(s-a)] = e^{at}f(t) \tag{5.5.4}$$

Likewise, from the slightly revised version of the second shifting property, we know that

$$\mathcal{L}[u(t-a)f(t-a)] = e^{-as}\mathcal{L}[f(t)] = e^{-as}F(s)$$

and therefore stated in inverse form,

$$\mathcal{L}^{-1}[e^{-as}F(s)] = u(t-a)f(t-a) \tag{5.5.5}$$

In our next example, we see how several of these fundamental concepts are employed in practice, specifically when step functions are involved.

Example 5.5.2   Use Laplace transforms to solve the initial-value problem

$$y' + y = 5u(t-1), \quad y(0) = 4$$

Solution.   Taking the Laplace transform of both sides of the differential equation and applying the initial condition,

$$s\mathcal{L}[y] - 4 + \mathcal{L}[y] = 5\mathcal{L}[u(t-1)]$$

Using the established fact that $\mathcal{L}[u(t-1)] = e^{-s}/s$ and letting $Y(s) = \mathcal{L}(y)$,

$$sY(s) - 4 + Y(s) = \frac{5e^{-s}}{s}$$

Solving for $Y(s)$,

$$Y(s) = \frac{4}{s+1} + 5e^{-s}\frac{1}{s(s+1)} \tag{5.5.6}$$

At this point, we need to use the inverse transform to solve for $y(t)$. Finding $\mathcal{L}^{-1}[4/(s+1)]$ is straightforward: by linearity and the first shifting property,[5]

$$\mathcal{L}^{-1}\left[\frac{4}{s+1}\right] = 4e^{-t} \tag{5.5.7}$$

To deal with the remaining term in (5.5.6), we note that with e^{-s} present we will need to use the second shifting property (5.5.5) in reverse. For this, it will be most useful to have the function

$$F(s) = \frac{1}{s(s+1)}$$

---

[5] We know $\mathcal{L}^{-1}[1/s] = 1$, and thus the first shifting property implies $\mathcal{L}^{-1}[1/(s+1)] = e^{-t} \cdot 1$

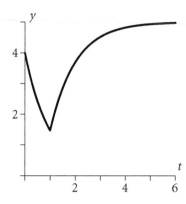

Figure 5.8 The solution to the IVP of example 5.5.2.

in a simpler form. Using its partial fraction decomposition, we observe that

$$F(s) = \frac{1}{s} - \frac{1}{s+1}$$

By (5.5.5), it now follows that

$$\mathcal{L}^{-1}\left[5e^{-s}\left(\frac{1}{s} - \frac{1}{s+1}\right)\right] = 5u(t-1)\left(1 - e^{-(t-1)}\right) \qquad (5.5.8)$$

Combining our work at (5.5.7) and (5.5.8) to determine $y(t)$ from (5.5.6), we have shown that

$$y(t) = 4e^{-t} + 5u(t-1) - 5u(t-1)e^{-(t-1)}$$

A plot of this solution curve is shown in figure 5.8, where we see qualitative behavior consistent with what we would expect from the forcing function in the IVP. In particular, the forcing function is $5u(t-1)$, which makes the forcing function behave as if the constant function 5 is turned on at $t = 1$ in the initial-value problem. For $t = 0$ to $t = 1$, we see the standard exponential decay that we would expect for the homogeneous equation $y' + y = 0$. But at $t = 1$, the solution function turns and begins to approach the equilibrium solution $y = 5$ that we expect in the nonhomogeneous equation $y' + y = 5$. We note specifically that the Laplace transform has successfully handled all of this at once, including the role of the initial condition $y(0) = 4$ and the corner in the solution function $y(t)$ at $t = 1$.

We next solve a second-order initial-value problem that involves the unit step function. Here, we will see how the higher order of the equation introduces additional complexity in determining the inverse Laplace transform needed to solve the IVP.

Example 5.5.3   Use the Laplace transform to solve the initial-value problem

$$y'' + 2y' + 5y = u(t - 2), \quad y(0) = 1, \ y'(0) = 0 \qquad (5.5.9)$$

Solution.   Taking the Laplace transform of both sides of (5.5.9) and writing $Y(s) = \mathcal{L}[y(t)]$, we observe that

$$s^2 Y(s) - sy(0) - y'(0) + 2(sY(s) - y(0)) + 5Y(s) = \frac{e^{-2s}}{s}$$

Substituting the given initial conditions and factoring on the left, we have

$$Y(s)(s^2 + 2s + 5) = s + 2 + \frac{e^{-2s}}{s}$$

Solving for $Y(s)$, we can write

$$Y(s) = Y_1(s) + Y_2(s) = \frac{s+2}{s^2 + 2s + 5} + e^{-2s} \frac{1}{s(s^2 + 2s + 5)} \qquad (5.5.10)$$

It remains for us to determine the function $y(t)$ whose transform is $Y(s)$. By linearity, it helps for us to break the function $Y(s)$ into the simplest pieces we can; we begin by determining the inverse transform of $Y_1(s)$. Because of shifting properties of the transform (and because of the fact that we cannot factor $s^2 + 2s + 5$ in an effort to apply partial fractions), it is useful to complete the square in expressions such as $s^2 + 2s + 5$. We instead write $(s+1)^2 + 4$, and seek to identify other parts of the expression that involve $(s + 1)$. Separating the numerator $(s + 2)$ into $(s + 1) + 1$, we can express the first term in (5.5.10) as

$$Y_1(s) = \frac{s+2}{s^2 + 2s + 5} = \frac{s+1}{(s+1)^2 + 4} + \frac{1}{(s+1)^2 + 4} \qquad (5.5.11)$$

Recalling that $\mathcal{L}[\cos 2t] = s/(s^2 + 4)$ and $\mathcal{L}[\sin 2t] = 2/(s^2 + 4)$, we know

$$\mathcal{L}^{-1}[s/(s^2 + 4)] = \cos 2t \ \text{ and } \ \mathcal{L}^{-1}[2/(s^2 + 4)] = \sin 2t$$

The inverse of the first shifting property, $\mathcal{L}^{-1}[F(s+1)] = e^{-t}f(t)$, now implies that

$$\mathcal{L}^{-1}\left[\frac{s+1}{(s+1)^2 + 4} + \frac{1}{(s+1)^2 + 4}\right] = e^{-t}\cos 2t + \frac{1}{2}e^{-t}\sin 2t \qquad (5.5.12)$$

Hence, the first term $Y_1(s)$ in (5.5.10) comes from taking the Laplace transform of the function $y_1(t) = e^{-t}\cos 2t + \frac{1}{2}e^{-t}\sin 2t$.

From (5.5.10), it remains for us to find the function $y_2(t)$ whose Laplace transform is

$$Y_2(s) = e^{-2s}\frac{1}{s(s^2 + 2s + 5)}$$

Using a partial fraction decomposition on the rational part of the function, we have

$$e^{-2s}\frac{1}{s(s^2 + 2s + 5)} = \frac{1}{5}e^{-2s}\left(\frac{1}{s} - \frac{s+2}{s^2 + 2s + 5}\right)$$

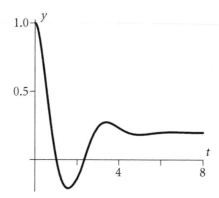

Figure 5.9 The solution $y(t)$ to the IVP in example 5.5.3.

Observe that we have already determined the inverse transform of the function $(s+2)/(s^2+2s+5)$ above at (5.5.12). Here, we must deal with the additional presence of the constant $1/5$, the multiplier e^{-2s}, and the basic function $1/s$. Recalling the inverse second shifting property, $\mathcal{L}^{-1}[e^{-as}F(s)] = u(t-a)f(t-a)$, and (5.5.12), we observe that

$$\mathcal{L}^{-1}\left[e^{-2s}\left(\frac{1}{s} - \frac{s+2}{s^2+2s+5}\right)\right]$$

$$= u(t-2)\left[1 - e^{-(t-2)}(\cos 2(t-2) + \frac{1}{2}\sin 2(t-2))\right] \qquad (5.5.13)$$

Combining (5.5.10), (5.5.12), and (5.5.13), we have shown that the solution $y(t)$ to the initial-value problem is

$$y(t) = e^{-t}\cos 2t + \frac{e^{-t}}{2}\sin 2t + \frac{1}{5}u(t-2)$$

$$\left[1 - e^{-(t-2)}\cos 2(t-2) + \frac{e^{-(t-2)}}{2}\sin 2(t-2)\right]$$

A plot of the function $y(t)$ is shown in figure 5.9. Here, we see evidence of the qualitative behavior we expect: until the unit step function turns on, the homogeneous equation should show damped oscillations so that $y(t) \to 0$. But once the step function turns on, the forcing function makes the equation nonhomogeneous with a constant forcing function, making $y = 1/5$ the stable equilibrium solution to which $y(t)$ tends.

To further explore the ideas that arise in computing inverse transforms, we next consider a slight modification of the preceding example, but in an applied setting where a more complicated forcing function is present. In particular, we examine a spring-mass system in which a periodic forcing function is introduced at $t = \pi$.

Example 5.5.4   Consider a mass of 1 kg attached to a spring with spring constant $k = 13$ such that the system has damping constant $c = 4$. Assume that the mass is displaced 1 m from equilibrium and released at $t = 0$; furthermore, at time $t = \pi$ the forcing function $f(t) = 2\sin 3t$ is applied. Assuming consistent units, set up an IVP that models this situation and solve the IVP using Laplace transforms.

Solution.   From our work with spring-mass systems, we know that the displacement $y(t)$ of the mass from equilibrium must satisfy the initial-value problem

$$y'' + 4y' + 13y = 2u(t - \pi)\sin 3t, \quad y(0) = 1, \; y'(0) = 0$$

Taking Laplace transforms, it follows that

$$s^2 Y(s) - sy(0) - y'(0) + 4(sY(s) - y(0)) + 13Y(s) = 2\mathcal{L}[u(t - \pi)\sin 3t] \tag{5.5.14}$$

We know that $\mathcal{L}[\sin 3t] = 3/(s^2 + 9)$, and by the second shifting property

$$\mathcal{L}[u(t - \pi)\sin 3t] = e^{-\pi s}\mathcal{L}[\sin 3(t + \pi)] \tag{5.5.15}$$

At this point, we observe by basic trigonometry that $\sin(3t + 3\pi) = \sin 3t \cos 3\pi + \cos 3t \sin 3\pi = -\sin 3t$. Hence, from (5.5.15) we have

$$\mathcal{L}[u(t - \pi)\sin 3t] = e^{-\pi s}\mathcal{L}[-\sin 3t] = -e^{-\pi s}\frac{3}{s^2 + 9}$$

Returning to (5.5.14) and using the given initial conditions, it follows that

$$s^2 Y(s) - s + 4sY(s) - 4 + 13Y(s) = -2e^{-\pi s}\frac{3}{s^2 + 9}$$

Factoring,

$$Y(s)(s^2 + 4s + 13) = s + 4 - 2e^{-\pi s}\frac{3}{s^2 + 9}$$

Solving for $Y(s)$,

$$Y(s) = Y_1(s) + Y_2(s)$$

$$= \frac{s + 4}{s^2 + 4s + 13} - 2e^{-\pi s}\frac{3}{(s^2 + 9)(s^2 + 4s + 13)} \tag{5.5.16}$$

It remains to find the inverse transform of $Y(s)$; we do so one piece at a time using the linearity of the inverse transform. In both $Y_1(s)$ and $Y_2(s)$, we will algebraically rearrange the expression in order to help us more easily determine the inverse Laplace transform, using an approach similar to our work in example 5.5.3.

Taking the first term in (5.5.16), we observe that since the denominator does not factor, we need to write it in a more familiar form. Completing the square and separating the numerator enables us to write

$$Y_1(s) = \frac{s + 4}{(s + 2)^2 + 9} = \frac{s + 2}{(s + 2)^2 + 9} + \frac{2}{(s + 2)^2 + 9}$$

and see the structure of Laplace transforms of basic functions. In particular, from the first shifting property and the known Laplace transforms of $\cos 3t$ and $\sin 3t$, it follows that

$$\mathcal{L}^{-1}[Y_1(s)] = \mathcal{L}^{-1}\left[\frac{s+2}{(s+2)^2+9} + \frac{2}{(s+2)^2+9}\right] = e^{-2t}\cos 3t + \frac{2}{3}e^{-2t}\sin 3t$$
(5.5.17)

Next we find the inverse transform of the term $Y_2(s)$ in (5.5.16). That is, we must determine

$$\mathcal{L}^{-1}[Y_2(s)] = \mathcal{L}^{-1}\left[-6e^{-\pi s}\frac{1}{(s^2+9)(s^2+4s+13)}\right] \qquad (5.5.18)$$

From the presence of $e^{-\pi s}$, we know the second shifting property will be used; in addition, we must algebraically rearrange the remaining part of the expression in order to find the inverse transform. Computing the partial fraction decomposition of the rational function in (5.5.18), we equivalently seek

$$\mathcal{L}^{-1}[Y_2(s)] = \mathcal{L}^{-1}\left[\frac{6}{40}e^{-\pi s}\left(\frac{s-1}{s^2+9} - \frac{s+3}{s^2+4s+13}\right)\right] \qquad (5.5.19)$$

One additional rearrangement will enable us to find the desired inverse transform. Completing the square in the second fraction and separating the numerator in each enables us to rewrite (5.5.19) as

$$\mathcal{L}^{-1}[Y_2(s)] = \frac{6}{40}\mathcal{L}^{-1}\left[e^{-\pi s}\left(\frac{s}{s^2+9} - \frac{1}{s^2+9} - \frac{s+2}{(s+2)^2+9} - \frac{1}{(s+2)^2+9}\right)\right]$$

Applying the inverse of the second shifting property to each of the terms in $\mathcal{L}^{-1}[Y_2(s)]$, it follows that

$$\mathcal{L}^{-1}[Y_2(s)] = \frac{6}{40}u(t-\pi)\left[\cos 3(t-\pi) - \frac{1}{3}\sin 3(t-\pi)\right.$$

$$\left. - e^{-2(t-\pi)}\cos 3(t-\pi) - \frac{1}{3}e^{-2(t-\pi)}\sin 3(t-\pi)\right] \qquad (5.5.20)$$

Noting that $\sin(3t - 3\pi) = -\sin 3t$ and $\cos(3t - 3\pi) = -\cos 3t$, we can simplify (5.5.20) to

$$\mathcal{L}^{-1}[Y_2(s)] = \frac{3}{20}u(t-\pi)\left[-\cos 3t + \frac{1}{3}\sin 3t - e^{-2(t-\pi)}(-\cos 3t + \frac{1}{3}\sin 3t)\right]$$

Combining our work with $\mathcal{L}^{-1}[Y_1(s)]$ and $\mathcal{L}^{-1}[Y_2(s)]$, we have therefore shown that $y(t) = \mathcal{L}^{-1}[Y(s)]$ is the function

$$y(t) = e^{-2t}(\cos 3t + \frac{2}{3}\sin 3t) + \frac{3}{20}u(t-\pi)[-\cos 3t$$

$$+ \frac{1}{3}\sin 3t - e^{-2(t-\pi)}(-\cos 3t + \frac{1}{3}\sin 3t)]$$

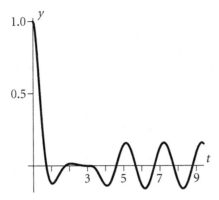

Figure 5.10 The solution to the IVP in example 5.5.4.

A plot of the function $y(t)$ is given in figure 5.10, where we see that until the forcing function activates at $t = \pi$, we see the standard damped oscillations decaying to zero. When the periodic forcing function turns on, the system demonstrates the repeating oscillations generated by this function.

At this point in our work, we have been exposed to most of the main ideas necessary for using the Laplace transform to solve initial-value problems. In addition to knowing the standard properties of the transform and its effects on basic functions, we must understand how to compute the inverse transform and the algebraic rearrangements that such inversion entails. Specifically, we have seen in several examples the need to determine partial fraction decompositions, complete the square, and separate the numerator in fractions. For example, the key computations necessary to find the inverse transform of the function

$$F(s) = \frac{11}{s(s^2 + 6s + 11)}$$

are to first determine the partial fraction decomposition and write

$$F(s) = \frac{1}{s} - \frac{s+6}{s^2 + 6s + 11}$$

The first term is straightforward to invert; but the second term requires further manipulation. Completing the square in the denominator, we see that $s^2 + 6s + 11 = (s+3)^2 + 2$, and therefore it is convenient to write the numerator as $s+6 = (s+3) + 3$. Doing so,

$$F(s) = \frac{1}{s} - \frac{s+3}{(s+3)^2 + 2} - \frac{3}{(s+3)^2 + 2}$$

It is at this point, together with the first shifting property, that we can finally compute $\mathcal{L}^{-1}[F(s)]$ and find

$$f(t) = \mathcal{L}^{-1}[F(s)] = 1 - e^{-3t} \cos\sqrt{2}t - \frac{3}{\sqrt{2}} e^{-3t} \sin\sqrt{2}t$$

Finally, we have also seen that the second shifting property also plays an important role. In the presence of the unit step function $u(t-a)$, the multiplier e^{-as} will arise in $F(s)$. In that case, we must invert $e^{-as}F(s)$; doing so, we get $u(t-a)f(t-a)$, as opposed to simply $f(t)$.

In light of these overall comments, we see the need to practice the computation of inverse Laplace transforms so that we can use these concepts in the solution of initial-value problems. In the next section, we will summarize key properties of the inverse transform, consider a few additional examples of more complicated inverse transforms, demonstrate the role technology plays in computations, and provide exercises for additional practice.

We close the current section with an example involving the Dirac delta function.

Example 5.5.5   Consider an undamped spring-mass system with spring constant $c = 4$. Suppose that the mass is displaced 1 unit from equilibrium and struck with a force to impart an initial velocity of $y'(0) = 1$. In addition, at times $t = 7$ and $t = 20$, a hammer delivers a one-unit impulse to the mass in the positive direction. Assuming consistent units, set up and solve an IVP that models this situation.

Solution.   We use the Dirac delta function to represent the impulse forces delivered at times $t = 7$ and $t = 20$. Coupled with the standard equation to represent the spring-mass system, we see that the displacement $y(t)$ of the mass at time t satisfies the initial-value problem

$$y'' + 4y = \delta(t-7) + \delta(t-20), \quad y(0) = 1, \ y'(0) = 1$$

To solve the IVP, we begin by taking Laplace transforms and find that

$$s^2 Y(s) - sy(0) - y'(0) + 4Y(s) = \mathcal{L}[\delta(t-7)] + \mathcal{L}[\delta(t-20)]$$

Recalling that $\mathcal{L}[\delta(t-a)] = e^{-as}$ and using the given initial conditions, $Y(s)$ must satisfy the equation

$$s^2 Y(s) - s - 1 + 4Y(s) = e^{-7s} + e^{-20s}$$

Factoring,

$$Y(s)(s^2 + 4) = s + 1 + e^{-7s} + e^{-20s}$$

and therefore

$$Y(s) = \frac{s}{s^2+4} + \frac{1}{s^2+4} + e^{-7s}\frac{1}{s^2+4} + e^{-20s}\frac{1}{s^2+4}$$

Using the second shifting property to find the inverse of the last two terms on the right, we find

$$y(t) = \mathcal{L}^{-1}[Y(s)]$$

$$= \cos 2t + \frac{1}{2}\sin 2t + \frac{1}{2}u(t-7)\sin 2(t-7) + \frac{1}{2}u(t-20)\sin 2(t-20)$$

A plot of the solution function $y(t)$ is shown in figure 5.11. We know that because the system is undamped, once it is set in motion it will oscillate at the

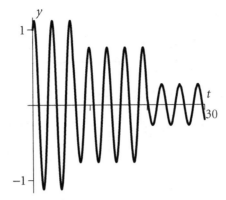

Figure 5.11 The solution to the IVP of example 5.5.5.

same amplitude indefinitely in the absence of other forces. When the hammer blows are delivered at $t = 7$ and $t = 20$, this will obviously change the amplitude of oscillation. At first the observed behavior may seem counterintuitive, as the hammer strikes are diminishing the amount of oscillation. However, if we note that the impulses are delivered in the positive direction at a time when the mass is traveling in the negative direction, then, indeed, the resulting solution accurately models the physical situation.

It is interesting to explore how delivering the impulses at other times impacts the system. Note that our work with Laplace transforms in example 5.5.5 is essentially unchanged by the times the impulses occur. In particular, if the hammer strikes occur at $t = a$ and $t = b$, then the solution will be

$$y(t) = \cos 2t + \frac{1}{2}\sin 2t + \frac{1}{2}u(t - a)\sin 2(t - a) + \frac{1}{2}u(t - b)\sin 2(t - b)$$

If we choose $a = 9$ and $b = 18$, we see substantially different behavior in the solution function due to the fact that these impulses occur in the same direction as the motion at the time they are delivered. A plot of the solution $y(t)$ in this case is shown in figure 5.12.

Exercises 5.5   In exercises 1–20, solve the stated initial-value problem using Laplace transforms. In each case, sketch a plot of your solution.

1. $y' + 5y = 20$, $y(0) = 3$
2. $y' + 3y = e^{2t}$, $y(0) = -2$
3. $y' - 2y = e^{2t}$, $y(0) = 1$
4. $y' + 4y = \sin 3t$, $y(0) = 5$
5. $y' + y = te^t$, $y(0) = -1$

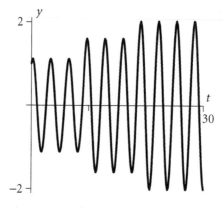

Figure 5.12 The solution to the IVP of example 5.5.5 where the impulses instead occur at $t = 9$ and $t = 18$.

6. $y' - 8y = u(t-1)$, $y(0) = -4$

7. $y' - 8y = u(t-3) \cdot t$, $y(0) = -4$

8. $y' - 8y = \delta(t-1)$, $y(0) = -4$

9. $y'' + 9y = 0$, $y(0) = 0$, $y'(0) = 5$

10. $y'' - 9y = 0$, $y(0) = 2$, $y'(0) = 0$

11. $y'' + 9y = 2$, $y(0) = 0$, $y'(0) = 1$

12. $y'' + 9y = 5\cos t$, $y(0) = 0$, $y'(0) = 0$

13. $y'' + 9y = 5\cos 3t$, $y(0) = 0$, $y'(0) = 0$

14. $y'' + 7y' + 12y = 0$, $y(0) = 0$, $y'(0) = 3$

15. $y'' + 6y' + 9y = 0$, $y(0) = 2$, $y'(0) = 0$

16. $y'' + 2y' + y = 3t$, $y(0) = 0$, $y'(0) = 0$

17. $y'' + 2y' + 5y = u(t-4)$, $y(0) = 1$, $y'(0) = 0$

18. $y'' - 2y' - 3y = u(t-3)$, $y(0) = 2$, $y'(0) = 0$

19. $y'' - 2y' - 3y = u(t-3)$, $y(0) = 2$, $y'(0) = 0$

20. $y'' + 2y' + 5y = \delta(t-1)$, $y(0) = 0$, $y'(0) = 0$

For exercises 21–26, solve the stated initial-value problem from exercises 1–20 by standard means developed in preceding chapters (i.e., without using Laplace transforms).

21. $y' + 3y = e^{2t}$, $y(0) = -2$

22. $y' + 4y = \sin 3t$, $y(0) = 5$

23. $y' + y = te^t$, $y(0) = -1$

24. $y'' + 9y = 2$, $y(0) = 0$, $y'(0) = 1$

25. $y'' + 9y = 5\cos 3t$, $y(0) = 0$, $y'(0) = 0$

26. $y'' + 2y' + y = 3t$, $y(0) = 0$, $y'(0) = 0$

In exercises 27–32, use Laplace transforms to determine the displacement $y(t)$ of the spring-mass system with spring constant $k = 72$ and mass $m = 2$ kg for the given forcing function $f(t)$. Assume each time the system starts from rest; solve for $y(t)$ in the cases where the spring constant c is (a) $c = 0$, (b) $c = 2$, (c) $c = 24$, and (d) $c = 40$, assuming consistent units. Sketch a plot of each solution.

27. $f(t) = 2$

28. $f(t) = 10\sin 2t$

29. $f(t) = 10\sin 6t$

30. $f(t) = 10[u(t) - u(t - 4\pi)]$

31. $f(t) = 10e^{-0.2t}$

32. $f(t) = 100\delta(t)$

In exercises 33–38, consider an RLC circuit for which an inductor of $L = 1$ H and capacitor $C = 0.01$ F are present. For each given forcing function $f(t)$, use Laplace transforms to determine the charge $Q(t)$ and current $I(t)$ in the circuit at time t if initially $Q(0) = 0$ and $I(0) = 0$. Determine the charge and current in the cases where the resistance is (a) $R = 0 \ \Omega$, (b) $R = 16 \ \Omega$, (c) $R = 20 \ \Omega$, and (d) $R = 25 \ \Omega$, assuming consistent units. Sketch a plot of each solution.

33. $f(t) = 10$

34. $f(t) = 10\sin 10t$

35. $f(t) = 5\sin 10t$

36. $f(t) = 10[u(t) - u(t - 2\pi)]$

37. $f(t) = 10\delta(t)$

38. $f(t) = 20e^{-t}$

5.6 More on the inverse Laplace transform

In this section, we provide an overall summary of properties of the inverse transform and present some further practice with computations. We close with a discussion of how transforms and inverse transforms may be found using a computer algebra system.

To begin, table 5.3 provides a list of familiar functions $F(s)$ and their inverse transforms, as well as several key general properties of the inverse transform.

Table 5.3

Inverse Laplace transforms of some basic functions and other fundamental properties.

$F(s)$	$f(t) = \mathcal{L}^{-1}[F(s)]$
$1/s^n$	$t^n/n!$
$1/(s-a)$	e^{at}
$s/(s^2+k^2)$	$\cos kt$
$k/(s^2+k^2)$	$\sin kt$
$s/(s^2-k^2)$	$\cosh kt$
$k/(s^2-k^2)$	$\sinh kt$
$aF(s)+bG(s)$	$af(t)+bg(t)$
$F(s-a)$	$e^{at}f(t)$
e^{-as}	$\delta(t-a)$
$e^{-as}F(s)$	$u(t-a)f(t-a)$

Most of the lines in the table are derived from taking the inverse perspective on statements in tables 5.1 and 5.2. While full tables of Laplace transforms typically number many pages, we present only a small collection for use in standard problems involving spring-mass systems and RLC circuits, leaving other examples for exploration in other sources or computer algebra systems.

The next several examples demonstrate standard techniques in the computation of inverse transforms.

Example 5.6.1   Determine $\mathcal{L}^{-1}[F(s)]$ for each of the following functions:

(a) $F(s) = \dfrac{e^{-2s}}{s(s+1)^2}$ (b) $F(s) = \dfrac{2}{s^4+4s^2}$ (c) $F(s) = \dfrac{4se^{-2\pi s}}{(s^2+2s+5)(s^2+9)}$

Solution.   (a) Because of the presence of e^{-2s} in $F(s)$, we will use the second shifting property. But first, we find the partial fraction decomposition

$$\frac{1}{s(s+1)^2} = \frac{1}{s} - \frac{1}{s+1} - \frac{1}{(s+1)^2}$$

and note that

$$\mathcal{L}^{-1}\left[\frac{1}{s(s+1)^2}\right] = \mathcal{L}^{-1}\left[\frac{1}{s} - \frac{1}{s+1} - \frac{1}{(s+1)^2}\right]$$

$$= 1 - e^{-t} - te^{-t}$$

Now, in order to compute the inverse transform of the given function, we use the second shifting property to address the presence of e^{-2s} in each term and thus find that

$$\mathcal{L}^{-1}\left[\frac{e^{-2s}}{s(s+1)^2}\right] = u(t-2)[1 - e^{-(t-2)} - (t-2)e^{-(t-2)}]$$

(b) Partial fractions shows that

$$F(s) = \frac{2}{s^4 + 4s^2} = \frac{1}{2}\left(\frac{1}{s^2} - \frac{1}{s^2+4}\right)$$

Using the inverses of familiar transforms of $f(t) = t$ and $f(t) = \sin 2t$, we see

$$\mathcal{L}^{-1}[F(s)] = \frac{1}{2}\left(t - \frac{1}{2}\sin 2t\right)$$

(c) Given the function

$$F(s) = \frac{4se^{-2\pi s}}{(s^2 + 2s + 5)(s^2 + 9)}$$

we see that the presence of $e^{-2\pi s}$ implies the inverse of the second shifting property will be used. As is now custom, we first use partial fractions to break the rational part of $F(s)$ into a sum of simpler expressions. Doing so and completing the square to re-express $s^2 + 2s + 5$,

$$\frac{4s}{(s^2+2s+5)(s^2+9)} = \frac{1}{13}\left(-\frac{4s-18}{s^2+9} + \frac{4s-10}{s^2+2s+5}\right)$$

$$= \frac{1}{13}\left(\frac{-4s}{s^2+9} + \frac{18}{s^2+9} + \frac{4(s+1)}{(s+1)^2+4} - \frac{14}{(s+1)^2+4}\right)$$

Letting $G(s) = 4s/(s^2 + 2s + 5)(s^2 + 9)$, it now follows from familiar rules with inverse transforms and the first shifting property that

$$\mathcal{L}^{-1}[G(s)] = -\frac{4}{13}\cos 3t + \frac{18}{39}\sin 3t + \frac{4}{13}e^{-t}\cos 2t - \frac{7}{13}e^{-t}\sin 2t$$

Finally, since $F(s) = e^{-2\pi s}G(s)$, the second shifting property implies

$$\mathcal{L}^{-1}[F(s)] = u(t-2\pi)\left(-\frac{4}{13}\cos 3(t-2\pi) + \frac{6}{13}\sin 3(t-2\pi)\right)$$

$$+ u(t-2\pi)\left(\frac{4}{13}e^{-t}\cos 2(t-2\pi) - \frac{7}{13}e^{-(t-2\pi)}\sin 2(t-2\pi)\right)$$

The 2π shift in each of the sine and cosine functions can be removed; for instance, $\cos 3(t - 2\pi) = \cos 3t$. Doing so throughout shows that

$$\mathcal{L}^{-1}[F(s)] = u(t - 2\pi)$$

$$\left(-\frac{4}{13}\cos 3t + \frac{6}{13}\sin 3t + \frac{4}{13}e^{-(t-2\pi)}\cos 2t - \frac{7}{13}e^{-(t-2\pi)}\sin 2t\right)$$

There are certainly other properties of the inverse Laplace transform that we could study. For example, theorem 5.3.4 in inverse form allows us to say that if $\mathcal{L}^{-1}[F(s)] = f(t)$ and $f(0) = 0$, then

$$\mathcal{L}^{-1}[sF(s)] = f'(t) \tag{5.6.1}$$

While results like this are theoretically interesting and can occasionally enable us to determine inverse transforms in alternate ways, they are less useful in pragmatic terms when we think of our overarching goal: using Laplace transforms to solve initial-value problems.

Indeed, our work throughout this chapter has given us a good overview of how Laplace transforms work, especially the role they play in solving initial-value problems. Of course, there are also many forcing functions we have not discussed for which Laplace transforms may be taken. There are books that contain lengthy tables of Laplace transforms and inverse transforms that we could, if necessary, consult. But because of the technology available to us, these tables have essentially been rendered obsolete. Most computer algebra systems are fully capable of computing Laplace transforms and their inverses, so we choose not to study methods for these more difficult calculations. The next example demonstrates one such function $F(s)$ which is beyond the methods we have developed but that can easily be handled by a computer algebra system.

Example 5.6.2   Find the inverse Laplace transform of

$$F(s) = \frac{9}{(s^2 + 1)^2(s^2 + 4)^2}$$

Solution.   The partial fraction decomposition of $F(s)$ is

$$F(s) = -\frac{2/3}{s^2 + 1} + \frac{1}{(s^2 + 1)^2} + \frac{2/3}{s^2 + 4} + \frac{1}{(s^2 + 4)^2} \tag{5.6.2}$$

Two of the terms in (5.6.2) are straightforward to invert, but the two involving squares of irreducible quadratic terms are not among familiar functions from our previous work. In the following subsection, we demonstrate how to use *Maple* to compute the inverse transform of such functions. These computations reveal that

$$\mathcal{L}^{-1}\left[\frac{1}{(s^2 + 1)^2}\right] = \frac{1}{2}\sin t - \frac{1}{2}t\cos t$$

and

$$\mathcal{L}^{-1}\left[\frac{1}{(s^2 + 4)^2}\right] = \frac{1}{16}\sin 2t - \frac{1}{8}t\cos 2t$$

From this work and (5.6.2), we find

$$\mathcal{L}^{-1}[F(s)] = -\frac{2}{3}\sin t + \frac{1}{2}\sin t - \frac{1}{2}t\cos t + \frac{1}{3}\sin 2t + \frac{1}{16}\sin 2t - \frac{1}{8}t\cos 2t$$

$$= -\frac{1}{6}\sin t - \frac{1}{2}t\cos t + \frac{19}{48}\sin 2t - \frac{1}{8}t\cos 2t$$

Further discussion of how to use *Maple* to compute transforms and inverse transform follows in the next subsection.

5.6.1 Laplace transforms and inverse transforms using *Maple*

As we have noted, while we have computed Laplace transforms for a range of functions, there are many more examples we have not considered. Moreover, even for familiar functions, certain combinations of them can lead to tedious, involved calculations. Computer algebra systems such as *Maple* are fully capable of computing Laplace transforms of functions, as well as inverse transforms. Here we demonstrate the syntax required in the solution of the initial-value problem from example 5.5.4:

$$y'' + 4y' + 13y = 2u(t - \pi)\sin 3t, \quad y(0) = 1, \ y'(0) = 0 \qquad (5.6.3)$$

To begin, we load the `inttrans` package in *Maple*.

```
> with(inttrans);
```

If, for example, we desire to use *Maple* to compute the Laplace transform of $2u(t - \pi)\sin 3t$, we use the syntax

```
> laplace(2*Heaviside(t-Pi)*sin(3*t),t,s);
```

This command results in the output

$$-\frac{6e^{-s\pi}}{s^2 + 9}$$

which is precisely the transform we expect.

After computing by hand the transform of the left-hand side of (5.6.3) and solving for $Y(s)$, as shown in detail in example 5.5.4, we have

$$Y(s) = \frac{s+4}{s^2 + 4s + 13} - 2e^{-\pi s}\frac{3}{(s^2 + 9)(s^2 + 4s + 13)}$$

Here, we may use *Maple*'s `invlaplace` command to determine $\mathcal{L}^{-1}[Y(s)]$. While we could choose to do so all at once, for simplicity of display we do so in two steps. First,

```
> invlaplace((s+4)/(s^2 + 4*s + 13),s,t);
```

results in the output

$$\frac{1}{3}e^{(-2t)}(3\cos(3t)+2\sin(3t)) \tag{5.6.4}$$

Similarly, for the second term in $Y(s)$, we compute

```
> invlaplace(2*exp(-Pi*s)*3/((s^2 + 9)*(s^2 + 4*s
  + 13)),s,t);
```

Maple produces the output

$$\frac{1}{3}\text{Heaviside}(t-\pi)(3\cos(3t)-\sin(3t)-e^{(-2t+2\pi)}(3\cos(3t)+\sin(3t))) \tag{5.6.5}$$

which corresponds to our work in example 5.5.4. The sum of the two functions of t that have resulted from inverse transforms in (5.6.4) and (5.6.5) is precisely the solution to the IVP.

Note that in computing the inverse transform (5.6.5), *Maple* has implicitly executed the partial fraction decomposition of the expression

$$\frac{3}{(s^2+9)(s^2+4s+13)}$$

If we wish to find this explicitly, we can use the command

```
> convert(3/((s^2 + 9)*(s^2 + 4*s + 13)),
  parfrac, s);
```

which produces the output

$$\frac{1}{40}\frac{3-3s}{s^2+9}+\frac{1}{40}\frac{9+3s}{s^2+4s+13}$$

In general, we see that to compute the Laplace transform of $f(t)$ in *Maple* we use the syntax

```
> laplace(f(t),t,s);
```

whereas to compute the inverse transform of $F(s)$, we enter

```
> invlaplace(F(s),s,t);
```

Exercises 5.6   In exercises 1–9, find the inverse Laplace transform of the given function $F(s)$ using familiar techniques or a computer algebra system.

1. $F(s) = \dfrac{2s}{(s+3)^2}$

2. $F(s) = \dfrac{4}{(s^2-4)^2}$

3. $F(s) = \dfrac{1}{s^2(s-2)}$

4. $F(s) = \dfrac{2}{(s^2-1)^2(s^2+1)}$

5. $F(s) = \dfrac{s^2+1}{(s+1)^2(s^2+4)}$

6. $F(s) = \dfrac{e^{-s}}{s^2(s-2)}$

7. $F(s) = e^{-3s}\dfrac{2}{(s^2-1)^2(s^2+1)}$

8. $F(s) = \dfrac{5s^2+20}{s(s-1)(s^2-5s+4)}$

9. $F(s) = e^{-\pi s}\dfrac{5s^2+20}{s(s-1)(s^2-5s+4)}$

In exercises 10–22, solve the stated initial-value problem using Laplace transforms (using a computer algebra system as necessary). Sketch a plot of each solution.

10. $y' + y = e^{-t} + te^{-t}$, $y(0) = 1$

11. $y'' + 4y = \sin 2t$, $y(0) = 0$, $y'(0) = 1$

12. $y'' + 4y = \sin 2t + \delta(t-6)$, $y(0) = 0$, $y'(0) = 1$

13. $y'' + 4y = \sin 2t + \delta(t-6) + \delta(t-12)$, $y(0) = 0$, $y'(0) = 1$

14. $y'' + 9y = \cos 3t + t\cos 3t$, $y(0) = 0$, $y'(0) = 1$

15. $y'' + 2y' + 5y = e^{-t}\sin 2t$, $y(0) = 0$, $y'(0) = 1$

16. $y'' + 2y' + 5y = e^{-t}\sin 2t + te^{-t}\sin 2t$, $y(0) = 0$, $y'(0) = 1$

17. $y'' + 2y' + 5y = e^{-t}\sin 2t + u(t-\pi)te^{-t}\sin 2t$, $y(0) = 0$, $y'(0) = 1$

18. $y'' + y' - 2y = 4e^t + 1$, $y(0) = 1$, $y'(0) = 0$

19. $y'' + y' - 2y = 4e^t + 1 + \delta(t-3)$, $y(0) = 1$, $y'(0) = 0$

20. $y'' + y' - 2y = 4e^t + u(t-3)$, $y(0) = 1$, $y'(0) = 0$

21. $y'' + 2y' + 5y = e^{-t}\sin 2t + te^{-t}\sin 2t + \delta(t-5)$, $y(0) = 0$, $y'(0) = 1$

22. $y'' + 2y' + 5y = 13e^t\sin t$, $y(0) = 0$, $y'(0) = 0$

5.7 For further study

5.7.1 Laplace transforms of infinite series

If $f(t)$ is a function of exponential order that is analytic[6] at $t = 0$ with an infinite radius of convergence, then $f(t)$ may be expressed as a power series and also has a Laplace transform. It therefore follows that if

$$f(t) = \sum_{n=0}^{\infty} a_n t^n$$

then its transform is

$$F(s) = \mathcal{L}[f(t)] = \sum_{n=0}^{\infty} a_n \mathcal{L}[t^n] = \sum_{n=0}^{\infty} n! a_n \frac{1}{s^{n+1}} \tag{5.7.1}$$

We begin by exploring the transforms of some familiar functions through the use of infinite series.

(a) Recall that $f(t) = e^t$ is analytic at $t = 0$ with series expansion

$$e^t = \sum_{n=0}^{\infty} \frac{t^n}{n!} = 1 + t + \frac{t^2}{2!} + \frac{t^3}{3!} + \cdots \tag{5.7.2}$$

By taking the Laplace transform of the series (5.7.2) term-wise,[7] show that

$$\mathcal{L}[e^t] = \sum_{n=0}^{\infty} \frac{1}{s^{n+1}} \tag{5.7.3}$$

Then, recognize (5.7.3) as a geometric series to show that

$$\mathcal{L}[e^t] = \frac{1}{s-1}$$

(b) Similarly, use the fact that $f(t) = \sin t$ has the series expansion

$$\sin t = t - \frac{t^3}{3!} + \frac{t^5}{5!} - \cdots$$

to show using infinite series that

$$\mathcal{L}[\sin t] = \frac{1}{s^2 + 1}$$

---

[6] More on power series expansions of functions and the meaning of terms such as "analytic" may be found in Section 8.2.

[7] While the Laplace transform of a finite sum is the sum of the Laplace transforms of the individual terms, it is not obvious that this property holds for infinite sums. The formal justification that this is valid in what follows is beyond the scope of this text; the reader may assume that this step is valid, and proceed as directed.

In addition, develop the Laplace transform of $f(t) = \cos t$ using the series expansion $\cos t = 1 - t^2/2! + t^4/4! - \cdots$.

While power series expansions of such familiar functions as e^t, $\sin t$, and $\cos t$ are important and offer a different perspective on the development of the transforms of these functions, power series are even more useful for working with functions that are more complicated. For example, if we seek the transform of

$$f(t) = \frac{e^{-t} - 1}{t} \tag{5.7.4}$$

none of the methods we have previously discussed apply. However, standard techniques[8] with infinite series may be used to address functions such as (5.7.4).

(c) Use the standard power series expansion for e^t to show that $f(t) = (e^{-t} - 1)/t$ has the series expansion

$$\frac{e^{-t} - 1}{t} = -1 + \frac{t}{2!} - \frac{t^2}{3!} + \frac{t^3}{4!} - \cdots = \sum_{n=1}^{\infty} \frac{(-1)^n}{n!} t^{n-1}$$

Then, compute the Laplace transform of the series expression to show that

$$\mathcal{L}\left[\frac{e^{-t} - 1}{t}\right] = -\frac{1}{s} + \frac{1}{2s^2} - \frac{1}{3s^3} + \cdots \tag{5.7.5}$$

(d) Even though the Laplace transform of an analytic function will result in an infinite sum involving negative powers of s, sometime we can recognize the transform as a familiar function. To see this in (5.7.5), use the known series expansion

$$\ln(1 + x) = x - \frac{1}{2}x^2 + \frac{1}{3}x^3 - \cdots$$

and the substitution $x = 1/s$ to show that

$$\mathcal{L}\left[\frac{e^{-t} - 1}{t}\right] = -\ln\left(1 + \frac{1}{s}\right)$$

(e) From the standard series expansion for the function $\sin t$, determine the Taylor series of

$$f(t) = \frac{\sin t}{t} \tag{5.7.6}$$

and hence compute the Laplace transform of (5.7.6). Then, use the expansion

$$\arctan x = x - \frac{1}{3}x^3 + \frac{1}{5}x^5 - \frac{1}{7}x^7 + \cdots$$

---

[8] A review of the development of power series of functions can be found in section 8.2.

and an appropriate substitution to show that

$$\mathcal{L}\left[\frac{\sin t}{t}\right] = \arctan\frac{1}{s}$$

(f) Use series techniques to show that

$$\mathcal{L}\left[\frac{\cos t - 1}{t}\right] = -\frac{1}{2}\ln\left(1 + \frac{1}{s^2}\right)$$

5.7.2 Laplace transforms of periodic forcing functions

Nonhomogeneous differential equations often involve periodic forcing functions. In section 4.5, we considered the effects of the forcing function $f(t) = \sin\omega t$ in connection with the natural frequency of a system. More generally, here we examine periodic forcing functions that are piecewise continuous. Such functions satisfy the relationship that for some value of a,

$$f(t) = f(t+a) + f(t+2a) + f(t+3a)$$
$$+ \cdots + f(t+na) + \cdots \qquad (5.7.7)$$

An example of such a function is shown in figure 5.13. Taking the Laplace transform of such a function f, we may write the transform as the infinite sum of integrals

$$\mathcal{L}[f(t)] = \int_0^\infty f(t)e^{-st}\,dt$$
$$= \int_0^a f(t)e^{-st}\,dt + \int_a^{2a} f(t)e^{-st}\,dt + \int_{2a}^{3a} f(t)e^{-st}\,dt + \cdots \qquad (5.7.8)$$

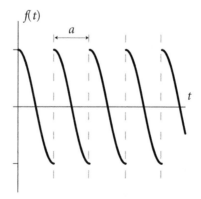

Figure 5.13 A periodic function with period a that is piecewise continuous.

(a) Using the change of variables $t = \tau + a$ in the second integral, $t = \tau + 2a$ in the third, and so on, show that

$$\mathcal{L}[f(t)] = \int_0^a f(t)e^{-st}\,dt + \int_0^a f(\tau + a)e^{-s(\tau+a)}\,d\tau$$
$$+ \int_0^a f(\tau + 2a)e^{-s(\tau+2a)}\,d\tau + \cdots \qquad (5.7.9)$$

(b) By replacing the integration variable τ with t in (5.7.9), show that

$$\mathcal{L}[f(t)] = [1 + e^{-as} + e^{-2as} + \cdots]\int_0^a f(t)e^{-st}\,dt \qquad (5.7.10)$$

Then, use the fact that the infinite series in (5.7.10) is geometric in order to conclude

$$\mathcal{L}[f(t)] = \frac{1}{1 - e^{-as}}\int_0^a f(t)e^{-st}\,dt \qquad (5.7.11)$$

(c) Use (5.7.11) to determine the Laplace transform of the square wave function shown in figure 5.14. (The vertical lines shown in the graph are not actually part of the function's graph; indeed, f is piecewise constant with value 3 on $[0, 2)$ and value -3 on $[-2, 4)$, and so on.) In particular, show that

$$\mathcal{L}[f(t)] = \frac{3}{s} \cdot \frac{1 - e^{-2s}}{1 + e^{-2s}}$$

where $f(t)$ is the function pictured in figure 5.14.

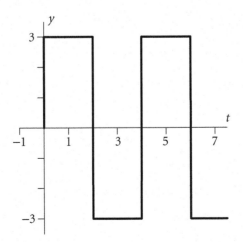

Figure 5.14 A square wave with amplitude 3 and period 4.

(d) Consider the periodic function with period 2π given by

$$f(t) = \begin{cases} \sin t, & \text{if } 0 < t < \pi \\ 0, & \text{if } \pi < t < 2\pi \end{cases}$$

This function is called the *half-rectified sine wave* since it only consists of the top-half of the standard sine function. Sketch a graph of this function and show that its Laplace transform is

$$\mathcal{L}[f(t)] = \frac{1 + e^{-\pi s}}{(1 - e^{-2\pi s})(s^2 + 1)}$$

(e) Let a slightly damped spring-mass system be given with $m = 1$, $c = 0.02$, and $k = 25$, and be driven by a square-wave periodic forcing function $f(t)$ with amplitude 5 and period 2π. We will use Laplace transforms to solve the initial-value problem that governs this system under the assumption that the system starts from rest.

(i) The stated problem is modeled by the initial-value problem

$$y'' + 0.02y' + 25y = f(t), \quad y(0) = 0, \ y'(0) = 0$$

Take Laplace transforms to show that $Y(s) = \mathcal{L}[y(t)]$ must satisfy the equation

$$Y(s) = \frac{F(s)}{s^2 + 0.02s + 25} \tag{5.7.12}$$

where $F(s) = \mathcal{L}[f(t)]$.

(ii) While we have learned in (c) how to write the transform of a square wave function without using infinite series in its expression, it turns out for this problem that a series expansion is necessary for finding the inverse transform when solving the IVP. By writing the square wave function given in this problem in the form

$$f(t) = 5u(t) - 10u(t - \pi) + 10u(t - 2\pi) - 10u(t - 3\pi) + \cdots$$

show that

$$F(s) = \mathcal{L}[f(t)] = \frac{5}{s}[1 - 2e^{-\pi s} + 2e^{-2\pi s} - 2e^{-3\pi s} + \cdots] \tag{5.7.13}$$

(iii) Explain why

$$\frac{1}{s^2 + 0.02s + 25} \approx \frac{1}{(s + 0.01)^2 + 5^2} \tag{5.7.14}$$

(iv) Combine (5.7.12), (5.7.13), and (5.7.14) in order to conclude that

$$y(t) = \mathcal{L}^{-1}[Y(s)]$$

$$= \mathcal{L}^{-1}\left[\frac{5}{s[(s + 0.01)^2 + 5^2]}[1 - 2e^{-\pi s} + 2e^{-2\pi s} - \cdots] \right] \tag{5.7.15}$$

Explain why we have to find the inverse transform in (5.7.15) term-by-term.

(v) Compute the inverse transform of the first term

$$y_1(t) = \mathcal{L}^{-1}\left[\frac{5}{s[(s+0.01)^2+5^2]}\right]$$

in (5.7.15) given the partial fraction decomposition

$$\frac{5}{s[(s+0.01)^2+5^2]} = \frac{0.2}{s} - \frac{0.2s+0.004}{(s+0.01)^2+5^2}$$

(Hint: $0.2s+0.004 = 0.2(s+0.01)+0.002$)

Conclude that

$$y_1(t) = 0.2 - e^{-0.01t}(0.2\cos 5t - 0.0004\sin 5t) \qquad (5.7.16)$$

(vi) Compute the inverse transform of the second term

$$y_2(t) = \mathcal{L}^{-1}\left[-2e^{-\pi s}\frac{5}{s[(s+0.01)^2+5^2]}\right]$$

in (5.7.15) using (5.7.16) and the second shifting property.

Using the fact that $\cos 5(t-\pi) = -\cos 5t$ and $\sin 5(t-\pi) = -\sin 5t$, conclude that

$$y_2(t) = -2u(t-\pi)\left\{0.2 + e^{-0.01(t-\pi)}(0.2\cos 5t + 0.0004\sin 5t)\right\}$$

$$= -2u(t-\pi)\{0.2 + e^{0.01\pi}[0.2 - y_0(t)]\} \qquad (5.7.17)$$

(vii) Compute the inverse transform of the third term

$$y_3(t) = \mathcal{L}^{-1}\left[2e^{-2\pi s}\frac{5}{s[(s+0.01)^2+5^2]}\right]$$

in (5.7.15) using (5.7.16) and the second shifting property.

Using the fact that $\cos 5(t-2\pi) = -\cos 5t$ and $\sin 5(t-2\pi) = -\sin 5t$, conclude that

$$y_3(t) = 2u(t-2\pi)\left\{0.2 - e^{-0.01(t-2\pi)}(0.2\cos 5t + 0.0004\sin 5t)\right\}$$

$$= 2u(t-2\pi)\{0.2 - e^{0.02\pi}[0.2 - y_0(t)]\} \qquad (5.7.18)$$

(viii) So far, we have found the formula for $y(t)$ valid up to $t = 3\pi$. In fact,

$$y(t) = y_1(t), \quad \text{if } 0 < t < \pi$$
$$y(t) = y_1(t) + y_2(t), \quad \text{if } \pi < t < 2\pi$$
$$y(t) = y_1(t) + y_2(t) + y_3(t), \quad \text{if } 2\pi < t < 3\pi$$

Using $y_1(t) = 0.2 - e^{0\pi}[0.2 - y_1(t)]$, together with (5.7.17) and (5.7.18), plus the fact that on $2\pi < t < 3\pi$ we know $u(t - \pi) = u(t - 2\pi) = 1$, show that on $2\pi < t < 3\pi$,

$$y(t) = 0.2 - [0.2 - y_1(t)]\{1 + 2e^{0.01\pi} + 2e^{0.02\pi}\}$$

(ix) Using the patterns established in (5.7.17) and (5.7.18), explain why

$$y(t) = y_1(t) + y_2(t) + \cdots + y_n(t)$$
$$= (-1)^n 0.2 - [0.2 - y_1(t)]$$
$$\{1 + 2e^{0.01\pi} + \cdots + 2e^{0.01n\pi}\} \quad (5.7.19)$$

is valid for $n\pi < t < (n+1)\pi$ for any positive integer n

(x) Letting $z(t) = e^{-0.01t}(\cos 5t + 0.002 \sin 5t)$ and using the fact that $1 - x^{n+1}/1 - x = 1 + x + x^2 + \cdots x^n$, show that on $n\pi < t < (n+1)\pi$,

$$y(t) = (-1)^n \left(\frac{1}{5} - \frac{2}{5(1 - e^{0.01\pi})}\right) z(t) + \frac{2e^{(n+1)0.01\pi}}{5(1 - e^{0.01\pi})} z(t) \quad (5.7.20)$$

Explain why as $t \to \infty$, it follows that $y(t) \to \infty$. Using a computer algebra system, graph the solution function on several consecutive large intervals of width π, such as $[200\pi, 201\pi]$, $[201\pi, 202\pi]$, etc., and discuss the behavior of the system.

5.7.3 Laplace transforms of systems

Recall that the standard initial-value problem for a system of first-order DEs is given in matrix form by

$$\mathbf{x}' = \mathbf{A}\mathbf{x} + \mathbf{f}(t), \qquad \mathbf{x}(0) = \mathbf{b} \quad (5.7.21)$$

In the event that $\mathbf{f}$ is a continuous function, the variation of parameters technique applies. But, if $\mathbf{f}$ is a step function or otherwise piecewise defined, our earlier methods fail, and Laplace transforms may be used. Regardless, the Laplace transform can be a useful tool for systems for many of the same reasons it is for single DEs, such as the fact that it treats all linear systems in a uniform manner and incorporates the initial conditions immediately into the process of finding the solution.

Since each of the three terms in the equation in (5.7.21) is a vector, Laplace transforms may be applied component-wise. For example,

$$\mathcal{L}[\mathbf{x}'(t)] = \mathcal{L}\begin{bmatrix} x_1'(t) \\ x_2'(t) \end{bmatrix} = \begin{bmatrix} \mathcal{L}[x_1'(t)] \\ \mathcal{L}[x_2'(t)] \end{bmatrix}$$
$$= \begin{bmatrix} sX_1(s) - x_1(0) \\ sX_2(s) - x_2(0) \end{bmatrix} = s\mathbf{X}(s) - \mathbf{x}(0)$$

where we let $X(s)$ denote the Laplace transform of the vector function $x(t)$. Letting $F(s)$ be the transform of the vector $f(t)$, we may deduce from (5.7.21) and theorem 5.3.4 that

$$sX(s) - x(0) = AX(s) + F(s) \tag{5.7.22}$$

(a) Solve (5.7.22) for $X(s)$ to show that

$$X(s) = Z(s)(F(s) + b) \tag{5.7.23}$$

where $Z(s) = (sI - A)^{-1}$ and $b = x(0)$. Explain why we must assume that s is not an eigenvalue of A when we write $X(s)$ in the form (5.7.23).

(b) Next we solve an example system in step-by-step fashion. Consider the IVP

$$x' = \begin{bmatrix} 1 & 0 \\ -1 & 3 \end{bmatrix} x + \begin{bmatrix} e^{2t} \\ 3 \end{bmatrix}, \qquad x(0) = \begin{bmatrix} 1 \\ 0 \end{bmatrix} \tag{5.7.24}$$

(i) Compute $F(s)$ and hence show that

$$F(s) + x(0) = \begin{bmatrix} \frac{1}{s-2} + 1 \\ \frac{3}{s} \end{bmatrix}$$

(ii) Use the given coefficient matrix A to compute $Z(s) = (sI - A)^{-1}$ and conclude[9] that

$$Z(s) = \frac{1}{(s-1)(s-3)} \begin{bmatrix} s-3 & 0 \\ -1 & s-1 \end{bmatrix}$$

(iii) Compute $X(s)$ using (5.7.23) to show that

$$X(s) = \frac{1}{s(s-2)} \begin{bmatrix} s \\ 2 \end{bmatrix}$$

(iv) Finally, use the inverse Laplace transform component-wise on $X(s)$ (using standard inverse transform techniques) to find

$$x(t) = \mathcal{L}^{-1}[X(s)] = \begin{bmatrix} e^{2t} \\ e^{2t-1} \end{bmatrix}$$

(c) Use Laplace transforms and the solution technique outlined in (b) above to find the solution of each system of IVPs below.

(i) $x' = \begin{bmatrix} 1 & 1 \\ -1 & 1 \end{bmatrix} x + \begin{bmatrix} \cos t \\ -\sin t \end{bmatrix}, \qquad x(0) = \begin{bmatrix} -1 \\ 0 \end{bmatrix}$

(ii) $x' = \begin{bmatrix} 0 & 2 \\ 1 & -1 \end{bmatrix} x + \begin{bmatrix} \sin t \\ \sin t \end{bmatrix}, \qquad x(0) = \begin{bmatrix} 1 \\ 0 \end{bmatrix}$

---

[9] Recall the shortcut $\begin{bmatrix} a & b \\ c & d \end{bmatrix}^{-1} = \frac{1}{ad - bc} \begin{bmatrix} d & -b \\ -c & a \end{bmatrix}$.

(iii) $\mathbf{x}' = \begin{bmatrix} 2 & 1 \\ 3 & 0 \end{bmatrix} \mathbf{x} + \begin{bmatrix} t \\ t \end{bmatrix}, \qquad \mathbf{x}(0) = \begin{bmatrix} 1 \\ 0 \end{bmatrix}$

(iv) $\mathbf{x}' = \begin{bmatrix} 2 & 1 \\ 3 & 0 \end{bmatrix} \mathbf{x} + \begin{bmatrix} t \\ t \end{bmatrix}, \qquad \mathbf{x}(0) = \begin{bmatrix} 0 \\ 0 \end{bmatrix}$

(v) $\mathbf{x}' = \begin{bmatrix} 2 & 1 & 0 \\ 0 & 2 & 0 \\ 0 & 0 & -1 \end{bmatrix} \mathbf{x} + \begin{bmatrix} e^t \\ 1 \\ 0 \end{bmatrix}, \qquad \mathbf{x}(0) = \begin{bmatrix} 0 \\ 0 \\ 0 \end{bmatrix}$

(vi) $\mathbf{x}' = \begin{bmatrix} 2 & 1 & 0 \\ 0 & 2 & 0 \\ 0 & 0 & -1 \end{bmatrix} \mathbf{x} + \begin{bmatrix} e^t \\ 1 \\ 0 \end{bmatrix}, \qquad \mathbf{x}(0) = \begin{bmatrix} 1 \\ 0 \\ 0 \end{bmatrix}$

6

Nonlinear systems of differential equations

6.1 Motivating problems

In our studies so far, we have seen that a variety of interesting physical situations can be modeled by linear systems of differential equations. Moreover, nearly all linear systems may be solved explicitly. But, many important phenomena are nonlinear in nature; in order to motivate our upcoming work with such systems, we consider two applications where nonlinear systems of equations arise.

A pendulum is a mesmerizing phenomenon. Whether on a grandfather clock or in the hand of a hypnotist, there is something fascinating about its motion. It turns out that a nonlinear second-order differential equation (and hence a system of nonlinear first-order equations) models its behavior. To develop this differential equation, let a rigid arm of length L be attached to a point from which it may swing freely. In this discussion, we will assume for simplicity that no damping is present. Similarly, to simplify the physics we assume that the arm itself has negligible mass. Finally, we attach a mass m to the end of the rigid arm and set the pendulum in motion, as shown in figure 6.1.

We are interested in how the mass travels along a circular arc once the mass is set in motion. The quantities of interest to us are noted in figure 6.1; the variable θ represents the angle (in radians) the arm makes with the vertical axis and s denotes the displacement of the center of the mass along the circular arc.

Because the mass is traveling along a circular arc, it follows that $s = L\theta$. Noting that both s and θ are implicit functions of t, we can differentiate with respect to t and find $s'(t) = L\theta'(t)$ and $s''(t) = L\theta''(t)$. In particular, the velocity of the center of the mass along the arc is $s'(t)$ and its acceleration is $s''(t)$.

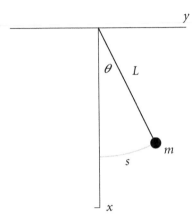

Figure 6.1 A simple pendulum.

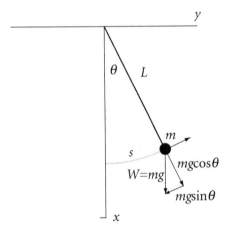

Figure 6.2 Component of gravity's force
along the pendulum's motion.

Since the acceleration $a(t)$ is given by $a(t) = s''(t)$, we have

$$a(t) = \frac{d^2 s}{dt^2} = L\frac{d^2\theta}{dt^2} \tag{6.1.1}$$

Since we have assumed that there is no damping present, once the mass
is set in motion the only force acting on the pendulum is gravity. Because we
are studying the displacement, velocity, and acceleration of the mass along its
path, we must consider the magnitude of the weight $W = mg$ in the direction
of motion. From figure 6.2, we see that gravity induces a force of magnitude
$W \sin\theta$ along the circular arc. Note, too, that this force opposes the motion of
the pendulum, assuming $s'(t)$ is positive.

From Newton's second law, $F = ma$, it now follows that $ma = -mg\sin\theta$, or

$$a(t) = -g\sin\theta(t) \tag{6.1.2}$$

Using the two equivalent expressions for acceleration in (6.1.1) and (6.1.2), it follows that

$$L\frac{d^2\theta}{dt^2} = -g\sin\theta \tag{6.1.3}$$

If we assume that an initial displacement angle $\theta(0) = \theta_0$ and initial angular velocity $\theta'(0) = \theta_0'$ are given, then after rearranging (6.1.3) it follows that θ satisfies the initial-value problem

$$\theta'' + \frac{g}{L}\sin\theta = 0, \quad \theta(0) = \theta_0, \; \theta'(0) = \theta_0' \tag{6.1.4}$$

Because of the presence of $\sin\theta$ in this equation, this second-order differential equation is nonlinear, which means that none of our previous solution methods apply. If we use the substitution $x_1 = \theta$ and $x_2 = \theta'$ to recast (6.1.4) as a nonlinear system of first-order differential equations, then it turns out that the system has a natural graphical interpretation through its slope field, just as we saw with linear systems of differential equations. Using this substitution, we observe that the pendulum is governed by the system

$$x_1' = x_2$$

$$x_2' = -\frac{g}{L}\sin x_1$$

with initial conditions $x_1(0) = \theta_0$ and $x_2(0) = \theta_0'$. Besides studying the associated slope field, we will also learn that it is possible to approximate this nonlinear system at key points with a linear system to better understand its behavior, particularly at any equilibrium points it may have. In subsequent sections, we will explore these issues in greater detail and return to this example involving the pendulum several times, including an investigation of what happens when friction is present.

In addition to the pendulum, another system of nonlinear differential equations arises in the study of population dynamics. Let us consider a population $W(t)$ of wolves (in hundreds) that prey upon a population $M(t)$ of moose (in hundreds), where t is time measured in years. A good example of such a situation, and one that biologists have studied in detail, occurs on Isle Royale in Lake Superior. On this remote island, wolves are the only predator of moose and moose are essentially the only prey of wolves.

Suppose that in the absence of moose, the wolves would die off at a rate proportional to their own number according to a differential equation such as

$$\frac{dW}{dt} = -0.75W$$

In the presence of moose, however, we expect more of the wolves to be able to survive, and to do so at a rate proportional to the moose–wolf interactions since these can result in food for the wolves. The number of moose–wolf interactions can be modeled by taking the product of M and W; only some fraction of such interactions will be beneficial to the wolves. Thus, the wolf population can be assumed to satisfy a differential equation of the form

$$\frac{dW}{dt} = -0.75W + 0.25MW \tag{6.1.5}$$

Likewise, in the absence of wolves, we would expect the number of moose to grow unencumbered (at least in the short term). We might, therefore, have a differential equation like

$$\frac{dM}{dt} = 0.5M$$

But with wolves around, some of the moose will die due to moose–wolf interactions, hence we assume the moose population satisfies an equation like

$$\frac{dM}{dt} = 0.5M - 0.1MW \tag{6.1.6}$$

Equations (6.1.5) and (6.1.6) lead to the system of nonlinear differential equations

$$\frac{dW}{dt} = -0.75W + 0.25MW$$

$$\frac{dM}{dt} = 0.5M - 0.1MW$$

Systems of this form (regardless of the values of the constants) are typically known as *predator–prey* or *Lotka–Volterra* equations. Factoring the right-hand side in each equation above, we see that the wolf and moose populations satisfy

$$\frac{dW}{dt} = W(-0.75 + 0.25M)$$

$$\frac{dM}{dt} = M(0.5 - 0.1W)$$

from which it is evident that the system of differential equations has not only the obvious equilibrium point at the origin, but also one at $(5, 3)$. What kind of behavior should we expect for the wolf and moose populations for initial conditions near $(5, 3)$? In particular, is this equilibrium point stable? Are there ways we can approximate this nonlinear system with a linear one? These questions and more are the focus of subsequent sections as we investigate nonlinear systems of DEs. Our in-depth study of linear systems of differential equations in chapter 3 will prove useful in the study of nonlinear systems: as we see in section 6.2, we can study the graphical behavior of solutions to nonlinear systems in the phase plane by plotting a direction field, just as we did with linear systems. Moreover, in section 6.3 we will study a process by which we can approximate the nonlinear system at a point by a linear system and use our understanding of the behavior of linear systems to make predictions about the nonlinear system.

6.2 Graphical behavior of solutions for 2 × 2 nonlinear systems

In our study of single first-order initial-value problems in chapter 2, we learned that every IVP associated with a linear differential equation with sufficiently well-behaved coefficient functions has a unique solution; moreover, we can determine an explicit formula for the solution. As we learned in chapter 3, essentially the same situation holds for linear systems of differential equations; those with constant coefficients and their corresponding IVPs can always be solved. However, in the case when the governing differential equation or system of equations is nonlinear, we are not guaranteed that solutions to initial-value problems exist, nor that they are unique when they do exist. In addition, as we now study nonlinear systems, we will find that even when unique solutions exist, we are usually unable to determine explicit formulas for them.

We therefore turn again to graphical and numerical investigations of the qualitative properties of solutions to nonlinear systems in order to understand their short- and long-term behavior. To begin, let us choose an example through which we can develop intuition. We consider the system given by

$$\begin{aligned} x_1' &= x_2 - x_1^3 \\ x_2' &= x_1 - x_2^3 \end{aligned} \qquad (6.2.1)$$

If we let

$$\mathbf{x}(t) = \begin{bmatrix} x_1(t) \\ x_2(t) \end{bmatrix}$$

and $\mathbf{F} : \mathbb{R}^2 \to \mathbb{R}^2$ be the function defined by

$$\mathbf{F}(\mathbf{x}) = \mathbf{F}(x_1, x_2) = (x_2 - x_1^3, x_1 - x_2^3)$$

then it follows that we may view (6.2.1) as having the form

$$\mathbf{x}' = \mathbf{F}(\mathbf{x}) \qquad (6.2.2)$$

This is analogous to our work with linear systems of differential equations that may be expressed in the form $\mathbf{x}' = \mathbf{A}\mathbf{x}$, where $\mathbf{A}$ is a matrix. In that setting, the right-hand side of the system is a linear function of $\mathbf{x}$, but in (6.2.2), $\mathbf{F}(\mathbf{x})$ is not linear. Nonetheless, a graphical interpretation of the system remains both possible and enlightening.

In section 3.4, we discussed the graphical behavior of a vector function. Here, we simply remind ourselves that for the system $\mathbf{x}' = \mathbf{F}(\mathbf{x})$ in (6.2.1), a solution $\mathbf{x}(t)$ is a vector function whose output lies in $\mathbb{R}^2$ and whose graph is the curve that is traced out by the vectors $\mathbf{x}(t)$ at various times t. Moreover, the derivative $\mathbf{x}'(t)$ of $\mathbf{x}(t)$ is itself a vector function that indicates the instantaneous velocity of a particle traveling along the curve traced out by $\mathbf{x}(t)$. In particular, scalar multiples of $\mathbf{x}'(t)$ tell us the direction of motion or *flow* along the solution curve as time increases.

We therefore turn again to direction fields to study the flow of the solution curves through the vector field generated by the system of differential equations. In particular, (6.2.2) indicates how, for any point (x_1, x_2) in the plane, we can easily compute $\mathbf{x}' = \mathbf{F}(x_1, x_2)$ at that point, and hence know the direction of the flow of the solution curve that passes through that point. Using a computer algebra system to execute these computations repeatedly at points sampled throughout the plane, we can view the *direction field* for the nonlinear system, which is analogous to the direction field for a linear system. A direction field for (6.2.1) is shown in figure 6.3.

The x_1–x_2 plane is again called the *phase plane*; the independent variable t remains implicit in the flow, while the behavior of the curve relative to the coordinate axes demonstrates the interrelationship among the components $x_1(t)$ and $x_2(t)$ of the solution $\mathbf{x}(t)$. Sample solution curves, such those plotted in figure 6.4 are typically called *trajectories*. In section 6.4 we will learn how to construct trajectories for systems through numerical approximation techniques such as Euler's method.

From figures 6.3 and 6.4, it appears that the system (6.2.1) has three equilibrium solutions. Specifically, the behavior of trajectories suggests the possibilities of equilibria at $(-1, -1)$, $(0, 0)$, and $(1, 1)$. We can confirm this algebraically by setting $\mathbf{x}' = \mathbf{0}$ and solving the resulting nonlinear system of equations

$$0 = x_2 - x_1^3 \tag{6.2.3}$$

$$0 = x_1 - x_2^3 \tag{6.2.4}$$

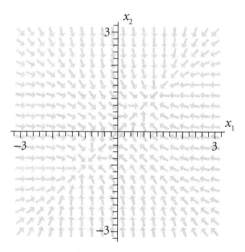

Figure 6.3 The direction field for the system $\mathbf{x}' = \mathbf{F}(\mathbf{x})$ given in (6.2.1).

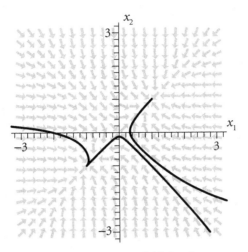

Figure 6.4 The direction field for the system $\mathbf{x'} = \mathbf{F}(\mathbf{x})$ given in (6.2.1) with three trajectories.

Equation (6.2.3) implies that $x_2 = x_1^3$. Substituting this result in (6.2.4), it follows that

$$0 = x_1 - (x_1^3)^3$$

Factoring, we see

$$0 = x_1(1 - x_1^8) = x_1(1 - x_1^4)(1 + x_1^4) = x_1(1 - x_1^2)(1 + x_1^2)(1 + x_1^4)$$

from which we determine that $x_1 = 0, 1$, or -1. Recalling that $x_2 = x_1^3$, the corresponding x_2-values are $x_2 = 0, 1$, and -1, and we have found that the equilibrium points of the system (6.2.1) are indeed $(-1, -1)$, $(0, 0)$, and $(1, 1)$.

Here, we see another distinction between linear and nonlinear systems of differential equations. For a linear system $\mathbf{x'} = \mathbf{Ax}$, the search for equilibrium solutions means we must solve $\mathbf{Ax} = \mathbf{0}$, which we know has either a unique solution or infinitely many solutions. With nonlinear systems, it is possible that any number of equilibrium solutions exist (from none to infinitely many). Moreover, there are no guarantees that we can even expect to analytically solve the resulting system of nonlinear algebraic equations to find such equilibria.

When we do find equilibrium solutions to a system, it is natural to ask about their stability. For example, for the equilibrium solution $(0, 0)$ to (6.2.1), we might observe from figure 6.3 that the origin seems to exhibit behavior similar to a saddle point and therefore may be unstable. To investigate this further, one option is to see if there is a linear system of differential equations to which we can compare (6.2.1). For x_1 and x_2 near zero, observe that both x_1^3 and x_2^3 are extremely small, so that in this region close to the origin it is reasonable for us to say that

$$\begin{aligned} x_1' &= x_2 - x_1^3 \approx x_2 \\ x_2' &= x_1 - x_2^3 \approx x_1 \end{aligned} \tag{6.2.5}$$

In particular, note that the approximate system is linear, and we can write $\mathbf{x}' = \mathbf{A}\mathbf{x}$, for $\mathbf{x}$ near $\mathbf{0}$ with

$$\mathbf{A} = \begin{bmatrix} 0 & 1 \\ 1 & 0 \end{bmatrix} \tag{6.2.6}$$

The eigenvalues of the matrix $\mathbf{A}$ are $\lambda_1 = -1$ and $\lambda_2 = 1$ with corresponding eigenvectors $\mathbf{v}_1 = [-1 \ 1]^T$ and $\mathbf{v}_2 = [1 \ 1]^T$. Due to the fact that the eigenvalues are real and of opposing signs, it follows that the origin is indeed a saddle point for this approximating linear system and is therefore unstable. The phase plane for the linear system corresponding to (6.2.6) near $\mathbf{0}$ is displayed in figure 6.5. This behavior is consistent with that observed near the origin in figure 6.3. We will call the system $\mathbf{x}' = \mathbf{A}\mathbf{x}$, where $\mathbf{A}$ is given by (6.2.6), the *linearization* of (6.2.1) near $\mathbf{0}$. In section 6.3, we will study this approximation to a nonlinear system of differential equations near any particular point of interest to us.

We close this section with two examples of nonlinear systems in which we determine all equilibrium solutions and examine the graphical behavior of solutions near the equilibria.

Example 6.2.1   Consider the system of differential equations given by

$$\begin{aligned} x_1' &= \sin x_2 \\ x_2' &= x_2 - x_1^2 \end{aligned} \tag{6.2.7}$$

Determine all equilibrium solutions of the system, plot the direction field, and discuss the behavior of solutions near at least two of the equilibrium solutions.

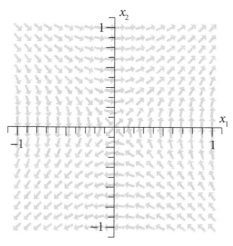

Figure 6.5 The direction field for the linear system $\mathbf{x}' = \mathbf{A}\mathbf{x}$ given in (6.2.5).

Solution.   To find the equilibrium solutions, we set $x_1' = x_2' = 0$ and solve the system of equations

$$0 = \sin x_2 \tag{6.2.8}$$

$$0 = x_2 - x_1^2 \tag{6.2.9}$$

Equation (6.2.8) implies that x_2 must be any integer multiple of π, while (6.2.9) shows that x_1 and x_2 must satisfy the relationship $x_1^2 = x_2$. This latter equation implies that x_2 must be non-negative, and therefore with $x_2 = k\pi$ for any non-negative integer k, it follows that $x_1 = \pm\sqrt{k\pi}$ and we have equilibrium solutions of the form $(\sqrt{k\pi}, k\pi), (-\sqrt{k\pi}, k\pi)$ for $k = 0, 1, 2, \ldots$.

An appropriate window in which to plot the direction field for this system might therefore be $[-3, 3] \times [-1, 8]$, as this will include the five equilibrium solutions $(0, 0), (-\sqrt{\pi}, \pi), (\sqrt{\pi}, \pi), (-\sqrt{2\pi}, 2\pi),$ and $(\sqrt{2\pi}, 2\pi)$. Plotting the direction field, as shown in figure 6.6, we see that the system appears to demonstrate familiar behavior around the equilibrium solutions. For example, at the solutions $(\sqrt{\pi}, \pi)$ and $(-\sqrt{2\pi}, 2\pi)$, each seems to be a saddle point, based on the behavior of trajectories nearby. In addition, at the equilibrium points $(-\sqrt{\pi}, \pi)$ and $(\sqrt{2\pi}, 2\pi)$, the system appears to demonstrate spiraling behavior where the equilibria might act as stable centers or possibly as unstable spiral sources. Based on the periodicity of the sine function, we can reasonably expect that we would see similar behavior demonstrated at other equilibrium points of the form $(\pm\sqrt{k\pi}, k\pi)$, for $k = 3, 4, \ldots$. Note further that all equilibria lie along the parabola $x_2 = x_1^2$, as dictated by (6.2.9). Finally, it is evident that $(0, 0)$ is an unstable equilibrium, though the precise behavior of solutions nearby is not entirely clear from the plot.

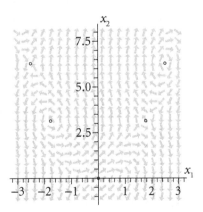

Figure 6.6 The direction field for the system (6.2.7) with equilibrium points $(0, 0), (-\sqrt{\pi}, \pi), (\sqrt{\pi}, \pi),$ $(-\sqrt{2\pi}, 2\pi),$ and $(\sqrt{2\pi}, 2\pi)$.

Indeed, it is apparent that we desire more precision, and not just in the vicinity of $(0, 0)$; our study of the linearization of a system of nonlinear differential equations in the next section will enable a much more rigorous understanding of a system's behavior near any equilibrium point.

Example 6.2.2   Consider the system of differential equations given by

$$\begin{aligned} x_1' &= -x_1 + x_1 x_2^2 \\ x_2' &= -2x_2 + x_2 x_1 \end{aligned} \tag{6.2.10}$$

Determine all equilibrium solutions of the system, plot the direction field, and discuss the behavior of solutions near at least two of the equilibrium solutions.

Solution.   In the standard way, to find the equilibrium solutions we set $x_1' = x_2' = 0$ and solve the nonlinear system of equations

$$0 = -x_1 + x_1 x_2^2 = x_1(-1 + x_2^2) \tag{6.2.11}$$

$$0 = -2x_2 + x_2 x_1 = x_2(-2 + x_1) \tag{6.2.12}$$

From (6.2.12), we see that either $x_2 = 0$ or $x_1 = 2$. If $x_2 = 0$, substituting this value for x_2 in (6.2.11), it follows that $x_1 = 0$, so one equilibrium solution is $(0, 0)$. If $x_1 = 2$, then (6.2.11) implies that $-1 + x_2^2 = 0$, which in turn shows that $x_2 = \pm 1$. Thus, two additional equilibrium solutions have been found: $(2, 1)$ and $(2, -1)$.

A reasonable window for plotting the direction field for this system is $[-2, 4] \times [-3, 3]$, since this will include the three equilibrium solutions we

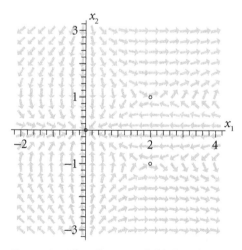

Figure 6.7 The direction field for the system (6.2.10) with equilibrium points $(0, 0)$, $(2, 1)$, and $(2, -1)$.

have found at $(0,0)$, $(2,1)$, and $(2,-1)$. As we see in figure 6.7, it appears that $(0,0)$ is a stable attracting fixed point and that both coordinate axes are straight-line solutions. This observation is not surprising if we also think about linear approximations: for x_1 and x_2 near zero, $x_1 x_2^2$ and $x_1 x_2$ will be extremely small, and thus for such values the nonlinear system (6.2.10) can be approximated by the linear system

$$\begin{aligned} x_1' &= -x_1 \\ x_2' &= -2x_2 \end{aligned}$$

(6.2.13)

The linear system (6.2.13) has the obvious solutions $x_1(t) = e^{-t}$ and $x_2(t) = e^{-2t}$, which lead to the observed behavior near $(0,0)$ in the nonlinear system. From figure 6.7, it also appears that the equilibrium points $(2,1)$ and $(2,-1)$ are saddle points.

From all of our work in this section, we see that equilibrium solutions remain a vital part of our understanding of any system, whether linear or not. In addition, the picture painted by the direction field is fundamental to understanding the behavior of solutions to a nonlinear system. And yet, we are left desiring more detail than the direction field can provide. In section 6.3 we will develop the concept of the linearization of a system in order to link our understanding of linear systems to the behavior of nonlinear systems near equilibrium points. Furthermore, in section 6.4, we will generalize Euler's method for single differential equations in order to apply it to systems to generate approximate solutions to solutions.

6.2.1 Plotting direction fields of nonlinear systems using *Maple*

The *Maple* syntax used to generate the plots in this section is essentially identical to that discussed for direction fields for linear systems in section 3.4.1. As always, we use the DEtools package, and load it with the command

```
> with(DEtools):
```

To define the system of differential equations from example 6.2.1 in *Maple*, we use the command

```
> sys := diff(x[1](t),t) = sin(x[2](t)),
diff(x[1](t),t) = x[2](t) - x[1](t)^2;
```

The system of differential equations of interest is now stored in "sys." The direction field may now be generated by the command

```
> DEplot([sys], [x[1](t),x[2](t)], t=-1..1,
x[1]=-3..3, x[2]=-1..8, arrows=large, color=gray);
```

In plots in section 6.2, we have also included the equilibrium points. These may be generated by the `pointplot` command, which requires us to load the `plots` package. For example, the syntax

```
> with(plots): pointplot([0,0], [sqrt(Pi),Pi],
[-sqrt(Pi),Pi], [sqrt(2*Pi),2*Pi], [-sqrt(2*Pi),
2*Pi], symbol=circle, symbolsize=7);
```

will produce a plot of just these five points in the plane. To superimpose these points on the direction field, we can assign names to each plot and then display them together. Giving the respective plots the names DF and EQsol, we can use the `display` command as follows. Note the use of colons, rather than semicolons, to suppress output when we assign names to the plots.

```
> DF := DEplot([sys], [x[1](t),x[2](t)], t=-1..1,
x[1]=-3..3, x[2]=-1..8, arrows=large, color=gray):
> EQsol := pointplot( [0,0], [sqrt(Pi),Pi],
[-sqrt(Pi),Pi], [sqrt(2*Pi),2*Pi], [-sqrt(2*Pi),
2*Pi], symbol=circle, symbolsize=7):
> display(DF, EQsol);
```

This combination of commands results in the output shown at left in figure 6.8.

If desired, we can now sketch trajectories by hand. *Maple* has the capacity to include such trajectories, given initial conditions. For example, if we are given the initial conditions $\mathbf{x}(0) = (2, 6)$ and $(-2, 6)$, we can modify the earlier DEplot

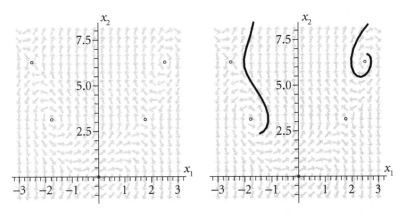

Figure 6.8 At left, the direction field for the system (6.2.7) with equilibrium points $(0,0)$, $(-\sqrt{\pi},\pi)$, $(\sqrt{\pi},\pi)$, $(-\sqrt{2\pi},2\pi)$, and $(\sqrt{2\pi},2\pi)$. At the right, the same direction field with trajectories through $(2,6)$ and $(-2,6)$ is included.

command to

```
> DEplot([sys], [x[1](t),x[2](t)], t=-2..2,
x[1]=-3..3, x[2]=-1..8, arrows=medium, color=gray),
[[x[1](0)=2,x[2](0)=6], [x[1](0)=-2,x[2](0)=6]]);
```

This most recent command, when saved and displayed simultaneously with the above plot of equilibrium solutions, results in the righthand plot in figure 6.8.

As a reminder, we always expect to experiment some with the window in which the plot is displayed: the range of x- and y-values certainly affects how clearly the direction field is revealed, and the range of t-values impacts how much of each trajectory is plotted. As the most recent section shows, a study of a system's equilibrium points is a helpful guide for choosing a window in which to display a plot.

Exercises 6.2
In exercises 1–7, (a) determine all equilibrium solutions, (b) use *Maple* to plot the direction field, and (c) from the direction field, visually estimate whether equilibrium solutions are stable or unstable and discuss the long-term behavior of solutions.

1. $x_1' = x_2 - 2x_1x_2$
   $x_2' = 4x_1x_2 - x_1$

2. $x_1' = 4 - x_2^2$
   $x_2' = 1 - x_1 + x_2$

3. $x_1' = \cos x_2$
   $x_2' = 1 - \sin x_1$

4. $x_1' = 2x_1 - x_2$
   $x_2' = -4x_1 + 2x_2$

5. $x_1' = e^{-x_2}$
   $x_2' = 1/(1 + x_1^2)$

6. $x_1' = \ln(2 + x_2)$
   $x_2' = x_1^2 + x_2$

7. $x_1' = x_2 - x_1^2$
   $x_2' = x_1 - 8x_2^2$

8. Recall from section 6.1 that the nonlinear system of differential equations

$$W' = -0.75W + 0.25MW$$

$$M' = 0.5M - 0.1MW$$

models the numbers of wolves and moose (each measured in hundreds) in a predator–prey situation. Determine all equilibrium solutions to this system, plot an appropriate direction field in a computer algebra system, and discuss the apparent long-term behavior of the wolf and moose populations.

9. Recall that if $x_1 = \theta$ is the angle that the arm of a pendulum forms with the positive x-axis (as shown in figure 6.2) and $x_2 = x_1' = \theta'$, then x_1 and x_2 satisfy the nonlinear system of differential equations

$$x_1' = x_2$$

$$x_2' = -\frac{g}{L} \sin x_1$$

Let $g = 9.8 \, \text{m/s}^2$ and assume that the length of the arm is $L = 2 \, \text{m}$. Determine all equilibrium solutions to this system, plot an appropriate direction field in a computer algebra system, and discuss the long-term behavior of solutions to the system. Be sure to relate your answers directly to the behavior of the pendulum and corresponding initial conditions.

6.3 Linear approximations of nonlinear systems

In our first look at nonlinear systems in the preceding section, we considered the system

$$\begin{aligned} x_1' &= x_2 - x_1^3 \\ x_2' &= x_1 - x_2^3 \end{aligned} \tag{6.3.1}$$

and observed informally that near the origin where $\mathbf{x} \approx \mathbf{0}$, we can drop the x_1^3 and x_2^3 terms so that (6.3.1) can be approximated by the linear system $\mathbf{x}' = A\mathbf{x}$ where

$$A = \begin{bmatrix} 0 & 1 \\ 1 & 0 \end{bmatrix} \tag{6.3.2}$$

In this section, we make this notion of linear approximation of nonlinear systems more precise and use this approach to classify the stability of equilibria of nonlinear systems.

An important idea in calculus is that all well-behaved functions are locally linear. That is, they appear linear when viewed up close; the line the function emulates is the tangent line to the curve at the point on which we focus. In particular, for a function $f(x)$ that is differentiable at the value $x = a$, $f(x) \approx L(x)$ for x near a, where

$$L(x) = f(a) + f'(a)(x - a) \tag{6.3.3}$$

The function $L(x)$ is usually called the *tangent line approximation* or *linearization* of f at $x = a$.

We encounter the very same ideas in multivariable calculus. For a differentiable vector function $\mathbf{r} : \mathbb{R} \to \mathbb{R}^3$ given by

$$\mathbf{r}(t) = \begin{bmatrix} f(t) \\ g(t) \\ h(t) \end{bmatrix}$$

for values of t near some fixed value a, the curve in space that $\mathbf{r}(t)$ generates can be approximated by the tangent line to the curve. In particular, $\mathbf{r}(t) \approx \mathbf{L}(t)$ where

$$\mathbf{L}(t) = \mathbf{r}(a) + \mathbf{r}'(a)(t-a) = \begin{bmatrix} f(a) + f'(a)(t-a) \\ g(a) + g'(a)(t-a) \\ h(a) + h'(a)(t-a) \end{bmatrix} \qquad (6.3.4)$$

for t near a. As in the case of the scalar function f, $\mathbf{L}$ is called the *tangent line approximation* or *linearization* of $\mathbf{r}$ at $t = a$.

Similarly for a differentiable real-valued function of several variables $F : \mathbb{R}^2 \to \mathbb{R}$ given by $z = F(x, y)$, $F(x, y)$ can be approximated by its tangent plane for (x, y) near some fixed point (a, b). That is, we have the approximation $F(x, y) \approx L(x, y)$ where

$$L(x, y) = f(a, b) + f_x(a, b)(x - a) + f_y(a, b)(y - b) \qquad (6.3.5)$$

L is called the *tangent plane approximation* or *linearization* of f at (a, b).

There is obviously a great deal of similarity in the algebraic forms of the linear approximations given in (6.3.3), (6.3.4), and (6.3.5). How can we apply these ideas to systems of nonlinear differential equations? The next example, in which we reconsider (6.3.1), suggests one approach. Because of the pending use of partial derivatives, we will temporarily use the notation $\mathbf{x} = [x_1\ x_2]^{\mathrm{T}} = [x\ y]^{\mathrm{T}}$.

Example 6.3.1   Consider the system of differential equations

$$\begin{aligned} x' &= f(x, y) = y - x^3 \\ y' &= g(x, y) = x - y^3 \end{aligned} \qquad (6.3.6)$$

Determine linear approximations to both $f(x, y)$ and $g(x, y)$ at the point $(1, 1)$. Then explain how these linear combinations may be combined to form an overall linear approximation of (6.3.6) near $(1, 1)$.

Solution.   In section 6.2, we considered this same system (using x_1 and x_2 for the functions, instead of x and y) and learned that the equilibrium solutions to the system are $(-1, -1)$, $(0, 0)$, and $(1, 1)$. As noted at the start of this section, we have already considered a linear approximation of the system at $(0, 0)$. Here, we focus on the behavior of solutions near the equilibrium solution $(1, 1)$.

To first approximate $x' = f(x, y) = y - x^3$ near $(1, 1)$, we use (6.3.5) to find the tangent plane approximation. Noting that $f_x(x, y) = -3x^2$ and $f_y(x, y) = 1$, it follows that $f_x(1, 1) = -3$ and $f_y(1, 1) = 1$. Moreover, $f(1, 1) = 0$ since $(1, 1)$ is an equilibrium solution of the system. Now, it follows that for (x, y) near $(1, 1)$,

$$f(x, y) \approx f(1, 1) + f_x(1, 1)(x - 1) + f_y(1, 1)(y - 1) = 0 - 3(x - 1) + 1(y - 1)$$
(6.3.7)

Similar ideas applied to $y' = g(x, y) = x - y^3$ show that for (x, y) near $(1, 1)$,

$$g(x, y) \approx g(1, 1) + g_x(1, 1)(x - 1) + g_y(1, 1)(y - 1) = 0 + 1(x - 1) - 3(y - 1)$$
(6.3.8)

If we now consider the overall system (6.3.6), for (x, y) near $(1, 1)$ we have the approximation

$$\begin{aligned} x' &= f(x, y) \approx -3(x - 1) + 1(y - 1) \\ y' &= g(x, y) \approx 1(x - 1) - 3(y - 1) \end{aligned}$$
(6.3.9)

Using the fact that both equations in (6.3.9) are linear and writing this system in matrix form with $\mathbf{x} = [x \; y]^{\mathrm{T}}$, we have

$$\mathbf{x}' \approx \begin{bmatrix} -3 & 1 \\ 1 & -3 \end{bmatrix} \begin{bmatrix} x - 1 \\ y - 1 \end{bmatrix} = \begin{bmatrix} -3 & 1 \\ 1 & -3 \end{bmatrix} \mathbf{x} + \begin{bmatrix} -3 & 1 \\ 1 & -3 \end{bmatrix} \begin{bmatrix} -1 \\ -1 \end{bmatrix}$$

$$= \begin{bmatrix} -3 & 1 \\ 1 & -3 \end{bmatrix} \mathbf{x} + \begin{bmatrix} 2 \\ 2 \end{bmatrix}$$
(6.3.10)

Hence we have approximated the original nonlinear system with a linear one by writing it in the form $\mathbf{x}' \approx \mathbf{A}(\mathbf{x} - \mathbf{a}) = \mathbf{A}\mathbf{x} + \mathbf{b}$, where $\mathbf{b} = -\mathbf{A}\mathbf{a}$, for $\mathbf{x}$ near $\mathbf{a}$.

Because we have found that we may approximate the system (6.3.6) with the linear system (6.3.10), we can now use our understanding of linear systems to determine the behavior of the nonlinear system near the chosen equilibrium point. Specifically, the fact that the eigenvalues of the matrix $\mathbf{A}$ in (6.3.10) are $\lambda = -2$ and $\lambda = -4$ tells us that the equilibrium solution $(1, 1)$ of (6.3.1) is a stable, attracting node, as we initially conjectured graphically from figure 6.4.

Moreover, the approach we have taken in example 6.3.1 may certainly be generalized. Any nonlinear system of two differential equations may be written in the form

$$\mathbf{x}' = \mathbf{F}(\mathbf{x})$$
(6.3.11)

where $\mathbf{F}$ is a function of the form $\mathbf{F}(\mathbf{x}) = \mathbf{F}(x, y) = (f(x, y), g(x, y))$. Given an equilibrium solution of (6.3.11) at $\mathbf{a} = (a, b)$, notice that $\mathbf{F}(\mathbf{a}) = \mathbf{0}$; in particular, $f(a, b) = g(a, b) = 0$.

If, as in example 6.3.1, we approximate f and g near (a, b) with

$$f(x, y) \approx f(a, b) + f_x(a, b)(x - a) + f_y(a, b)(y - b)$$
$$= f_x(a, b)(x - a) + f_y(a, b)(y - b)$$
$$g(x, y) \approx g(a, b) + g_x(a, b)(x - a) + g_y(a, b)(y - b)$$
$$= g_x(a, b)(x - a) + g_y(a, b)(y - b)$$

we observe that in matrix form we have

$$\mathbf{x}' = \mathbf{F}(\mathbf{x})$$

$$= \begin{bmatrix} f(x, y) \\ g(x, y) \end{bmatrix}$$

$$\approx \begin{bmatrix} f_x(a, b)(x - a) + f_y(a, b)(y - b) \\ g_x(a, b)(x - a) + g_y(a, b)(y - b) \end{bmatrix}$$

$$= \begin{bmatrix} f_x(a, b) & f_y(a, b) \\ g_x(a, b) & g_y(a, b) \end{bmatrix} \begin{bmatrix} x - a \\ y - b \end{bmatrix}$$

In matrix notation, we have written that $\mathbf{x}' = \mathbf{F}(\mathbf{x}) \approx \mathbf{J}(\mathbf{a})(\mathbf{x} - \mathbf{a})$ for $\mathbf{x}$ near $\mathbf{a}$, where $\mathbf{a}$ is an equilibrium point of the original system and $\mathbf{J}(\mathbf{a})$ is a matrix with constant entries. The matrix $\mathbf{J}(\mathbf{a})$, which is defined by

$$\mathbf{J}(\mathbf{a}) = \begin{bmatrix} f_x(a, b) & f_y(a, b) \\ g_x(a, b) & g_y(a, b) \end{bmatrix} \tag{6.3.12}$$

is known as the *Jacobian* matrix of the function $\mathbf{F}$ evaluated at the point (a, b).

More generally, for any differentiable function $\mathbf{F} : \mathbb{R}^n \to \mathbb{R}^m$ given by $\mathbf{F}(\mathbf{x}) = \mathbf{F}(x_1, \ldots, x_n) = (f_1(x_1, \ldots, x_n), \ldots, f_m(x_1, \ldots, x_n))$, the Jacobian matrix $\mathbf{J}(\mathbf{x})$ is given by

$$\mathbf{J}(\mathbf{x}) = \begin{bmatrix} \partial f_1/\partial x_1 & \partial f_1/\partial x_2 & \cdots & \partial f_1/\partial x_n \\ \partial f_2/\partial x_1 & \partial f_2/\partial x_2 & \cdots & \partial f_2/\partial x_n \\ \vdots & \vdots & \vdots & \vdots \\ \partial f_m/\partial x_1 & \partial f_m/\partial x_2 & \cdots & \partial f_m/\partial x_n \end{bmatrix} \tag{6.3.13}$$

The Jacobian enables us to write the linearization of any differentiable function $\mathbf{F}$ for $\mathbf{x}$ near a point $\mathbf{a}$ as

$$\mathbf{F}(\mathbf{x}) \approx \mathbf{F}(\mathbf{a}) + \mathbf{J}(\mathbf{a})(\mathbf{x} - \mathbf{a}) \tag{6.3.14}$$

which is remarkably similar to the tangent line approximation (6.3.3). Note that we must evaluate the Jacobian matrix at the point $\mathbf{a}$ of interest; moreover, if we are working with a nonlinear system of differential equations with equilibrium point $\mathbf{a}$, it follows that $\mathbf{F}(\mathbf{a}) = \mathbf{0}$, so that we have

$$\mathbf{x}' = \mathbf{F}(\mathbf{x}) \approx \mathbf{J}(\mathbf{a})(\mathbf{x} - \mathbf{a}) \tag{6.3.15}$$

This entire discussion of linearizing nonlinear systems is important for several reasons. One is that it demonstrates how we can take a problem we do not fully understand (the nonlinear system) and gain more knowledge of it by approximating the system near a point of interest with a simpler (linear) system that we do understand. Moreover, because we have completely classified the stability of equilibria of linear systems through the eigenvalues of the system's matrix, we classify the equilibria of nonlinear systems by doing so for the corresponding linearization. We will use the same terminology and classification scheme for equilibria of nonlinear systems that we established for linear ones in sections 3.4 and 3.5. Two examples now follow to demonstrate these ideas in greater detail.

Example 6.3.2   Given the system of differential equations

$$x_1' = 9x_2 - x_2^2$$

$$x_2' = x_1$$

determine all equilibrium points of the system, evaluate the Jacobian at each equilibrium point, and find a corresponding linearization of the system in order to analyze the behavior of trajectories near each equilibrium point and the stability of equilibria. Finally, plot the direction field of the given system to confirm the observations made.

Solution.   First, we observe that $x' = F(x)$ for

$$F(x) = F(x_1, x_2) = (f(x_1, x_2), g(x_1, x_2)) = (9x_2 - x_2^2, x_1)$$

Setting $x' = 0$, it follows that $x_1 = 0$ and $x_2(9 - x_2) = 0$, so that the equilibrium points of the system are $(0, 0)$ and $(0, 9)$.

Taking the appropriate partial derivatives, the Jacobian of F is

$$J(x) = \begin{bmatrix} 0 & 9 - 2x_2 \\ 1 & 0 \end{bmatrix}$$

Therefore, for values of x_1 and x_2 near the equilibrium point $a = (0, 0) = 0$, we have that $x' = F(x) \approx J(0)(x - 0)$, or

$$x' \approx \begin{bmatrix} 0 & 9 \\ 1 & 0 \end{bmatrix} x$$

For this linear system, the eigenvalues of the matrix $J(0)$ are $\lambda = 3$ and $\lambda = -3$, so the origin is a saddle point and therefore unstable. Moreover, we expect there to be two approximately straight-line solutions (along the respective eigenvectors of $J(0)$) that pass through the origin, along one of which the solution tends toward $(0, 0)$ while on the other the solution is repelled away from $(0, 0)$.

For x_1 and x_2 near the equilibrium point $a = (0, 9)$, we have that $x' = F(x) \approx J(a)(x - a)$, or

$$x' \approx \begin{bmatrix} 0 & -9 \\ 1 & 0 \end{bmatrix} \begin{bmatrix} x_1 - 0 \\ x_2 - 9 \end{bmatrix} = \begin{bmatrix} 0 & -9 \\ 1 & 0 \end{bmatrix} x + \begin{bmatrix} 81 \\ 0 \end{bmatrix}$$

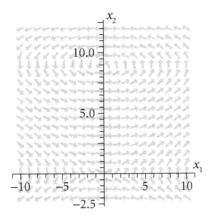

Figure 6.9 The direction field for Example 6.3.2.

For this nonhomogeneous linear system, the eigenvalues of the matrix $\mathbf{J}(0,9)$ are $\lambda = 3i$ and $\lambda = -3i$. Because the eigenvalues are purely imaginary, it follows that the equilibrium point $(0, 9)$ is a stable center. Nearby this point, we expect to see trajectories orbit the point in approximately elliptical loops.

All of our observations are confirmed by the graphical behavior evidenced in figure 6.9.

Example 6.3.3   For the system of differential equations

$$x_1' = \sin x_2$$
$$x_2' = x_2 - x_1^2 \qquad (6.3.16)$$

determine all equilibrium points of the system, evaluate the Jacobian at each equilibrium point, and find a corresponding linearization of the system in order to analyze the behavior of trajectories near each equilibrium point and the stability of equilibria. Finally, plot the direction field of the given system to confirm the observations made.

Solution.   The given system is the same one that we studied in example 6.2.1 in the preceding section. There we discovered that for any equilibrium solution $\mathbf{x} = (x_1, x_2)$, x_2 must be any integer multiple of π and $x_1^2 = x_2$, so that x_2 must be non-negative. Thus, the equilibrium solutions have the form $(\sqrt{k\pi}, k\pi)$, $(-\sqrt{k\pi}, k\pi)$ for $k = 0, 1, 2, \ldots$.

Letting $\mathbf{x}' = \mathbf{F}(\mathbf{x}) = (\sin x_2, x_2 - x_1^2)$, it follows that the Jacobian of $\mathbf{F}$ is

$$\mathbf{J}(\mathbf{x}) = \begin{bmatrix} 0 & \cos x_2 \\ -2x_1 & 1 \end{bmatrix}$$

For values of x_1 and x_2 near the equilibrium point $\mathbf{a} = (0,0) = \mathbf{0}$, we have that $\mathbf{x}' = \mathbf{F}(\mathbf{x}) \approx \mathbf{J}(\mathbf{0})(\mathbf{x} - \mathbf{0})$, or

$$\mathbf{x}' \approx \begin{bmatrix} 0 & 1 \\ 0 & 1 \end{bmatrix} \mathbf{x}$$

The eigenvalues of the matrix $\mathbf{J}(\mathbf{0})$ are $\lambda = 0$ and $\lambda = 1$, so the origin is unstable, because the real eigenvalue $\lambda = 1 > 0$ will drive solutions away from the origin as $t \to \infty$. Moreover, because $\lambda = 0$ is an eigenvalue of $\mathbf{J}(\mathbf{0})$, it also follows that all solutions near $\mathbf{0}$ are approximately straight-line solutions.

For x_1 and x_2 near the equilibrium point $\mathbf{a} = (\sqrt{\pi}, \pi)$, we have that $\mathbf{x}' = \mathbf{F}(\mathbf{x}) \approx \mathbf{J}(\mathbf{a})(\mathbf{x} - \mathbf{a})$, or

$$\mathbf{x}' \approx \begin{bmatrix} 0 & -1 \\ -2\sqrt{\pi} & 1 \end{bmatrix} \begin{bmatrix} x_1 - \sqrt{\pi} \\ x_2 - \pi \end{bmatrix} = \begin{bmatrix} 0 & -1 \\ -2\sqrt{\pi} & 1 \end{bmatrix} \mathbf{x} + \begin{bmatrix} \pi \\ \pi \end{bmatrix}$$

The eigenvalues of the matrix $\mathbf{J}(\sqrt{\pi}, \pi)$ are approximately $\lambda = 2.448$ and $\lambda = -1.448$, and so the equilibrium point $(\sqrt{\pi}, \pi)$ is a saddle point and unstable. However, if we consider the equilibrium point $\mathbf{a} = (-\sqrt{\pi}, \pi)$, we have that $\mathbf{x}' = \mathbf{F}(\mathbf{x}) \approx \mathbf{J}(\mathbf{a})(\mathbf{x} - \mathbf{a})$, or

$$\mathbf{x}' \approx \begin{bmatrix} 0 & -1 \\ 2\sqrt{\pi} & 1 \end{bmatrix} \begin{bmatrix} x_1 + \sqrt{\pi} \\ x_2 - \pi \end{bmatrix} = \begin{bmatrix} 0 & -1 \\ 2\sqrt{\pi} & 1 \end{bmatrix} \mathbf{x} + \begin{bmatrix} \pi \\ \pi \end{bmatrix}$$

In this case, the eigenvalues of the matrix $\mathbf{J}(-\sqrt{\pi}, \pi)$ are approximately $\lambda = 0.5 \pm 1.815i$. Because these complex eigenvalues have positive real parts, it follows that the equilibrium solution $(-\sqrt{\pi}, \pi)$ is a spiral source and is unstable.

If we continue exploring equilibrium points of the form $(\pm\sqrt{k\pi}, k\pi)$, we can show through the Jacobian that whenever k is odd, the point $(\sqrt{k\pi}, k\pi)$ is a saddle point and the point $(-\sqrt{k\pi}, k\pi)$ is a spiral source. Conversely, whenever k is even, $(\sqrt{k\pi}, k\pi)$ is a spiral source and the point $(-\sqrt{k\pi}, k\pi)$ is a saddle. In particular, every equilibrium point of the system is unstable.

These observations are all confirmed in the direction field shown in figure 6.10.

Through linear approximation, the tools we developed for linear systems enable us to understand and classify the stability of equilibria and behavior of solutions near equilibrium points for nonlinear systems. In the next section, we will explore how to actually compute approximate solutions via Euler's method for systems.

Exercises 6.3
In exercises 1–6, find the Jacobian of the given function, $\mathbf{F}$.

1. $\mathbf{F}(x_1, x_2) = (x_1^2 + x_2, x_1 - x_2^2)$

2. $\mathbf{F}(x_1, x_2) = (e^{2x_1 x_2}, \cos x_1 + \sin x_2)$

3. $\mathbf{F}(x_1, x_2) = (x_2 - 2x_1 x_2, 4x_1 x_2 - x_1)$

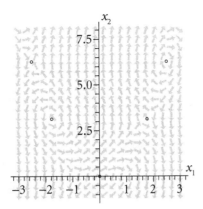

Figure 6.10 The direction field for the system (6.3.16) with equilibrium points $(0,0)$, $(-\sqrt{\pi},\pi)$, $(\sqrt{\pi},\pi)$, $(-\sqrt{2\pi},2\pi)$, and $(\sqrt{2\pi},2\pi)$.

4. $\mathbf{F}(x_1, x_2) = (4 - x_2^2, 1 - x_1^2)$

5. $\mathbf{F}(x_1, x_2, x_3) = (1/(1 + x_1^2 + x_2^2 + x_3^2), e^{-x_1^2 - x_2^2 - x_3^2}, 2x_1 - 3x_2^2 + x_3^4)$

6. $\mathbf{F}(x_1, x_2, x_3) = (3x_1 - x_2 + 4x_3, x_1 + x_2 - 2x_3, -2x_1 + 5x_2 - x_3)$

In exercises 7–10, find the linearization of the given function, $\mathbf{F}(x_1, x_2)$, at the given point **a**.

7. $\mathbf{F}(x_1, x_2) = (x_1^2 + x_2, x_1 - x_2^2)$, $\mathbf{a} = (1, -1)$

8. $\mathbf{F}(x_1, x_2) = (x_2 e^{2x_1}, \cos x_1 + \sin x_2)$, $\mathbf{a} = (\pi/2, 0)$

9. $\mathbf{F}(x_1, x_2) = (x_2 - 2x_1 x_2, 4x_1 x_2 - x_1)$, $\mathbf{a} = (1/2, 1/4)$

10. $\mathbf{F}(x_1, x_2) = (4 - x_2^2, 1 - x_1^2)$, $\mathbf{a} = (-1, 2)$

In exercises 11–17, find all equilibrium points of the system, determine the linearization of the given system near each equilibrium point, classify the stability of each equilibrium point, and compare your work to a plot of the direction field for the system.[1]

11. $x_1' = x_2 - 2x_1 x_2$

   $x_2' = 4x_1 x_2 - x_1$

12. $x_1' = 4 - x_2^2$

   $x_2' = 1 - x_1 + x_2$

---

[1] Note that in the exercises of section 6.2, equilibrium solutions were found and direction fields were plotted in exercises 1–7, which correspond to the same systems of differential equations given here.

13. $x_1' = \cos x_2$
    $x_2' = 1 - \sin x_1$

14. $x_1' = 2x_1 - x_2$
    $x_2' = -4x_1 + 2x_2$

15. $x_1' = e^{-x_2}$
    $x_2' = 1/(1 + x_1^2)$

16. $x_1' = \ln(2 + x_2)$
    $x_2' = x_1^2 + x_2$

17. $x_1' = x_2 - x_1^2$
    $x_2' = x_1 - 8x_2^2$

18. Recall from section 6.1 that the nonlinear system of differential equations

$$W' = -0.75W + 0.25MW$$

$$M' = 0.5M - 0.1MW$$

models the numbers of wolves and moose (each measured in hundreds) in a predator-prey situation. Determine the linearization of the system near the nonzero equilibrium solution, classify the stability of this equilibrium, and discuss the long-term behavior of the wolf and moose populations.[2]

19. Recall that if $x_1 = \theta$ is the angle that the arm of a pendulum forms with the positive x-axis (as shown in figure 6.2) and $x_2 = x_1' = \theta'$, then x_1 and x_2 satisfy the nonlinear system of differential equations

$$x_1' = x_2$$

$$x_2' = -\frac{g}{L} \sin x_1$$

Let $g = 9.8 \text{ m/s}^2$ and $L = 2$ m. Determine the linearization of the system near the equilibrium solution at zero and at least one other equilibrium solution, classify the stability of these equilibria, and discuss the long-term behavior of the pendulum. Be sure to relate your answers directly to the behavior of the pendulum and corresponding initial conditions.

20. In example 6.2.2, we considered the system of differential equations given by

$$x_1' = -x_1 + x_1 x_2^2$$

$$x_2' = -2x_2 + x_2 x_1$$

Determine the linearization of the system near each equilibrium solution, classify the stability of each equilibrium point, and discuss the behavior of solutions nearby.

---

[2] In the exercises of section 6.2, equilibrium solutions were found and the direction field was plotted for this system in exercise 8; similarly, see the results of exercise 9 in section 6.2 for use in the problem 19 below.

6.4 Euler's method for nonlinear systems

Just as we experienced with single nonlinear initial-value problems such as

$$y' = ye^{-y} + 1, \qquad y(0) = 1 \tag{6.4.1}$$

or

$$y' = t^2 + y^2 + 1, \qquad y(0) = -1 \tag{6.4.2}$$

that we could not solve explicitly, in the past two sections we have encountered systems of nonlinear differential equations for which solutions to corresponding initial-value problems cannot be determined analytically. We therefore desire to explore ways to estimate solutions to these problems.

For IVPs such as (6.4.1) and (6.4.2), we know that we may estimate a solution to the problem through Euler's method. Recall from Section 2.6 that for any first-order IVP in the form $y' = f(t, y)$, $y(t_0) = y_0$, given a step-size h we are able to generate the sequence of points $(t_1, y_1), \ldots, (t_n, y_n)$ such that

$$t_{n+1} = t_n + h \quad \text{and} \quad y_{n+1} = y_n + hf(t_n, y_n), \quad \text{for } n \geq 0 \tag{6.4.3}$$

where $y_n \approx y(t_n)$. That is, y_n approximates the solution y to the initial-value problem at the point where $t = t_n$.

To explore how we can extend Euler's method to systems of differential equations, let us consider the initial-value problem given by

$$\begin{aligned} x' &= 9y - y^2, & x(0) &= 1 \\ y' &= x, & y(0) &= 8 \end{aligned} \tag{6.4.4}$$

Here, we choose to use the notation $\mathbf{x} = [x \ y]^T$ rather than $[x_1 \ x_2]^T$ due to the fact that we will be using subscripts to label approximations to the component solutions $x(t)$ and $y(t)$. Keeping in mind that x and y are each implicit functions of t, we can view (6.4.4) as being of the form

$$\begin{aligned} x' &= f(x, y, t), & x(t_0) &= x_0 \\ y' &= g(x, y, t), & y(t_0) &= y_0 \end{aligned} \tag{6.4.5}$$

To see how to approximate solutions to this system of IVPs, let us reconsider our earlier studies of single differential equations. In section 2.6, we considered the equation $y' = f(t, y)$ in a first-order IVP and emphasized the fact that Euler's method relies on following the tangent line approximation to $y(t)$ at each step. In particular, if we have some approximation y_n to the solution y at the t-value t_n, then to move along the tangent line to the next approximation (t_{n+1}, y_{n+1}), it follows that

$$y_{n+1} = y_n + \Delta y$$

$$= y_n + \frac{\Delta y}{\Delta t} \cdot \Delta t$$

$$= y_n + m \cdot \Delta t \tag{6.4.6}$$

where m is the slope at each step of our approximation given by $m = y' = f(t, y)$ in the differential equation that we are attempting to solve. Specifically, given the approximation y_n at time t_n, the slope of the tangent line to the solution curve at this point is $f(t_n, y_n)$. Therefore, using this value for m in (6.4.6), letting $h = \Delta t$ be the step size, we have

$$y_{n+1} = y_n + hf(t_n, y_n) \qquad (6.4.7)$$

An essentially identical approach will work for the system (6.4.5). In particular, given the initial condition (x_0, y_0) and a step-size h, we can generate the approximate solution $(x(t_1), y(t_1)) \approx (x_1, y_1)$ by taking

$$\begin{aligned} x_1 &= x_0 + h \cdot f(t_0, x_0, y_0) \\ y_1 &= y_0 + h \cdot g(t_0, x_0, y_0) \end{aligned} \qquad (6.4.8)$$

The only difference between this approach and our experience with Euler's method for a single equation is that we obviously have to update two approximations at once, as estimates of both $x(t_n)$ and $y(t_n)$ are needed to generate approximations of $x(t_{n+1})$ and $y(t_{n+1})$. We generalize our latest observation in (6.4.8) for a step from the approximation (x_n, y_n) to the approximation (x_{n+1}, y_{n+1}) by

$$\begin{aligned} x_{n+1} &= x_n + h \cdot f(t_n, x_n, y_n) \\ y_{n+1} &= y_n + h \cdot g(t_n, x_n, y_n) \end{aligned} \qquad (6.4.9)$$

At the end of this section, we will discuss the implementation of Euler's method for systems in Excel. For now, we simply report the results of such an implementation here to see the approximations generated. For the original system we considered above,

$$\begin{aligned} x' &= 9y - y^2, & x(0) &= 1 \\ y' &= x, & y(0) &= 8 \end{aligned} \qquad (6.4.10)$$

recall that this system was also studied in example 6.3.2 in section 6.3. There we observed that the equilibrium solution $(0, 9)$ is a stable center of the system and that we expect elliptical orbits nearby. If, for the IVP (6.4.10), we choose a step-size of $h = 0.1$ and take enough steps to complete the expected loop in the orbit, we see the abbreviated data in table 6.1.

In particular, we notice that after taking a sufficient number of steps to loop back around to near the initial condition $(1, 8)$, we have in fact not returned to this point; in fact, we have missed it appreciably with the two nearest approximations being $(0.527, 6.259)$ and $(2.243, 6.312)$.

If we decrease the step size h and take more steps, we can improve the accuracy of the approximation. Doing so with $h = 0.01$ results in the values in table 6.2.

We see that the approximate trajectory has completed one full loop and has nearly returned to pass through the point $(1, 8)$ where the trajectory began. This behavior is more consistent with what we expected based on the classification of the equilibrium point $(0, 9)$ as a stable center through linearization in the preceding section.

Table 6.1
Euler's method applied to (6.4.10) with step-size $h = 0.1$

t_n	x_n	y_n
0	1	8
0.1	1.8	8.1
0.2	2.529	8.28
0.3	3.12516	8.5329
⋮	⋮	⋮
2	−1.146540202	6.373703158
2.1	0.527383445	6.259049138
2.2	2.242958058	6.311787483
2.3	3.93970067	6.536083289

Table 6.2
Euler's method applied to (6.4.10) with step-size $h = 0.01$

t_n	x_n	y_n
0	1	8
0.01	1.08	8.01
0.02	1.159299	8.0208
0.03	1.237838674	8.03239299
⋮	⋮	⋮
2.09	0.934286677	7.878994865
2.1	1.022610614	7.888337731
2.11	1.110302289	7.898563837
2.12	1.197299927	7.90966686

In the first example with Euler's method we just completed, we observe one of the major weaknesses of the method: when a large number of steps are needed and some of the changes in x and y are large, a substantial amount of roundoff error enters the calculations. While more sophisticated numerical methods exist (and are studied in chapter 7), for now we limit ourselves to Euler's method in order to first get an intuitive feel for the numerical behavior of approximate solutions. Another example follows.

Example 6.4.1   For the system of initial-value problems given by

$$x' = y - x^3, \quad x(0) = 2$$
$$y' = x - y^3, \quad y(0) = -1 \tag{6.4.11}$$

estimate the solution to the IVP up to $t = 5$ using $h = 0.1$ and comment on the behavior of the trajectory.

Solution.   In the given problem, if we take the perspective that $x' = f(t, x, y)$ and $y' = g(t, x, y)$, then it follows that $f(t, x, y) = y - x^3$ and $g(t, x, y) = x - y^3$. Applying (6.4.9) with $h = 0.1$, we have

$$x_{n+1} = x_n + 0.1 \cdot (y_n - x_n^3)$$
$$y_{n+1} = y_n + 0.1 \cdot (x_n - y_n^3)$$

Beginning this iteration with $x_0 = 2$ and $y_0 = -1$, we generate the following table.

t_n	x_n	y_n
0	2	-1
0.1	1.1	-0.7
0.2	0.8969	-0.5557
0.3	0.769180708	-0.448849846
$\vdots$	$\vdots$	$\vdots$
4.7	0.994536765	0.994533281
4.8	0.995620126	0.995618024
4.9	0.996490144	0.996488877
5	0.997188297	0.997187534

In the table, we see behavior consistent with the fact that the equilibrium point $(1, 1)$ of the system is a stable attracting node. In addition, the numerical data is in agreement with the graphical behavior we expect based on the direction field in figure 6.4 where we first considered the given nonlinear system. This behavior is also seen in the following plot in figure 6.11, which shows the (x_n, y_n) data from $n = 0, \ldots, 50$ generated by Excel.

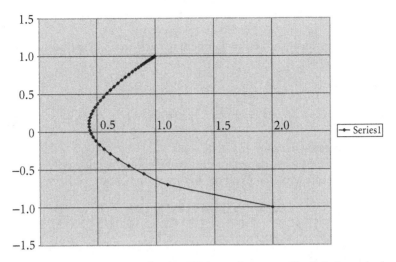

Figure 6.11 The trajectory for the IVP (6.4.11) generated by Euler's method with $h = 0.1$.

Example 6.4.1 shows that when small changes in t lead to very small changes in $x(t)$ and $y(t)$, such as near a stable, attracting node, Euler's method produces reasonable approximations without having to resort to extremely small h-values. We also see the importance of having a theoretical understanding of the expected behavior in advance of executing computations in order to check the reasonableness of our results.

6.4.1 Implementing Euler's method for systems in Excel

Just as we did for single initial-value problems in section 2.6.1, we will use Excel to generate approximate solutions to system IVPs. In this setting, given an initial value problem

$$\begin{aligned} x' &= f(x, y, t), & x(t_0) &= x_0 \\ y' &= g(x, y, t), & y(t_0) &= y_0 \end{aligned} \tag{6.4.12}$$

we seek approximations $x_1, x_2, \ldots$ and $y_1, y_2, \ldots$ such that $(x_n, y_n) \approx (x(t_n), y(t_n))$, where $t_{n+1} = t_n + h$ for some chosen step-size h. In particular, we have shown that these approximations are generated using Euler's method by the rule

$$\begin{aligned} x_{n+1} &= x_n + h \cdot f(t_n, x_n, y_n) \\ y_{n+1} &= y_n + h \cdot g(t_n, x_n, y_n) \end{aligned} \tag{6.4.13}$$

In a spreadsheet, we will view the following data: step number n, stepsize h, t_n, x_n, y_n, $f(t_n, x_n, y_n)$, and $g(t_n, x_n, y_n)$, where t_n is the value of the independent variable and $(x_n, y_n) \approx (x(t_n), y(t_n))$ is an estimate to the solution to the IVP at the value t_n. This data will appear in a given row where the row contains all these values for the corresponding n-value. From this, we naturally build subsequent approximations (x_{n+1}, y_{n+1}) based on the preceding row.

We will demonstrate the development of such an Excel spreadsheet for the particular example

$$x' = y - x^3, \quad x(0) = 2$$
$$y' = x - y^3, \quad y(0) = -1 \tag{6.4.14}$$

that we investigated in example 6.4.1.

To begin, we establish names for the various columns, say in cells A1 through G1, and see on our screen in Excel the information below.

	A	B	C	D	E	F	G
1	n	h	t__n	x__n	y__n	f(x__n,y__n)	g(x__n,y__n)

In most of the examples we consider with Euler's method, the system will be autonomous (i.e., t is implicit in the functions f and g), and therefore we choose to omit t from the column labels for $f(t_n, x_n, y_n)$ and $g(t_n, x_n, y_n)$.

In the subsequent row 2, we now enter the given data at step zero. In particular, in cell A2 we enter the step number ("0"), in B2 the chosen stepsize ("0.1"), in C2 the starting t-value ("0"), in D2 the starting x-value ("2"), and in E2 the starting y-value ("-1"). Next, in F2, we apply the function $f(t, x, y)$ to get the slope at the point at this step. That is, since in this IVP $f(t, x, y) = y - x^3$, we enter in F2 the command "= E2 - D2^3". Similarly, since $g(t, x, y) = x - y^3$, in G2 we enter "= D2 - E2^3". Now our spreadsheet appears as follows.

	A	B	C	D	E	F	G
1	n	h	t__n	x__n	y__n	f(x__n,y__n)	g(x__n,y__n)
2	0	0.1	0	2	-1	-9	3

In the next row, row 3, we may now build subsequent entries based on existing data. To increase the step number, in A3 we enter "= A2 + 1". Since the step-size stays constant throughout, in B3 we input "= B2". Since the next t-value will be the preceding t-value plus the stepsize ($t_1 = t_0 + h$), we enter in C3 the command "= C2 + B2".

To compute the next x-value in cell D3 from Euler's method, we know that $x_1 = x_0 + hf(t_0, x_0, y_0)$. Hence, in D3 we write "= D2 + B2*F2". Similarly, to compute $y_1 = y_0 + hg(t_0, x_0, y_0)$, in cell E3 we enter "= E2 + B2*G2".

Finally, we also need values of $f(t_1, x_1, y_1)$ and $g(t_1, x_1, y_1)$ for use in the following step. This involves simply updating the functions $f(t, x, y)$ and $g(t, x, y)$ at the given t-, x-, and y-values, so we select cell F2, copy it, and paste it into cell F3. Equivalently, we can directly enter in F3 "= E3 - D3^3".

We can similarly copy G2 into G3, or in G3 enter "= D3 - E3^3". Below is the current state of our spreadsheet.

	A	B	C	D	E	F	G
1	n	h	t_n	x_n	y_n	f(x_n,y_n)	g(x_n,y_n)
2	0	0.1	0	2	-1	-9	3
3	1	0.1	0.1	1.1	-0.7	-2.031	1.443

Now we can harness the power of Excel to compute as many subsequent steps as we like. By using the mouse to highlight row 3, and then placing the cursor on the bottom right corner of cell E3, we can click and drag downward to fill subsequent rows with similar calculations. For example, doing so through row 7 yields the following.

	A	B	C	D	E	F	G
1	n	h	t_n	x_n	y_n	f(x_n,y_n)	g(x_n,y_n)
2	0	0.1	0	2	-1	-9	3
3	1	0.1	0.1	1.1	-0.7	-2.031	1.443
4	2	0.1	0.2	0.8969	-0.5557	-1.2771929	1.0685015
5	3	0.1	0.3	0.7691807	-0.4488498	-0.9039271	0.8596087
6	4	0.1	0.4	0.6787879	-0.3628889	-0.6756426	0.7265762
7	5	0.1	0.5	0.6112237	-0.2902313	-0.5185811	0.6356711

As we have noted previously, besides the relative simplicity of these computations, there are further advantages Excel offers. One is that changing one appropriately chosen cell will update all of our computations. For example, if we are interested in the change induced by a different step-size, say $h = 0.01$, all we need to do is enter "0.01" in cell B2, and every other cell will update accordingly. In addition, if we desire to see the graphical results of our work, we can use Excel's Chart Wizard.

To plot the trajectory generated by our approximations, we can simultaneously highlight the x and y columns in our chart above (cells C2 through C7 and D2 through D7), and then go to Insert menu and select Chart (alternatively, we may click on the Chart Wizard icon on the toolbar). In the prompt window that arises, we choose "XY (Scatter)" and select one of the graph style options at the right. By clicking "Next" in a few subsequent windows (in which advanced users can avail themselves of more options), we eventually get to a final window where our graph appears and the option to "Finish." Clicking on "Finish," the graph will appear in the spreadsheet and may be moved around by clicking and dragging it accordingly. We see the resulting plot displayed as in figure 6.12.

Exercises 6.4

In exercises 1–7, use Euler's method with the stated h-value to estimate the solution of the given system of IVPs at the given t-value. Compare your work to

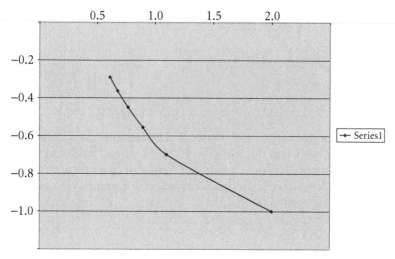

Figure 6.12   An Excel plot of an approximate solution to the IVP (6.4.14).

a plot of the direction field for the system and the classification of any relevant
equilibrium solutions.[3]

1. $x' = y - 2xy$, $x(0) = 0.75$ $t = 1, h = 0.1$
   $y' = 4xy - x$, $y(0) = 0.5$

2. $x' = 4 - y^2$, $x(0) = -2$ $t = 1, h = 0.05$
   $y' = 1 - x + y$, $y(0) = -1$

3. $x' = \cos y$, $x(0) = 2$ $t = 1, h = 0.1$
   $y' = 1 - \sin x$, $y(0) = 3$

4. $x' = 2x - y$, $x(0) = 1$ $t = 1, h = 0.1$
   $y' = -4x + 2y$, $y(0) = 1$

5. $x' = e^{-y}$, $x(0) = 0$ $t = 1, h = 0.05$
   $y' = 1/(1 + x^2)$, $y(0) = 0$

6. $x' = \ln(2 + y)$, $x(0) = -1$ $t = 1, h = 0.1$
   $y' = x^2 + y$, $y(0) = -0.5$

7. $x' = y - x^2$, $x(0) = 1$ $t = 1, h = 0.05$
   $y' = x - 8y^2$, $y(0) = 0.75$

---

[3] In the exercises of section 6.2, equilibrium solutions were found and direction fields were plotted in
exercises 1–7, which correspond to the same systems of differential equations given here. Similarly, in
section 6.3, equilibrium solutions were classified through linearization in exercises 11–17, which also
correspond to these systems.

8. Recall from section 6.1 that the nonlinear system of differential equations

$$W' = -0.75W + 0.25MW$$

$$M' = 0.5M - 0.1MW$$

models the numbers of wolves and moose (each measured in hundreds), in a predator-prey model where time is measured in years. Assume that at time $t = 0$ there are 250 moose and 550 wolves. Estimate the numbers of moose and wolves present at $t = 3, 6,$ and 9 years using a step-size of (a) $h = 0.1$, and (b) $h = 0.01$. Discuss your findings and describe the behavior of the trajectory.[4]

6.5 For further study

6.5.1 The damped pendulum

In our development of the pendulum equation, we learned that for a pendulum with an arm of length L and bob of mass m, the angle θ that the arm forms with the positive x-axis at time t satisfies the IVP

$$L\theta'' = -g\sin\theta, \quad \theta(0) = \theta_0, \ \theta'(0) = \theta_0' \tag{6.5.1}$$

provided that we assume no friction is present in the screw from which the pendulum hangs and there is no air drag on the bob. Here, we investigate the effects of such resistance on the pendulum's behavior.

(a) Under the natural assumption that the friction or damping that is present is directly proportional to the velocity of the bob along the arc of motion, explain why it follows the pendulum is governed by the IVP

$$L\theta'' = -g\sin\theta - c\theta', \quad \theta(0) = \theta_0, \ \theta'(0) = \theta_0' \tag{6.5.2}$$

where c is the damping constant.

(b) Using the standard change of variables, convert the nonlinear second-order IVP (6.5.2) to a nonlinear system of first-order IVPs. Write the system in the form $\mathbf{x}' = \mathbf{F}(\mathbf{x})$ for an appropriate function $\mathbf{F}$.

(c) Determine all equilibrium solutions of the system in (b). Are the equilibria different from those of the undamped pendulum?

(d) Let a given pendulum have an arm of length $L = 1$ m, and recall that $g = 9.8$ m/sec^2. For each of the c-values $c = 0.5, c = 1, c = 2$, and $c = 5$, plot the direction field for the system in (b) as well as trajectories that correspond to the stated initial conditions below. For each plot, discuss the

---

[4] In the exercises of section 6.2, equilibrium solutions were found and the direction field was plotted for this system in exercise 8.

behavior of the pendulum over time and how damping affects the observed behavior.

   (i) $\theta(0) = 2,\ \theta'(0) = 0$
   (ii) $\theta(0) = 4,\ \theta'(0) = 0$
   (iii) $\theta(0) = 2,\ \theta'(0) = 10$
   (iv) $\theta(0) = 2,\ \theta'(0) = -10$

In addition, be sure to discuss the physical interpretation of each set of initial conditions and how these conditions affect the trajectories.

(e) Using $c = 1$, find the linear approximation of the system in (b) at two different equilibrium points, one that is stable and another that is unstable. Discuss the graphical behavior of the two linear systems you find near the equilibrium points and how this compares to the plot of the corresponding direction field in (d).

(f) Again using $c = 1$ and $L = 1$, apply Euler's method with $h = 0.01$ to the system in (b) with the initial conditions $\theta(0) = 2,\ \theta'(0) = 10$. Experiment with how many steps are needed in order to have the approximations approach the stable equilibrium $(2\pi, 0)$, plot the approximations you compute, and compare the results to the appropriate direction field in (d).

6.5.2 Competitive species

In our development of the predator–prey equations, we used the fundamental assumption that the prey population would, in the absence of a predator, grow according to an exponential model, and similarly that the predator would decay exponentially if no prey is available. These hypotheses led us to equations of the form

$$\begin{aligned} x' &= ax - cxy \\ y' &= -by + dxy \end{aligned} \tag{6.5.3}$$

where x is the prey population and y represents the number of predators. Recall that the terms $-cxy$ and dxy represent a fraction of the number of predator–prey interactions that are, respectively, harmful or beneficial to the two species.

    In what follows, we consider a similar scenario where, instead of one species preying on the other, two species are competing for resources. In this setting, species interactions (modeled by "xy") are harmful to both species. In addition, rather than assuming exponential growth or decay for the individual populations, we explore the affects of the assumption that each population on its own grows logistically.

(a) Assume that in the absence of another species competing for resources, the population $x(t)$ grows according to the logistic model

$$x' = ax\left(1 - \frac{x}{A}\right)$$

where a and A are positive constants (a is the population's growth constant and A is its carrying capacity). Similarly, for a second population $y(t)$, assume that without another competing species present $y(t)$ is governed by the model

$$y' = by\left(1 - \frac{y}{B}\right)$$

where b and B are positive constants.

By viewing a fraction of the interactions xy as harmful, we can subtract from each of the above differential equations a term proportional to xy – say αxy from x' and βxy from y' – to account for this competition. Do so, and show that the populations $x(t)$ and $y(t)$ satisfy the system of equations given by

$$\begin{aligned} x' &= ax(1 - \tfrac{1}{A}x - \tfrac{\alpha}{a}y) \\ y' &= by(1 - \tfrac{1}{B}y - \tfrac{\beta}{b}x) \end{aligned} \qquad (6.5.4)$$

(b) Throughout the remaining questions, we assume that x and y represent populations measured in thousands. We explore the impact of different constants in the equations, as well as various initial conditions. In (6.5.4), let $a = 0.5$, $b = 0.25$, $A = 5$, $B = 2$, $\alpha = 0.04$, and $\beta = 0.02$. Find all equilibrium points of the system. (Hint: there are more than two equilibria.)

(c) At each of the equilibrium points determined in (b), compute the linearization of the system (6.5.4), and hence determine the stability of the equilibrium point.

(d) In an appropriate window, plot the direction field for the system (6.5.4) and discuss how the direction field supports your conclusions regarding the stability of various equilibrium points in (c). Discuss the long-term behavior of the two populations for several different initial conditions.

(e) With the initial conditions $x(0) = 2$, $y(0) = 2$, use Euler's method for systems to estimate the values of the populations at a range of time values. Use a step size of $h = 0.1$ and compare your results to the plot in (d).

(f) In (6.5.4), use the parameter values given in (b), except change the carrying capacity of the second population to $B = 15$. Respond to prompts (b), (c), (d), and (e) for this scenario and compare and contrast the updated system with the first one considered. In the new situation, which population will dominate in the long run? Why do you think this is the case?

(g) In (6.5.4), let $a = 0.5$, $b = 0.25$, $A = 5$, and $B = 2$, but now adjust the parameters α and β to reflect greater competition for resources by setting $\alpha = 0.4$, and $\beta = 0.2$. Respond to prompts (b), (c), (d), and (e) for this scenario and compare and contrast the updated system with the first one considered. In the new situation, which population is more likely to dominate in the long run? For which initial conditions is the weaker population able to survive?

(h) Suppose there are three different species x, y, and z, all competing for resources. Under the assumption that population interactions xy and xz are harmful to x, and so on, what system of differential equations models the behavior of the three species?

7

Numerical methods for differential equations

7.1 Motivating problems

In previous chapters, we have learned to solve a wide range of differential equations. Primarily, our focus has been on linear differential equations: first-order linear equations, higher order linear equations with constant coefficients, and systems of linear equations with constant coefficients. Indeed, we have learned through a variety of techniques that under the proviso that a differential equation or system is linear, we can almost always find a solution.

The situation is much more complicated for nonlinear equations. For example, while we can use an integrating factor to solve the linear first-order differential equation $y' + y = t$, if we replace y by y^2, the differential equation

$$y' + y^2 = t \tag{7.1.1}$$

is no longer linear. In addition, (7.1.1) is not separable, nor is it exact. With none of our established analytical methods available, it appears that we cannot solve this differential equation. If faced with the related initial-value problem

$$y' + y^2 = t, \qquad y(0) = 1 \tag{7.1.2}$$

we know that we can visually approximate a solution by plotting the direction field that corresponds to the differential equation. Moreover, we learned in section 2.6 that we can generate a sequence of estimates of the values of the solution $y(t)$ at discrete t-values separated by a step-size h according to the rule

$$t_{n+1} = t_n + h \quad \text{and} \quad y_{n+1} = y_n + hf(t_n, y_n), \text{ for } n \geq 0 \tag{7.1.3}$$

The algorithm that generates this sequence of approximations is called Euler's method.

We encounter the same difficulties with higher order differential equations. While we can solve almost any higher order linear equation with constant coefficients, such as

$$y'' + a_1 y' + a_0 y = f(t)$$

nonlinear equations are much more difficult. For instance, as discussed in section 6.1, a simple pendulum may be modeled by the nonlinear second-order initial-value problem

$$\theta'' + \frac{g}{L}\sin\theta = 0, \qquad \theta(0) = \theta_0, \quad \theta'(0) = \theta_1 \qquad (7.1.4)$$

where $\theta(t)$ is the angle the arm of the pendulum forms with a vertical axis at time t. In chapter 6, we introduced several different approaches to approximate the solution to (7.1.4); each was based on converting the second-order equation to a system of first-order equations and approximating the solution to the resulting system.

Finally, nonlinear systems of differential equations are important in their own right. A prominent example is the predator–prey equations, discussed in detail in section 6.1, where two populations $M(t)$ and $W(t)$ (in hundreds) are modeled by the following system of nonlinear first-order initial-value problems:

$$\begin{aligned} W' &= W(-0.75 + 0.25M), & W(0) &= 3 \\ M' &= M(0.5 - 0.1W), & M(0) &= 7 \end{aligned} \qquad (7.1.5)$$

As with the pendulum, the nonlinearity of these equations makes determining an analytical solution (i.e., formulas for $W(t)$ and $M(t)$) impossible, and therefore we must instead be content to find approximate solutions. In section 6.4, we introduced an extension of Euler's method that can be used to produce some basic approximations to the solution of a system of nonlinear initial-value problems such as (7.1.5).

But through a variety of examples considered in sections 2.6 and 6.4, we have seen that Euler's method has a big downside: each step produces significant error, and each step compounds the error from the preceding step. To get an accurate approximation using Euler's method, a very small step-size h is usually needed. With modern computing power so readily available, we might be tempted to simply take very small h-values in this approach and be content to do thousands of computations to get estimates of solutions. But taking smaller and smaller values of h proves to be an unsatisfactory approach for many reasons, perhaps most significantly because of the fact that as numbers get extremely small, computers have great difficulty distinguishing them from zero and major round-off errors can result.

Instead, we will seek to develop approaches in the spirit of Euler's method, but more sophisticated in that they naturally reduce the error that comes from using a step of $h = \triangle t$. Our goal is to develop numerical methods for initial-value problems (for first-order, higher order, and systems) that, given a step-size h, produce an accurate approximate solution to the initial-value problem. We desire that the methods give reasonably good approximations for small (but not too small) values of h, while at the same time not requiring too many calculations. In the upcoming sections, we will discuss problems of the nature of (7.1.2), (7.1.4), (7.1.5), and more, and develop and apply algorithms that produce acceptable approximations to solutions.

7.2 Beyond Euler's method

To approach an initial-value problem that we cannot solve by standard techniques, such as separation of variables or integrating factors, we have learned that one option is to use Euler's method. Given the IVP

$$y' = f(t, y), \qquad y(t_0) = y_0$$

this algorithm generates a sequence of points (t_1, y_1), (t_2, y_2), ..., (t_n, y_n) according to the rule

$$y_{n+1} = y_n + hf(t_n, y_n) \quad \text{for } n \geq 0 \tag{7.2.1}$$

where $t_{n+1} = t_n + h$. Each y_n is an approximation to the value of the actual solution y at the value t_n. That is, $y(t_n) \approx y_n$.

Euler's method is developed by using the standard tangent line approximation in calculus. While this is instructive and intuitive, the method is the least accurate of many other available methods. In this section, we begin to develop algorithms beyond Euler's method in an effort to increase the accuracy of our approximations while actually decreasing the number of computations we execute.

Before we develop new approaches, we first revisit some important concepts from numerical integration in calculus. These ideas not only remind us of key issues in approximation techniques, but also inform our efforts to approximate solutions to initial-value problems. Given a continuous function $f(t)$ on an interval $[t_0, t_0 + h]$, there are several basic approximations to $\int_{t_0}^{t_0+h} f(t)\, dt$. Specifically,

$$\int_{t_0}^{t_0+h} f(t)\, dt \approx h \cdot f(t_0) \qquad \text{(left endpoint rule)}$$

$$\int_{t_0}^{t_0+h} f(t)\, dt \approx h \cdot f(t_0 + h) \qquad \text{(right endpoint rule)}$$

$$\int_{t_0}^{t_0+h} f(t)\, dt \approx h \cdot \frac{f(t_0)+f(t_0+h)}{2} \qquad \text{(trapezoid rule)}$$

$$\int_{t_0}^{t_0+h} f(t)\, dt \approx h \cdot f\left(t_0 + \frac{h}{2}\right) \qquad \text{(midpoint rule)}$$

It is a standard exercise in calculus to show that the left and right endpoint rules are the least accurate approximations of the four, while the midpoint rule is the best. While one can make sophisticated arguments using Taylor series to justify claims about the size of the error in such an approximation, visual arguments are just as convincing: sampling f at the midpoint of the interval usually balances the behavior of the function and leads to the best approximation of the integral of the four options above.

There is a direct link between the numerical approximation of definite integrals and numerical methods to estimate solutions to initial-value problems such as Euler's method. Given the IVP

$$y'(t) = f(t, y), \qquad y(t_0) = y_0$$

if we integrate both sides of the differential equation with respect to t from $t = t_0$ to $t = t_0 + h$ for some $h > 0$, then

$$\int_{t_0}^{t_0+h} y'(t)\, dt = \int_{t_0}^{t_0+h} f(t, y(t))\, dt \qquad (7.2.2)$$

Integrating the left side of (7.2.2), we have

$$y(t_0 + h) - y(t_0) = \int_{t_0}^{t_0+h} f(t, y(t))\, dt$$

or equivalently

$$y(t_0 + h) = y(t_0) + \int_{t_0}^{t_0+h} f(t, y(t))\, dt \qquad (7.2.3)$$

Estimating the integral in (7.2.3) with the left endpoint rule,

$$y(t_0 + h) \approx y(t_0) + hf(t_0, y(t_0)) \qquad (7.2.4)$$

Using the initial condition $y(t_0) = y_0$, it follows that

$$y(t_0 + h) \approx y_0 + hf(t_0, y_0) \qquad (7.2.5)$$

which is precisely the first step in Euler's method. That is, we have shown in our efforts to step from $t = t_0$ to $t = t_0 + h$ along the solution $y(t)$ that this process can be equivalently achieved by estimating the value of a definite integral. Moreover, Euler's method can be viewed as arising naturally from estimating the required definite integral through a left endpoint rule.

As such, it is not surprising that Euler's method is not an accurate approach, for neither is the left endpoint rule for approximating integrals. The availability of the trapezoid and midpoint rules as better approximations leads us to consider two improvements upon Euler's method.

7.2.1 Heun's method

To improve on Euler's method, we return to (7.2.3), and instead estimate the definite integral on the right-hand side with the trapezoid rule. Doing so, we find

$$y(t_0 + h) \approx y(t_0) + h \cdot \frac{f(t_0, y(t_0)) + f(t_0 + h, y(t_0 + h))}{2} \qquad (7.2.6)$$

The difficulty in (7.2.6) is that the last term in the approximation on the right-hand side involves $y(t_0 + h)$, the very quantity we are trying to estimate. One way to view what is occurring in this approach is that we are trying to use not only the slope at (t_0, y_0), computed as $f(t_0, y_0)$, but also the slope at $(t_0 + h, y(t_0 + h))$. While we do not know $y(t_0 + h)$ exactly, we can estimate this value using Euler's method. In particular, if we use the fact that $y(t_0) = y_0$ and employ the Euler approximation $y(t_0 + h) \approx y_0 + hf(t_0, y_0)$, then from (7.2.6) we find that

$$y(t_0 + h) \approx y_0 + h \cdot \frac{f(t_0, y_0) + f(t_0 + h, y_0 + hf(t_0, y_0))}{2} \qquad (7.2.7)$$

Generalizing (7.2.7) to the situation where we are moving from the known approximation $y(t_n) \approx y_n$ at point (t_n, y_n) to a new approximation (t_{n+1}, y_{n+1}) with $t_{n+1} = t_n + h$, we have developed *Heun's method* given by

$$y_{n+1} = y_n + h \cdot \frac{f(t_n, y_n) + f(t_{n+1}, y_n + hf(t_n, y_n))}{2} \qquad (7.2.8)$$

Because this algorithm is more complicated than Euler's method, some additional notation can assist us in its implementation. We first let

$$a_n = f(t_n, y_n) \qquad (7.2.9)$$

which is the slope of the solution curve at (t_n, y_n) given by the IVP. We observe that the expression a_n arises twice in (7.2.8), and that we also have to compute $f(t_{n+1}, y_n + ha_n)$. We therefore let

$$b_n = f(t_{n+1}, y_n + ha_n) \qquad (7.2.10)$$

It follows that Heun's method is then executed by computing

$$y_{n+1} = y_n + h \cdot \frac{a_n + b_n}{2} \qquad (7.2.11)$$

In this light, we see that Heun's method uses the average of two slopes (the slope at (t_n, y_n) and the approximate slope at (t_{n+1}, y_{n+1})) in order to predict the next value of the solution $y(t)$. We consider an example to demonstrate how Heun's method is implemented and to contrast its results with those from Euler's method.

Example 7.2.1   Execute ten steps of Heun's method with $h = 0.1$ to find an approximate solution of the initial-value problem

$$y' = 2t(2 - y), \qquad y(0) = 1$$

Compare the results to Euler's method as well as the exact solution of the IVP.

Solution.   Note first that the given differential equation is both linear and separable. The exact solution of the IVP is $y(t) = 2 - e^{-t^2}$.

To apply Heun's method, we must compute a_n, b_n, and y_n at each step. To begin, $a_0 = f(t_0, y_0)$. From the stated IVP, $f(t, y) = 2t(2 - y)$ and $(t_0, y_0) = (0, 1)$. Thus,

$$a_0 = 2 \cdot 0 \cdot (2 - 1) = 0$$

In addition, $b_0 = f(t_1, y_0 + ha_0)$, so

$$b_0 = 2 \cdot 0.1 \cdot (2 - (1 + 0.1 \cdot 0)) = 0.2$$

With both a_0 and b_0 calculated, we can now determine y_1 to be

$$y_1 = y_0 + \frac{h}{2}(a_0 + b_0) = 1 + \frac{0.1}{2}(0 + 0.2) = 1.01$$

Repeating these same steps to determine y_2, we find that

$$a_1 = f(t_1, y_1) = f(0.1, 1.01) = 2 \cdot 0.1 \cdot (2 - 1.01) = 0.198$$

and

$$b_1 = f(t_2, y_1 + ha_1) = f(0.2, 1.01 + 0.1 \cdot 0.198)$$
$$= 2 \cdot 0.2 \cdot (2 - 1.0298) = 0.38808$$

so that

$$y_2 = y_1 + \frac{0.1}{2}(a_1 + b_1) = 1.01 + 0.05(0.198 + 0.38808) = 1.039304$$

Implementing the remaining computations in a program such as Excel, it follows that we can generate the values shown in table 7.1. Included in the table are the approximations generated by Euler's method, as well as the errors resulting from both methods which are computed by comparison to the exact solution of the IVP. For simplicity, we report the results from every other step in each algorithm.

Table 7.1
Euler's method and Heun's method applied to the IVP $y' = 2t(2 - y)$, $y(0) = 1$, using $h = 0.1$

	Euler	Heun	Solution	Euler error	Heun error
t_n	y_n	y_n	$y(t_n)$	$|y(t_n) - y_n|$	$|y(t_n) - y_n|$
0	1	1	1	0	0
0.2	1.02	1.039304	1.039210561	0.019989439	0.000093439
0.4	1.115648	1.147959794	1.147856211	0.038539949	0.000103583
0.6	1.267756544	1.302226785	1.302323674	0.053302085	0.000096889
0.8	1.445838152	1.472149858	1.472707576	0.061796472	0.000557718
1	1.618293319	1.630946606	1.632120559	0.062514097	0.001173953

Obviously, Heun's method is a major improvement over Euler's method. In fact, given that we use the Euler approximation at each step to help forecast the next slope encountered, it is somewhat remarkable how accurate Heun's method is. It can be shown rigorously that the error in Heun's method is a significant improvement over Euler's method by relating the error in the approximation to the step-size h; it turns out[1] that the error in Euler's method is proportional to h^2, while the error in Heun's method is proportional to h^3. Finally, we might observe that it appears unusual that the error in Heun's method actually drops from $t_4 = 0.4$ to $t_6 = 0.6$, and that the growth in the error slows in Euler's method at the same stage. This is due to the fact that the solution function $y(t) = 2 - e^{-t^2}$ is an increasing function whose concavity changes (from concave up to concave down) at the point $t = 1/2$; the change in concavity allows the linear approximations to temporarily catch up, instead of having the error continue to increase at an increasing rate.

We have seen that Heun's method is developed using an application of the trapezoid rule in numerical integration. We consider another similar method (based on the midpoint rule) before introducing more sophisticated techniques in section 7.3.

7.2.2 Modified Euler's method

The midpoint rule is normally more accurate than the trapezoid rule.[2] Given our experience with Heun's method and its connection to the trapezoid rule, it makes sense to see if we can develop a related method that uses the perspective of the midpoint rule.

Recalling (7.2.3),

$$y(t_0 + h) = y(t_0) + \int_{t_0}^{t_0+h} f(t, y(t)) \, dt$$

if we use the midpoint rule to estimate the integral, then we have to evaluate the integrand at the midpoint $t_0 + h/2$ of the interval $[t_0, t_0 + h]$. Doing so,

$$y(t_0 + h) \approx y(t_0) + hf\left(t_0 + \frac{h}{2}, y\left(t_0 + \frac{h}{2}\right)\right) \qquad (7.2.12)$$

As with Heun's method, in the context of trying to solve the IVP $y' = f(t, y)$, $y(t_0) = y_0$, only $y(t_0)$ is known. Thus, we do not know—and therefore have to estimate—the value of $y(t_0 + h/2)$ in (7.2.12). We again employ Euler's method and write

$$y\left(t_0 + \frac{h}{2}\right) \approx y(t_0) + \frac{h}{2}f[t_0, y(t_0)] \qquad (7.2.13)$$

---

[1] A more formal analysis of errors that shows the dependence on powers of h is discussed in section 7.3.

[2] On an interval where $f(x)$ has consistent concavity, the midpoint rule is approximately twice as accurate as the trapezoid rule.

Substituting (7.2.13) in (7.2.12) and replacing $y(t_0)$ with y_0,

$$y(t_0 + h) \approx y_0 + hf\left[t_0 + \frac{h}{2}, y_0 + \frac{h}{2}f(t_0, y_0)\right] \qquad (7.2.14)$$

Generalizing (7.2.14) to the situation where we are moving from a known approximation $y(t_n) \approx y_n$ at point (t_n, y_n) to the next approximation at (t_{n+1}, y_{n+1}), we have developed the *Modified Euler method* given by

$$y_{n+1} = y_n + hf\left[t_n + \frac{h}{2}, y_n + \frac{h}{2}f(t_n, y_n)\right] \qquad (7.2.15)$$

As with Heun's method, some additional notation assists us in tracking our computations. Let $a_n = f(t_n, y_n)$ and

$$c_n = y_n + \frac{h}{2}a_n$$

so that

$$y_{n+1} = y_n + hf\left(t_n + \frac{h}{2}, c_n\right) \qquad (7.2.16)$$

We consider an example in order to see the implementation of the Modified Euler method and to compare its results to those of Heun's method. We again employ an IVP that we can solve exactly in order to compare the errors of the two methods.

Example 7.2.2   Consider the initial-value problem $y' = e^{2t} - y, y(0) = 1$. Apply the Modified Euler method to estimate the value of $y(1)$ using $h = 0.1$ and compare the results with Heun's method and the exact solution.

Solution.   Since $y' = e^{2t} - y$ is a linear first-order differential equation, we can find the general solution $y(t) = Ce^{-t} + \frac{1}{3}e^{2t}$, and hence the exact solution to the IVP is

$$y(t) = \frac{2}{3}e^{-t} + \frac{1}{3}e^{2t}$$

To begin the Modified Euler method, we know from the given IVP that $f(t, y) = e^{2t} - y$ and that $(t_0, y_0) = (0, 1)$. Thus, $a_0 = f(t_0, y_0) = e^{2 \cdot 0} - 1 = 0$. Next, we observe that $c_0 = y_0 + \frac{h}{2}a_0 = 1 + 0.05 \cdot 0 = 1$. To compute y_1, by (7.2.16) we have

$$y_1 = y_0 + hf\left(t_0 + \frac{h}{2}, c_0\right) = 1 + 0.1 \cdot (\exp 2(0 + 0.05) - 1)$$

$$= 1 + 0.1 \cdot 0.105170918 = 1.010517092$$

Continuing to the next step, $a_1 = f(t_1, y_1) = \exp(2 \cdot 0.1) - 1.010517092 = 0.210885666$. Next,

$$c_1 = y_1 + \frac{h}{2}a_1 = 1 + 0.05 \cdot 0.210885666 = 1.021061375$$

Table 7.2
Heun's method and Modified Euler's method (ME) applied to the IVP $y' = e^{2t} - y$, $y(0) = 1$ with $h = 0.1$

	Heun	ME	Solution	Heun error	ME error
t_n	y_n	y_n	$y(t_n)$	$\lvert y(t_n) - y_n \rvert$	$\lvert y(t_n) - y_n \rvert$
0	1	1	1	0	0
0.2	1.044572834	1.043396835	1.043095401	0.001477433	0.000301434
0.4	1.192009094	1.189291538	1.188727007	0.003282087	0.000564531
0.6	1.478251184	1.473408204	1.472580065	0.005671119	0.000828139
0.8	1.959569856	1.951698881	1.950563451	0.009006405	0.00113543
1	2.722082435	2.70981115	2.70827166	0.013810775	0.001539489

Finally,

$$y_2 = y_1 + hf\left(t_1 + \frac{h}{2}, c_1\right) = 1.010517092 + 0.1 \cdot (\exp2(0.1 + 0.05) - 1.021061375)$$

$$= 1.010517092 + 0.1 \cdot 0.328797432 = 1.043396835$$

Executing eight more steps using a computer, we find the results in table 7.2. We also show the results from Heun's method in order to make a comparison between the two approaches we have developed beyond Euler's method, again reporting the results from every other step.

From the table, we see that the Modified Euler method is an improvement over Heun's method. This is not too surprising since the former stems from the midpoint rule for integration, while the latter from the trapezoid rule. In addition, if we plot the exact solution function, we see that the solution is always increasing and concave up over the interval of interest; in the presence of such consistent concavity in the solution function, the midpoint rule will generate noticeably more accurate approximations than will the trapezoid rule.

Obviously Heun's method and the Modified Euler method are substantial improvements over the standard Euler's method. Not only are their errors much smaller, but the errors grow less quickly. To better understand why this is so, observe that Euler's method relies solely on presently available data in generating its estimates. That is, the method takes an approach that relies on just one data point in order to proceed to the next approximation. Our two newest methods instead look into the future: rather than using the current point and the slope at that location, they use the current point and an estimate of the slope at a point that is ahead of our current location. We create these estimates using only the currently available data, but the approaches lead to a substantial increase in accuracy that makes us hopeful for significant improvements through other predictive approximation techniques that we are yet to investigate.

Exercises 7.2   In exercises 1–10, use (a) Euler's method, (b) Heun's method, and (c) the Modified Euler method to estimate $y(1)$ using $h = 0.1$, and compare the approximations generated by the three methods. In exercises 1–6, compare the approximations with the exact solution.

1. $y' + 2ty = 0$, $y(0) = -2$

2. $y' = 2y - 1$, $y(0) = 2$

3. $y' - y = 0$, $y(0) = 2$

4. $(y')^2 - 2y = 0$, $y(0) = 2$

5. $y' - y^2 = 1$, $y(0) = 0$

6. $tyy' = -1 - y^2$, $y(0) = 2$

7. $y' + ty = t^2$, $y(0) = 1$

8. $y' + y^2 = t$, $y(0) = 1$

9. $y' + \sin y = 2e^{-t}$, $y(0) = 0$

10. $y' = 2e^{t/2} \sin \sqrt{y}$, $y(0) = 0$

7.3 Higher order methods

In calculus, we learn that if $F(x)$ is a function with $n+1$ derivatives in an interval surrounding a value $x = a$, then F has a Taylor polynomial expansion that obeys the relationship

$$F(x) = F(a) + F'(a)(x-a) + \frac{F''(a)}{2!}(x-a)^2 + \cdots + \frac{F^{(n)}(a)}{n!}$$

$$+ \frac{F^{(n+1)}(\zeta_x)}{(n+1)!}(x-a)^{n+1} \tag{7.3.1}$$

which is valid for x-values in an interval surrounding a and ζ_x is a number within that interval that depends on x. If we think of our interest in the solution $y(t)$ of an initial-value problem, assuming that y is sufficiently differentiable, the Taylor series expansion of y provides insight into errors that arise in approximation schemes. In (7.3.1), if we replace F by y, a by t_0, and x by $t_0 + h$, noting that $x - a = h$, it follows that

$$y(t_0 + h) = y(t_0) + hy'(t_0) + \frac{h^2}{2!}y''(t_0) + \cdots + \frac{h^n}{n!}y^{(n)}(t_0) + O(h^{n+1}) \tag{7.3.2}$$

where by "$O(h^{n+1})$" we mean "of order h^{n+1} or "proportional to h^{n+1}."

From (7.3.2), we can discern the so-called *truncation error* of certain methods. For example, if we use the approximation

$$y(t_0 + h) \approx y(t_0) + h y'(t_0) \tag{7.3.3}$$

which corresponds to Euler's method,[3] we see that the truncation error is proportional to h^2 from the equation $y(t_0 + h) = y(t_0) + h y'(t_0) + O(h^2)$. We therefore say that Euler's method is *first-order*, in reference to the highest power of h present in (7.3.3).

Since we use a small step-size h, it is evident that higher order methods are superior: in the error due to truncation, higher powers of h will approach zero faster. In what follows, we will investigate second-, third-, and fourth-order approaches. The first two arise through using the Taylor series expansion directly, and are therefore called *Taylor methods*.

7.3.1 Taylor methods

To employ a second-order Taylor method, from (7.3.2) we must be able to compute

$$y(t_0 + h) \approx y(t_0) + h y'(t_0) + \frac{h^2}{2} y''(t_0) \tag{7.3.4}$$

In a standard initial-value problem, we are given $y' = f(t, y)$ (plus an initial condition), so we can compute y'' from the form of the differential equation. In particular, since

$$y'(t) = f(t, y(t))$$

the chain rule for functions of two variables,[4] implies that

$$
\begin{aligned}
y''(t) &= \frac{d}{dt}\big[f(t, y(t))\big] \\
&= f_t(t, y)\frac{d}{dt}[t] + f_y(t, y)\frac{d}{dt}[y] \\
&= f_t(t, y) + f_y(t, y)y' \\
&= f_t(t, y) + f_y(t, y)f(t, y) \tag{7.3.5}
\end{aligned}
$$

Combining (7.3.5) with (7.3.4), we have developed the *second-order Taylor method* given by

$$y(t_0 + h) \approx y(t_0) + h f(t_0, y_0) + \frac{h^2}{2}[f_t(t_0, y_0) + f_y(t_0, y_0)f(t_0, y_0)] \tag{7.3.6}$$

Generalizing (7.3.6) to the step from y_n to y_{n+1}, we find that

$$y_{n+1} = y_n + h f(t_n, y_n) + \frac{h^2}{2}[f_t(t_n, y_n) + f_y(t_n, y_n)f(t_n, y_n)] \tag{7.3.7}$$

---

[3] Observe that we are writing $y'(t_0)$, which is given by $f(t_0, y_0)$ in Euler's method.

[4] We are using the rule that if $f(x, y)$ is a differentiable function of x and y, and x and y are each differentiable functions of t, then $d/dt[f(x, y)] = f_x(x, y)dx/dt + f_y(x, y)dy/dt$.

where $y_n \approx y(t_n)$. We consider an example to demonstrate the implementation of this method and compare it to results previously considered.

Example 7.3.1   Execute ten steps of the second-order Taylor series method with $h = 0.1$ to find an approximate solution of the initial-value problem

$$y' = e^{2t} - y, \qquad y(0) = 1$$

Compare the results to those of Heun's method and to the exact solution.

Solution.   This is the same IVP that we considered in example 7.2.2 with Heun's method and the Modified Euler method. To employ (7.3.7), we first must compute $f_t(t, y)$ and $f_y(t, y)$. Since $f(t, y) = e^{2t} - y$, we know that $f_t(t, y) = 2e^{2t}$ and $f_y(t, y) = -1$. In addition, to simplify the implementation of the method, we use notation similar to Heun's method. We let $a_n = f(t_n, y_n)$, $r_n = f_t(t_n, y_n)$, and $s_n = f_y(t_n, y_n)$, so that

$$y_{n+1} = y_n + h a_n + \frac{h^2}{2}[r_n + s_n a_n]$$

Beginning with $t_0 = 0$ and $y_0 = 1$, observe that

$$a_0 = f(0, 1) = e^{2 \cdot 0} - 1 = 0$$

$$r_0 = f_t(0, 1) = 2e^{2 \cdot 0} = 2$$

$$s_0 = f_y(0, 1) = -1$$

We then have

$$y_1 = y_0 + h a_0 + \frac{h^2}{2}[r_0 + s_0 a_0]$$

$$= 1 + 0.1 \cdot 0 + \frac{0.1^2}{2}[2 - 1 \cdot 0]$$

$$= 1.01$$

Similarly, we can compute

$$a_1 = f(0.1, 1.01) = e^{2 \cdot 0.1} - 1.01 = 0.211402758$$

$$r_1 = f_t(0.1, 1.01) = 2e^{2 \cdot 0.1} = 2.442805516$$

$$s_1 = f_y(0.1, 1.01) = -1$$

and thus

$$y_2 = y_1 + h a_1 + \frac{h^2}{2}[r_1 + s_1 a_1]$$

$$= 1.01 + 0.1 \cdot 0.211402758 + \frac{0.1^2}{2}[2.442805516 - 1 \cdot 0.211402758]$$

$$= 1.04229729$$

Continuing these computations through ten steps, we find the results noted in table 7.3, which are listed for every other step. Note, too, that we have included

Table 7.3

Taylor's method and Heun's method applied to the IVP $y' = 2t(2-y)$, $y(0) = 1$ using $h = 0.1$

	Taylor	Heun	Solution	Taylor error	Heun error
t_n	y_n	y_n	$y(t_n)$	$|y(t_n) - y_n|$	$|y(t_n) - y_n|$
0	1	1	1	0	0
0.2	1.04229729	1.044572834	1.043095401	0.000798112	0.001477433
0.4	1.186750654	1.192009094	1.188727007	0.001976353	0.003282087
0.6	1.468880073	1.478251184	1.472580065	0.003699992	0.005671119
0.8	1.944339609	1.959569856	1.950563451	0.006223842	0.009006405
1	2.698337638	2.722082435	2.70827166	0.009934023	0.013810775

the results of Heun's method from its application to the same IVP with the same step-size $h = 0.1$.

From table 7.3, we can see that the errors in Heun's method and the second-order Taylor method are roughly proportionate and seem to grow at the same rate. This suggests that Heun's method may also be a second-order method—an assertion that may be proved by studying related higher order methods. In particular, Heun's method can be viewed as one of a collection of algorithms known as Runge–Kutta methods, which we will consider after some additional work with Taylor methods.

Having shown that we can use the Taylor series (7.3.2) to motivate the development of the second-order method (7.3.7), it is natural to wonder if we could extend this work further to a third-order method. This is desirable since if the error in our method is proportionate to h^4, then the method will be more accurate without having to use smaller values of h.

It is indeed possible to develop a third-order method, provided that the function $f(t, y)$ from the given IVP is sufficiently differentiable. In particular, in order to write

$$y(t_0 + h) \approx y(t_0) + hy'(t_0) + \frac{h^2}{2}y''(t_0) + \frac{h^3}{3!}y'''(t_0) \qquad (7.3.8)$$

we must compute the third derivative of y. From our earlier work (7.3.5), we know that

$$y'' = f_t(t, y) + f_y(t, y)f(t, y) \qquad (7.3.9)$$

Applying the chain rule to the first term in (7.3.9), along with the fact that $y' = f(t, y)$,

$$\frac{d}{dt}\left[f_t(t, y)\right] = f_{tt}(t, y)\frac{d}{dt}[t] + f_{ty}(t, y)\frac{d}{dt}[y]$$

$$= f_{tt}(t, y) + f_{ty}(t, y)f(t, y) \tag{7.3.10}$$

where the final step follows from using $y' = f(t, y)$. Using both the product rule and the chain rule on the second term in (7.3.9) and suppressing the "(t, y)" argument of each function present,

$$\frac{d}{dt}[f_y f] = f_y \frac{d}{dt}[f] + \frac{d}{dt}[f_y]f$$

$$= f_y(f_t + f_y f) + (f_{yt} + f_{yy}f)f$$

$$= f_y f_t + f_y^2 f + f_{yt}f + f_{yy}f^2 \tag{7.3.11}$$

Combining (7.3.10) and (7.3.11) and using the fact that $f_{ty} = f_{yt}$, we have shown that

$$y''' = f_{tt} + f_{ty}f + f_y f_t + f_y^2 f + f_{yt}f + f_{yy}f^2$$

$$= f_{tt} + 2f_{ty}f + f_y f_t + f_y^2 f + f_{yy}f^2 \tag{7.3.12}$$

From (7.3.12), we understand why we normally do not use third-order Taylor methods in practice: the computations are extremely cumbersome. Were we to attempt to write

$$y(t_0 + h) \approx y(t_0) + hy'(t_0) + \frac{h^2}{2}y''(t_0) + \frac{h^3}{3!}y'''(t_0)$$

in terms of the function f from the given IVP, we would have to compute

$$y(t_0 + h) \approx y_0 + hf + \frac{h^2}{2}(f_t + f_y f) + \frac{h^3}{3!}(f_{tt} + 2f_{ty}f + f_y f_t + f_y^2 f + f_{yy}f^2)$$

where each appearance of the function f or one of its partial derivatives is also being evaluated at the point (t_0, y_0). This combination of the determination of a large number of functions and the evaluation of each at every stage of an algorithm makes Taylor methods of orders higher than two unreasonable to use. Hence, we next introduce one of the most popular and effective numerical methods for the solution of IVPs (known as *Runge–Kutta methods*) that enable us to achieve higher order approximations without the difficulty of computing multiple partial derivatives and evaluating these functions repeatedly.

7.3.2 Runge–Kutta methods

Where higher order Taylor methods require finding partial derivatives of $y' = f(t, y)$ and evaluating these derivatives at each stage of the algorithm, Runge–Kutta methods seek to avoid using partial derivatives altogether, while still achieving the desired higher order accuracy. Instead, in Runge–Kutta methods the function f is evaluated at a greater number of points, essentially seeking to compute the slope at the current and future points in an effort to make as accurate a prediction as possible.

Formally, Runge–Kutta methods can be viewed as a generalization of Heun's method. Recall that in Heun's method we write

$$y_{n+1} = y_n + \frac{h}{2}(a_n + b_n)$$

where

$$a_n = f(t_n, y_n) \quad \text{and} \quad b_n = f(t_{n+1}, y_n + ha_n)$$

Rather than prescribing that we compute or estimate slopes at the points (t_n, y_n) and (t_{n+1}, y_{n+1}) and simply average them, a *two-stage Runge–Kutta method* takes an arbitrary combination of the function values $f(t_n, y_n)$ and $f(t_n + \alpha h, y_n + \beta h f(t_n, y_n))$. Specifically, we set

$$y_{n+1} = y_n + c_1 h f(t_n, y_n) + c_2 h f(t_n + \alpha h, y_n + \beta h f(t_n, y_n)) \qquad (7.3.13)$$

and then determine conditions on c_1, c_2, α, and β that guarantee the approximation generated by (7.3.13) is second-order through a comparison to the Taylor expansion of $y(t_n + h)$. It can be shown that among the infinitely many possible valid choices for c_1, c_2, α, and β, taking $\alpha = \beta = 1$ and $c_1 = c_2 = 1/2$ results in Heun's method, which justifies the fact that Heun's method is second-order.

Heun's method is an example of a *two-stage* Runge–Kutta method; two-stage refers to the fact that slopes are evaluated or estimated at two points. It is possible to achieve even higher order Runge–Kutta methods by generalizing the idea in (7.3.13). In particular, we can take arbitrary combinations of the values (or estimated values) of $f(t, y)$ at points in the interval $t_n \leq t \leq t_{n+1}$ and select the weights so that the approximation agrees with the Taylor series expansion for $y(t_n + h)$ up to, and including, the term involving h^4, h^5, or whatever accuracy we desire. The details of the rigorous development of such methods are complicated and unenlightening. But, a more intuitive approach can help us gain a better sense of why the Runge–Kutta method works so well and where the formulas used in the algorithm come from.

If we recall our development of Heun's method and the Modified Euler method, each was linked to the idea of numerically approximating a definite integral. Specifically, Heun's method is analogous to the trapezoid rule, and the Modified Euler method corresponds to the midpoint rule. The trapezoid rule and midpoint rule both give the exact value of the definite integral of any linear function; in addition, when a function has consistent concavity over an interval, the midpoint rule is roughly twice as accurate as the trapezoid rule and the errors in the midpoint and trapezoid rules have opposite signs. As such, it makes sense to take a weighted average of the two rules in an effort to cancel out the error of each. Computing the weighted average

$$\frac{2 \cdot MID + TRAP}{3}$$

results in a new method known as Simpson's rule that is a remarkably accurate approximation of the definite integral. In fact, it can be shown that Simpson's rule is exact for every cubic polynomial.

This same increase in accuracy can be accomplished through similar ideas in the numerical approximation of solutions to initial-value problems. Recalling our work with Heun's method (H) and the Modified Euler method (ME),

$$\text{H:} \quad y_{n+1} = y_n + \frac{f(t_n, y_n) + f(t_{n+1}, y_n + hf(t_n, y_n))}{2} \tag{7.3.14}$$

$$\text{ME:} \quad y_{n+1} = y_n + hf\left(t_n + \frac{h}{2}, y_n + \frac{h}{2} \cdot f(t_n, y_n)\right) \tag{7.3.15}$$

we note that each uses a different expression for $\triangle y$, the approximate change in $y(t)$ in moving from t_n to t_{n+1}. If we let

$$\triangle y_H = \frac{h}{2}[f(t_n, y_n) + f(t_{n+1}, y_n + hf(t_n, y_n))]$$

and

$$\triangle y_{ME} = hf\left(t_n + \frac{h}{2}, y_n + \frac{h}{2} \cdot f(t_n, y_n)\right)$$

then the analogy to Simpson's Rule for approximating the solution y to the IVP $y' = f(t, y)$, $y(t_0) = y_0$ is given by

$$y_{n+1} = y_n + \frac{2\triangle y_{ME} + \triangle y_H}{3} \tag{7.3.16}$$

Using (7.3.14) and (7.3.15) and letting $a_n = f(t_n, y_n)$, we have the approximation rule given by $y_{n+1} = y_n + \triangle y_S$ where

$$\triangle y_S = \frac{2}{3}hf\left(t_n + \frac{h}{2}, y_n + \frac{h}{2}a_n\right) + \frac{1}{3} \cdot \frac{h}{2}[a_n + f(t_{n+1}, y_n + ha_n)]$$

$$= \frac{h}{6}\left[a_n + 4f\left(t_n + \frac{h}{2}, y_n + \frac{h}{2}a_n\right) + f(t_{n+1}, y_n + ha_n)\right] \tag{7.3.17}$$

If we slightly modify this expression for $\triangle y_S$ in recognition of the fact that as we proceed across the interval we have more and more information available (and hence a better approximation of the slope to use), the fourth-order Runge–Kutta rule emerges. In particular, rather than rely on the value a_n at every stage in (7.3.17), we recognize that we are attempting to compute approximate slopes at not just the left endpoint, but also at the midpoint and right endpoint. It makes sense that we should use these approximations as they become available to us; for instance, when we compute the approximate slope at the right endpoint, we ought to use the approximate slope at the midpoint to do so. Furthermore, given that the midpoint slope is weighted at 4 and the others at 1 in the average given by (7.3.17), it is reasonable to invest additional effort ensuring that the midpoint slope is as accurate as possible.

As in Heun's method, the computations are easier to understand, track, and implement if we introduce some additional notation. In particular, letting

$$
\begin{aligned}
a_n &= f(t_n, y_n) && \text{slope at left endpoint} \\
b_n &= f(t_n + \tfrac{1}{2}h, y_n + \tfrac{1}{2}ha_n) && \text{slope at midpoint} \\
c_n &= f(t_n + \tfrac{1}{2}h, y_n + \tfrac{1}{2}hb_n) && \text{updated slope at midpoint} \\
d_n &= f(t_n + h, y_n + hc_n) && \text{slope at right endpoint}
\end{aligned}
\tag{7.3.18}
$$

we can replace the expression $4f(t_n + h/2, y_n + h/2 a_n)$ in (7.3.17) with the more accurate estimate $2b_n + 2c_n$, and replace $f(t_{n+1}, y_n + ha_n)$ with $f(t_{n+1}, y_n + hc_n)$; each of these updates takes advantage of the most recent calculation of the approximate slope at points nearby. We thus arrive at the *fourth-order Runge–Kutta method* by setting $y_{n+1} = y_n + \triangle y$ to find

$$
y_{n+1} = y_n + \frac{h}{6}(a_n + 2b_n + 2c_n + d_n)
\tag{7.3.19}
$$

where a_n, b_n, c_n, and d_n are defined as at (7.3.18).

Again, through a lengthy development involving complicated calculations, it can be established rigorously that (7.3.19) is a fourth-order approximation technique: the resulting truncation error in the approximation is proportional to h^5. The next example demonstrates the remarkable accuracy of the Runge–Kutta method.

Example 7.3.2 Execute ten steps of the fourth-order Runge–Kutta method with $h = 0.1$ to find an approximate solution of the initial-value problem

$$
y' = e^{2t} - y, \qquad y(0) = 1
$$

Compare the results to those of the second-order Taylor method.

Solution. This is the same IVP as we considered in example 7.3.1. Recall that the exact solution to the problem is $y(t) = 2/3 e^{-t} + 1/3 e^{2t}$.

To implement the Runge–Kutta method, we use $f(t, y) = e^{2t} - y$ and compute a_n, b_n, c_n, and d_n as given by (7.3.18). Using the initial condition $(t_0, y_0) = (0, 1)$, we compute

$$
a_0 = f(t_0, y_0) = f(0, 1) = e^{2 \cdot 0} - 1 = 0
$$

$$
b_0 = f\left(t_0 + \frac{h}{2}, y_0 + \frac{ha_0}{2}\right) = f(0.05, 1 + 0.05 \cdot 0) = f(0.05, 1)
$$

$$
= e^{2 \cdot 0.05} - 1 = 0.105170918
$$

$$
c_0 = f\left(t_0 + \frac{h}{2}, y_0 + \frac{hb_0}{2}\right) = f(0.05, 1 + 0.05 \cdot 0.105170918)
$$

$$
= f(0.05, 1.005258546) = e^{2 \cdot 0.05} - 1.005258546 = 0.099912372
$$

$$
d_0 = f(t_1, y_0 + hc_0) = f(0.1, 1 + 0.1 \cdot 0.099912372) = f(0.1, 1.009991237)
$$

$$
= e^{2 \cdot 0.1} - 1.009991237 = 0.211411521
$$

Table 7.4
Fourth-order Runge–Kutta method and second-order Taylor's method applied to the IVP $y' = 2t(2 - y)$, $y(0) = 1$ using $h = 0.1$

	Runge–Kutta (RK)	Solution	RK error	Taylor error
t_n	y_n	$y(t_n)$	$|y(t_n) - y_n|$	$|y(t_n) - y_n|$
0	1	1	0	0
0.2	1.043096313	1.043095401	0.000000912	0.000798112
0.4	1.188729047	1.188727007	0.000002040	0.001976353
0.6	1.472583611	1.472580065	0.000003546	0.003699992
0.8	1.950569107	1.950563451	0.000005656	0.006223842
1	2.708280362	2.70827166	0.000008701	0.009934023

and therefore

$$y_1 = y_0 + \frac{h}{6}(a_0 + 2b_0 + 2c_0 + d_0)$$

$$= 1 + \frac{0.1}{6}(0 + 0.210341836 + 0.199824744 + 0.211411521)$$

$$= 1.010359635$$

Implementing these same calculations for subsequent steps, we can generate the output displayed in table 7.4, where again we report the results from every other step. The error from Taylor's method is being reported from table 7.3.

In table 7.4 we can see the exceptional accuracy of the fourth-order Runge–Kutta method. In one sense, this is not surprising. Being a fourth-order method, we expect the error in the first step to be proportional to $h^5 = (0.1)^5 = 0.00001$, which is in contrast to the second-order Taylor's method with error proportional to $h^3 = 0.001$. In each method, the errors are in fact much smaller; one reason why this is so can be understood by thinking about the coefficient $1/5! = 1/120$ that arises in the Taylor expansion of $y(t_0 + h)$ and multiplies h^5.

What can be considered surprising about the Runge–Kutta method is that it generates such significant accuracy through a relatively limited number of computations and by only evaluating the function $f(t, y)$ from the IVP at a select number of points, without the need to compute higher order derivatives. Fundamentally, the method takes four actual or approximate slopes and computes a weighted average of them in order to predict the next value of the solution function $y(t)$. This fourth-order Runge–Kutta method is so accurate that it is used as the standard plotting tool in *Maple* when using the DEplot command. In addition, if we command *Maple* to produce a numerical estimate to the solution of a stated IVP, the standard option in the `dsolve` command is a slightly more sophisticated algorithm known as the Runge–Kutta–Fehlberg method.

Exercises 7.3   In exercises 1–10, use (a) the second-order Taylor's method and (b) the fourth-order Runge–Kutta method to estimate $y(1)$ using $h = 0.1$, and compare the approximations generated by the methods. In exercises 1–6, compare the approximations with the exact solution. Each IVP in exercises 1–10 is identical to those in exercises 1–10 in section 7.2.

1. $y' + 2ty = 0, \quad y(0) = -2$

2. $y' = 2y - 1, \quad y(0) = 2$

3. $y' - y = 0, \quad y(0) = 2$

4. $(y')^2 - 2y = 0, \quad y(0) = 2$

5. $y' - y^2 = 1, \quad y(0) = 0$

6. $tyy' = -1 - y^2, \quad y(0) = 2$

7. $y' + ty = t^2, \quad y(0) = 1$

8. $y' + y^2 = t, \quad y(0) = 1$

9. $y' + \sin y = 2e^{-t}, \quad y(0) = 0$

10. $y' = 2e^{t/2} \sin \sqrt{y}, \quad y(0) = 0$

7.4 Methods for systems and higher order equations

In section 6.4, we introduced an extension of Euler's method for estimating the solution to nonlinear IVPs such as

$$\begin{aligned} x' &= 9y - y^2, & x(0) &= 1 \\ y' &= x, & y(0) &= 8 \end{aligned} \tag{7.4.1}$$

We again choose to use the notation $\mathbf{x} = [x \ y]^{\mathrm{T}}$ rather than $[x_1 \ x_2]^{\mathrm{T}}$ because we will be using subscripts to label approximations to the component solutions $x(t)$ and $y(t)$: for instance, $x_1 \approx x(t_1)$, where $t_1 = t_0 + h$. Recalling that x and y are each implicit functions of t, we can view (7.4.1) in the form

$$\begin{aligned} x' &= f(x, y, t), & x(t_0) &= x_0 \\ y' &= g(x, y, t), & y(t_0) &= y_0 \end{aligned} \tag{7.4.2}$$

For a single initial-value problem $y' = f(t, y)$, $y(0) = y_0$, we have developed a variety of methods for estimating the solution, including Euler's method, Heun's method, and Runge–Kutta, in order of increasing accuracy. We will generalize each of these methods to the situation for systems, leaving it as an exercise for the reader to consider other alternatives, such as the Modified Euler method. Throughout, we keep in mind that for a single IVP, every method has the form

$$y_{n+1} = y_n + \Delta y$$

where Δy is an estimate that is obtained by taking the step-size h times some approximation of the slope of the solution y at or near (t_n, y_n).

Because Euler's method is the simplest, we begin there.

7.4.1 Euler's method for systems

Recall that for a single IVP $y' = f(t, y)$, $y(0) = y_0$, Euler's method is given by the algorithm

$$y_{n+1} = y_n + hf(t_n, y_n) \qquad (7.4.3)$$

where $t_{n+1} = t_n + h$, given a step-size h. As was shown in section 6.4, to implement Euler's method for a system of two IVPs in the form (7.4.2), for the step from the approximation (x_n, y_n) to the approximation (x_{n+1}, y_{n+1}), we compute

$$\begin{aligned} x_{n+1} &= x_n + h \cdot f(t_n, x_n, y_n) \\ y_{n+1} &= y_n + h \cdot g(t_n, x_n, y_n) \end{aligned} \qquad (7.4.4)$$

Viewed from a vector perspective, if we let

$$\mathbf{x} = \begin{bmatrix} x \\ y \end{bmatrix} \text{ and } \mathbf{F}(t, \mathbf{x}) = \begin{bmatrix} f(t, x, y) \\ g(t, x, y) \end{bmatrix}$$

it follows that Euler's method for systems is given by the rule

$$\mathbf{x}^{(n+1)} = \mathbf{x}^{(n)} + h\mathbf{F}(t_n, \mathbf{x}^{(n)}) \qquad (7.4.5)$$

We use the superscript $\mathbf{x}^{(n)} \approx \mathbf{x}(t_n)$ to denote the approximation since subscripts on vectors often indicate particular entries in the vector.

In section 6.4, we saw evidence that Euler's method is not very effective because of the errors that arise. To demonstrate this further, we consider an example involving a linear system whose solution we know exactly.

Example 7.4.1   Use Euler's method with $h = 0.1$ to estimate the solution $\mathbf{x}(1)$ to the initial-value problem

$$\mathbf{x}' = \begin{bmatrix} -1 & 2 \\ -2 & -1 \end{bmatrix} \mathbf{x}, \quad \mathbf{x}(0) = \begin{bmatrix} 2 \\ 0 \end{bmatrix}$$

Compare the results to the exact solution.

Solution.   Using established methods from chapter 3, it is straightforward to show that the solution to the given IVP is

$$\mathbf{x}(t) = 2e^{-t} \begin{bmatrix} \cos 2t \\ \sin 2t \end{bmatrix}$$

To estimate this solution via Euler's method, we first observe that

$$\mathbf{x}' = \mathbf{F}(t, \mathbf{x}) = \begin{bmatrix} -1 & 2 \\ -2 & -1 \end{bmatrix} \begin{bmatrix} x \\ y \end{bmatrix} = \begin{bmatrix} -x + 2y \\ -2x - y \end{bmatrix}$$

To compute $\mathbf{x}^{(1)} \approx \mathbf{x}(t_1)$, we use (7.4.5) and write

$$\mathbf{x}^{(1)} = \mathbf{x}^{(0)} + h\mathbf{F}(0, \mathbf{x}^{(0)}) = \begin{bmatrix} 2 \\ 0 \end{bmatrix} + 0.1 \begin{bmatrix} -1 & 2 \\ -2 & -1 \end{bmatrix} \begin{bmatrix} 2 \\ 0 \end{bmatrix}$$

$$= \begin{bmatrix} 2 \\ 0 \end{bmatrix} + 0.1 \begin{bmatrix} -2 \\ -4 \end{bmatrix} = \begin{bmatrix} 1.8 \\ -0.4 \end{bmatrix}$$

Continuing Euler's method in this manner for the subsequent nine steps with $h = 0.1$ to estimate $\mathbf{x}(1)$, we find the results shown in table 7.5, where the values from every other step are reported.

The final column in table 7.5 merits some discussion. Since our exact solution is a vector function and the approximate solutions are also vectors, the error at each stage is given by the vector $\mathbf{e}^{(n)} = |\mathbf{x}(t_n) - \mathbf{x}^{(n)}|$, where $|\cdot|$ denotes the absolute value function. The size of a vector can be measured by a single number, its *length* (or *magnitude* or *norm*), which is computed by taking the square root of the sum of the squares of its entries. For a vector $\mathbf{x} \in \mathbb{R}^3$, its length is $\|\mathbf{x}\| = \sqrt{(x_1^2 + x_2^2 + x_3^2)}$, where x_1, x_2, and x_3 are the entries in $\mathbf{x}$. The entries in

Table 7.5
Euler's method applied to the IVP in example 7.4.1 using $h = 0.1$

t_n	Euler's method $\mathbf{x}^{(n)}$	Exact solution $\mathbf{x}(t_n)$	Euler error $\|\mathbf{x}(t_n) - \mathbf{x}^{(n)}\|$
0	$\begin{bmatrix} 2 \\ 0 \end{bmatrix}$	$\begin{bmatrix} 2 \\ 0 \end{bmatrix}$	0.000000000
0.2	$\begin{bmatrix} 1.54 \\ -0.72 \end{bmatrix}$	$\begin{bmatrix} 1.508201923 \\ -0.637657545 \end{bmatrix}$	0.088268894
0.4	$\begin{bmatrix} 0.9266 \\ -1.1088 \end{bmatrix}$	$\begin{bmatrix} 0.934032947 \\ -0.961716336 \end{bmatrix}$	0.147271358
0.6	$\begin{bmatrix} 0.314314 \\ -1.187352 \end{bmatrix}$	$\begin{bmatrix} 0.397732304 \\ -1.023027791 \end{bmatrix}$	0.184285265
0.8	$\begin{bmatrix} -0.18542494 \\ -1.02741408 \end{bmatrix}$	$\begin{bmatrix} -0.026240382 \\ -0.898274743 \end{bmatrix}$	0.204979735
1	$\begin{bmatrix} -0.512646273 \\ -0.724355863 \end{bmatrix}$	$\begin{bmatrix} -0.306183731 \\ -0.669023658 \end{bmatrix}$	0.213748529

the final column in table 7.5 are computed by taking the length of the vector $\mathbf{e}^{(n)}$ which is the difference between the exact solution and the approximate solution at step n. For example, the error that is present at the second step is

$$\mathbf{e}^{(1)} = \left\| \begin{bmatrix} 1.54 \\ -0.72 \end{bmatrix} - \begin{bmatrix} 1.5082 \\ -0.6376 \end{bmatrix} \right\| = \left\| \begin{bmatrix} 0.03180 \\ -0.08234 \end{bmatrix} \right\|$$

$$= \sqrt{(0.03180)^2 + (-0.08234)^2} = 0.08827$$

which is the second entry in the third column of table 7.5.

Clearly, the errors in Euler's method are significant. From our earlier work with Heun's method and the Runge–Kutta method, we expect that we can attain much better approximations by using analogous approaches for systems. We consider Heun's method next.

7.4.2 Heun's method for systems

From our most recent work, we know that if we view a system of IVPs from the perspective of vector functions, we are trying to estimate the solution to

$$\mathbf{x}' = \mathbf{F}(t, \mathbf{x}), \quad \mathbf{x}(t_0) = \mathbf{x}_0$$

and that from this point of view, the vector version of Euler's method is

$$\mathbf{x}^{(n+1)} = \mathbf{x}^{(n)} + h\mathbf{F}(t_n, \mathbf{x}^{(n)})$$

Recalling that Heun's method for a single differential equation is given by the rule

$$y_{n+1} = y_n + \frac{h}{2}(a_n + b_n) \tag{7.4.6}$$

where $a_n = f(t_n, y_n)$ and $b_n = f(t_{n+1}, y_n + ha_n)$, we realize that the vector analog of (7.4.6) is

$$\mathbf{x}^{(n+1)} = \mathbf{x}^{(n)} + \frac{h}{2}(\mathbf{a}^{(n)} + \mathbf{b}^{(n)}) \tag{7.4.7}$$

where $\mathbf{a}^{(n)}$ and $\mathbf{b}^{(n)}$ are given by

$$\mathbf{a}^{(n)} = \mathbf{F}(t_n, \mathbf{x}^{(n)}) \quad \text{and} \quad \mathbf{b}^{(n)} = \mathbf{F}(t_{n+1}, \mathbf{x}^{(n)} + h\mathbf{a}^{(n)}) \tag{7.4.8}$$

In order to compare and contrast the vector version of Heun's method with Euler's method, we consider the following example which builds upon example 7.4.1.

Example 7.4.2   Use Heun's method with $h = 0.1$ to estimate the solution $\mathbf{x}(1)$ to the initial-value problem

$$\mathbf{x}' = \begin{bmatrix} -1 & 2 \\ -2 & -1 \end{bmatrix} \mathbf{x}, \quad \mathbf{x}(0) = \begin{bmatrix} 2 \\ 0 \end{bmatrix}$$

Compare the results to the exact solution and to those from Euler's method in example 7.4.1.

Solution.   We are considering the IVP

$$\mathbf{x}' = \mathbf{F}(t, \mathbf{x}) = \begin{bmatrix} -1 & 2 \\ -2 & -1 \end{bmatrix} \begin{bmatrix} x \\ y \end{bmatrix} = \begin{bmatrix} -x + 2y \\ -2x - y \end{bmatrix}, \quad \mathbf{x}(0) = \begin{bmatrix} x_0 \\ y_0 \end{bmatrix} = \begin{bmatrix} 2 \\ 0 \end{bmatrix}$$

To compute $\mathbf{x}^{(1)} \approx \mathbf{x}(0.1)$ by Heun's method, we first compute

$$\mathbf{a}^{(1)} = \mathbf{F}(t_0, \mathbf{x}^{(0)}) = \begin{bmatrix} -1 & 2 \\ -2 & -1 \end{bmatrix} \mathbf{x}^{(0)}$$

$$= \begin{bmatrix} -1 & 2 \\ -2 & -1 \end{bmatrix} \begin{bmatrix} 2 \\ 0 \end{bmatrix} = \begin{bmatrix} -2 \\ -4 \end{bmatrix}$$

Next, to determine $\mathbf{b}^{(1)}$ we write

$$\mathbf{b}^{(1)} = \mathbf{F}(t_0, \mathbf{x}^{(0)} + h\mathbf{a}^{(0)}) = \begin{bmatrix} -1 & 2 \\ -2 & -1 \end{bmatrix} (\mathbf{x}^{(0)} + h\mathbf{a}^{(0)})$$

$$= \begin{bmatrix} -1 & 2 \\ -2 & -1 \end{bmatrix} \begin{bmatrix} 2 + 0.1 \cdot (-2) \\ 0 + 0.1 \cdot (-4) \end{bmatrix} = \begin{bmatrix} -2.6 \\ -3.2 \end{bmatrix}$$

Finally, we determine $\mathbf{x}^{(1)} = \mathbf{x}^{(0)} + h/2(\mathbf{a}^{(1)} + \mathbf{b}^{(1)})$ to find

$$\mathbf{x}^{(1)} = \begin{bmatrix} 2 \\ 0 \end{bmatrix} + 0.05 \left(\begin{bmatrix} -2 \\ -4 \end{bmatrix} + \begin{bmatrix} -2.6 \\ -3.2 \end{bmatrix} \right) = \begin{bmatrix} 1.77 \\ -0.36 \end{bmatrix}$$

Updating our work and computing the subsequent approximations results in the values for $\mathbf{x}^{(2)}, \ldots, \mathbf{x}^{(10)}$ shown in table 7.6, where we also display the errors computed in table 7.5 for Euler's method applied to the same IVP.

It is apparent from table 7.6 that just as Heun's method for a single IVP is a substantial improvement over Euler's method, it is also better for systems. At the same time, knowing that even higher order methods such as Runge–Kutta are available, we aspire to develop even more accurate methods for systems by converting the Runge–Kutta method for a single DE to one for systems.

7.4.3 Runge–Kutta method for systems

Recall that for the single first-order IVP $y' = f(t, y)$, $y(t_0) = y_0$, the fourth-order Runge–Kutta method is given by

$$y_{n+1} = y_n + \frac{h}{6}(a_n + 2b_n + 2c_n + d_n) \tag{7.4.9}$$

where

$$\begin{aligned} a_n &= f(t_n, y_n) \\ b_n &= f\left(t_n + \tfrac{1}{2}h, y_n + \tfrac{1}{2}ha_n\right) \\ c_n &= f\left(t_n + \tfrac{1}{2}h, y_n + \tfrac{1}{2}hb_n\right) \\ d_n &= f(t_n + h, y_n + hc_n) \end{aligned} \tag{7.4.10}$$

Table 7.6
Heun's method applied to the IVP in example 7.4.2 using $h = 0.1$

	Heun	Solution	Heun error	Euler error
t_n	$\mathbf{x}^{(n)}$	$\mathbf{x}(t_n)$	$\|\mathbf{x}(t_n) - \mathbf{x}^{(n)}\|$	$\|\mathbf{x}(t_n) - \mathbf{x}^{(n)}\|$
0	$\begin{bmatrix} 2 \\ 0 \end{bmatrix}$	$\begin{bmatrix} 2 \\ 0 \end{bmatrix}$	0	0
0.2	$\begin{bmatrix} 1.50165 \\ -0.6372 \end{bmatrix}$	$\begin{bmatrix} 1.508201923 \\ -0.637657545 \end{bmatrix}$	0.006567879	0.088268894
0.4	$\begin{bmatrix} 0.924464441 \\ -0.95685138 \end{bmatrix}$	$\begin{bmatrix} 0.934032947 \\ -0.961716336 \end{bmatrix}$	0.010734249	0.147271358
0.6	$\begin{bmatrix} 0.389258164 \\ -1.012962308 \end{bmatrix}$	$\begin{bmatrix} 0.397732304 \\ -1.023027791 \end{bmatrix}$	0.013157697	0.184285265
0.8	$\begin{bmatrix} -0.03046503 \\ -0.884575076 \end{bmatrix}$	$\begin{bmatrix} -0.026240382 \\ -0.898274743 \end{bmatrix}$	0.014336266	0.204979735
1	$\begin{bmatrix} -0.304699526 \\ -0.654454923 \end{bmatrix}$	$\begin{bmatrix} -0.306183731 \\ -0.669023658 \end{bmatrix}$	0.014644143	0.213748529

Just as with Euler's method and Heun's method, we can develop the vector analog of the Runge–Kutta method. We do so by letting

$$\mathbf{x}^{(n+1)} = \mathbf{x}^{(n)} + \frac{h}{6}\left(\mathbf{a}^{(n)} + 2\mathbf{b}^{(n)} + 2\mathbf{c}^{(n)} + \mathbf{d}^{(n)}\right) \qquad (7.4.11)$$

where

$$\begin{aligned}
\mathbf{a}^{(n)} &= \mathbf{F}\left(t_n, \mathbf{x}^{(n)}\right) \\
\mathbf{b}^{(n)} &= \mathbf{F}\left(t_n + \tfrac{1}{2}h, \mathbf{x}^{(n)} + \tfrac{1}{2}h\mathbf{a}^{(n)}\right) \\
\mathbf{c}^{(n)} &= \mathbf{F}\left(t_n + \tfrac{1}{2}h, \mathbf{x}^{(n)} + \tfrac{1}{2}h\mathbf{b}^{(n)}\right) \\
\mathbf{d}^{(n)} &= \mathbf{F}\left(t_n + h, \mathbf{x}^{(n)} + h\mathbf{c}^{(n)}\right)
\end{aligned} \qquad (7.4.12)$$

The computations for the Runge–Kutta method for systems can be implemented in a way very similar to those for Heun's method. Doing so and applying the Runge–Kutta method to the IVP stated in examples 7.4.1 and 7.4.2 results in the values shown in table 7.7; we also display the error from Heun's method by way of contrast.

As with single IVPs, the results of the Runge–Kutta method for systems are impressive. This is again due to the fact that the Runge–Kutta method is fourth-order, while Heun's method is only second-order.

We close this section by recalling the important link between higher order differential equations and systems of first-order equations.

Table 7.7
Runge–Kutta method applied to the IVP in example 7.4.2 using $h = 0.1$

	RK	Solution	RK error	Heun error
t_n	$\mathbf{x}^{(n)}$	$\mathbf{x}(t_n)$	$\|\mathbf{x}(t_n) - \mathbf{x}^{(n)}\|$	$\|\mathbf{x}(t_n) - \mathbf{x}^{(n)}\|$
0	$\begin{bmatrix} 2 \\ 0 \end{bmatrix}$	$\begin{bmatrix} 2 \\ 0 \end{bmatrix}$	0	0
0.2	$\begin{bmatrix} 1.508211151 \\ -0.637671316 \end{bmatrix}$	$\begin{bmatrix} 1.508201923 \\ -0.637657545 \end{bmatrix}$	0.00001658	0.006567879
0.4	$\begin{bmatrix} 0.934038085 \\ -0.96174299 \end{bmatrix}$	$\begin{bmatrix} 0.934032947 \\ -0.961716336 \end{bmatrix}$	0.00002714	0.010734249
0.6	$\begin{bmatrix} 0.397725368 \\ -1.023060398 \end{bmatrix}$	$\begin{bmatrix} 0.397732304 \\ -1.023027791 \end{bmatrix}$	0.00003334	0.013157697
0.8	$\begin{bmatrix} -0.026261217 \\ -0.89830458 \end{bmatrix}$	$\begin{bmatrix} -0.026240382 \\ -0.898274743 \end{bmatrix}$	0.00003639	0.014336266
1	$\begin{bmatrix} -0.306215262 \\ -0.66904348 \end{bmatrix}$	$\begin{bmatrix} -0.306183731 \\ -0.669023658 \end{bmatrix}$	0.00003724	0.014644143

7.4.4 Methods for higher order IVPs

We have repeatedly used the fact that any linear nth-order differential equation can be converted to a system of linear first-order equations. For example, given a second-order equation such as $y'' + 2y' - 3y = \sin t$, we know that with the substitution $x_1 = y$, $x_2 = y'$, it follows that $\mathbf{x} = [x_1 \ x_2]^T$ is a solution to the system of differential equations

$$x_1' = x_2$$
$$x_2' = 3x_1 - 2x_2 + \sin t$$

Given our current interest in approximating solutions to initial-value problems, we are particularly focused on nonlinear equations, including

$$\theta'' + \frac{g}{L}\sin\theta = 0, \qquad \theta(0) = a, \ \theta'(0) = b$$

which governs the motion of a simple undamped pendulum, as developed in section 6.1. In this setting, we are unable to determine an exact solution, and thus wish to generate an approximate one. More generally, we want to be able to develop an approximate solution to any nonlinear IVP. In the second-order case, we can view this problem as having the form

$$y'' = f(t, y, y'), \quad y(0) = a, \ y'(0) = b \tag{7.4.13}$$

We introduce the substitution $z = y'$, then $z' = y'' = f(t, y, y') = f(t, y, z)$, so that (7.4.13) may be rewritten as the system of IVPs

$$y' = z, \qquad y(0) = a$$
$$z' = f(t, y, z), \quad z(0) = b \tag{7.4.14}$$

Letting $\mathbf{x} = [y \; z]^T$ and $\mathbf{F}(t, \mathbf{x}) = [z \; f(t, y, z)]^T$, we may rewrite (7.4.14) in the form

$$\mathbf{x}' = \mathbf{F}(t, \mathbf{x}), \quad \mathbf{x}(0) = \begin{bmatrix} a \\ b \end{bmatrix}$$

which is precisely the form we considered for Euler's method, Heun's method, and the Runge–Kutta method for systems. That is, once we have converted a higher order IVP to a system of first-order IVPs, we may choose from any of our existing approximation methods for systems of DEs. We demonstrate this for a particular example using Heun's method.

Example 7.4.3   Use Heun's method to estimate the solution $y(t)$ from $t = 0$ to $t = 1$ to the second-order IVP

$$y'' + 0.1y' + 4\sin y = 0, \qquad y(0) = 1, \; y'(0) = 0$$

with step-size $h = 0.1$.

Solution.   We begin by letting $z = y'$, so that $z' = y'' = -4\sin y - 0.1y' = -4\sin y - 0.1z$. Writing $\mathbf{x} = [y \; z]^T$, it follows that

$$\mathbf{x}' = \begin{bmatrix} z \\ -4\sin y - 0.1z \end{bmatrix} = \mathbf{F}(t, \mathbf{x})$$

Recalling Heun's method, we must compute

$$\mathbf{x}^{(n+1)} = \mathbf{x}^{(n)} + \frac{h}{2}(\mathbf{a}^{(n)} + \mathbf{b}^{(n)})$$

where

$$\mathbf{a}^{(n)} = \mathbf{F}(t_n, \mathbf{x}^{(n)}) \text{ and } \mathbf{b}^{(n)} = \mathbf{F}(t_{n+1}, \mathbf{x}^{(n)} + h\mathbf{a}^{(n)})$$

With the initial condition $\mathbf{x}^{(0)} = [1 \; 0]$, we first find that

$$\mathbf{a}^{(0)} = \begin{bmatrix} 0 \\ -4\sin(1) - 0.1 \cdot 0 \end{bmatrix} = \begin{bmatrix} 0 \\ -3.366 \end{bmatrix}$$

from which it follows that

$$\mathbf{b}^{(0)} = \mathbf{F}(0.1, \mathbf{x}^{(0)} + h\mathbf{a}^{(0)}) = \begin{bmatrix} -0.3366 \\ -3.332 \end{bmatrix}$$

Therefore, $\mathbf{x}^{(1)}$ is given by

$$\mathbf{x}^{(1)} = \mathbf{x}^{(0)} + \frac{h}{2}(\mathbf{a}^{(0)} + \mathbf{b}^{(0)})$$

$$= \begin{bmatrix} 1 \\ 0 \end{bmatrix} + \frac{0.1}{2}\left(\begin{bmatrix} 0 \\ -3.366 \end{bmatrix} + \begin{bmatrix} -0.3366 \\ -3.332 \end{bmatrix} \right)$$

$$= \begin{bmatrix} 0.98317 \\ -0.33490 \end{bmatrix}$$

Table 7.8
Heun's method applied to the second-order IVP in example 7.4.3 using $h = 0.1$

n	$\mathbf{x}^{(n)}$	$\mathbf{a}^{(n)}$	$\mathbf{b}^{(n)}$	$\mathbf{x}^{(n+1)}$
0	$\begin{bmatrix} 1 \\ 0 \end{bmatrix}$	$\begin{bmatrix} 0 \\ -3.365883939 \end{bmatrix}$	$\begin{bmatrix} -0.336588394 \\ -3.3322251 \end{bmatrix}$	$\begin{bmatrix} 0.98317058 \\ -0.334905452 \end{bmatrix}$
2	$\begin{bmatrix} 0.933202302 \\ -0.659006349 \end{bmatrix}$	$\begin{bmatrix} -0.659006349 \\ -3.148220455 \end{bmatrix}$	$\begin{bmatrix} -0.973828394 \\ -2.952961961 \end{bmatrix}$	$\begin{bmatrix} 0.851560565 \\ -0.96406547 \end{bmatrix}$
4	$\begin{bmatrix} 0.740589862 \\ -1.240510452 \end{bmatrix}$	$\begin{bmatrix} -1.240510452 \\ -2.574842511 \end{bmatrix}$	$\begin{bmatrix} -1.497994703 \\ -2.163059344 \end{bmatrix}$	$\begin{bmatrix} 0.603664604 \\ -1.477405545 \end{bmatrix}$
6	$\begin{bmatrix} 0.445309489 \\ -1.663161107 \end{bmatrix}$	$\begin{bmatrix} -1.663161107 \\ -1.556632719 \end{bmatrix}$	$\begin{bmatrix} -1.818824379 \\ -0.919669919 \end{bmatrix}$	$\begin{bmatrix} 0.271210214 \\ -1.786976239 \end{bmatrix}$
8	$\begin{bmatrix} 0.088048126 \\ -1.840715689 \end{bmatrix}$	$\begin{bmatrix} -1.840715689 \\ -0.167666053 \end{bmatrix}$	$\begin{bmatrix} -1.857482294 \\ 0.569252015 \end{bmatrix}$	$\begin{bmatrix} -0.096861773 \\ -1.820636391 \end{bmatrix}$
10	$\begin{bmatrix} -0.276080886 \\ -1.728307853 \end{bmatrix}$	$\begin{bmatrix} -1.728307853 \\ 1.263178986 \end{bmatrix}$	$\begin{bmatrix} -1.601989955 \\ 1.896140173 \end{bmatrix}$	$\begin{bmatrix} -0.442595776 \\ -1.570341895 \end{bmatrix}$

Executing similar computations for the remaining nine steps to approximate $\mathbf{x}(1)$, we find the results shown in table 7.8.

From the results of table 7.8, we see that

$$\mathbf{x}(1) \approx \mathbf{x}^{(10)} = \begin{bmatrix} -0.276080886 \\ -1.728307853 \end{bmatrix}$$

Recalling that $\mathbf{x}(t) = [y(t) \ z(t)]^{\mathrm{T}}$ and that our ultimate goal is to estimate the solution $y(t)$ to the stated IVP, it follows that $y(1) \approx -0.2761$.

The approach in example 7.4.3 can be implemented for higher order initial-value problems through a substitution to convert a given higher order equation to a system of first-order ones. More accurate results may be obtained through applying the fourth-order Runge–Kutta method for systems. We note particularly that not only can we estimate solutions to nonlinear equations, but even those with non-constant coefficients. For example, solutions to IVPs like

$$y'' + ty = 10\sin 2t, \quad y(0) = y'(0) = 0$$

can now be approximated.

Exercises 7.4 In exercises 1–6, (a) use Euler's method for systems with $h = 0.1$ to estimate the solution $\mathbf{x}(1)$ to the initial-value problem, (b) use Heun's method for systems with $h = 0.1$ to estimate the solution $\mathbf{x}(1)$ to the initial-value problem, and (c) if possible, compare the results to the exact solution.

1. $\mathbf{x}' = \begin{bmatrix} 0 & -1 \\ 1 & 0 \end{bmatrix} \mathbf{x}, \quad \mathbf{x}(0) = \begin{bmatrix} 1 \\ 1 \end{bmatrix}$

2. $\mathbf{x}' = \begin{bmatrix} -1 & 3 \\ 3 & 1 \end{bmatrix} \mathbf{x}, \quad \mathbf{x}(0) = \begin{bmatrix} 1 \\ 1 \end{bmatrix}$

3. $\mathbf{x}' = \begin{bmatrix} -1 & 2 \\ 2 & -4 \end{bmatrix} \mathbf{x}, \quad \mathbf{x}(0) = \begin{bmatrix} 1 \\ 1 \end{bmatrix}$

4. $\mathbf{x}' = \begin{bmatrix} t & -1 \\ 1 & 0 \end{bmatrix} \mathbf{x}, \quad \mathbf{x}(0) = \begin{bmatrix} 1 \\ 0 \end{bmatrix}$

5. $\mathbf{x}' = \begin{bmatrix} 0 & -1 \\ 1 & 0 \end{bmatrix} \mathbf{x} + \begin{bmatrix} 1 \\ t \end{bmatrix}, \quad \mathbf{x}(0) = \begin{bmatrix} 1 \\ 0 \end{bmatrix}$

6. $\mathbf{x}' = \begin{bmatrix} 1 & -1 \\ t & 0 \end{bmatrix} \mathbf{x} + \begin{bmatrix} 1 \\ 1 \end{bmatrix}, \quad \mathbf{x}(0) = \begin{bmatrix} 0 \\ 0 \end{bmatrix}$

In exercises 7–13, (a) use Heun's method and (b) use the Runge–Kutta method to estimate the solution of the system of IVPs at the given t-value using the stated h-value.

7. $x' = y - 2xy, \quad x(0) = 0.75 \qquad t = 1, h = 0.1$
   $y' = 4xy - x, \quad y(0) = 0.5$

8. $x' = 4 - y^2, \qquad x(0) = -2 \qquad t = 3, h = 0.05$
   $y' = 1 - x + y, \quad y(0) = -1$

9. $x' = \cos y, \qquad x(0) = 2 \qquad t = 1.5, h = 0.1$
   $y' = 1 - \sin x, \quad y(0) = 3$

10. $x' = 2x - y, \qquad x(0) = 1 \qquad t = 1.5, h = 0.1$
    $y' = -4x + 2y, \quad y(0) = 1$

11. $x' = e^{-y}, \qquad\qquad x(0) = 0 \qquad t = 2, h = 0.05$
    $y' = 1/(1 + x^2), \quad y(0) = 0$

12. $x' = \ln(2 + y), \quad x(0) = -1 \qquad t = 2, h = 0.1$
    $y' = x^2 + y, \qquad y(0) = -0.5$

13. $x' = y - x^2, \qquad x(0) = 1 \qquad t = 1, h = 0.05$
    $y' = x - 8y^2, \quad y(0) = 0.75$

14. Recall from section 6.1 that the nonlinear system of differential equations

$$W' = -0.75W + 0.25MW$$
$$M' = 0.5M - 0.1MW$$

models the numbers of wolves and moose (each measured in hundreds) in a predator–prey model, where time is measured in years. Assume that at time $t = 0$ there are 250 moose and 550 wolves present. Estimate the numbers of moose and wolves present at $t = 3$, 6, and 9 years using a step-size of (a) $h = 0.1$, and (b) $h = 0.05$ with both Euler's method and Heun's method.

In exercises 15–18, (a) convert the given second-order IVP to a system of first-order IVPs, (b) use Euler's method for systems with $h = 0.1$ to estimate the solution $y(1)$ to the initial-value problem, (c) use Heun's method for systems with $h = 0.1$ to estimate the solution $y(1)$ to the initial-value problem, and (d) if possible, compare the results to the exact solution.

15. $y'' + 16y = 2t + 1$, $\quad y(0) = y'(0) = 0$

16. $y'' + 16y = 2\sin 2t$, $\quad y(0) = y'(0) = 0$

17. $y'' + 16y^2 = 2\sin 2t$, $\quad y(0) = y'(0) = 0$

18. $y'' + 0.2(y')^2 + 2y^2 = 4e^{-t}\sin t$, $\quad y(0) = y'(0) = 0$

7.5 For further study

7.5.1 Predator–prey equations

Recall that a predator–prey scenario is modeled by the equations

$$\begin{aligned} x' &= 0.6x - 0.3xy & x(0) &= 2 \\ y' &= -0.9x + 0.6xy & y(0) &= 3 \end{aligned} \qquad (7.5.1)$$

(a) Determine the nontrivial equilibrium solution of (7.5.1) and use a computer algebra system to plot the direction field of the system in a suitable window containing the equilibrium solution and the given initial condition.

(b) Use a computer to implement Heun's method to estimate the solution $(x(t), y(t))$ of (7.5.1) on the interval $0 \le t \le 20$ using $h = 0.1$.

(c) Use your data from (b) to generate two plots: one a parametric plot of the approximate curve $(x(t), y(t))$ and the other a simultaneous plot of the separate functions $x(t)$ and $y(t)$ on the same coordinate axes. Discuss the behavior of the populations $x(t)$ and $y(t)$ over time.

(d) Modify your calculations in (b) appropriately to investigate the impact of changing the parameter '0.3' in the first equation to each of the values 0.1, 0.2, 0.4, 0.5, and 0.9. In each case, generate the same plots as instructed in (c). What impact does this have on the behavior of the populations?

(e) Modify your calculations in (b) in order to consider the following different initial conditions: $x(0) = 1.7$, $y(0) = 1.8$; $x(0) = 2.5$, $y(0) = 3.6$; $x(0) = 5$,

$y(0) = 1$. In each case, generate the same plots as instructed in (c). What impact do the initial conditions have on the behavior of the populations?

7.5.2 Competitive species

In section 6.5.2, we developed the model

$$x' = ax\left(1 - \tfrac{1}{A}x - \tfrac{\alpha}{a}y\right)$$
$$y' = by\left(1 - \tfrac{1}{B}y - \tfrac{\beta}{b}x\right) \tag{7.5.2}$$

where a, A, and α are positive constants (a is the population $x(t)$'s growth constant, A its carrying capacity, and α a parameter that reflects the competition for resources from population $y(t)$). The constants b, B, and β play the same roles for the second population.

(a) In (7.5.2), let $a = 0.5$, $b = 0.25$, $A = 5$, $B = 2$, $\alpha = 0.04$, and $\beta = 0.02$. Find all equilibrium points of the system and plot a direction field in a computer algebra system of this system that contains all the equilibrium solutions.

(b) Apply Heun's method to estimate the solution $(x(t), y(t))$ of (7.5.2) on the interval $0 \leq t \leq 20$ using $h = 0.1$. Plot the trajectory of the approximate solution.

(c) Leaving all other parameters the same, change the value of B to $B = 8$. Repeat questions (a) and (b) and discuss the differences between the results for the two B-values.

(d) Repeat question (c) with $B = 15$.

(e) What is the largest value of B for which the two populations can coexist with a stable equilibrium in which each population tends to a nonzero value as $t \to \infty$? What value(s) of B ensure that population $y(t)$ will dominate as $t \to \infty$ and force $x(t) \to 0$?

(f) For each of the three values of B above, experiment with the impact of the following different sets of initial conditions: $x(0) = 1$, $y(0) = 1$; $x(0) = 5$, $y(0) = 1$; $x(0) = 1$, $y(0) = 5$; $x(0) = 5$, $y(0) = 5$. How do the different initial conditions impact the behaviors of the two populations?

7.5.3 The damped pendulum

In section 6.5.1, it was shown that for a pendulum with an arm of length L, bob of mass m, and damping constant c, the angle θ that the arm forms with the vertical axis at time t satisfies the IVP

$$L\theta'' = -g\sin\theta - c\theta', \quad \theta(0) = \theta_0, \; \theta'(0) = \theta_0' \tag{7.5.3}$$

(a) Using the change of variables $x = \theta$, $y = x'$, show that the nonlinear second-order IVP (6.5.2) is equivalent to the system

$$x' = y$$
$$y' = -\frac{g}{L}\sin x - \frac{c}{L}y \qquad\qquad (7.5.4)$$

(b) Apply Heun's method to estimate the solution $(x(t), y(t))$ of (7.5.4) with $g = 9.8$, $L = 1$, and $c = 1$ with initial conditions $x(0) = 2$, $y(0) = 2$ on the interval $0 \le t \le 10$ using $h = 0.1$. Plot the trajectory of the approximate solution.

(c) Repeat question (b) using $c = 0.1$ and $c = 5$. Discuss the differences in the results.

(d) Investigate the effects of changing the initial conditions to the following: $x(0) = 2$, $y(0) = 5$; $x(0) = 2$, $y(0) = 15$; $x(0) = 2$, $y(0) = -5$. Do so for each of the three c-values noted above and discuss the differences among the results and the physical interpretation that explains how the pendulum is behaving.

8

Series solutions for differential equations

8.1 Motivating problems

In more sophisticated courses in mathematical physics or special functions, a different type of linear differential equation frequently arises from those we have studied to date. From several perspectives, we have thoroughly analyzed the behavior of linear differential equations with constant coefficients of the form

$$y'' + a_1 y' + a_0 y = f(t)$$

But there are other important and well-known equations with non-constant coefficients. We list some of these here in anticipation of more in-depth study in subsequent sections.

Airy's equation is a linear second-order equation that arises in physics in the study of light refraction. While it can be stated in a slightly more general form, a good example to begin with is

$$y'' + ty = 0 \tag{8.1.1}$$

The explicit presence of the coefficient "t" in (8.1.1) makes this equation substantially different from those (such as $y'' + y = 0$) we have already solved.

If we recall the initial approach to solving $y'' + y = 0$, we can gain intuition for how to proceed with (8.1.1). We know that guessing $y = e^{rt}$ in $y'' + y = 0$ leads to the characteristic equation $r^2 + 1 = 0$, so that $y = e^{it}$ or $y = e^{-it}$. We then know from Euler's formula that both $y = \sin t$ and $y = \cos t$ arise as linearly independent solutions to $y'' + y = 0$. One key characteristic the exponential, sine, and cosine functions have in common is that they can be expressed as infinite power series; indeed, this fact was used to justify the validity of Euler's formula.

In particular, we can write

$$e^t = 1 + t + \frac{t^2}{2!} + \frac{t^3}{3!} + \cdots + \frac{t^n}{n!} + \cdots \tag{8.1.2}$$

$$\sin t = t - \frac{t^3}{3!} + \frac{t^5}{5!} - \cdots + (-1)^{n+1}\frac{t^{2n+1}}{(2n+1)!} + \cdots \tag{8.1.3}$$

$$\cos t = 1 - \frac{t^2}{2!} + \frac{t^4}{4!} - \cdots + (-1)^n\frac{t^{2n}}{(2n)!} + \cdots \tag{8.1.4}$$

Each of these expressions for e^t, $\sin t$, and $\cos t$ is of the form $\sum_{n=0}^{\infty} a_n t^n$ and is valid for every real number t.

In the upcoming chapter, rather than making guesses of the form $y = e^{rt}$, we instead assume much more generally that y is a nice enough function to have a power series expansion of the form $y = \sum_{n=0}^{\infty} a_n t^n$, and then substitute this form of the potential solution function y into the differential equation in order to deduce the coefficients a_n.

Other well-known differential equations that we will consider include the *Hermite equation*

$$y'' - 2ty' + 2qy = 0 \tag{8.1.5}$$

where q is a constant, the *Laguerre equation*

$$ty'' + (1 - t)y' + qy = 0 \tag{8.1.6}$$

(again where q is constant), and the *Bessel equation*

$$t^2 y'' + ty' + (t^2 - n^2)y = 0 \tag{8.1.7}$$

where n is a constant.

Again, in each of (8.1.5), (8.1.6), and (8.1.7), it is the presence of non-constant coefficient(s) involving t that makes us seek new ways to find solutions. Finally, recalling an elementary differential equation from calculus further motivates the importance of infinite series representations of functions. Among the simplest of all first-order differential equations are those of the form $y' = f(t)$; these can be solved (in theory) by integrating. But if we consider an example such as

$$y' = e^{-t^2}$$

we are immediately stuck since the function e^{-t^2} lacks an elementary anti-derivative.

If we use (8.1.2) and replace t with $-t^2$, then we can write

$$y' = e^{-t^2} = 1 - t^2 + \frac{t^4}{2!} - \frac{t^6}{3!} + \cdots + (-1)^{n+1}\frac{t^{2n}}{n!} + \cdots$$

Integrating, it follows that

$$y = C + t - \frac{t^3}{3} + \frac{t^5}{5 \cdot 2!} - \frac{t^7}{7 \cdot 3!} + \cdots + (-1)^{n+1}\frac{t^{2n+1}}{(2n+1) \cdot n!} + \cdots$$

Hence we are able to determine the general solution function y, although we must be content to leave y in its series representation. Discovering solutions in this power series form will be typical of the results we obtain in our work in this chapter.

8.2 A review of Taylor and power series

From calculus, we know that if a function has a derivative at a given point $t = a$, then the function is approximately linear near $t = a$. Indeed, the existence of the first derivative ensures that the function is smooth: the function must be continuous at a and it's graph cannot have a corner there. Of course, if having one derivative is a good thing, having several derivatives is even better. The best possible scenario of all is that the function is *infinitely differentiable* at $t = a$. That is, $f^{(k)}(a)$ exists for every $k = 0, 1, 2, \ldots$. A function that is infinitely differentiable at $t = a$ and at all points in some small open interval containing a is said to be *analytic*[1] at $t = a$. If a function fails to be analytic at a given point, we say that f is *singular* at that point. For example, the rational function

$$f(t) = \frac{t}{(t^2 + 9)(t - 4)}$$

is singular at $t = 4$ and $t = \pm 3i$ since it is undefined at these values (as are each of its derivatives). At every other value of t, $f(t)$ is analytic.

Much of the theory of analytic functions is a natural extension of the ideas of Taylor polynomials and Taylor series from calculus. Here our intention is not to develop a complete theory of analytic functions, but rather to remind the reader of important results on Taylor series and extend this perspective slightly in order to suit our purposes. Most results will be stated without proof.

To begin, we assume that f is an analytic function at $a = 0$ and recall that the polynomial functions

$$P_0(t) = f(0)$$

$$P_1(t) = f(0) + f'(0)t$$

$$P_2(t) = f(0) + f'(0)t + \frac{f''(0)}{2!}t^2$$

$$\vdots$$

$$P_k(t) = f(0) + f'(0)t + \frac{f''(0)}{2!}t^2 + \cdots + \frac{f^{(k)}(0)}{k!}t^k \qquad (8.2.1)$$

are called the Taylor polynomials of f at $a = 0$ and form the *sequence of partial sums* of the infinite series

$$P(t) = f(0) + f'(0)t + \frac{f''(0)}{2!}t^2 + \cdots + \frac{f^{(k)}(0)}{k!}t^k + \cdots \qquad (8.2.2)$$

---

[1] Usually when analytic functions are discussed, we allow the function to have complex inputs and consider a disk of a given radius around a complex point. For our purposes, a discussion restricted to real values is sufficient.

In particular, the function $P_k(t)$ in (8.2.1) is the kth *Taylor polynomial* of f at $a = 0$, and the infinite series (8.2.2) is called the *Taylor series* of f centered at $a = 0$; the series converges in (8.2.2) if and only if the sequence of partial sums converges. That is, $P(t)$ is defined if and only if

$$\lim_{k \to \infty} P_k(t)$$

exists. If this limit fails to exist, we say that the Taylor series diverges at this point. What is perhaps most remarkable is the fact that wherever the series (8.2.2) converges, it does so to the value of the given analytic function f; moreover, the Taylor series converges in an interval centered at $t = 0$ that extends to the nearest singular point. Formally, we have the following theorem.

Theorem 8.2.1   Suppose that $f(t)$ is an analytic function at 0 and R is the distance from 0 to the nearest singular point of $f(t)$. Then the Taylor series of $f(t)$ centered at $t = 0$ converges to $f(t)$ in the interval $|t| < R$ and diverges in the interval $|t| > R$.

The number R is called the *radius of convergence* of the Taylor series. We note, too, that it is possible for singular points to be complex, so R is not necessarily the distance from 0 to the nearest real singular point. We also observe specifically that for any t such that $|t| < R$, we know

$$f(t) = f(0) + f'(0)t + \frac{f''(0)}{2!}t^2 + \cdots + \frac{f^{(k)}(0)}{k!}t^k + \cdots$$

We consider an example to see many of these ideas at work.

Example 8.2.1   Find the Taylor series of $f(t) = \ln(1 + t)$ centered at $t = 0$ and determine the radius of convergence of the series.

Solution.   We begin by taking the first several derivatives of f and evaluating them at 0:

$$f(t) = \ln(1 + t) \qquad\qquad f(0) = \ln(1) = 0$$
$$f'(t) = (1 + t)^{-1} \qquad\qquad f'(0) = 1$$
$$f''(t) = (-1)(1 + t)^{-2} \qquad\qquad f''(0) = -1$$
$$f'''(t) = (-2)(-1)(1 + t)^{-3} \qquad f'''(0) = 2!$$
$$f^{(4)}(t) = (-3)(-2)(-1)(1 + t)^{-4} \quad f^{(4)}(0) = -3!$$

From these calculations, we see that the fourth Taylor polynomial is

$$P_4(t) = 0 + 1t - \frac{1}{2!}t^2 + \frac{2!}{3!}t^3 - \frac{3!}{4!}t^4$$

$$= t - \frac{1}{2}t^2 + \frac{1}{3}t^3 - \frac{1}{4}t^4$$

The established pattern implies that the Taylor series of $f(t) = \ln(1+t)$ is

$$P(t) = t - \frac{1}{2}t^2 + \frac{1}{3}t^3 - \frac{1}{4}t^4 + \cdots = \sum_{n=1}^{\infty}(-1)^{n+1}\frac{1}{n}t^n$$

From calculus, the standard way to test a power series for convergence is to use the Ratio Test. Doing so here with $a_n = (-1)^{n+1}(1/n)t^n$, we observe that

$$\lim_{n\to\infty}\left|\frac{a_{n+1}}{a_n}\right| = \lim_{n\to\infty}\left|\frac{(-1)^{n+2}(1/n+1)\,t^{n+1}}{(-1)^{n+1}(1/n)\,t^n}\right|$$

$$= \lim_{n\to\infty}\left|-1\cdot\frac{n}{n+1}\cdot t\right|$$

$$= |t|$$

The Ratio Test states that a given series converges if $\lim_{n\to\infty}|a_{n+1}/a_n| < 1$. Thus, if $|t| < 1$, it follows that

$$\ln(1+t) = t - \frac{1}{2}t^2 + \frac{1}{3}t^3 - \frac{1}{4}t^4 + \cdots = \sum_{n=1}^{\infty}(-1)^{n+1}\frac{1}{n}t^n \qquad (8.2.3)$$

converges.

The result of example 8.2.1 makes further sense in light of theorem 8.2.1 since we know that $f(t) = \ln(1+t)$ has a singularity at $t = -1$. If we substitute $t = -1$ in (8.2.3), the opposite of the harmonic series arises $(-1 - \frac{1}{2} - \frac{1}{3} - \frac{1}{4} - \cdots)$, which diverges. However, it can be shown by the alternating series test that (8.2.3) does converge when $t = 1$; indeed, for any power series that converges for $|t| < R$, it is possible for the series to converge at both $t = \pm R$, neither, or just one of the points. While this is an interesting mathematical topic in its own right, it is largely irrelevant in our discussion of series solutions to differential equations.

We next state several prominent Taylor series expansions along with their respective radii of convergence and leave the development and testing of these series for convergence to the exercises at the end of this section.

$$e^t = 1 + t + \frac{t^2}{2!} + \frac{t^3}{3!} + \cdots + \frac{t^n}{n!} + \cdots \qquad\qquad R = \infty$$

$$\sin t = t - \frac{t^3}{3!} + \frac{t^5}{5!} - \cdots + (-1)^{n+1}\frac{t^{2n+1}}{(2n+1)!} + \cdots \qquad R = \infty$$

$$\qquad\qquad\qquad\qquad\qquad\qquad\qquad\qquad\qquad\qquad\qquad (8.2.4)$$

$$\cos t = 1 - \frac{t^2}{2!} + \frac{t^4}{4!} - \cdots + (-1)^n\frac{t^{2n}}{(2n)!} + \cdots \qquad R = \infty$$

$$\frac{1}{1-t} = 1 + t + t^2 + t^3 + \cdots + t^n + \cdots \qquad\qquad R = 1$$

From these fundamental Taylor series, the series expansions of other related functions may often be easily found. The following example demonstrates one way in which this may be accomplished.

Example 8.2.2   Find the Taylor series expansion of

$$f(t) = \frac{t}{1 + 4t^2}$$

as well as its radius of convergence.

Solution.   If we first omit the t in the numerator of $f(t)$, we can use the final result from (8.2.4) and substitute $-4t^2$ for t, writing

$$\frac{1}{1 - (-4t^2)} = 1 + (-4t^2) + (-4t^2)^2 + (-4t^2)^3 + \cdots + (-4t^2)^n + \cdots$$

$$= 1 - 4t^2 + 16t^4 - 64t^6 + \cdots + (-4)^n t^{2n} + \cdots \qquad (8.2.5)$$

To get the Taylor series of $f(t)$, we now multiply both sides of (8.2.5) by t, and have

$$f(t) = \frac{t}{1 + 4t^2} = t - 4t^3 + 16t^5 + 64t^7 + \cdots + (-4)^n t^{2n+1} + \cdots \qquad (8.2.6)$$

Since the original series from (8.2.4) converges for $|t| < 1$ and we replaced t with $-4t^2$, it follows that (8.2.5) converges for $|-4t^2| < 1$, or in other words for $|t| < 1/2$. Multiplying (8.2.5) by t has no effect on the radius of convergence of the series, and therefore (8.2.6) converges for $|t| < 1/2$. Note further that the denominator $1 + 4t^2$ of $f(t)$ is zero at $t = \pm i/2$; each of these complex numbers lies a distance of $1/2$ unit away from the origin and is a singular point of f. This observation is additional evidence that $R = 1/2$ is the radius of convergence of the series expansion of $f(t)$.

Similar reasoning may be used to find expansions for such functions as e^{-t^2}, $t \sin 4t$, and $(\cos t - 1)/t^2$. In each case, the approach of example 8.2.2 is far simpler than using the definition of Taylor series directly and computing derivatives of the given function.

One reason why the development of Taylor series for functions similar to those in (8.2.4) is so straightforward is the fact that Taylor series are unique. Said differently, if we can find a power series expression for a given function, it must be the Taylor series. This is stated formally in the following theorem.

Theorem 8.2.2   The series $\sum_{k=0}^{\infty} b_k t^k$ converges in the interval $|t| < R$ to the function $f(t)$ if and only if $f(t)$ is analytic for all t such that $|t| < R$ and

$$b_k = \frac{1}{k!} f^{(k)}(0)$$

An immediate consequence of theorem 8.2.2 is that if $\sum_{k=0}^{\infty} b_k t^k = 0$ for $|t| < R$, then $b_k = 0$ for all t in the interval. We will use this result frequently when we solve differential equations by *equating like coefficients* of two equal power series.

If we cannot use substitution to find a Taylor series expansion (as we did in example 8.2.2), it may be possible to use differentiation or integration to do so. The following example introduces this approach.

Example 8.2.3   Find the Taylor series expansion and radius of convergence of $f(t) = \arctan t$.

Solution.   If we were to attempt to find the series via the definition by taking derivatives, we would find that the process becomes laborious after computing $f'(t) = 1/(1+t^2)$, since differentiating will involve both the chain and quotient rules. Instead, we observe that

$$f'(t) = \frac{1}{1+t^2}$$

itself has a series expansion that is not difficult to find. Similar to our work in example 8.2.2, we use the final result in (8.2.4) and substitute $-t^2$ for t to write

$$f'(t) = \frac{1}{1-(-t^2)} = 1 + (-t^2) + (-t^2)^2 + (-t^2)^3 + \cdots + (-t^2)^n + \cdots$$

$$= 1 - t^2 + t^4 - t^6 + \cdots + (-1)^n t^{2n} + \cdots \qquad (8.2.7)$$

Because we now have a series expansion for $f'(t)$, it is natural to integrate both sides of (8.2.7) to find the series for $f(t)$. Doing so, we see that

$$f(t) = \arctan t = C + t - \frac{1}{3} t^3 + \frac{1}{5} t^5 - \frac{1}{7} t^7 + \cdots + \frac{(-1)^n}{2n+1} t^{2n+1} \cdots \qquad (8.2.8)$$

It is a straightforward exercise to use the Ratio Test to show that (8.2.8) converges for all t such that $|t| < 1$. Moreover, since $\arctan(0) = 0$, it follows that $C = 0$.

While intuition guides our work in example 8.2.3, and we certainly know that we can integrate any finite polynomial, the one step that is perhaps questionable is when we say we will integrate both sides of (8.2.7) to find the series for $f(t)$. That this step is legitimate (and that it preserves the radius of convergence) is the conclusion of our next formal result, the Taylor series Differentiation and Integration Theorem.

Theorem 8.2.3   If $f(t)$ has the Taylor series expansion

$$f(t) = \sum_{k=0}^{\infty} b_k t^k, \quad |t| < R$$

then its antiderivative $F(t) = \int_0^t f(x)\,dx$ and its derivative $f'(t)$ have the respective Taylor series expansions

$$F(t) = \sum_{k=0}^{\infty} b_k \int_0^t x^k\,dx = \sum_{k=0}^{\infty} \frac{b_k}{k+1} t^{k+1}, \quad |t| < R \qquad (8.2.9)$$

$$f'(t) = \sum_{k=0}^{\infty} b_k \frac{d}{dt}[t^k]\,dx = \sum_{k=1}^{\infty} k b_k t^{k-1}, \quad |t| < R \qquad (8.2.10)$$

That is, theorem 8.2.3 states that any power series may be differentiated or integrated term-wise and that doing so does not change the radius of convergence of the power series. This fact makes more reasonable our plan to solve differential equations by letting y be an unknown power series, taking its appropriate derivative(s), and substituting into the differential equation to determine the coefficients in the series.

Finally, it is not always possible to determine an explicit expression for the nth coefficient of the Taylor series expansion of a function in terms of n. In this situation, we must be content with knowing the values of the first few coefficients. For this type of computation, we sometimes abbreviate the tail end of a power series by writing

$$O(t^n) = c_n t^n + c_{n+1} t^{n+1} + \cdots \qquad (8.2.11)$$

where we read the notation $O(t^n)$ as "order of t^n". For instance, we could write

$$e^t = 1 + t + \frac{t^2}{2} + O(t^3)$$

The next example emphasizes the fact that we cannot always explicitly determine a formula for the general nth term in the Taylor expansion of a function.

Example 8.2.4   Find the first four terms of the Taylor series expansion about $t = 0$ of the function

$$f(t) = \frac{t}{e^t + 1}$$

Solution.   Because f is the quotient of two functions that are analytic everywhere and the denominator is never zero, it follows that f is analytic everywhere. In particular, f is analytic at $a = 0$ and, therefore, has a Taylor series expansion there of the form

$$\frac{t}{e^t + 1} = b_0 + b_1 t + b_2 t^2 + b_3 t^3 + \cdots \qquad (8.2.12)$$

We know from the standard expansion of e^t that

$$e^t + 1 = 2 + t + \frac{t^2}{2!} + \frac{t^3}{3!} + \cdots$$

Multiplying both sides of (8.2.12) by this expression for $e^t + 1$, we obtain the identity

$$t = \left(2 + t + \frac{t^2}{2!} + \frac{t^3}{3!} + \cdots \right) \left(b_0 + b_1 t + b_2 t^2 + b_3 t^3 + \cdots \right)$$

Distributing to multiply these two series, we find that

$$t = 2b_0 + (2b_1 + b_0)t + \left(2b_2 + b_1 + \frac{b_0}{2} \right) t^2 + \left(2b_3 + b_2 + \frac{b_1}{2} + \frac{b_0}{6} \right) t^3 + \cdots$$

In order for this identity to hold, the uniqueness of Taylor series expansions established in theorem 8.2.2 implies that all of the coefficients of powers of t on the left must equal the corresponding coefficients of powers of t on the right. In particular, it must be the case that

$$0 = 2b_0$$

$$1 = 2b_1 + b_0$$

$$0 = 2b_2 + b_1 + \frac{1}{2} b_0$$

$$0 = 2b_3 + b_2 + \frac{1}{2} b_1 + \frac{1}{6} b_0$$

From this sequence of equalities, it follows that $b_0 = 0$, $b_1 = 1/2$, $b_2 = -1/4$, and $b_3 = 0$, so that

$$f(t) = \frac{t}{e^t + 1} = \frac{1}{2} t - \frac{1}{4} t^2 + 0t^3 + \cdots$$

Exercises 8.2

In exercises 1–4, determine the radius of convergence of the stated power series.

1. $\displaystyle\sum_{n=1}^{\infty} \frac{t^n}{n}$

2. $\displaystyle\sum_{n=1}^{\infty} \frac{2^n t^n}{n!}$

3. $\displaystyle\sum_{n=1}^{\infty} \frac{n^2 (t-2)^n}{5^n}$

4. $\displaystyle\sum_{n=1}^{\infty} \frac{(n!)^2 (t+3)^n}{(2n)!}$

In exercises 5–17, find the first four nonzero coefficients of the Taylor series expansion for each function $f(t)$ about $a = 0$. In addition, state the radius of convergence of the series expansion. Wherever possible, use known expansions and the techniques of examples 8.2.2, 8.2.3, and 8.2.4.

5. $f(t) = \sqrt{t+1}$

6. $f(t) = t^3 + 5t^2 - 3t + 8$

7. $f(t) = \dfrac{1}{1 + t^4}$

8. $f(t) = e^{-t^2}$

9. $f(t) = \dfrac{e^{2t} - 1}{2t}$

10. $f(t) = \dfrac{\sin t}{t}$

11. $f(t) = t^3 \sin t^2$

12. $f(t) = \cos t^3$

13. $f(t) = \cos t \sin t$

14. $f(t) = \cos^2(t)$

15. $f(t) = e^{-t} \sin t$

16. $f(t) = \dfrac{e^t}{1 + t}$

17. $f(t) = \arctan t^2$

In exercises 18–24, find the first four nonzero coefficients of the Taylor series expansion for each integral by first finding the expansion of the integrand and then integrating term by term.[2]

18. $\displaystyle\int_0^t \dfrac{1}{1 + s^4}\, ds$

19. $\displaystyle\int_0^t e^{-s^2}\, ds$

20. $\displaystyle\int_0^t \dfrac{e^{2s} - 1}{2s}\, ds$

21. $\displaystyle\int_0^t \dfrac{\sin s}{s}\, ds$

22. $\displaystyle\int_0^t s^3 \sin s^2\, ds$

––––––––

[2] Your work in exercises 5–17 will be helpful.

23. $\displaystyle\int_0^t \cos s^3 \, ds$

24. $\displaystyle\int_0^t \arctan s^2 \, ds$

8.3 Power series solutions of linear equations

In this section, we begin solving linear differential equations by assuming that the solution function may be expressed as a power series. To motivate our work, we revisit a familiar first-order equation (which we can solve easily by other means) to explore how series can be used in this way.

Example 8.3.1 By assuming that y has a power series expansion of the form $y(t) = a_0 + a_1 t + a_2 t^2 + a_3 t^3 + \cdots$, determine the solution to the initial-value problem

$$y' = y, \qquad y(0) = 1$$

Solution. Writing $y(t) = a_0 + a_1 t + a_2 t^2 + a_3 t^3 + \cdots$, we know

$$y'(t) = a_1 + 2a_2 t + 3a_3 t^2 + 4a_4 t^3 + \cdots$$

Equating y and y', we observe that

$$a_0 + a_1 t + a_2 t^2 + a_3 t^3 + \cdots = a_1 + 2a_2 t + 3a_3 t^2 + 4a_4 t^3 + \cdots \qquad (8.3.1)$$

Because of the uniqueness of Taylor series expansions (theorem 8.2.2), we may equate like coefficients of powers of t in (8.3.1), from which we deduce that the following recurrence relation among the coefficients a_i must hold:

$$a_0 = a_1$$

$$a_1 = 2a_2$$

$$a_2 = 3a_3$$

$$\vdots$$

$$a_n = (n+1)a_{n+1}$$

Provided that we know a_0, we can find all of the remaining values of a_i. Clearly, $a_0 = y(0)$, so using the initial condition $y(0) = 1$,

$$a_0 = 1, \quad a_1 = 1, \quad a_2 = \frac{1}{2}, \quad a_3 = \frac{1}{3}a_2 = \frac{1}{3\cdot 2}, \quad \cdots$$

From this sequence of coefficients and the general recurrence relation $a_{n+1} = \frac{1}{n+1}a_n$, we observe that $a_n = \frac{1}{n!}$, and therefore

$$y(t) = 1 + t + \frac{1}{2!}t^2 + \frac{1}{3!}t^3 + \cdots + \frac{1}{n!}t^n + \cdots$$

which we recognize as the familiar power series expansion of $y(t) = e^t$, the solution to the IVP $y' = y$, $y(0) = 1$.

Obviously there is no need to use power series to solve the IVP given in example 8.3.1, as it is a standard linear first-order equation. However, given our desire to solve higher order equations that are linear, but for which we currently lack a method for obtaining an analytic solution, this example is important since we hope to generalize from the simpler first-order constant coefficient case to the more difficult second-order non-constant coefficient one. For example, a linear second-order differential equation such as

$$y'' - 2ty' + y = 0 \tag{8.3.2}$$

in which the coefficients of y, y', and y'' are not all constant is not among the collection of equations whose solutions we can currently determine. Equations such as (8.3.2) belong to a family of equations of the general form

$$y'' + p(t)y' + q(t)y = f(t) \tag{8.3.3}$$

that we now aspire to solve.

Before we solve equations of form (8.3.3), we consider one more familiar example that introduces other critical ideas that arise when solving linear second-order equations through power series expansions. Because we already know the solution to the equation we consider, we will be able to check our work appropriately and better see the role that series expansions play.

Example 8.3.2   Solve the initial-value problem

$$y'' + y = 0, \quad y(0) = 1, \quad y'(0) = 1$$

by assuming that y has a power series expansion $y(t) = a_0 + a_1 t + a_2 t^2 + a_3 t^3 + a_4 t^4 + \cdots$.

Solution.   Since $y = a_0 + a_1 t + a_2 t^2 + a_3 t^3 + a_4 t^4 + \cdots$, it follows that

$$y' = a_1 + 2a_2 t + 3a_3 t^2 + 4a_4 t^3 + \cdots \text{ and}$$

$$y'' = 2a_2 + 3 \cdot 2a_3 t + 4 \cdot 3a_4 t^2 + 5 \cdot 4a_5 t^3 + \cdots$$

Substituting for y and y'' in the given equation $y'' + y = 0$, we have

$$(a_0 + a_1 t + a_2 t^2 + a_3 t^3 + a_4 t^4 + \cdots) + (2a_2 + 6a_3 t + 12a_4 t^2 + 20a_5 t^3 + \cdots) = 0$$

Gathering terms with like coefficients,

$$(a_0 + 2a_2) + (a_1 + 6a_3)t + (a_2 + 12a_4)t^2 + (a_3 + 20a_5)t^3 + \cdots = 0 \tag{8.3.4}$$

Setting each coefficient of powers of t in (8.3.4) equal to zero implies that the following sequence of equalities holds:

$a_0 = -2a_2$	$a_1 = -6a_3$
$a_2 = -12a_4$	$a_3 = -20a_5$
$a_4 = -30a_6$	$a_5 = -42a_7$
$\vdots$	$\vdots$
$a_{2n} = -(2n+2)(2n+1)a_{2n+2}$	$a_{2n+1} = -(2n+3)(2n+2)a_{2n+3}$

We group these equations into the two columns shown for the natural reason that the coefficients with even indices depend recursively on one another, as do the coefficients with odd indices. Furthermore, we see that if we can identify both a_0 and a_1 (which we can through the two stated initial conditions), then we can determine all of the remaining coefficients.

Specifically, since $y(0) = 1$ and $a_0 = y(0)$, it follows that $a_0 = 1$. Similarly, with the given condition $y'(0) = 1$ and the fact that $a_1 = y'(0)$, we know $a_1 = 1$. Thus, from the sequence of equalities with even indices above,

$$a_0 = 1, \quad a_2 = -\frac{1}{2}, \quad a_4 = -\frac{1}{4 \cdot 3} a_2 = \frac{1}{4 \cdot 3 \cdot 2} = \frac{1}{4!},$$

$$\text{and} \quad a_6 = -\frac{1}{30} a_4 = -\frac{1}{6 \cdot 5 \cdot 4!} = -\frac{1}{6!}$$

From this and the stated recurrence relation for a_{2n} and a_{2n+2}, we observe that

$$a_{2n} = (-1)^n \frac{1}{(2n)!}, \quad n = 0, 1, 2, \ldots. \tag{8.3.5}$$

The formula (8.3.5) implies that the portion of the series expansion for y in which all of the powers of t are even will be

$$y_1 = 1 - \frac{1}{2!} t^2 + \frac{1}{4!} t^4 - \frac{1}{6!} t^6 + \cdots \tag{8.3.6}$$

which we recognize as the familiar series expansion for $\cos t$.

Returning to the recurrence relation involving the coefficients with odd indices, nearly identical work to that with the even coefficients shows that

$$a_1 = 1, \quad a_3 = -\frac{1}{3!}, \quad a_5 = -\frac{1}{5 \cdot 4} a_3 = \frac{1}{5!}, \quad \text{and } a_7 = -\frac{1}{42} a_4 = -\frac{1}{7!}$$

These observations imply that the part of the expansion of y involving odd coefficients has form

$$y_2 = t - \frac{1}{3!} t^3 + \frac{1}{5!} t^5 - \frac{1}{7!} t^7 + \cdots \tag{8.3.7}$$

which is $\sin t$.

Hence our work with series expansions at (8.3.6) and (8.3.7) has shown that

$$y = 1 + t - \frac{1}{2!} t^2 - \frac{1}{3!} t^3 + \frac{1}{4!} t^4 + \frac{1}{5!} t^5 - \frac{1}{6!} t^6 - \frac{1}{7!} t^7 + \cdots$$

$$= 1 - \frac{1}{2!} t^2 + \frac{1}{4!} t^4 - \frac{1}{6!} t^6 + \cdots + t - \frac{1}{3!} t^3 + \frac{1}{5!} t^5 - \frac{1}{7!} t^7 + \cdots$$

$$= \cos t + \sin t \tag{8.3.8}$$

Again, it is no surprise that $y = \cos t + \sin t$ is the solution to the IVP $y'' + y = 0, y(0) = 1, y'(0) = 1$. We know from our work in several different contexts that the general solution to this differential equation is $y = c_1 \cos t + c_2 \sin t$, and can easily see that the given two initial conditions lead to $c_1 = c_2 = 1$. Even without the initial conditions, we could have determined from our work in example 8.3.2 that $y = a_0 \cos t + a_1 \sin t$. Regardless, there is a great deal we can learn about series solutions to differential equations by thinking carefully about our work in this familiar example.

First, we saw that in order to get the recurrence relations started, we needed to know the values of a_0 and a_1. This reinforces the fact that the solution space to the second-order equation is two dimensional, and suggests that the power series expansion has the property that it detects the need for two linearly independent solutions. Next, we observe from our work in example 8.3.2 that two different unlinked series solutions arose in the solution; these turned out to be the expansions for the cosine and sine functions, respectively, each of which has an infinite radius of convergence. This led to the overall solution series being convergent for every value of t. Finally, we note that normally we will need to be content with expressions that state the first few nonzero terms of a power series expansion, as we cannot expect in general to be able to recognize familiar power series expansions within solutions, as we did at (8.3.8).

In general, we will be interested in linear differential equations of the form

$$y'' + p(t)y' + q(t)y = 0 \tag{8.3.9}$$

If $p(t)$ and $q(t)$ are both analytic functions at $t = a$ (that is, both have a Taylor expansion at a), then we call $t = a$ an *ordinary point* of the DE (8.3.9). Otherwise, $t = a$ is a *singular point* of (8.3.9). The following theorem tells us that if $t = 0$ is an ordinary point of (8.3.9), then there exist two linearly independent solutions to the DE that may be represented by Taylor series centered at $t = 0$.

Theorem 8.3.1   If $t = 0$ is an ordinary point of (8.3.9), then there exist two linearly independent solutions

$$y_1(t) = \sum_{n=0}^{\infty} a_n t^n \quad \text{and} \quad y_2(t) = \sum_{n=0}^{\infty} b_n t^n \tag{8.3.10}$$

Both series converge in a disk $|t| < R$, where R is at least as large as the distance from the origin to the nearest singular point of the functions $p(t)$ and $q(t)$.

In example 8.3.2, the coefficient functions of y' and y in the DE were simply the constant functions 0 and 1, which are each analytic everywhere. Theorem 8.3.1 implies that the two series expansions we found (which were those of the cosine and sine functions) must therefore converge everywhere. We see from this result that anytime the coefficient functions $p(t)$ and $q(t)$ are constant, the solution functions that arise must converge everywhere. This is not surprising, given our experience that in the case of linear differential equations with constant coefficients, solutions essentially consist of the functions e^{kt}, $\sin kt$, and $\cos kt$. More generally, we can now state that if $p(t)$ and $q(t)$ are polynomial functions, which are also analytic everywhere, then the series in (8.3.10) must both converge everywhere.

We now consider an example involving a differential equation that we are unable to solve by other means in order to gain more understanding of the role played by infinite series in its solution.

Example 8.3.3   Consider the linear second-order differential equation

$$y'' - 2ty' + y = 0 \qquad (8.3.11)$$

Determine two linearly independent series solutions to this equation. Then, solve the initial-value problem given by this DE along with the initial conditions $y(0) = 2$, $y'(0) = -1$.

Solution.   We begin by assuming that $y = a_0 + a_1 t + a_2 t^2 + a_3 t^3 + \cdots$. From this, it follows

$$y' = a_1 + 2a_2 t + 3a_3 t^2 + 4a_4 t^3 + \cdots \qquad = \sum_{n=1}^{\infty} n a_n t^{n-1}$$

$$-2ty' = -2at - 4a_2 t^2 - 6a_3 t^3 - 8a_4 t^3 + \cdots = -\sum_{n=1}^{\infty} 2n a_n t^n$$

$$y'' = 2a_2 + 6a_3 t + 12a_4 t^2 + 20a_5 t^3 + \cdots = \sum_{n=2}^{\infty} n(n-1) a_n t^{n-2}$$

In many instances, it will be most convenient to work with power series represented in the shorthand sigma (Σ) notation, which is how we will proceed from here. Substituting in (8.3.11) with the series expressions for y'', $-2ty'$, and y, we find

$$\sum_{n=2}^{\infty} n(n-1) a_n t^{n-2} - \sum_{n=1}^{\infty} 2n a_n t^n + \sum_{n=0}^{\infty} a_n t^n = 0 \qquad (8.3.12)$$

In order to equate the coefficients of like powers of t, it is helpful to write each series in (8.3.12) using the same indices for the sum. Replacing n with $n+2$ allows us to write

$$\sum_{n=2}^{\infty} n(n-1) a_n t^{n-2} = \sum_{n=0}^{\infty} (n+2)(n+1) a_{n+2} t^n$$

In addition, observe that

$$\sum_{n=1}^{\infty} 2n a_n t^n = \sum_{n=0}^{\infty} 2n a_n t^n$$

because the term $-2n a_n$ vanishes when $n = 0$. Therefore we can revise (8.3.12) to have the form

$$\sum_{n=0}^{\infty} (n+2)(n+1) a_{n+2} t^n + \sum_{n=0}^{\infty} -2n a_n t^n + \sum_{n=0}^{\infty} a_n t^n = 0 \qquad (8.3.13)$$

Now that each series is indexed from $n = 0$ with corresponding powers of t, we can combine the three sums into one and write

$$\sum_{n=0}^{\infty} [(n+2)(n+1) a_{n+2} - 2n a_n + a_n] t^n = 0 \qquad (8.3.14)$$

Because (8.3.14) implies that every coefficient of the series must be zero, we see that the constants a_n must satisfy the recurrence relation

$$(n+2)(n+1)a_{n+2} - 2na_n + a_n = 0$$

or equivalently

$$a_{n+2} = \frac{2n-1}{(n+2)(n+1)}a_n, \quad n = 0, 1, 2, \ldots. \tag{8.3.15}$$

Here it is essential to observe that since the subscripts differ by two in (8.3.15), we can obtain two distinct series solutions to the original equation (8.3.11), one involving all of the even terms and the other all of the odd ones. In particular, considering $n = 0, 2, 4, \ldots$, we have from (8.3.15) that

$$a_2 = -\frac{1}{2}a_0, \quad a_4 = \frac{3}{3 \cdot 4}a_2 = \frac{-1 \cdot 3}{2 \cdot 3 \cdot 4}a_0, \quad \text{and } a_6 = \frac{7}{6 \cdot 5}a_4 = \frac{-1 \cdot 3 \cdot 7}{2 \cdot 3 \cdot 4 \cdot 5 \cdot 6}a_0$$

More generally, the pattern

$$a_{2n} = \frac{-1 \cdot 3 \cdot 7 \cdots (4n-5)}{(2n)!}$$

holds and therefore

$$y_1(t) = a_0 - \frac{1}{2}a_0 t^2 - \frac{1}{8}a_0 t^4 - \frac{7}{240}a_0 t^6 + \cdots$$

$$= a_0 - a_0 \sum_{n=1}^{\infty} \frac{1 \cdot 3 \cdot 7 \cdots (4n-5)}{(2n)!} t^{2n} \tag{8.3.16}$$

Similarly, if we examine the odd terms for $n = 1, 3, 5, \ldots$ in (8.3.15), we see

$$a_3 = \frac{1}{2 \cdot 3}a_1, \quad a_5 = \frac{5}{4 \cdot 5}a_3 = \frac{1 \cdot 5}{2 \cdot 3 \cdot 4 \cdot 5}a_1, \quad \text{and } a_7 = \frac{9}{6 \cdot 7}a_5 = \frac{1 \cdot 5 \cdot 9}{2 \cdot 3 \cdot 4 \cdot 5 \cdot 6 \cdot 7}a_1$$

Thus, we find

$$a_{2n+1} = \frac{1 \cdot 5 \cdot 9 \cdots (4n-3)}{(2n+1)!}a_1$$

and therefore

$$y_2(t) = a_1 t + \frac{1}{6}a_1 t^3 + \frac{1}{24}a_1 t^5 + \cdots$$

$$= a_1 t + a_1 \sum_{n=1}^{\infty} \frac{1 \cdot 5 \cdot 9 \cdots (4n-3)}{(2n+1)!} t^{2n+1} \tag{8.3.17}$$

Because y_1 only involves even powers of t and y_2 only involves odd powers of t, it is obvious that y_1 and y_2 must be linearly independent functions: it is impossible for one to be a scalar multiple of the other. Hence we have found the two basic solutions to the given DE and the general solution is

$$y = a_0 y_1 + a_1 y_2$$

$$= a_0 \left(1 - \sum_{n=1}^{\infty} \frac{1 \cdot 3 \cdot 7 \cdots (4n-5)}{(2n)!} t^{2n}\right) + a_1 \left(t + \sum_{n=1}^{\infty} \frac{1 \cdot 5 \cdot 9 \cdots (4n-3)}{(2n+1)!} t^{2n+1}\right)$$

Moreover, since $p(t) = -2t$ and $q(t) = 1$ are analytic everywhere, it follows from theorem 8.3.1 that both y_1 and y_2 converge for all values of t, as must the general solution (8.3.18).

Finally, if we desire to solve the initial-value problem with $y(0) = 2$ and $y'(0) = -1$, we need only observe from our beginning assumption regarding the series expansion of y that $y(0) = a_0 = 2$ and $y'(0) = a_1 = -1$. Therefore, the solution to the IVP is

$$y = 2\left(1 - \sum_{n=1}^{\infty} \frac{1 \cdot 3 \cdot 7 \cdots (4n-5)}{(2n)!} t^{2n}\right) - \left(t + \sum_{n=1}^{\infty} \frac{1 \cdot 5 \cdot 9 \cdots (4n-3)}{(2n+1)!} t^{2n+1}\right)$$

In the recurrence relation that arises from assuming that $y = a_0 + a_1 t + a_2 t^2 + \cdots$, it is not always obvious that two linear solutions to the original linear second-order equation arise. Often, we must content ourselves with finding the first several terms of the overall general solution and rely on theorem 8.3.1 to tell us that both have been found. We close this section with an example that demonstrates this fact through connections to earlier material we have studied.

Example 8.3.4   Use infinite series to determine the solution to the initial-value problem

$$y'' - 2y' - 3y = 0, \quad y(0) = 4, \quad y'(0) = 0 \qquad (8.3.18)$$

Compare your result to the known solution to this IVP which can be found without using series.

Solution.   Considering the series expansions for y, y', and y'', we observe that

$$y = a_0 + a_1 t + a_2 t^2 + a_3 t^3 + \cdots + a_n t^n + \cdots$$

$$y' = a_1 + 2a_2 t + 3a_3 t^2 + 4a_4 t^3 + \cdots + (n+1)a_{n+1} t^n + \cdots$$

$$y'' = 2a_2 + 6a_3 t + 12a_4 t^2 + 20a_5 t^3 + \cdots + (n+2)(n+1)a_{n+2} t^n + \cdots$$

From the differential equation $y'' - 2y' - 3y = 0$, we know that $y'' = 2y' + 3y$. Equating like coefficients from the expressions for y'' and $2y' + 3y$, we find the recurrence relation

$$2a_2 = 2a_1 + 3a_0$$

$$6a_3 = 4a_2 + 3a_1$$

$$12a_4 = 6a_3 + 3a_2$$

$$20a_5 = 8a_4 + 3a_3$$

$$\vdots$$

More generally, we can state that for any $n \geq 2$,

$$a_n = \frac{(2n-2)a_{n-1} + 3a_{n-2}}{n(n-1)}$$

Using the given initial conditions, we find that $a_0 = y(0) = 4$ and $a_1 = y'(0) = 0$, and subsequently that

$$a_2 = \frac{2a_1 + 3a_0}{2} = \frac{0+12}{2} = 6$$

$$a_3 = \frac{4a_2 + 3a_1}{6} = \frac{24+0}{6} = 4$$

$$a_4 = \frac{6a_3 + 3a_2}{12} = \frac{24+18}{12} = \frac{7}{2}$$

and therefore the solution to the IVP is

$$y = 4 + 6t^2 + 4t^3 + \frac{7}{2}t^4 + \cdots$$

We can confirm that this is in fact the correct solution by solving the IVP through another approach and considering power series expansions of the basic solution functions. In particular, since the characteristic equation of (8.3.18) is $r^2 - 2r - 3 = 0$ with roots $r = 3$ and $r = -1$, the general solution of the DE is

$$y = c_1 e^{3t} + c_2 e^{-t}$$

It is a standard exercise to show that the values of the constants that satisfy the initial conditions are $c_1 = 1$ and $c_2 = 3$, so that

$$y = e^{3t} + 3e^{-t}$$

If we now employ the standard power series expansion for e^t to write series expansions for the two solutions present in y, and then combine like terms, we observe that

$$y = e^{3t} + 3e^{-t}$$

$$= \left(1 + 3t + \frac{9t^2}{2!} + \frac{27t^3}{3!} + \frac{81t^4}{4!} + \cdots\right) + \left(3 - 3t + \frac{3t^2}{2!} - \frac{3t^3}{3!} + \frac{3t^4}{4!} - \cdots\right)$$

$$= 4 + \frac{12t^2}{2!} + \frac{24t^3}{3!} + \frac{84t^4}{4!} + \cdots$$

$$= 4 + 6t^2 + 4t^3 + \frac{7}{2}t^4 + \cdots$$

which is precisely the power series expansion of the solution we found at the outset.

Example 8.3.4 demonstrates that although the series form of the solution can hide some of the inherent structure in the solution, this approach is nonetheless straightforward to apply and will effectively lead us to the power series expansion of the solution to a stated IVP.

Exercises 8.3
In exercises 1–13, find the first four terms in the Taylor series representation of the general solution to the stated DE.

1. $y'' + ty' = 0$

2. $y'' + 4y' = 0$

3. $y'' + 4y = 0$

4. $y'' + ty = 0$

5. $y'' + 6y' + 5y = 0$

6. $y'' + y' + 4y = 0$

7. $y'' - y' - 6y = 0$

8. $y'' + t^2 y = 0$

9. $(1 - t)y'' + y = 0$

10. $(t^2 - 1)y'' - 4y = 0$

11. $y'' + 3ty' + 3y = 0$

12. $(t^2 + 1)y'' - 2y = 0$

13. $(1 - t^2)y'' - 12ty' - 18y = 0$

In exercises 14–17, find the first four nonzero coefficients of the Taylor series expansion for the solution to the stated IVP.

14. $(4 - t^2)y'' + 2y = 0$, $\quad y(0) = 0$, $\quad y'(0) = 1$

15. $y'' + (1 - t)y = 0$, $\quad y(0) = 1$, $\quad y'(0) = 0$

16. $y'' - t^2 y' + y\sin t = 0$, $\quad y(0) = 0$, $\quad y'(0) = 1$

17. $y' + y\sin t = 0$, $\quad y(0) = 1$, $\quad y'(0) = 0$

8.4 Legendre's equation

A differential equation that arises naturally in physics, particularly when using spherical coordinates, is the *Legendre equation*,

$$(1 - t^2)y'' - 2ty' + \lambda(\lambda + 1)y = 0 \tag{8.4.1}$$

The parameter λ is often a positive integer, though it is allowed to be any real, non-negative constant. If we divide both sides of (8.4.1) by $1 - t^2$ to write the equation in standard form $y'' + p(t)y' + q(t)y = 0$, we have

$$y'' - \frac{2t}{1-t^2}y' + \frac{\lambda(\lambda+1)}{1-t^2}y = 0 \tag{8.4.2}$$

With

$$p(t) = -\frac{2t}{1-t^2} \quad \text{and} \quad q(t) = \frac{\lambda(\lambda+1)}{1-t^2}$$

it follows that the origin is an ordinary point of Legendre's equation and the nearest singularities lie at $t = \pm 1$. We therefore expect that we can find Taylor series expansions about $t = 0$ for each of the two linearly independent solutions of (8.4.1), and the radius of convergence of each such series will be at least 1.

To solve the Legendre equation, we assume that

$$y(t) = \sum_{n=0}^{\infty} a_n t^n$$

and consider the three terms present in the DE: $(1-t^2)y''$, $-2ty'$, and $\lambda(\lambda+1)y$. Letting $\alpha = \lambda(\lambda + 1)$ and writing each of these expressions in their series expansion, we have

$$(1-t^2)y'' = (1-t^2)\sum_{n=2}^{\infty} n(n-1)a_n t^{n-2} = \sum_{n=2}^{\infty} n(n-1)a_n t^{n-2} - \sum_{n=2}^{\infty} n(n-1)a_n t^n$$

$$= \sum_{n=0}^{\infty} (n+2)(n+1)a_{n+2} t^n - \sum_{n=0}^{\infty} n(n-1)a_n t^n \tag{8.4.3}$$

$$-2ty' = -2t\sum_{n=1}^{\infty} na_n t^{n-1} = \sum_{n=1}^{\infty} -2na_n t^n = \sum_{n=0}^{\infty} -2na_n t^n \tag{8.4.4}$$

$$\alpha y = \sum_{n=0}^{\infty} \alpha a_n t^n \tag{8.4.5}$$

To achieve the final expression for $(1 - t^2)y''$ in (8.4.3), we re-indexed the first sum by letting n be replaced by $n+2$ and lowering the index, and re-indexed the second sum by noting that when $n = 0$ and $n = 1$, the coefficient $n(n-1)$ vanishes, so starting at $n = 0$ is the same as starting at $n = 2$. Likewise, for the expression for $-2ty'$, the term $na_n t^n$ is zero when $n = 0$, so we can start the sum at $n = 0$ instead of $n = 1$ in (8.4.4). Thus, all three series are written in terms of powers of t^n starting at $n = 0$.

Next, to satisfy Legendre's equation (8.4.1), we take the series expressions in (8.4.3), (8.4.4), and (8.4.5) and set their collective sum to zero. Doing so,

$$0 = (1 - t^2)y'' - 2ty' + \alpha y$$

$$= \sum_{n=0}^{\infty} (n+2)(n+1)a_{n+2} t^n - \sum_{n=0}^{\infty} n(n-1)a_n t^n + \sum_{n=0}^{\infty} -2na_n t^n + \sum_{n=0}^{\infty} \alpha a_n t^n$$

$$= \sum_{n=0}^{\infty} [(n+2)(n+1)a_{n+2} - (n(n-1) + 2n - \alpha)a_n] t^n$$

$$= \sum_{n=0}^{\infty} [(n+2)(n+1)a_{n+2} - (n^2 + n - \alpha)a_n] t^n \qquad (8.4.6)$$

We thus observe (8.4.6) implies the recurrence relation

$$(n+2)(n+1)a_{n+2} - (n^2 + n - \alpha)a_n = 0 \qquad (8.4.7)$$

Recalling that $\alpha = \lambda(\lambda + 1) = \lambda^2 + \lambda$, we may write

$$n^2 + n - \alpha = n^2 + n - \lambda^2 - \lambda = (n - \lambda)(n + \lambda + 1) \qquad (8.4.8)$$

Hence, (8.4.7) and (8.4.8) together show

$$a_{n+2} = \frac{(n - \lambda)(n + \lambda + 1)}{(n+2)(n+1)} a_n \qquad (8.4.9)$$

As we have seen in certain other DEs, the recurrence relation (8.4.9) makes all of the even coefficients in the expansion for y depend on a_0, and all of the odd coefficients depend on a_1. Assuming that $a_0 = 1$ and computing the first few even coefficients, we find that

$$a_0 = 1, \quad a_2 = \frac{(-\lambda)(\lambda + 1)}{2 \cdot 1} a_0, \quad a_4 = \frac{(2 - \lambda)(3 + \lambda)}{4 \cdot 3} a_2$$

so that one solution to the Legendre equation is

$$y_1(t) = 1 - \frac{1}{2!}\lambda(\lambda + 1)t^2 + \frac{1}{4!}\lambda(\lambda + 1)(\lambda - 2)(\lambda + 3)t^4 + \cdots \qquad (8.4.10)$$

Similar computations for the odd coefficients with $a_1 = 1$ results in the function

$$y_2(t) = t - \frac{1}{3!}(\lambda - 1)(\lambda + 2)t^3 + \frac{1}{5!}(\lambda - 1)(\lambda - 3)(\lambda + 2)(\lambda + 4)t^5 + \cdots \qquad (8.4.11)$$

The solutions y_1 and y_2 are clearly linearly independent and therefore form a basis for the set of all solutions to the Legendre equation. Note particularly that each depends directly on the parameter λ, as the Legendre equation is actually a family of equations where each equation depends on λ. In our development of y_1 and y_2, note that we assumed $a_0 = 1$ and $a_1 = 1$, which is equivalent to assuming that $y(0) = 1$ and $y'(0) = 1$. The general solution of the Legendre equation is $y = a_0 y_1 + a_1 y_2$, where y_1 and y_2 are given by 8.4.10 and 8.4.11, respectively.

The case when λ is a non-negative integer is particularly interesting. From the recurrence relation (8.4.9), whenever $\lambda = n$, it follows that $a_{n+2} = 0$ and hence $a_{n+4}, a_{n+6}, \dots$ are all zero. Since this causes the series expansion of y_1 or y_2 to terminate, one of the resulting solutions to the differential equation is a polynomial. In particular, if λ is an even integer, say $\lambda = 2m$, then $y_1(t)$ is a polynomial of degree $2m$. For example,

$$\lambda = 0: \qquad y_1(t) = 1$$

$$\lambda = 2: \qquad y_1(t) = 1 - 3t^2$$

$$\lambda = 4: \qquad y_1(t) = 1 - 10t^2 + \frac{35}{3}t^4$$

Similarly, in the case where $\lambda = 2m + 1$ is an odd integer, $y_2(t)$ is a polynomial of degree $2m + 1$. The first few examples for small values of λ are

$$\lambda = 1: \qquad y_2(t) = t$$

$$\lambda = 3: \qquad y_2(t) = t - \frac{5}{3}t^3$$

$$\lambda = 5: \qquad y_2(t) = t - \frac{14}{3}t^3 + \frac{21}{5}t^5$$

These polynomials demonstrate that when λ is non-negative integer, at least one basic solution of the Legendre equation is a polynomial function. Moreover, since the Legendre equation is linear, any scalar multiple of a solution is also a solution, so we can scale these polynomials however we like. Doing so to make the polynomial's value 1 when $t = 1$ results in the family of polynomials

$$P_0(t) = 1$$

$$P_1(t) = t$$

$$P_2(t) = \frac{3}{2}t^2 - \frac{1}{2}$$

$$P_3(t) = \frac{5}{2}t^3 - \frac{3}{2}$$

$$P_4(t) = \frac{35}{8}t^4 - \frac{30}{8}t^2 + \frac{3}{8}$$

$$P_5(t) = \frac{63}{8}t^5 - \frac{70}{8}t^3 + \frac{15}{8}$$

The polynomials $P_n(t)$, which can also be described through a recurrence relation linking P_{n+2} to P_{n+1} and P_n, are known as the *Legendre polynomials* and form a well-known class of so-called *orthogonal polynomials*. The Legendre polynomials have many interesting properties, including the fact that each has n real, distinct roots that lie in the interval $(-1, 1)$ and demonstrate an oscillatory behavior similar to the graph of $P_{11}(t)$ shown in figure 8.1. The study of orthogonal polynomials has important ramifications in many areas of mathematics and physics, but lies beyond the scope of this text.

Regardless of whether λ is a non-negative integer or not, the two infinite series expansions for y_1 and y_2 in (8.4.10) and (8.4.11) are the two linearly independent solutions of the Legendre equation. In the case where λ is a non-negative integer, we have shown that one of these two infinite series terminates

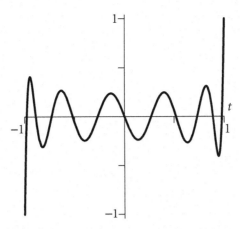

Figure 8.1 The degree 11 Legendre polynomial, $P_{11}(t)$.

to form a polynomial, one of the Legendre polynomials. The other solution turns out to have recognizable structure as well.

For instance, when $\lambda = 0$, we know that one solution to the Legendre equation comes from $y_1(t) = 1 = P_0(t)$. Setting $\lambda = 0$ in $y_2(t)$, it follows

$$y_2(t) = t - \frac{-1 \cdot 2}{3!}t^3 + \frac{-1 \cdot (-3) \cdot 2 \cdot 4}{5!}t^5 + \cdots$$

$$= t + \frac{1}{3}t^3 + \frac{1}{5}t^5 + \cdots \qquad (8.4.12)$$

It can be shown from this expansion that

$$y_2(t) = \frac{1}{2}\ln\left(\frac{1+t}{1-t}\right)$$

Thus, when $\lambda = 0$, a second linearly independent solution is given by $Q_0(t) = \frac{1}{2}\ln\left(\frac{1+t}{1-t}\right)$ and we write $y = c_1 P_0 + c_2 Q_0$. More generally, it can be shown that for any non-negative integer $\lambda = n$, a related expression involving Q_0 exists for the second linearly independent solution Q_n that is not a polynomial. In particular, these functions are known as *Legendre functions of the second kind*; the first several of these functions are given by

$$Q_0(t) = \frac{1}{2}\ln\frac{1+t}{1-t}$$

$$Q_1(t) = P_1(t)Q_0(t) - 1$$

$$Q_2(t) = P_2(t)Q_0(t) - \frac{3}{2}t$$

$$Q_3(t) = P_3(t)Q_0(t) - \frac{5}{2}t^2 + \frac{2}{3}$$

$$Q_4(t) = P_3(t)Q_0(t) - \frac{35}{8}t^3 + \frac{55}{24}$$

Note that the presence of $Q_0(t)$ in each solution highlights the fact that singularities are present in the Legendre equation at $t = \pm 1$. The functions $P_1(t)$, $P_2(t)$, ... are the previously noted Legendre polynomials. Further, the general solution of the Legendre equation with $\lambda = n \geq 0$ is therefore

$$y(t) = c_1 P_n(t) + c_2 Q_n(t) \qquad (8.4.13)$$

We close this section with an example.

Example 8.4.1   Find the solution of the initial-value problem

$$(1 - t^2)y'' - 2ty' + 12y = 0, \quad y(0) = 1, \quad y'(0) = 1$$

Solution.   First, observe that the given DE is Legendre's equation with $\lambda = 3$, since $3(3 + 1) = 12$. From our earlier work in this section, we know that the general solution is

$$y(t) = c_1 P_3(t) + c_2 Q_3(t)$$

$$= c_1 P_3(t) + c_2 \left(P_3(t)Q_0(t) - \frac{5}{2}t^2 + \frac{2}{3} \right)$$

$$= P_3(t)(c_1 + c_2 Q_0(t)) + c_2 \left(-\frac{5}{2}t^2 + \frac{2}{3} \right)$$

$$= \left(\frac{5}{2}t^3 - \frac{3}{2}t \right)\left(c_1 + \frac{c_2}{2}\ln\frac{1-t}{1+t} \right) + c_2 \left(-\frac{5}{2}t^2 + \frac{2}{3} \right) \qquad (8.4.14)$$

Applying the initial conditions $y(0) = 1$ and $y'(0) = 1$ to 8.4.14, we can show that $c_1 = -2/3$ and $c_2 = 3/2$, and thus

$$y = \left(\frac{5}{2}t^3 - \frac{3}{2}t \right)\left(-\frac{2}{3} + \frac{3}{4}\ln\frac{1-t}{1+t} \right) - \frac{15}{4}t^2 + 1$$

is the solution to the given IVP.

Exercises 8.4

1. Verify by direct substitution that the Legendre equation is satisfied by the polynomials $P_2(t)$ and $P_3(t)$ when $\lambda = 2$ and $\lambda = 3$, respectively.

2. Verify by direct substitution that $Q_0(t) = \frac{1}{2}\ln(1+t)/(1-t)$ is a solution of Legendre's equation with $\lambda = 0$.

3. Determine the Taylor series expansion about $a = 0$ of
$f(t) = \frac{1}{2}\ln(1+t)/(1-t)$ and confirm that this matches (8.4.12).

4. Determine expressions for $P_6(t)$ and $P_7(t)$.

In exercises 5–7, find the general solution of the stated differential equation in terms of $P_n(t)$ and $Q_n(t)$. (*Hint:* Use the method of undetermined coefficients in the standard way to find a particular solution of each equation.)

5. $(1-t^2)y'' - 2ty' + 6y = 6$

6. $(1-t^2)y'' - 2ty' + 20y = 36t$

7. $(1-t^2)y'' - 2ty' + 30y = 12t^2$

In exercises 8–17, find the first four nonzero coefficients of the Taylor series expansion (about $t = 0$) for the solution to the stated IVP.

8. $(1-t^2)y'' - 2ty' + 2y = 0$, $\quad y(0) = 1$, $y'(0) = 0$

9. $(1-t^2)y'' - 2ty' + 3y = 0$, $\quad y(0) = 1$, $y'(0) = 0$

10. $(1-t^2)y'' - 2ty' + 20y = 18t$, $\quad y(0) = 0$, $y'(0) = 1$

11. $9(1-t^2)y'' - 18ty' + 4y = 0$, $\quad y(0) = 0$, $y'(0) = 1$

12. $(1-t^2)y'' - 2ty' + 20y = 0$, $\quad y(0) = 1$, $y'(0) = 1$

13. $(1-t^2)y'' - 2ty' + 20y = 14t^2$, $\quad y(0) = 3$, $y'(0) = 1$

8.5 Three important examples

In this penultimate section on series solutions to differential equations, we consider and discuss three examples that arise in applied physics.

8.5.1 The Hermite equation

The *Hermite equation* is the linear second-order differential equation given by

$$y'' - 2ty' + 2qu = 0 \tag{8.5.1}$$

where q is a real constant. Using the Taylor series expansions for y, y', and y'' in the usual way with $y = a_0 + a_1 t + a_2 t^2 + \cdots$, it can be shown that

$$\sum_{n=0}^{\infty}[(n+2)(n+1)a_{n+2} - 2(n-q)a_n]t^n = 0 \tag{8.5.2}$$

from which follows the recurrence relation

$$a_{n+2} = \frac{2(n-q)}{(n+1)(n+2)}a_n, \quad n = 0, 1, 2, \ldots . \tag{8.5.3}$$

As we have seen in previous examples, the even-subscripted coefficients depend on $y(0) = a_0$, and the odd-subscripted coefficients involve $y'(0) = a_1$.

To calculate the first few nonzero terms in the expansions for the solution $y_1(t)$ involving even powers of t, we observe that

$$a_2 = \frac{2(0-q)}{1 \cdot 2} a_0 = -2\frac{q}{2!} a_0$$

$$a_4 = \frac{2(2-q)}{3 \cdot 4} a_2 = -2^2 \frac{q(2-q)}{4!} a_0$$

$$a_6 = \frac{2(4-q)}{5 \cdot 6} a_2 = -2^3 \frac{q(2-q)(4-q)}{6!} a_0$$

More generally, it follows that

$$a_{2k} = -2^k \frac{q(2-q)\cdots(2k-2-q)}{(2k)!} a_0 \tag{8.5.4}$$

If we elect to use the initial conditions $y(0) = 1$ and $y'(0) = 0$, this implies that $a_0 = 1$ and $a_1 = 0$; the latter condition and the recurrence relation (8.5.3) imply that all odd-subscripted coefficients are zero, and hence one solution to the Hermite differential equation is

$$y_1(t) = a_0 + a_1 t + a_2 t^2 + \cdots$$

$$= 1 - \frac{2q}{2!}t^2 - \frac{2^2 q(2-q)}{4!}t^4 - \cdots$$

$$= 1 - \sum_{n=1}^{\infty} 2^n \frac{q(2-q)\cdots(2n-2-q)}{(2n)!}t^{2n} \tag{8.5.5}$$

Using similar reasoning with odd-subscripted coefficients, (8.5.3) implies

$$a_3 = \frac{2(1-q)}{2 \cdot 3} a_1$$

$$a_5 = \frac{2(3-q)}{4 \cdot 5} a_3 = 2^2 \frac{(1-q)(3-q)}{5!} a_1$$

$$a_7 = \frac{2(5-q)}{6 \cdot 7} a_5 = 2^3 \frac{(1-q)(3-q)(5-q)}{7!} a_1$$

From this, we can deduce that the general odd coefficient is given by

$$a_{2k+1} = 2^k \frac{(1-q)(3-q)\cdots(2k-1-q)}{(2k+1)!} a_1 \tag{8.5.6}$$

Using the initial conditions $y(0) = 0 = a_0$ and $y'(0) = 1 = a_1$, a second solution to the Hermite equation is

$$y_2(t) = t + \sum_{n=1}^{\infty} 2^n \frac{(1-q)(3-q)\cdots(2n-1-q)}{(2n+1)!}t^{2n+1} \tag{8.5.7}$$

Since $y_1(t)$ and $y_2(t)$ are linearly independent, the general solution to the Hermite equation is

$$y = c_1 y_1 + c_2 y_2$$

$$= c_1 \left(1 - \sum_{n=1}^{\infty} 2^n \frac{q(2-q)\cdots(2n-2-q)}{(2n)!} t^{2n} \right)$$

$$+ c_2 \left(t + \sum_{n=1}^{\infty} 2^n \frac{(1-q)(3-q)\cdots(2n-1-q)}{(2n+1)!} t^{2n+1} \right) \qquad (8.5.8)$$

Just as we experienced with Legendre's equation, there are values for the constant q in the Hermite equation that lead to polynomial solutions. In particular, the presence of the factor $(2n-2-q)$ in $y_1(t)$ implies that whenever q is an even, non-negative integer, then $y_1(t)$ is a polynomial. Specifically, from (8.5.5), when $q=0$, $q=2$, and $q=4$, it follows that

$$q=0: \qquad y_1(t) = 1$$

$$q=2: \qquad y_1(t) = 1 - 2t^2 \qquad (8.5.9)$$

$$q=4: \qquad y_1(t) = 1 - 4t^2 + \frac{4}{3}t^4$$

Similarly, for $q=1$, $q=3$, and $q=5$, the function $y_2(t)$ that is a solution to the Hermite equation is found to be

$$q=1: \qquad y_2(t) = t$$

$$q=3: \qquad y_2(t) = t - \frac{2}{3}t^3 \qquad (8.5.10)$$

$$q=5: \qquad y_2(t) = t - \frac{4}{3}t^3 + \frac{4}{15}t^5$$

The polynomial solutions to Hermite's equation given in (8.5.9) and (8.5.10) are usually called the *Hermite polynomials* $H_n(t)$ when scaled such that the coefficient of the highest power of t is 2^n. The first four Hermite polynomials are

$$H_0(t) = 1$$

$$H_1(t) = 2t$$

$$H_2(t) = 4t^2 - 2$$

$$H_3(t) = 8t^3 - 12t$$

The Hermite polynomials are another example of a family of orthogonal polynomials; Hermite polynomials are orthogonal on $(-\infty, \infty)$ with respect to the weighting function $w(t) = e^{-t^2}$. Like Legendre polynomials, they have a wide range of interesting properties and the possibilities they present for further study go well beyond the scope of this text. A plot of $H_{11}(t)$ is shown in figure 8.2. The Hermite polynomials have large oscillations; the degree 11

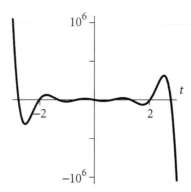

Figure 8.2 The degree 11 Hermite polynomial, $H_{11}(t)$, plotted on the interval $[-3, 3]$.

polynomial has two more zeros, located at approximately ± 3.7, which are not shown in figure 8.2.

8.5.2 The Laguerre equation

The *Laguerre equation* is given by

$$ty'' + (1-t)y' + qy = 0 \tag{8.5.11}$$

where q is, once again, a real constant. If we divide through by t, Laguerre's equation is equivalently expressed as

$$y'' + \frac{1-t}{t}y' + \frac{q}{t}y = 0$$

Since the coefficient functions $p(t)$ of y' and $q(t)$ of y are each undefined at $t = 0$, the Laguerre equation has a singular point at the origin. Nonetheless, it turns out that we can find a series expansion for a solution at the origin.

Letting $y = a_0 + a_1 t + a_2 t^2 + \cdots$ and substituting for y, y', and y'' in (8.5.11) it can be shown that the coefficients a_n must satisfy

$$\sum_{n=1}^{\infty} \left[(n+1)^2 a_{n+1} + (q-n)a_n \right] t^n = 0 \tag{8.5.12}$$

It follows from (8.5.12) that

$$(n+1)^2 a_{n+1} + (q-n)a_n = 0$$

and therefore

$$a_{n+1} = -\frac{q-n}{(n+1)^2} a_n \tag{8.5.13}$$

Note that this recurrence relation relies only on the value of a_0, and therefore only leads to one solution to the Laguerre equation.[3] Applying (8.5.13), we see

$$a_1 = -\frac{q}{1^2}a_0$$

$$a_2 = -\frac{q-1}{2^2}a_1 = \frac{1}{(1\cdot 2)^2}(q-1)qa_0$$

$$a_3 = -\frac{q-2}{3^2}a_2 = -\frac{1}{(1\cdot 2\cdot 3)^2}(q-2)(q-1)qa_0$$

More generally,

$$a_n = -\frac{q-n-1}{n^2}a_{n-1} = (-1)^n\frac{(q-n+1)\cdots(q-1)q}{n!^2}a_0$$

Taking $a_0 = 1$, we have found that one solution to the Laguerre equation is

$$y_1(t) = 1 + \sum_{n=1}^{\infty}(-1)^n\frac{(q-n+1)\cdots(q-1)q}{n!^2}t^n \qquad (8.5.14)$$

When q is a non-negative integer, we see from (8.5.14) that $y_1(t)$ is a polynomial of degree q. Recalling the binomial coefficient $\binom{q}{n}$ given by

$$\binom{q}{n} = \frac{q!}{n!(q-n)!} = \frac{q(q-1)\cdots(q-n+1)}{n!} \qquad (8.5.15)$$

we are able to find a relatively simple expression for these polynomial solutions. The *Laguerre polynomial* of degree q is given by

$$L_q(t) = 1 + \sum_{n=1}^{q}\frac{(-1)^n}{n!}\binom{q}{n}t^n \qquad (8.5.16)$$

and these functions turn out to be the only solutions (up to scalar multiples) of the Laguerre equation that are analytic at $t = 0$. The Laguerre polynomials are yet another family of orthogonal polynomials. The first few of these polynomials are given below, followed by a graph of $L_{11}(t)$ in figure 8.3.

$$L_1(t) = 1 - t$$

$$L_2(t) = 1 - 2t + \frac{1}{2}t^2$$

$$L_3(t) = 1 - \frac{3}{2}t + \frac{3}{2}t^2 - \frac{1}{6}t^3$$

$$L_4(t) = 1 - 4t + 3t^2 - \frac{2}{3}t^3 + \frac{1}{24}t^4$$

---

[3] A second solution can be found by more sophisticated techniques that lie beyond the scope of this book.

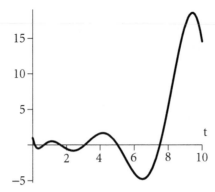

Figure 8.3 The degree 11 Laguerre polynomial $L_{11}(t)$ plotted on the interval $[0, 10]$.

8.5.3 The Bessel equation

The *Bessel equation*

$$t^2 y'' + ty' + (t^2 - \lambda^2)y = 0 \tag{8.5.17}$$

is a very important DE in mathematical physics. The properties of its solutions have been well studied; the equation often appears in the process of solving certain partial differential equations that appear when using cylindrical coordinates.

The parameter λ in (8.5.17) is a real constant. Like the Laguerre equation, the Bessel equation has a singular point at $t = 0$, so we cannot expect to find solutions to the equation with Taylor series centered at $a = 0$. Nonetheless, as we will show shortly, a solution analytic at $t = 0$ exists when λ is a non-negative integer. While a second linearly independent solution to the Bessel equation can be found, the techniques required are beyond the scope of this text.

Here we only explore the series solutions that do exist for the Bessel equation. Let $\lambda = m$ be a non-negative integer and assume that $y_1(t) = a_0 + a_1 t + a_2 t^2 + \cdots$. Substituting directly in (8.5.17) leads to

$$-m^2 a_0 + (1 - m^2)a_1 t + \sum_{k=2}^{\infty} [(k^2 - m^2)a_k + a_{k-2}]t^k = 0 \tag{8.5.18}$$

Since each coefficient of powers of t in (8.5.18) must be zero, it follows that $m^2 a_0 = 0$, $(1 - m^2)a_1 = 0$, and

$$(k^2 - m^2)a_k + a_{k-2} = 0, \quad k \geq 2 \tag{8.5.19}$$

If $k < m$, then it follows $a_k = 0$ for each such k by the three preceding equalities. When $k = m$, the coefficient $k^2 - m^2$ of a_k vanishes and thus (8.5.19) becomes the identity, rendering the value of a_m arbitrary. Note further that $a_{m+1} = a_{m+3} = \cdots = 0$ is another consequence of (8.5.19). Thus, a_m can be any constant, and subsequent terms must satisfy the recurrence

$$a_{k+2} = -\frac{1}{(k+2)^2 - m^2}a_k = -\frac{1}{(k+2-m)(k+2+m)}, \quad k = m, m+2, m+4, \ldots$$

$$(8.5.20)$$

Hence, given a positive integer $\lambda = m$ and a value for a_m, we can determine all of the coefficients of the Taylor expansion of an analytic solution to the Bessel equation. In particular, these coefficients a_{m+2j} for $j \geq 0$ must satisfy the recurrence relation (8.5.20), from which using $a_m = 1$ we find the closed formula

$$a_{m+2j} = 2^{-2j}\frac{(-1)^j}{j!(m+1)(m+2)\cdots(m+j)} \tag{8.5.21}$$

Hence, one solution of Bessel's equation (again, when $\lambda = m$ is a positive integer) is

$$y_1(t) = \sum_{j=0}^{\infty} 2^{-2j}\frac{(-1)^j}{j!(m+1)(m+2)\cdots(m+j)}t^{m+2j} \tag{8.5.22}$$

The *Bessel function of the first kind of order n* (it is standard to use n rather than m for the order of the Bessel function) is the scalar multiple of $y_1(t)$ given by

$$J_n(t) = \frac{2^{-n}}{n!}y_1(t) = \sum_{j=0}^{\infty} 2^{-2j-n}\frac{(-1)^j}{j!(n+j)!}t^{n+2j} \tag{8.5.23}$$

For example, the first two Bessel functions are

$$J_0(t) = \sum_{j=0}^{\infty} 2^{-2j}\frac{(-1)^j}{j!j!}t^{2j} \tag{8.5.24}$$

and

$$J_1(t) = \sum_{j=0}^{\infty} 2^{-2j-1}\frac{(-1)^j}{j!(j+1)!}t^{2j+1} \tag{8.5.25}$$

The graph of $J_0(t)$ in figure 8.4 shows that the Bessel function exhibits damped oscillation.

In this section, through the Hermite, Laguerre, and Bessel equations, we have encountered examples not only of three important DEs, but also of the various types of important functions that arise as solutions to these equations. Hermite polynomials, Laguerre polynomials, and Bessel functions are often studied in courses on special functions and demonstrate a wide range of interesting properties that mathematicians, engineers, and physicists have studied.

Exercises 8.5

1. Determine the degree 4 and 5 Hermite polynomials, $H_4(t)$ and $H_5(t)$.

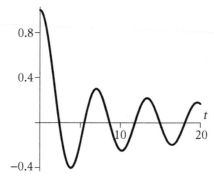

Figure 8.4 The Bessel function of the first kind, $J_0(t)$.

In exercises 2–4, find the first three nonzero terms in the Taylor series representation of the general solution to the given Hermite equation.

2. $y'' - 2ty' + 6y = 0$

3. $y'' - 2ty' + 10y = 0$

4. $y'' - 2ty' + 4y = 0$

In exercises 5–7, find the first three nonzero terms in the Taylor series representation of the general solution to the given IVP.

5. $y'' - 2ty' + 6y = 0$, $y(0) = 2$, $y'(0) = 10$

6. $y'' - 2ty' + 10y = 0$, $y(0) = 1$, $y'(0) = 0$

7. $y'' - 2ty' + 4y = 8t$, $y(0) = 1$, $y'(0) = 0$

8. Determine the degree 5 and 6 Laguerre polynomials, $L_5(t)$ and $L_6(t)$.

Given that a general solution of Laguerre's equation is $c_1 L_q(t) + c_2 u_2(t)$, where $u_2(t)$ is singular at the origin, in exercises 9–11, determine the solution to the given IVP.

9. $ty'' + (1 - t)y' + 3y = 0$, $y(0) = $ finite, $y(1) = 1$

10. $ty'' + (1 - t)y' + 4y = 0$, $y(0) = $ finite, $y(2) = 2$

11. $ty'' + (1 - t)y' + 4y = 3t$, $y(0) = $ finite, $y(1) = 4$

12. Determine the first five nonzero terms in the series expansion of $J_2(t)$ about $t = 0$. In addition, state the form of $J_2(t)$ in sigma notation.

It can be shown that a second linearly independent solution to the Bessel equation when $\lambda = n$ (called the *Bessel function of the second kind of* order n is given by

$$Y_n(t) = \frac{2}{\pi} J_n(t) \left(\ln \frac{t}{2} + \gamma \right) + R(t) + u(t)$$

where $R(t)$ is a rational function, $\gamma \approx 0.577215665$ is *Euler's constant*, and $u(t)$ is a power series convergent for all t. Note that $Y_n(t)$ is singular at the origin. In exercises 13–15, determine the general solution to the given equation.

13. $t^2 y'' + t y' + (t^2 - 4) y = 0$

14. $t^2 y'' + t y' + (t^2 - 9) y = 0$

15. $t^2 y'' + t y' + (t^2 - 16) y = 0$

In exercises 16–18, determine the solution to the given IVP.

16. $t^2 y'' + t y' + (t^2 - 4) y = 0$, $y(0) = $ finite, $y(1) = 1$

17. $t^2 y'' + t y' + (t^2 - 9) y = 0$, $y(0) = $ finite, $y(1) = -3$

18. $t^2 y'' + t y' + (t^2 - 16) y = 0$, $y(0) = $ finite, $y(1) = 2$

8.6 The Method of Frobenius

Some second-order linear DEs that appear in physical applications do not have two linearly independent analytic solutions about $t = 0$. Perhaps the most important and well-studied example is the Bessel equation (8.5.17). A somewhat simpler example is

$$t^2 y'' + \frac{3}{2} t y' - \frac{1}{2} y \qquad (8.6.1)$$

which is a Cauchy–Euler equation (on which more information can be found in section 4.7.3). It is a straightforward exercise to show that for all $t > 0$, $y_1(t) = t^{-1}$ and $y_2(t) = \sqrt{t}$ are linearly independent solutions of (8.6.1). Note that neither y_1 nor y_2 has a derivative at the origin, and therefore neither is analytic at $t = 0$; thus, each lacks a Taylor series expansion at the origin.

F. Georg Frobenius (1847–1917) showed that a certain class of linear second-order DEs with a singular point at the origin can be represented in series form by a slight generalization of a Taylor series. In particular, he showed that these series solutions have the form

$$y = t^r \sum_{k=0}^{\infty} b_k t^k = \sum_{k=0}^{\infty} b_k t^{k+r} \qquad (8.6.2)$$

where r is a real number and $\sum_{k=0}^{\infty} b_k t^k$ converges in some open interval containing the origin. The series (8.6.2) is called a *Frobenius series*, and the following method we will discuss for obtaining r and the coefficients b_k is known as the *Method of Frobenius*.

The Cauchy–Euler equation and the Bessel equation both belong to this class of equations that can be solved by the Method of Frobenius. In what follows, we focus particularly on equations of the form

$$t^2 y'' + tp(t)y' + q(t)y = 0 \qquad (8.6.3)$$

where $p(t)$ and $q(t)$ are low-degree polynomials. Note that p and q are analytic at the origin, and therefore each has a convergent Taylor series there. Any linear second-order DE with this property is said to have a *regular singular point* at the origin. The Method of Frobenius applies to all such equations. Finally, observe that if $p(t)$ and $q(t)$ are constant polynomials, then (8.6.3) reduces to a Cauchy–Euler equation.

To begin, we suppose that there is a solution of (8.6.3) that has a series expansion of the form

$$y = \sum_{k=0}^{\infty} b_k t^{k+r} \qquad (8.6.4)$$

where $b_0 \neq 0$ and $\sum_{k=0}^{\infty} b_k t^k$ converges in $0 < |t| < R$. From this, it follows that

$$y' = \sum_{k=0}^{\infty} (k+r)b_k t^{k+r-1} \qquad (8.6.5)$$

and

$$y'' = \sum_{k=0}^{\infty} (k+r)(k+r-1)b_k t^{k+r-2} \qquad (8.6.6)$$

Furthermore, we suppose that $p(t)$ and $q(t)$ have the expansions

$$p(t) = p_0 + p_1 t + p_2 t + \cdots + p_n^n + \cdots$$

$$q(t) = q_0 + q_1 t + q_2 t + \cdots + q_n^n + \cdots$$

Substituting these expressions for y, y', y'', p, and q in (8.6.3) and gathering like terms, we find that

$$0 = t^2 y'' + tp(t)y' + q(t)y$$

$$= \sum_{k=0}^{\infty} (k+r)(k+r-1)b_k t^{k+r} + (p_0 + p_1 t + p_2 t + \cdots + p_n^n + \cdots)$$

$$\times \sum_{k=0}^{\infty} (k+r)b_k t^{k+r} + (q_0 + q_1 t + q_2 t + \cdots + q_n^n + \cdots) \sum_{k=0}^{\infty} b_k t^{k+r}$$

$$= (r(r-1) + p_0 r + q_0)b_0 + c_1 t + c_2 t^2 + \cdots \qquad (8.6.7)$$

where the general term c_n depends on n and all earlier coefficients for each $n \geq 1$. A general formula for c_n turns out to be complicated and not particularly useful for the examples we wish to study, so we choose not to derive such a formula.

The most important conclusion to draw from (8.6.7) comes from the fact that each coefficient of the general power series expansion must be zero, so that since $b_0 \neq 0$,

$$r(r-1) + p_0 r + q_0 = 0 \tag{8.6.8}$$

Equation (8.6.8) is called the *indicial equation* for the Method of Frobenius. Note that this equation is quadratic in r; its two roots are the values of r that are used in (8.6.2). At this point, it is useful for us to turn our attention to two specific example of the Method of Frobenius at work.

Example 8.6.1   Find a Frobenius series solution for the Bessel–Clifford equation

$$t^2 y'' + (1-a)ty' + ty = 0 \tag{8.6.9}$$

where a is a constant.

Solution.   With a being a constant, we have $p(t) = 1 - a$, so in the series expansion for p, $p_0 = 1 - a$. Moreover, $q(t) = t$, so $q_0 = 0$. Thus, for the given DE the indicial equation is

$$r(r-1) + (1-a)r = 0$$

Rearranging, we see that $r(r - 1 + 1 - a) = r(r - a) = 0$, and thus the roots of the indicial equation are $r = 0$ and $r = a$.

In the case that $r = 0$, the Method of Frobenius is providing an analytic solution to (8.6.9) of the form

$$y_1 = \sum_{k=0}^{\infty} b_k t^k$$

Dividing both sides of (8.6.9) by t and substituting this expression for y using the standard series methods we have already discussed, it follows that

$$\sum_{k=0}^{\infty} [(k+1)(k+1-a)b_{k+1} + b_k]t^k$$

from which we obtain the recurrence relation

$$b_{k+1} = \frac{-1}{(k+1)(k+1-a)} b_k \tag{8.6.10}$$

It follows from (8.6.10) that the closed form expression for b_k is

$$b_k = \frac{(-1)^k}{k!(1-a)(2-a)\cdots(k-a)} b_0, \quad k \geq 1$$

so we find that

$$y_1(t) = b_0 \left(1 + \sum_{k=1}^{\infty} \frac{(-1)^k}{k!(1-a)(2-a)\cdots(k-a)} t^k \right) \tag{8.6.11}$$

which is valid for all t provided that $a \neq 1, 2, \ldots$. Note that from this recurrence relation, every b_n is a function of b_0, and thus there cannot be two linearly independent solutions to the Bessel–Clifford equation that are analytic at 0. Indeed, every solution linearly independent of $y_1(t)$ must be singular at 0. And while the equation has a singular point at the origin, there is an analytic solution there for every a except when a is a positive integer. We now turn to the other root of the indicial equation in search of a second solution to the Bessel–Clifford equation.

Using $r = a$, we have

$$ty(t) = \sum_{k=0}^{\infty} b_k t^{k+a+1}$$

$$(1-a)ty'(t) = \sum_{k=0}^{\infty} (1-a)(k+a)b_k t^{k+a}$$

$$t^2 y''(t) = \sum_{k=0}^{\infty} (k+a)(k+a-1)b_k t^{k+a}$$

Adding these equations forms the left side of the differential equation we aspire to solve; doing so and simplifying, we find that

$$0 = t^2 y''(t) + (1-a)ty'(t) + ty(t) = \sum_{k=0}^{\infty} k(k+a)b_k t^{k+a} + \sum_{k=0}^{\infty} b_k t^{k+a+1}$$

Since the first term in the first sum is zero, if we adjust the index of the summation in the second sum and combine, we have

$$\sum_{k=1}^{\infty} [k(k+a)b_k + b_{k-1}]t^{k+a} = 0$$

from which it follows that

$$k(k+a)b_k + b_{k-1} = 0, \quad k \geq 1$$

This standard recurrence relation can be solved to write every b_k in terms of b_0. Indeed, we see

$$b_k = \frac{(-1)^k}{k!(1+a)(2+a)\cdots(k+a)} b_0, \quad k \geq 1$$

so that the Frobenius series representation of the solution is

$$y_2(t) = b_0 t^a \left(1 + \sum_{k=1}^{\infty} \frac{(-1)^k}{k!(1+a)(2+a)\cdots(k+a)} t^k \right) \qquad (8.6.12)$$

We close this example with a few important observations. First, if $a = 0$, then the Frobenius solution $y_2(t)$ is identical to the earlier obtained $y_1(t)$. Moreover, if a is a non-negative integer, then the Method of Frobenius produces a Taylor series expansion that is analytic at $t = 0$. Thus, the cases for a valid analytic solution excluded by our approach in finding $y_1(t)$ are here reconciled. Finally, if a is not an integer, then $y_2(t)$ is singular at $t = 0$ and, together with the analytic $y_1(t)$ given by (8.6.11), we have found a linearly independent set of solutions for the Bessel–Clifford equation valid for $t > 0$.

To complete this section, we consider a second example.

Example 8.6.2   Find a Frobenius series solution of Bessel's equation,

$$t^2 y'' + ty' + (t^2 - \lambda^2) y = 0 \tag{8.6.13}$$

Solution.   In section 8.5.3, we derived a solution to (8.6.13) in the case where λ is an integer. Thus, in what follows we assume that $\lambda > 0$ is not an integer.
   Since $p(t) = 1$ and $q(t) = -\lambda^2 + t^2$, we have $p_0 = 1$ and $q_0 = -\lambda^2$, which tells us that the indicial equation is

$$r(r-1) + r - \lambda^2 = r^2 - \lambda^2 = 0$$

Thus, $r = \pm\lambda$. Choosing $r = \lambda$ and using (8.6.4), (8.6.5), and (8.6.6), we find that the three relevant series for the differential equation (8.6.13) are

$$(t^2 - \lambda^2) y(t) = \sum_{k=0}^{\infty} b_k t^{k+\lambda+2} - \sum_{k=0}^{\infty} b_k t^{k+\lambda+2}$$

$$ty'(t) = \sum_{k=0}^{\infty} (k+\lambda) b_k t^{k+\lambda}$$

$$t^2 y''(t) = \sum_{k=0}^{\infty} (k+\lambda)(k+\lambda-1) b_k t^{k+\lambda}$$

From the form of Bessel's equation, the sum of these three expressions vanishes; adding and simplifying, we observe that

$$\sum_{k=0}^{\infty} k(k+2\lambda) b_k t^{k+\lambda} - \sum_{k=0}^{\infty} b_k t^{k+\lambda+2} = 0$$

To combine the sums, we step up the index in the second summation by 2 and find

$$(1+2\lambda) b_1 t^{1+\lambda} - \sum_{k=2}^{\infty} [k(k+2\lambda) b_k + b_{k-2}] t^{k+\lambda} = 0$$

So, $(1+2\lambda) b_1 = 0$, and

$$k(k+2\lambda) b_k + b_{k-2} = 0, \quad k \geq 2 \tag{8.6.14}$$

One solution to this recurrence relation is obtained by setting $b_0 = 1$ and $b_1 = 0$. Then, since we are assuming that λ is not an integer and $b_1 = 0$, (8.6.14) implies that all odd-subscripted coefficients are zero and that

$$b_k = \frac{-1}{k(2\lambda + k)} b_{k-2}, \quad k = 2, 4, \ldots$$

Therefore, it follows that in closed form we have

$$b_{2k} = \frac{(-1)^k 2^{-2k}}{k!(1+\lambda)(2+\lambda)\cdots(k+\lambda)}$$

and thus a Frobenius solution to the Bessel equation is

$$y(t) = t^\lambda + \sum_{k=1}^{\infty} \frac{(-1)^k 2^{-2k}}{k!(1+\lambda)(2+\lambda)\cdots(k+\lambda)} t^{2k+\lambda}$$

Note that since $\lambda > 0$, the ratio test can be applied to show that this series converges for all values of t.

A more detailed study of the Method of Frobenius is beyond the scope of this text. (For further discussion, see Potter and Goldberg, *Mathematical Methods*, second edition, Great Lakes Press 1995.)

Exercises 8.6

In exercises 1–10, find the indicial equation and use the root that either is not an integer or that is the larger integer to find the first three nonzero coefficients in a Frobenius series solution to the given DE.

1. $2t^2 y'' - ty' + (1+t)y = 0$

2. $2ty'' + y' + ty = 0$

3. $ty'' + (t-2)y' + y = 0$

4. $2ty'' + (1+4t)y' + y = 0$

5. $t^2 y'' - t(t+5)y' + (t+5)y = 0$

6. $2t^2 y'' - ty' + (t-5)y = 0$

7. $4t^2 y'' + 6ty' + (t-2)y = 0$

8. $2ty'' + (1-t)y' - y = 0$

9. $t^2 y'' + ty' + (t-3)y = 0$

10. $3t^2 y'' - ty' - 4y = 0$

11. Find the indicial equation for the Cauchy–Euler equation

$$t^2 y'' + pty' + qy = 0$$

12. Show that the roots of the indicial equation are equal for the Laguerre equation

$$ty'' + (1-t)y' + qu = 0$$

8.7 For further study

8.7.1 Taylor series for first-order differential equations

Let $y(t) = \sum_{n=0}^{\infty} a_n t^n$ be the Taylor series of a solution of

$$ty' + \lambda y = f(t) \qquad (8.7.1)$$

where λ is constant and $f(t) = \sum_{n=0}^{\infty} f_n t^n$.

(a) Show that

$$y(t) = \sum_{n=0}^{\infty} \frac{f_n}{n+\lambda} t^n$$

(b) In terms of the infinite series derived in (a), what is the general solution to (8.7.1)?

(c) Using series expansions appropriately and your work in (a), determine the general solution to each of the following DEs.

  (i) $ty' + 2y = e^t$
  (ii) $ty' + 3y = \sin t$
  (iii) $ty' + 4y = \arctan t$

(d) Show that

$$\sum_{n=0}^{\infty} \frac{f_n}{n+\lambda} t^n = t^{-\lambda} \sum_{n=0}^{\infty} \frac{f_n}{n+\lambda} t^{n+\lambda} = t^{-\lambda} \int_0^t \sum_{n=0}^{\infty} f_n x^{n+\lambda-1} \, dx$$

$$ = t^{-\lambda} \int_0^t x^{\lambda-1} \sum_{n=0}^{\infty} f_n x^n \, dx = t^{-\lambda} \int_0^t x^{\lambda-1} f(x) \, dx$$

(e) Substitute directly in (8.7.1) to show that

$$y(t) = t^{-\lambda} \int_0^t x^{\lambda-1} f(x) \, dx$$

is indeed a solution.

(f) Solve (8.7.1) by use of an integrating factor (see section 2.3) and compare your result to $y(t)$ as given in (e).

8.7.2 The Gamma function

The Gamma function $\Gamma(x)$, like Bessel functions and families of orthogonal polynomials, is a special function that plays an important role in many areas of mathematics. The Gamma function is defined by

$$\Gamma(s+1) = \int_0^{\infty} e^{-t} t^s \, dt, \qquad s > -1 \qquad (8.7.2)$$

(a) Show that $\Gamma(1) = 1$.

(b) Use integration by parts to show that $\Gamma(s+1) = s\Gamma(s)$.

(c) Show that if s is a positive integer, then $\Gamma(s) = s!$.

(d) Let $r > 0$ be given and recall that $\mathcal{L}[t^r] = \int_0^\infty e^{-st} t^r \, dt$. Hence show that

$$\mathcal{L}[t^{r-1}] = \frac{\Gamma(r)}{s^r}$$

(e) Show that

$$\Gamma\left(\frac{1}{2}\right) = 2\int_{-\infty}^{\infty} e^{-x^2} \, dx = \sqrt{\pi}$$

(f) Use (b) to show that

$$h^n \frac{\Gamma(h+x/h)}{\Gamma(x/h)} = x(x+h)(x+2h)\cdots(x+(n-1)h)$$

Hence, show that

$$1 \cdot 3 \cdot 5 \cdots (2n-1) = 2^n \frac{\Gamma(n+1/2)}{\Gamma(1/2)} = \frac{2^n}{\sqrt{\pi}} \Gamma(n+1/2)$$

(g) Finally, explain why $1 \cdot 3 \cdot 5 \cdots (2n-1) = (2n)!/(2^n n!)$ and therefore show

$$\Gamma\left(n+\frac{1}{2}\right) = \frac{(2n)!}{2^n n!} \sqrt{\pi}$$

A

Review of integration techniques

Several standard solution techniques for differential equations require us to integrate functions. Here we briefly review some fundamentals from calculus.

u-substitution

For integrals of the form

$$\int f(g(t))g'(t)\, dt$$

we can evaluate the integral by undoing the chain rule through a change of variables. Letting $u = g(t)$, it follows $du = g'(t)\, dt$, and thus

$$\int f(g(t))g'(t)\, dt = \int f(u)\, du$$

If we can evaluate the new, simpler integral in u, all that remains is to substitute back to the variable t. For instance, to evaluate

$$\int t \sin t^2\, dt$$

we let $u = t^2$ and $du = 2t\, dt$. We note that $t\, dt = \frac{1}{2} du$. Thus, substituting for t^2 and $t\, dt$, we find that the given integral is equivalently

$$\int \frac{1}{2} \sin u\, du$$

Evaluating the integral in u and substituting back to t,

$$\int t \sin t^2\, dt = \int \frac{1}{2} \sin u\, du = -\frac{1}{2} \cos u + C = -\frac{1}{2} \cos t^2 + C$$

Overall, u-substitution is particularly relevant for working with composite functions. In attempting to use u-substitution, we should search the integrand for an *inside* function, and then hope that its derivative (up to a constant multiple) is present *outside* the composite function.

Examples for further practice:

1. $\displaystyle\int te^{-t^2}\,dt$

2. $\displaystyle\int t^{21}(4t^{22}-13)^{20}\,dt$

3. $\displaystyle\int 6e^{1/t}\cdot t^{-2}\,dt$

4. $\displaystyle\int \frac{\sin t}{1+\cos^2 t}\,dt$

5. $\displaystyle\int (\sin t)^3\,dt$ Hint: $\sin^2 t = 1-\cos^2 t$.

Integration by parts

As u-substitution is used to undo the chain rule, integration by parts undoes the product rule. It is particularly applicable to integrals that involve products of basic functions such as $\int te^t\,dt$.

Recall that the product rule states

$$\frac{d}{dt}[u(t)v(t)] = u(t)v'(t) + v(t)u'(t) \tag{A.1}$$

Integrating both sides of (A.1), it follows that

$$u(t)v(t) = \int u(t)v'(t)\,dt + \int v(t)u'(t)\,dt \tag{A.2}$$

Solving for $\int u(t)v'(t)\,dt$, we have

$$\int u(t)v'(t)\,dt = u(t)v(t) - \int v(t)u'(t)\,dt \tag{A.3}$$

Writing $dv = v'(t)\,dt$ and $du = u'(t)\,du$ and suppressing the presence of t, we see in (A.3) the standard statement of the integration by parts rule:

$$\int u\,dv = uv - \int v\,du \tag{A.4}$$

For example, let's evaluate $\int te^t\,dt$. Letting $u=t$ and $dv = e^t\,dt$, we observe that $du = dt$ and $v = e^t$. Thus, integrating by parts,

$$\int te^t\,dt = te^t - \int e^t\,dt = te^t - e^t + C$$

A good way to think of integration by parts is to view it as integrating the product $u\,dv$ by trading u for its derivative and trading dv for its antiderivative. In particular, once we have decided to use integration by parts, we must make appropriate choices for u and dv. One guideline is that dv should be fairly easy to antidifferentiate; another is that the derivative of u should not be significantly more complicated than u itself. Overall, we generally want the integral of $v\,du$ to be somehow simpler (or at least not more complicated) than the integral of $u\,dv$.

Examples for further practice:

1. $\displaystyle\int t^4 \ln t\,dt$

2. $\displaystyle\int 5t \sin t\,dt$

3. $\displaystyle\int 3te^{2t}\,dt$

4. $\displaystyle\int t\sqrt{7t+5}\,dt$

5. $\displaystyle\int \ln t\,dt$ Hint: Try $dv = 1$.

6. $\displaystyle\int t^2 e^t\,dt$

7. $\displaystyle\int e^t \cos t\,dt$

Partial fractions

A remarkable fact is that any rational function (that is, any quotient of two polynomials) may be integrated. The standard method for approaching an integration problem of the form

$$\int \frac{p(t)}{q(t)}\,dt$$

is the technique known as *partial fractions*. It is necessary to assume (or apply long division so) that the degree of p is less than the degree of q. While partial fractions is an important technique for integration, it is also a useful tool in its own right. For example, we frequently use it when working with the Laplace transform; see sections 5.5 and 5.6.

The method is best understood through a sequence of examples.

Example A.1   Evaluate the integral

$$\int \frac{t}{t^2 + 5t + 6}\,dt \tag{A.5}$$

Solution.   Factoring the integrand, we can write

$$\frac{t}{t^2 + 5t + 6} = \frac{t}{(t+2)(t+3)} \tag{A.6}$$

If we view the righthand side as the result of adding two simpler fractions, we can make the reasonable assumption that two fractions of the form $A/(t+2)$ and $B/(t+3)$ had to be combined by getting a common denominator to form (A.6). Thus we assume

$$\frac{t}{(t+2)(t+3)} = \frac{A}{t+2} + \frac{B}{t+3} \tag{A.7}$$

and seek values of A and B which make this relationship hold for all values of t. Multiplying both sides of (A.7) by $(t+2)(t+3)$, we find

$$t = A(t+3) + B(t+2) \tag{A.8}$$

Since (A.8) must be valid for every value of t, we can choose t-values that make it especially easy to identify A and B. Choosing $t = -2$, we see that $-2 = A(-2+1) = A$. Choosing $t = -3$, it follows $-3 = B(-3+2)$, so $B = 3$. Thus, we have determined

$$\frac{t}{(t+2)(t+3)} = -\frac{2}{t+2} + \frac{3}{t+3} \tag{A.9}$$

Having completed the partial fraction decomposition, we can now integrate. In particular,

$$\int \frac{t}{t^2 + 5t + 6}\, dt = -\frac{2}{t+2} + \frac{3}{t+3}$$

$$= -2\ln(t+2) + 3\ln(t+3) + C$$

The approach of example A.1 works any time the denominator $q(t)$ can be written as a product of distinct linear terms. That is, if $q(t) = (t - r_1)(t - r_2)\cdots(t - r_n)$, then we can write

$$\frac{p(t)}{q(t)} = \frac{A_1}{t - r_1} + \frac{A_2}{t - r_2} + \cdots + \frac{A_n}{t - r_n}$$

and use algebra similar to our work above to determine $A_1, \ldots, A_n$.

Example A.2   Evaluate the integral

$$\int \frac{t^2 - 4}{t^3 + t^2}\, dt$$

Solution.   Factoring the denominator of the integrand, we have

$$\frac{t^2 - 4}{t^3 + t^2} = \frac{t^2 - 4}{t^2(t+1)}$$

If we think of the possible simpler fractions from which the given one can arise, we see that it is possible for terms of the form

$$\frac{A}{t}, \ \frac{B}{t^2}, \ \text{and} \ \frac{C}{t+1}$$

to be present. In particular, we must include A/t since this denominator is included in the necessary B/t^2. Thus we write

$$\frac{t^2-4}{t^2(t+1)} = \frac{A}{t} + \frac{B}{t^2} + \frac{C}{t+1} \tag{A.10}$$

Multiplying both sides of (A.10) by the least common denominator $t^2(t+1)$, we find

$$t^2 - 4 = At(t+1) + B(t+1) + Ct^2$$

Setting $t = 0$ implies $-4 = B$; using $t = -1$ shows $-3 = C$. To find A, we may use any other value of t, along with the established values of B and C. With $t = 1$,

$$-3 = A(1)(2) + (-4)(2) + (-3)1^2$$

and therefore $A = 4$. We now apply the partial fractions decomposition and integrate:

$$\int \frac{t^2-4}{t^3+t^2}\, dt = \left(\int \frac{4}{t} - \frac{4}{t^2} - \frac{3}{t+1} \right) dt$$

$$= 4\ln t + 4t^{-1} - 3\ln(t+1) + C$$

In any rational function where the denominator contains a repeated factor, we use a similar form of partial fraction decomposition. For instance,

$$\frac{t^3 - 2t + 1}{(t+4)^3(t-2)^2(t-5)} = \frac{A}{t+4} + \frac{B}{(t+4)^2} + \frac{C}{(t+4)^3} + \frac{D}{t-2} + \frac{E}{(t-2)^2} + \frac{F}{t-5}$$

so that each repeated factor is represented once for each possible order up to the highest power.

Example A.3 Evaluate the integral

$$\int \frac{t-5}{t^3+t}\, dt$$

Solution. When we factor the integrand, we observe that a quadratic term is present that cannot be factored further. In particular,

$$\frac{t-5}{t^3+t} = \frac{t^2-4}{t(t^2+1)}$$

In this case, we assume that the right hand fraction may be decomposed into the sum

$$\frac{t-5}{t(t^2+1)} = \frac{A}{t} + \frac{Bt+C}{t^2+1} \tag{A.11}$$

The linear term $Bt + C$ in the numerator of the last fraction is necessary; if we used only C as the numerator of the second fraction, a contradiction may arise in attempting to find A and C. Multiplying both sides of (A.11) by $t(t^2 + 1)$,

$$t - 5 = A(t^2 + 1) + t(Bt + C) \tag{A.12}$$

Besides $t = 0$, there are no obvious real values of t that enable us to easily deduce the values of A, B, and C. Choosing any three distinct values of t will lead to a system of three linear equations in A, B, and C which may be solved. Alternatively, we can expand and equate like coefficients in (A.12). Specifically, since

$$t - 5 = At^2 + A + Bt^2 + Ct$$

equating constant terms implies $A = -5$, equating linear terms shows $C = 1$, and the quadratic terms require that $A + B = 0$, thus $B = 5$. We have now found the partial fraction decomposition and are ready to integrate. Doing so,

$$\int \frac{t - 5}{t(t^2 + 1)}\, dt = \int \left(-\frac{5}{t} + \frac{5t + 1}{t^2 + 1} \right) dt$$

$$= \int \left(-\frac{5}{t} + \frac{5t}{t^2 + 1} + \frac{1}{t^2 + 1} \right) dt \tag{A.13}$$

$$= -5\ln t + \frac{5}{2}\ln(t^2 + 1) + \arctan t + C$$

Note that from the first step to (A.13) we performed the key algebraic separation

$$\frac{5t + 1}{t^2 + 1} = \frac{5t}{t^2 + 1} + \frac{1}{t^2 + 1}$$

so that we could integrate the first term by u-substitution ($u = t^2 + 1$) and recognize the integral of the second as the familiar arctangent function.

When a rational function's denominator is factored, any time a term of the form $s^2 + a^2$ arises, we must include a linear term in the numerator of the proposed partial fraction decomposition. For instance, if we were decomposing

$$\frac{t}{(s^2 + 9)(s^2 + 25)}$$

the appropriate form to assume for the sum of simpler fractions would be

$$\frac{t}{(s^2 + 9)(s^2 + 25)} = \frac{At + B}{s^2 + 9} + \frac{Ct + D}{s^2 + 25}$$

The observations we have made for the cases of distinct linear terms, repeated linear terms, and irreducible quadratic terms may be combined, as need, in any problem where a partial fraction decomposition is sought.

Examples for further practice:

1. $\displaystyle\int \frac{t^2+1}{(t-2)(t-1)(t+3)}\,dt$

2. $\displaystyle\int \frac{t^3+t+1}{t^4-1}\,dt$

3. $\displaystyle\int \frac{t^2-t-1}{t^3-6t^2+11t-6}\,dt$

4. $\displaystyle\int \frac{t^2-t-1}{(t-3)^3}\,dt$

5. $\displaystyle\int \frac{t+2}{t^4+4t^2}\,dt$

6. $\displaystyle\int \frac{e^t}{e^{2t}-e^t-6}\,dt$

Tables and computer algebra systems

In addition to the methods of u-substitution, integration by parts, and partial fractions, there are other standard integration techniques that enable us to deduce a wide range of results. Students normally learn a handful of integration techniques in calculus; it is also the case that entire books exist that are filled with tables of integrals and almost every calculus book includes at least a short table of integrals, typically a few pages long.

It is common for integral tables to include results such as

$$\int \sin mt \sin nt\,dt = \frac{1}{2(n-m)}\sin(n-m)t - \frac{1}{2(n+m)}\sin(n+m)t, \quad m \neq \pm n$$

Given an integral that aligns with this form, say

$$\int \sin 5t \sin 3t\,dt$$

it is a straightforward exercise to identify m and n and thus evaluate the integral.

In other cases, the identification of the appropriate rule in a table is more subtle and involved. In table A.1, we see that for the given collection of examples, even a slight change in the integrand leads to a major difference in the result.

In addition, we note that it takes some care in order to correctly identify which line in an integral table to use in certain examples. For instance, if we wish to evaluate the integral

$$\int \frac{dt}{5t\sqrt{4t^2+9}} \tag{A.14}$$

Table A.1
Integrals involving $u^2 \pm a^2$

Function	Antiderivative
$\int \dfrac{du}{a^2+u^2}$	$\dfrac{1}{a}\arctan\dfrac{u}{a}$
$\int \dfrac{du}{\sqrt{u^2\pm a^2}}$	$\ln\lvert u+\sqrt{u^2\pm a^2}\rvert$
$\int \sqrt{u^2\pm a^2}\,du$	$\dfrac{u}{2}\sqrt{u^2\pm a^2}\pm\dfrac{a^2}{2}\ln\lvert u+\sqrt{u^2\pm a^2}\rvert$
$\int \dfrac{u^2\,du}{\sqrt{u^2\pm a^2}}$	$\dfrac{u}{2}\sqrt{u^2\pm a^2}\mp\dfrac{a^2}{2}\ln\lvert u+\sqrt{u^2\pm a^2}\rvert$
$\int \dfrac{du}{u\sqrt{u^2+a^2}}$	$-\dfrac{1}{a}\ln\left\lvert\dfrac{a+\sqrt{u^2+a^2}}{u}\right\rvert$
$\int \dfrac{du}{u\sqrt{u^2-a^2}}$	$\dfrac{1}{a}\sec^{-1}\dfrac{u}{a}$

it appears that (A.14) most resembles (5) in table A.1. To use this statement in the table, it is necessary that we execute a u-substitution. We see that letting $u=2t$ implies $u^2=4t^2$, $t=u/2$, and $dt=du/2$. Replacing the three appearances of t in (A.14), we have

$$\int \frac{dt}{5t\sqrt{4t^2+9}} = \int \frac{\frac{1}{2}\,du}{\frac{5}{2}u\sqrt{u^2+9}} = \frac{1}{5}\int \frac{du}{u\sqrt{u^2+9}}$$

Applying (5) in table A.1 to our most recent result (with $a=3$) and then substituting back to t, we find

$$\int \frac{dt}{5t\sqrt{4t^2+9}} = \frac{1}{5}\left(-\frac{1}{3}\ln\left\lvert\frac{3+\sqrt{u^2+9}}{u}\right\rvert\right) + C$$

$$= -\frac{1}{15}\ln\left\lvert\frac{3+\sqrt{4t^2+9}}{2t}\right\rvert + C$$

An available option in the consideration of any integral is the use of a computer algebra system. In *Maple*, the syntax > int(f(t), t); results in the program attempting to evaluate the integral. For example,

```
>int(exp(sqrt(t))/sqrt(t), t);
```

produces the output

$$2e^{\sqrt{t}}$$

which shows that

$$\int \frac{e^{\sqrt{t}}}{\sqrt{t}}\, dt = 2e^{\sqrt{t}} + C$$

There are integrals that *Maple* evaluates but that produce unusual output, such as

```
> int(exp(t^2), t);
```

which results in

$$\frac{1}{2}\sqrt{\pi}\,\mathrm{erf}(t)$$

The function *erf* is the so-called *error function* which arises frequently in probability and statistics and is itself defined by a definite integral. The notation $\mathrm{erf}(t)$ is used since e^{-t^2} lacks an elementary antiderivative.

Other integrals, some of which may be evaluated with human intervention, *Maple* is unable to execute. For instance, the integral

$$\int (1+t)e^t \sqrt{1 + (te^t)^2}\, dt \tag{A.15}$$

cannot be evaluated by *Maple* (when entered and executed, the program simply returns the integral unevaluated). However, if we recognize that the u-substitution $u = te^t$ leads to (A.15) being equivalently expressed as the integral

$$\int \sqrt{1 + u^2}\, du$$

then we observe that this integral in u may be easily evaluated by *Maple* or found in any standard table.

Overall, the reader is advised to be well versed in the standard integration methods, to practice them as needed, and to realize that even with lengthy tables and the availability of computer algebra systems, evaluating integrals if often both a challenging and involved task.

B

Complex numbers

Complex numbers arise naturally in the solution of quadratic (and other polynomial) equations. For example, the equation

$$t^2 + 1 = 0$$

has no real number solutions. But, if we want any quadratic equation to have two solutions, it is natural to say that $t^2 = -1$ and therefore $t = \pm\sqrt{-1}$. We denote $\sqrt{-1}$ by the symbol "i", and thus say that $t = \pm i$ are solutions to $t^2 + 1 = 0$.

Similarly, if we have the equation $t^2 + 2t + 5 = 0$ and we apply the quadratic formula, it follows

$$t = \frac{-2 \pm \sqrt{2^2 - 4 \cdot 1 \cdot 5}}{2} = \frac{-2 \pm \sqrt{-16}}{2}$$

Using $i = \sqrt{-1}$, we have

$$t = \frac{-2 \pm 4i}{2} = -1 \pm 2i$$

In general, a *complex number* z is any number of the form

$$z = a + bi$$

where a and b are both real numbers and i satisfies $i^2 = -1$. Complex numbers are naturally represented as points in the so-called *complex plane*, which corresponds to $\mathbb{R}^2$; the set of all complex numbers is denoted by $\mathbb{C}$. In particular, given any complex number $z = a + bi$ we can associate z with the point (a, b), as shown in figure B.1, where we see the particular example $z = 3 + 2i$.

In the complex plane, the horizontal axis is known as the *real* axis, denoted Re, and the vertical axis is the *imaginary* axis, labelled Im. For the complex

503

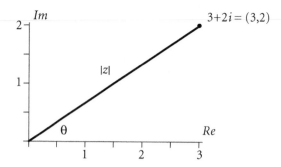

Figure B.1   The complex number $z = 3 + 2i$.

number $z = a + bi$, the *real part* of z is a, and we write $Re(z) = a$, while the *imaginary part* of z is b, denoted $Im(z) = b$.

This geometric interpretation of complex numbers leads to other natural concepts. The *modulus* $|z|$ of $z = 3 + 2i$ is defined to be the length of the line segment from the origin to the point $(3, 2)$, or $|z| = \sqrt{3^2 + 2^2} = \sqrt{13}$. Similarly, to each complex number we associate an angle θ, as shown in figure B.1, which is known as the *argument* of z. For $z = 3 + 2i$, $\theta = \arctan 2/3$. In general, for $z = a + bi$, $|z| = \sqrt{a^2 + b^2}$ and $\theta = \arctan b/a$. The modulus and argument essentially give us the polar representation of z, while a and b provide its rectangular coordinates.

Just like with real numbers, we can add, subtract, multiply, and divide complex numbers. For example, if $w = 2 + i$ and $z = 3 + 2i$, then

$$w + z = (2 + i) + (3 + 2i) = 5 + 3i$$

Complex addition, much like vector addition, is performed component-wise. Subtraction is executed in the same manner. For multiplication, the distributive law enables us to compute products of complex numbers. Specifically,

$$w \cdot z = (2 + i)(3 + 2i) = 6 + 4i + 3i + 2i^2 = 6 + 7i - 2 = 4 + 7i$$

To divide, we use the *complex conjugate* of the denominator to convert the division problem to one of multiplication. The complex conjugate of $z = a + bi$ is $\bar{z} = a - bi$. For instance,

$$\frac{w}{z} = \frac{2 + i}{3 + 2i} = \frac{2 + i}{3 + 2i} \cdot \frac{3 - 2i}{3 - 2i}$$

$$= \frac{6 + 3i - 2i - 2i^2}{9 - 4i^2} = \frac{4 + i}{5}$$

$$= \frac{4}{5} + \frac{1}{5}i$$

Using basic trigonometry and Euler's formula[1], we can gain a particularly nice geometric perspective on the multiplication of complex numbers.

---

[1] Euler's formula, $e^{i\theta} = \cos\theta + i\sin\theta$, is introduced in section 3.5.

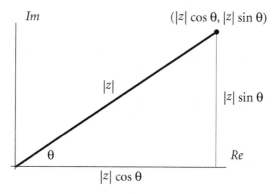

Figure B.2 The complex number $z = |z|\cos\theta + i|z|\sin\theta$.

Given a complex number z with modulus $|z|$ and argument θ, we may write z in its rectangular form as

$$z = |z|\cos\theta + i|z|\sin\theta$$

as demonstrated in figure B.2. From Euler's formula, we see that it is equivalent to write z in the form

$$z = |z|\cos\theta + i|z|\sin\theta$$
$$= |z|(\cos\theta + i\sin\theta)$$
$$= |z|e^{i\theta}$$

Note particularly that the complex number $e^{i\theta} = \cos\theta + i\sin\theta$ has modulus 1; that is, $e^{i\theta}$ lies on the unit circle in the complex plane.

Given another complex number w with modulus $|w|$ and argument α, we may write $w = |w|e^{i\alpha}$, from which the product $w \cdot z$ is

$$w \cdot z = (|w|e^{i\alpha}) \cdot (|z|e^{i\theta}) = |w||z|e^{i(\alpha+\theta)} \tag{B.1}$$

The expression (B.1) for $w \cdot z$ shows that when two complex numbers are multiplied, the modulus of the product is the product of the two numbers' moduli, while the argument of the product is the sum of the arguments of the two numbers. This is shown geometrically in figure B.3.

Finally, it is important to note that because the complex numbers have so much in common with the real numbers, it makes sense to work with them in functions, too. For instance, we can consider a function such as

$$P(z) = z^6 - 3z^5 + (5 - 2i)z^3 + iz^2 - 21z + 3 - 5i.$$

P is a function for which we can input any complex number z; the output will also be a complex number $P(z)$. For our work with solving differential equations, it will sometimes be the case that we can find a complex solution

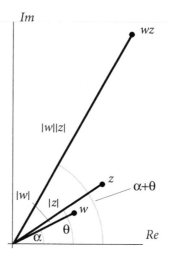

Figure B.3 The product of complex numbers $w = |w|e^{i\alpha}$ and $z = |z|e^{i\theta}$.

to a real differential equation, and that certain parts of the complex function (in fact, its real and imaginary parts) will themselves be real solutions to the differential equation. Our exposure to complex functions will be largely limited to doing some algebraic work with them; when studied in depth these functions lead to a rich area of mathematics known as complex analysis, where one can discover how calculus can be extended from working with real functions to complex ones.

Examples for further practice:

1. For each complex number, identify its real and imaginary parts, determine its complex conjugate, and write the number in the form $z = |z|e^{i\theta}$.

   (a) $z = 3 - 2i$
   (b) $z = -4 + 9i$
   (c) $z = 5$
   (d) $z = 4i$

2. Evaluate the stated sum, difference, product, or quotient, and write the result in the form $z = a + bi$.

   (a) $(1 - 3i) + (4 + 7i)$
   (b) $(2 - 5i) - (10 - i)$
   (c) $(1 - 2i)i$
   (d) $(5 - 2i)(i - 3)$

(e) $(1+i)(1-i)$

(f) $\dfrac{1}{1-i}$

(g) $\dfrac{3-i}{2+3i}$

3. For any complex numbers z and w, show that

   (a) $\overline{z+w} = \overline{z}+\overline{w}$ and $\overline{zw} = \overline{z}\,\overline{w}$

   (b) $z\overline{z} = |z|^2 \geq 0$

   (c) $\mathrm{Re}(z+w) = \mathrm{Re}z + \mathrm{Re}w$

   (d) $\mathrm{Re}(zw) = \mathrm{Re}z\,\mathrm{Re}w - \mathrm{Im}z\,\mathrm{Im}w$

   (e) $\mathrm{Im}(zw) = \mathrm{Im}z\,\mathrm{Re}w + \mathrm{Re}z\,\mathrm{Im}w$

4. Using the fact that $e^{i\theta} = \cos\theta + i\sin\theta$, determine the real and imaginary parts of

   (a) $e^{i\pi}$

   (b) $e^{i\pi/6}$

   (c) e^{2-3i}

C

Roots of polynomials

Polynomials are the most basic functions in all of mathematics. A *real polynomial of degree n* is a function $p(t)$ of the form

$$p(t) = a_n t^n + a_{n-1} t^{n-1} + \cdots + a_1 t + a_0 \tag{C.1}$$

where $a_0, \ldots, a_n$ are real numbers. A number r is a *root* or *zero* of a polynomial p if and only if $p(r) = 0$. In addition, we note that r is a root of p if and only if $(t - r)$ is a *factor* of $p(t)$, which means that we can express $p(t)$ in the form $p(t) = (t - r)q(t)$, where q is a polynomial of degree one less than p.

The roots of polynomials find important applications in many settings; in our study of differential equations and linear algebra, we must find polynomial zeros when solving the eigenvalue problem, as well as when determining fundamental solutions to higher order linear differential equations and linear systems of DEs. Here we briefly review some of the most important facts about the zeros of polynomial functions.

From quadratic polynomials of the form $p(t) = at^2 + bt + c$, we know that there are three possibilities for the zeros: p may have two distinct real zeros, one repeated real zero, or no real zeros. This can be observed in a variety of ways, but a graphical perspective is compelling: if a quadratic function p opens upward (that is, its coefficient $a > 0$), then the function either its vertex lies above the t-axis, on the t-axis, or below the t-axis, thus leading to the three noted possibilities, as shown in figure C.1.

We can see the three cases from an algebraic perspective as well. From the quadratic formula, we know the zeros are given by

$$t = \frac{-b \pm \sqrt{b^2 - 4ac}}{2a} \tag{C.2}$$

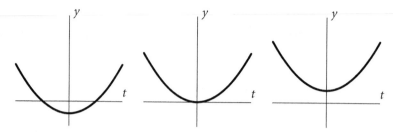

Figure C.1 Three concave up quadratic functions whose vertices lie, respectively, below, on, and above the t-axis.

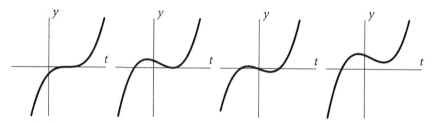

Figure C.2 Four cubic polynomials that demonstrate possible arrangements of the zeros of cubic functions.

Thus, if $b^2 - 4ac > 0$, it follows that $p(t)$ has two distinct real roots. In the case that $b^2 - 4ac = 0$, $p(t)$ has one repeated real root; here we say that $p(t)$ has a root of *multiplicity* 2. Finally, if $b^2 - 4ac < 0$, then although the term $\sqrt{b^2 - 4ac}$ permits no real solutions, if we use complex numbers and write $\sqrt{b^2 - 4ac} = i\sqrt{4ac - b^2}$, we find that $p(t)$ has two distinct complex roots. Note from (C.2) that these two complex roots are complex conjugates of one another; more on complex numbers can be found in appendix B.

The factored form of quadratic polynomials is also important. If $p(t)$ has two real roots, say $t = -1$ and $t = 1$, then $p(t)$ can be written in the form $p(t) = a(t+1)(t-1)$, where $p(t)$ is the product of two real linear terms. If $p(t)$ has a repeated root, say $t = 1$, then we have $p(t) = a(t-1)^2$. Finally, if $p(t)$ has complex roots, then $p(t)$ cannot be factored into a product of real linear terms. For example, $p(t) = t^2 + 1$ is a quadratic function with roots $t = \pm i$; we say that the quadratic term $t^2 + 1$ is *irreducible*.

For polynomials of degree greater than 2, many similar properties hold. For example, for polynomials of degree 3, we can see graphically several possibilities in figure C.2. In particular, a cubic polynomial can have a single real, repeated root of multiplicity 3, such as the function $p(t) = (t-1)^3$ shown at left in figure C.2. Alternatively, it is possible for the function to have algebraic form $p(t) = (t-1)^2(t+1)$, which leads to two real roots, one of which has multiplicity 2, which corresponds to the left center function in figure C.2. Likewise, a cubic function such as $p(t) = t(t-1)(t+1)$ can have three distinct real zeros—see the right center graph in the figure—or have only a single real root (which leaves the remaining two roots to be complex) as shown in the right-most graph in figure C.2.

Because a cubic function will have one end tend to $+\infty$ and the other to $-\infty$, this guarantees that every cubic function will have at least one real zero. It follows that we can write p in the form $p(t) = (t - r)q(t)$ where q is quadratic, and from this we can deduce the four possible cases for the zeros of p discussed in the preceding paragraph. In fact, there even exists a cubic formula analogous to the quadratic formula that explicitly provides the zeros of $p(t) = at^3 + bt^2 + ct + d$ in terms of formulas involving the coefficients a, b, c, and d. This formula is sufficiently complicated that we choose not to state it here.

The patterns we have observed for quadratic and cubic polynomials can be proved to hold for real polynomials of any degree. In particular, we have seen so far that for any degree-2 polynomial, the function has two zeros provided we allow them to be complex and count them according to their multiplicity. Similarly, for any degree-3 polynomial, the function has exactly three zeros under the same proviso. The Fundamental Theorem of Algebra, first proved by Carl Friedrich Gauss in 1799, beautifully summarizes the situation.

Theorem C.1   (The Fundamental Theorem of Algebra) If $p(t)$ is a real polynomial of degree n, then $p(t)$ has exactly n zeros provided we include complex zeros and count all zeros according to their multiplicity.

Theorem C.1 can be proved using methods of complex analysis. Its main purpose for our work is that we are always guaranteed that n roots of a polynomial of degree n exist. Through the methods established in chapters 3 and 4 for dealing with complex and repeated roots of characteristic equations, the Fundamental Theorem of Algebra ultimately enables us to find all solutions to any homogeneous linear higher order DE or system of linear first-order DEs.

We also note that it is possible to use standard ideas in complex analysis to show that if r is a complex root of a real polynomial p, then its complex conjugate $\bar{r}$ is also a root of p. This guarantees that for real polynomials, complex roots will always appear in conjugate pairs, just as we saw for the case of quadratic functions.

While the Fundamental Theorem of Algebra guarantees the existence of n zeros to a polynomial of degree n, it unfortunately does not provide an algorithm for finding them. In fact, though formulas exist for quadratic and cubic equations, as well as the degree four case, mathematicians have shown that there exists no general formula to provide the roots of a polynomial of degree 5 or greater. For higher degree polynomial equations, this leads us to resort to numerical methods or computer algebra systems; see section 4.6.1 for more on how to use *Maple* to compute the roots of polynomial functions.

Examples for further practice:

1. For each of the following polynomial functions, state the degree, determine all of the zeros, and state the multiplicity of each zero.

(a) $p(t) = t(t+2)(t+5)^2(t-3)(t-\pi)(t^2+1)$
(b) $p(t) = t^4 - 1$
(c) $p(t) = t^4 + 1$
(d) $p(t) = (t^2+1)^3(t-3)^5(t^2-t-12)$
(e) $p(t) = t^5 + 6t^3 + 9t$

2. Determine a formula for a real polynomial function of the least possible degree that satisfies the given criteria. State the degree of the function you find. If no such function is possible, explain why.

(a) distinct zeros at $t = -3, -1, 2$ and a zero of multiplicity 3 at $t = 0$
(b) complex zeros $t = \pm 3i$, each of multiplicity 2, and a single real zero of multiplicity 1 at $t = 4$
(c) a zero of multiplicity 2 at $t = -1$, a zero of multiplicity 3 at $t = 2$, and a zero of multiplicity 4 at $t = 5$
(d) a polynomial of even degree with exactly one real zero of multiplicity 1 at $t = 0$

D

Linear transformations

The notion of function is central to mathematics. Given any two collections of objects A and B, a function $f : A \rightarrow B$ is a rule that associates each element of A with one and only one element of B. Sometimes, we use the terms *mapping* or *transformation* in place of the word *function*. Among all functions, certain types stand out for their important properties and/or simplicity. In what follows, we focus on the property of *linearity*.

In many different areas of our study of linear algebra and differential equations, we find that linear combinations of objects play a key role. Similarly, we encounter important functions that transform a certain group of objects into another collection. The combination of these ideas makes us naturally interested in transformations that preserve linear combinations. Let us consider three familiar examples.

(1) For any $m \times n$ matrix $\mathbf{A}$, any vectors $\mathbf{x}$ and $\mathbf{y}$ in $\mathbb{R}^n$, and any real number c,
$$\mathbf{A}(\mathbf{x}+\mathbf{y}) = \mathbf{A}\mathbf{x}+\mathbf{A}\mathbf{y} \quad \text{and} \quad \mathbf{A}(c\mathbf{x}) = c\mathbf{A}\mathbf{x}$$

(2) From calculus, if we let D denote the differential operator, then for any differentiable functions f and g and any real number c, we know by the sum and constant multiple rules that
$$D(f+g) = D(f) + D(g) \quad \text{and} \quad D(cf) = cD(f)$$

(3) In our studies of the Laplace transform $\mathcal{L}$ in chapter 5, we found that the transform satisfies the property that for any acceptable functions f and g and any real constant c,
$$\mathcal{L}[f(t)+g(t)] = \mathcal{L}[f(t)] + \mathcal{L}[g(t)] \quad \text{and} \quad \mathcal{L}[cf(t)] = c\mathcal{L}[f(t)]$$

Matrix–vector multiplication, differentiation, and the Laplace transform are all examples of transformations: they take a given input (a vector or a function) and transform that input to a (unique) output, a new vector or function. Moreover, each satisfies the property that the transformation preserves sums and scalar multiples: the transformation applied to a sum is the same as the sum of the results of the transformation applied to the individual objects, and the transformation applied to a scalar multiple of an input is identical to the same scalar multiple of the output that results from the transformation applied to the original object. Viewing these inputs as belonging to a vector space[1], we arrive at the following formal definition.

> **Definition D.1**   Let U and V be vector spaces. A transformation $T : U \to V$ is a *linear transformation* provided that for any vectors $\mathbf{u}$ and $\mathbf{v}$ in U and any scalar c, T satisfies the properties $T(\mathbf{u}+\mathbf{v}) = T(\mathbf{u}) + T(\mathbf{v})$ and $T(c\mathbf{u}) = cT(\mathbf{u})$.

Two consequences of the definition are immediate: $T(\mathbf{0}) = \mathbf{0}$ and $T(a\mathbf{u} + b\mathbf{v}) = aT(\mathbf{u}) + bT(\mathbf{v})$ for all scalars a, b and vectors $\mathbf{u}$, $\mathbf{v}$. Note that in the equation $T(\mathbf{0}) = \mathbf{0}$, the zero vector on the left is from U while the one on the right is from V, and thus these may not be the same zero vectors.

Linear transformations play a key structural role in linear algebra and in the theory of linear DEs. We first turn to a discussion of the matrix of a linear transformation of finite dimensional vector spaces.

Matrix transformations

In section 1.3, we first saw that Property (1) above holds for matrix–vector multiplication. That is, given an $m \times n$ matrix $\mathbf{A}$, for any two vectors $\mathbf{x}$ and $\mathbf{y}$ in $\mathbb{R}^n$ and any constant c,

$$\mathbf{A}(\mathbf{x}+\mathbf{y}) = \mathbf{A}\mathbf{x} + \mathbf{A}\mathbf{y} \quad \text{and } \mathbf{A}(c\mathbf{x}) = c\mathbf{x}$$

Thus, if we define the transformation $T : \mathbb{R}^n \to \mathbb{R}^m$ by the rule $T(\mathbf{x}) = \mathbf{A}\mathbf{x}$, then it follows immediately that $T(\mathbf{x}+\mathbf{y}) = T(\mathbf{x}) + T(\mathbf{y})$ and $T(c\mathbf{x}) = cT(\mathbf{x})$, which means that T is a linear transformation. Said differently, the natural *multiplication function* associated with a given matrix $\mathbf{A}$ always generates a linear transformation. We usually call $\mathbf{A}$ the *matrix of the transformation* T. Consider the following particular example.

Example D.1   Let $\mathbf{A} = \begin{bmatrix} 3 & -2 & 5 \\ -1 & 0 & -7 \end{bmatrix}$, and let $T(\mathbf{x}) = \mathbf{A}\mathbf{x}$. Determine $T(\mathbf{e}_1)$, $T(\mathbf{e}_2)$, and $T(\mathbf{e}_3)$ where $\{\mathbf{e}_1, \mathbf{e}_2, \mathbf{e}_3\}$ is the standard basis of $\mathbb{R}^3$, and then use properties of linearity to determine $T(\mathbf{z})$ when $\mathbf{z} = [-5\ 2\ -6]^T$.

Solution.   First, we observe that

$$T(\mathbf{e}_1) = \mathbf{A}\mathbf{e}_1 = \begin{bmatrix} 3 & -2 & 5 \\ -1 & 0 & -7 \end{bmatrix} \begin{bmatrix} 1 \\ 0 \\ 0 \end{bmatrix} = \begin{bmatrix} 3 \\ -1 \end{bmatrix} \tag{D.1}$$

---

[1] This appendix assumes that the reader is familiar with basic concepts in sections 1.11 and 1.12. If the Laplace transform has not yet been studied, references to it may simply be skipped.

Similarly,

$$T(\mathbf{e}_2) = \begin{bmatrix} -2 \\ 0 \end{bmatrix} \quad \text{and} \quad T(\mathbf{e}_3) = \begin{bmatrix} 5 \\ -7 \end{bmatrix} \tag{D.2}$$

Next, to compute $T(\mathbf{z})$, we observe that

$$\mathbf{z} = \begin{bmatrix} -5 \\ 2 \\ -6 \end{bmatrix} = -5\begin{bmatrix} 1 \\ 0 \\ 0 \end{bmatrix} + 2\begin{bmatrix} 0 \\ 1 \\ 0 \end{bmatrix} - 6\begin{bmatrix} 0 \\ 0 \\ 1 \end{bmatrix} = -5\mathbf{e}_1 + 2\mathbf{e}_2 - 6\mathbf{e}_3$$

and thus by the linearity of T and (D.1) and (D.2), we have

$$\begin{aligned} T(\mathbf{z}) &= T(-5\mathbf{e}_1 + 2\mathbf{e}_2 - 6\mathbf{e}_3) \\ &= -5T(\mathbf{e}_1) + 2T(\mathbf{e}_2) - 6T(\mathbf{e}_3) \\ &= -5\begin{bmatrix} 3 \\ -1 \end{bmatrix} + 2\begin{bmatrix} -2 \\ 0 \end{bmatrix} - 6\begin{bmatrix} 5 \\ -7 \end{bmatrix} \\ &= \begin{bmatrix} -49 \\ 47 \end{bmatrix} \end{aligned}$$

There are at least two important observations to make from example D.1. The first is that, due to linearity, we can find the result of applying T to any vector if we first know the results of applying T to the basis vectors in the domain of T. Since any vector in the domain can be uniquely expressed as a linear combination of basis elements and T preserves linear combinations, we can easily apply T to the linear combination that generates the vector of our choice. This holds not just for the transformation in the example, but indeed for any linear transformation on a vector space.

Furthermore, (D.1) and (D.2) indicate that there is a key relationship between the values of the transformation applied to the domain's basis vectors and the matrix of the transformation. Specifically, $T(\mathbf{e}_1)$ is the first column of $\mathbf{A}$, and $T(\mathbf{e}_2)$ and $T(\mathbf{e}_3)$ are the second and third columns of $\mathbf{A}$. That this result holds in general is the following theorem.

Theorem D.1 If $T : \mathbb{R}^n \to \mathbb{R}^m$ is a linear transformation, then $T(\mathbf{x}) = \mathbf{Ax}$ where $\mathbf{A}$ is the $m \times n$ matrix

$$\mathbf{A} = [T(\mathbf{e}_1) \ T(\mathbf{e}_2) \ \cdots \ T(\mathbf{e}_n)]$$

and $\mathbf{e}_j$ is the jth standard basis vector of $\mathbb{R}^n$. Moreover, the matrix $\mathbf{A}$ is unique.

Example D.2 Let $T : \mathbb{R}^2 \to \mathbb{R}^3$ be a linear transformation such that

$$T(\mathbf{e}_1) = \begin{bmatrix} -2 \\ 3 \\ 9 \end{bmatrix} \text{ and } T(\mathbf{e}_2) = \begin{bmatrix} 4 \\ -2 \\ 0 \end{bmatrix}$$

Determine the matrix $\mathbf{A}$ of the transformation T and use $\mathbf{A}$ to compute $T(\mathbf{z})$ where $\mathbf{z} = [-3 \ -2]^{\mathrm{T}}$.

Solution.   By theorem D.1, it follows that

$$T(\mathbf{x}) = \mathbf{Ax} = \begin{bmatrix} -2 & 4 \\ 3 & -2 \\ 9 & 0 \end{bmatrix} \mathbf{x}$$

Thus, we can compute $T(\mathbf{z})$ as

$$T(\mathbf{z}) = \mathbf{Az} = \begin{bmatrix} -2 & 4 \\ 3 & -2 \\ 9 & 0 \end{bmatrix} \begin{bmatrix} -3 \\ -2 \end{bmatrix} = \begin{bmatrix} -2 \\ -5 \\ -27 \end{bmatrix}$$

Linear differential equations

In chapter 4, we solve higher order linear differential equations with constant coefficients of the form

$$y^{(n)} + a_{n-1}y^{(n-1)} + \cdots + a_1 y' + a_0 y = f(t) \tag{D.3}$$

In this setting, we can take a sophisticated perspective through linearity to see how solving an equation such as

$$y'' + 2y' + 3y = 0$$

is very similar to solving the homogeneous linear system of algebraic equations given by $\mathbf{Ax} = \mathbf{0}$ where $\mathbf{A}$ is an $m \times n$ matrix.

Recall that the derivative operator, D, is linear. The same is true of the second derivative operator, D^2, since

$$D^2(f + g) = (f + g)'' = f'' + g'' = D^2(f) + D^2(g)$$

and $D^2(cf) = (cf)'' = cf'' = cD^2(f)$. This alternate notation for derivatives permits a new perspective on DEs. Consider that $y'' + 2y' + 3y = 0$ can now be expressed as

$$D^2(y) + 2D(y) + 3y = 0 \tag{D.4}$$

In this setting, we observe that the left side of (D.4) appears as if a function or process is being applied to the input y. If we let L be the transformation defined by

$$L(y) = D^2(y) + 2D(y) + 3y$$

then we see that (D.4) can be written equivalently as the equation

$$L(y) = 0$$

Moreover, this new transformation L is linear. Observe that

$$\begin{aligned} L(f + g) &= D^2(f + g) + 2D(f + g) + 3(f + g) \\ &= D^2(f) + D^2(g) + 2D(f) + 2D(g) + 3f + 3g \\ &= D^2(f) + 2D(f) + 3f + D^2(g) + 2D(g) + 3g \\ &= L(f) + L(g) \end{aligned}$$

Similarly, it is straightforward to show that for any constant c, $L(cf) = D^2(cf) + 2D(cf) + 3(cf) = cD^2(f) + 2cD(f) + 3cf = cL(f)$. Hence, we see that solving the second-order equation (D.4) is equivalent to solving the linear homogeneous equation $L(y) = 0$, where L is the linear transformation just discussed. More generally, solving equations of the form (D.3) is equivalent to solving the linear equation

$$L(y) = f$$

where L is the linear transformation defined by $L(y) = D^n(y) + a_{n-1}D^{n-1}(y) + \cdots + a_1 D(y) + a_0 y$. While this perspective does not contribute substantially to our methods for solving such equations, it does further emphasize why these equations are classified as *linear* and why the characteristic polynomial $r^n + a_{n-1}r^{n-1} + \cdots a_1 r + a_0$ arises so naturally.

Furthermore, the linearity of differential equations of form (D.3) together with the fact that the Laplace transform is a linear operator is part of what enables the Laplace transform to be such an effective tool. For example, to solve

$$y'' + 2y' + 3y = \delta(t - 3)$$

we take the Laplace transform of both sides of the equation to find

$$\mathcal{L}[y'' + 2y' + 3y] = \mathcal{L}[\delta(t - 3)]$$

and thus by linearity

$$\mathcal{L}[y''] + 2\mathcal{L}[y'] + 3\mathcal{L}[y] = \mathcal{L}[\delta(t - 3)]$$

From there, properties of the transform discussed in sections 5.3 and 5.4 enable us to proceed to where we only need to use the inverse Laplace transform to solve the equation, which brings us to yet another class of important linear transformations.

Invertible transformations

A function or transformation $T : U \to V$ is invertible provided that there exists a function $T^{-1} : V \to U$ that satisfies the properties that

$$T^{-1}[T(\mathbf{u})] = \mathbf{u} \text{ for all } \mathbf{u} \in U \text{ and } T[T^{-1}(\mathbf{v})] = \mathbf{v} \text{ for all } \mathbf{v} \in V$$

Equivalently, in order for T to be invertible, there must exist a function T^{-1} that when composed with T results in the appropriate identity mapping: $T^{-1} \circ T = I_U$ and $T \circ T^{-1} = I_V$, where $I_U(\mathbf{u}) = \mathbf{u}$ for every $\mathbf{u} \in U$. Loosely, the transformation T is invertible whenever there exists a function T^{-1} that reverses the work of T.

Any time a matrix $\mathbf{A}$ is invertible, the resulting matrix transformation $T(\mathbf{x}) = \mathbf{Ax}$ is an invertible transformation. Consider the following example.

Example D.3   Let $A = \begin{bmatrix} 3 & 2 \\ -2 & -1 \end{bmatrix}$ and let $T(\mathbf{x}) = A\mathbf{x}$. Show that T is an invertible transformation and determine a formula for T^{-1}.

Solution.   We first observe that since $\det(A) = 1 \neq 0$, the matrix A is invertible. In addition, we can compute A^{-1} according to the standard algorithm, finding via row-reduction that

$$\begin{bmatrix} 3 & 2 & 1 & 0 \\ -2 & -1 & 0 & 1 \end{bmatrix} \rightarrow \begin{bmatrix} 1 & 0 & -1 & -2 \\ 0 & 1 & 2 & 3 \end{bmatrix}$$

Thus, the inverse of A is

$$A^{-1} = \begin{bmatrix} -1 & -2 \\ 2 & 3 \end{bmatrix}$$

Letting $T^{-1}(\mathbf{x}) = A^{-1}\mathbf{x}$, it follows that $T^{-1}(T(\mathbf{x})) = A^{-1}(A\mathbf{x}) = I\mathbf{x} = \mathbf{x}$ and $T(T^{-1}(\mathbf{x})) = A(A^{-1}\mathbf{x}) = I\mathbf{x} = \mathbf{x}$, which demonstrates that T is invertible and its inverse is given by the formula

$$T^{-1}(\mathbf{x}) = \begin{bmatrix} -1 & -2 \\ 2 & 3 \end{bmatrix} \mathbf{x}$$

Invertible matrix transformations find many important applications, including a prominent role in computer graphics. When matrix transformations are used to move a graphical image in a particular way, the inverse transformation is needed to move the object back. More on such transformations can be studied in section 1.8.1 and in the project found at the end of chapter 1 in 1.13.1.

In the study of differential equations, two other invertible linear transformations are important. One is found in the integral operator

$$S(f(x)) = \int_0^x f(t)\, dt$$

which is closely linked to the differential operator, $D(f(x)) = f'(x)$. Specifically, since a typical differential equation involves an unknown function and one or more of its derivatives, a natural approach is to attempt to integrate. In fact, for first-order equations that are separable, integration is the standard approach; with some care, integration also works well for linear first-order equations as well as exact equations. In these approaches, as well as in others used to solve differential equations, we use the fact that integration essentially reverses the process of differentiation. Here, we take care to be more precise about this fact.

Let U be the vector space of all continuously differentiable functions f such that $f(0) = 0$, and V the vector space of all continuous functions.[2] Then, we see

---

[2] The choices of U and V can be made considerably broader; doing so involves some subtleties from real analysis that are beyond the scope of this course. See, for instance, *Real Analysis*, by Bruckner, Bruckner, and Thomson, 1996, for a discussion on which functions have the property that they are differentiable and their derivative is integrable, as well as which functions are integrable.

that $D : U \to V$ and $S : V \to U$. Furthermore, for any f in U,

$$S(D(f)) = S(f') = \int_0^x f'(t)\,dt = f(x) - f(0) = f(x)$$

since $f(0) = 0$, and for any function g in V,

$$D(S(g)) = D\left(\int_0^x g(t)\,dt \right) = g(x)$$

by the Fundamental Theorem of Calculus. In each case, $S(D(f)) = f$ and $D(S(g)) = g$ for all relevant functions. This shows that D and S are each invertible transformations, and moreover that they are each other's respective inverses. Moreover, as we have noted on several occasions and is studied in calculus, both D and S are linear transformations.

Finally, the Laplace transform is a key example of an invertible linear transformation, and its invertibility ultimately is what makes it such a useful tool in the solution of linear differential equations. To emphasize several of the important properties, we consider an example of a fundamental initial-value problem and discuss the role of the Laplace transform in its solution. Specifically, we examine the role of the Laplace transform in the solution of the IVP

$$y'' + 3y' + 2y = 0, \quad y(0) = 1, \quad y'(0) = -1$$

First, recall that $\mathcal{L}$ is a linear transformation on the vector space of acceptable functions and that $\mathcal{L}$ transforms a given acceptable function $y(t)$ to a new function $Y(s)$. If we now apply the transform to both sides of the differential equation, the linearity of $\mathcal{L}$ implies that

$$\mathcal{L}[y''] + 3\mathcal{L}[y'] + 2\mathcal{L}[y] = 0 \tag{D.5}$$

From properties of $\mathcal{L}$ developed in chapter 5, we know that $\mathcal{L}[y''] = s^2 \mathcal{L}[y] - sy(0) - y'(0)$ and $\mathcal{L}[y'] = s\mathcal{L}[y] - y(0)$. Therefore, (D.5) can be updated to the equation

$$s^2 \mathcal{L}[y] + s - 1 + 3(s\mathcal{L}[y] - 1) + 2\mathcal{L}[y] = 0 \tag{D.6}$$

Observe that (D.6) is now an algebraic (rather than differential) equation in $Y(s) = \mathcal{L}[y(t)]$. Moreover, whereas before the equation we were trying to solve was a differential equation with three unknowns (y, y', and y''), now there is only one unknown, $\mathcal{L}[y]$, in (D.6). Solving for $\mathcal{L}[y]$, we find

$$\mathcal{L}[y](s^2 + 3s + 2) = 4 - s$$

and therefore

$$\mathcal{L}[y] = \frac{4 - s}{s^2 + 2s + 3} \tag{D.7}$$

At this point, the natural remaining step to solve for y becomes evident. Since $\mathcal{L}$ is an invertible transformation, $\mathcal{L}^{-1}[\mathcal{L}[y]] = y$, and thus we want to take the inverse Laplace transform of both sides of (D.7). One key computation must be performed first, as it turns out that a different algebraic form of the right-hand side is useful. A partial fraction decomposition of $(4 - s)/(s^2 + 2s + 3)$ reveals that (D.7) can be equivalently expressed as

$$\mathcal{L}[y] = \frac{5}{s+1} - \frac{6}{s+2} \tag{D.8}$$

Now we are ready to use the inverse Laplace transform; it, like the transform itself, is linear, and thus we find that

$$\mathcal{L}^{-1}[\mathcal{L}[y]] = \mathcal{L}^{-1}\left[\frac{5}{s+1} - \frac{6}{s+2}\right] \tag{D.9}$$

and therefore

$$y = 5\mathcal{L}^{-1}\left[\frac{1}{s+1}\right] - 6\mathcal{L}^{-1}\left[\frac{1}{s+2}\right] \tag{D.10}$$

A standard fact about the Laplace transform is that for any real number a, $\mathcal{L}[e^{at}] = 1/(s-a)$. From this, (D.10) implies that

$$y = 5e^{-t} - 6e^{-2t}$$

which is the solution to the original initial-value problem.

As we have noted throughout our discussion, the Laplace transform's linearity and invertibility play essential roles in the application of this tool to initial-value problems. These fundamental ideas demonstrate the valuable nature of the properties of linearity and invertibility, not just with the Laplace transform, but indeed in any setting.

Examples for further practice:

1. For the given linear transformation T from $\mathbb{R}^n$ to $\mathbb{R}^m$, find the matrix of the transformation T, and hence compute $T(\mathbf{z})$, where $\mathbf{z}$ is the given vector

    (a) $T : \mathbb{R}^2 \to \mathbb{R}^3$ with the property that $T(\mathbf{e}_1) = [1 \ -3 \ 4]^T$ and $T(\mathbf{e}_2) = [-2 \ 1 \ 0]^T$; $\mathbf{z} = [3 \ -2]$.

    (b) $T : \mathbb{R}^3 \to \mathbb{R}^2$ with the property that $T(\mathbf{e}_1) = [-2 \ -1]^T$, $T(\mathbf{e}_2) = [5 \ 1]^T$, and $T(\mathbf{e}_3) = [3 \ 4]^T$; $\mathbf{z} = [6 \ -1 \ 3]$.

    (c) $T : \mathbb{R}^2 \to \mathbb{R}^2$ with the property that $T(\mathbf{e}_1) = [7 \ 5]^T$ and $T(\mathbf{e}_2) = [-11 \ 3]^T$; $\mathbf{z} = [3 \ -2]$.

2. Let $T : \mathbb{P}_2 \to \mathbb{R}^3$ be a linear mapping such that

$$T(t^2) = \begin{bmatrix} 1 \\ 0 \\ -1 \end{bmatrix}, \ T(t) = \begin{bmatrix} 0 \\ -2 \\ 1 \end{bmatrix}, \text{ and } T(1) = \begin{bmatrix} -3 \\ 4 \\ 0 \end{bmatrix}$$

Determine $T(3t^2 - 4t + 7)$. (Recall that the standard basis of $\mathbb{P}_2$ is $\{1, t, t^2\}$.)

3. For each given linear transformation T below, find the matrix A of the transformation.

(a) $T(x, y) = (2x + y, -3x + 2y)$
(b) $T(x, y, z) = (x + y - z, -x + 2y + 3z)$
(c) $T(x, y) = (-x + 4y, x - 2y, 3x + 7y)$

4. Let D denote the differential operator and D^2 the second derivative. Use this notation to recast the following differential equations as equations involving linear transformations, as shown in (D.4).

(a) $y'' - 6y' + 5y = 0$
(b) $y'' + 4y = 0$
(c) $y' + 5y = 10$

5. Again, let D denote the differential operator. Let $L(y) = D^2(y) + 5D(y) + 4y$. Show that L is a linear operator. In addition, find all polynomial solutions to the equation $L(y) = 2t + 3$.

6. For each linear transformation T given below, determine whether or not the transformation is invertible and, if so, find a formula for its inverse.

(a) $T(x, y) = (2x + y, -3x + 2y)$
(b) $T(x, y) = (2x + y, -4x - 2y)$
(c) $T : \mathbb{R}^2 \to \mathbb{R}^2$ with the property that $T(e_1) = [7 \ 5]^T$ and $T(e_2) = [-11 \ 3]^T$
(d) $T : \mathbb{R}^2 \to \mathbb{R}^2$ with the property that $T(e_1) = [7 \ -5]^T$ and $T(e_2) = [-14 \ 10]^T$
(e) T is the mapping that takes each point (x, y) in the plane and reflects the point in the line $y = x$.
(f) T is the mapping that rotates each point (x, y) in the plane by $90°$ counterclockwise about the origin.

Solutions to selected exercises

Section 1.2

1. The unique solution to the system is $(-1, 1)$.

3. The system has no solution.

5. The system is consistent with unique solution $(4, -2, 3)$.

7. The system is consistent with infinitely many solutions given parametrically by $(-3 - 2t, -2 - t, t)$, $t \in \mathbb{R}$.

9. The system is consistent with infinitely many solutions given parametrically by $(-1 + 2t - 4s, t, 2 - 3s, s, -5)$, $t, s \in \mathbb{R}$.

11. No solution exists.

13. There are infinitely many solutions given parametrically by $(1 - 19t, s, 1 + 4t, t)$, $s, t \in \mathbb{R}$.

15. The system is consistent if $h = -21$ and inconsistent otherwise.

17. The system is consistent for all values of h; if $h \neq 0$, the solution is unique.

19. The system is consistent with unique solution $(53/3, -8/3, -46/3)$.

21. The system is consistent with infinitely many solutions given parametrically by $(5/3 - 1/6t, -13/3 + 5/6t, t)$, $t \in \mathbb{R}$.

23. The system is consistent with infinitely many solutions given parametrically by $(19/2 - 9t, -5/2 + 17/4t, 2 - 3/2t, t)$, $t \in \mathbb{R}$.

25. No.

27. Yes. $(2, 1, 2)$.

29. Yes. Consider the system $x_1 + x_2 + x_3 = 0$, $x_1 + x_2 + x_3 = 1$.

31. The number of pivot columns must equal the number of variables, so that no free variables are present.

33. $4 = a_2 1^2 + a_1 1 + a_0$, $7 = a_2 2^2 + a_1 2 + a_0$, $6 = a_2 3^2 + a_1 3 + a_0$, so $a_0 = -3$, $a_1 = 9$, and $a_2 = -2$.

35. $I_1 = 10/41$, $I_2 = 80/41$, and $I_3 = 70/41$.

Section 1.3

1. The product is not defined.

3. $\mathbf{Ax} = [19 \; 5 \; -13]^T$.

5. To get each entry in $\mathbf{Ax}$, we take the dot product of the corresponding row in A with the column vector $\mathbf{x}$.

7. $\begin{bmatrix} x_1' \\ x_2' \end{bmatrix} = \begin{bmatrix} -1/20 & 1/80 \\ 1/40 & -1/40 \end{bmatrix} \begin{bmatrix} x_1 \\ x_2 \end{bmatrix} + \begin{bmatrix} 2250 \\ 3750 \end{bmatrix}$.

9. Yes, $\mathbf{b}$ is a linear combination of the vectors $\mathbf{a}_1$, $\mathbf{a}_2$, $\mathbf{a}_3$; infinitely many weights work. For example, $x_1 = 3$, $x_2 = 1$, $x_3 = 0$.

11. The system has infinitely many solutions, so $\mathbf{b}$ is a linear combination of the columns of $\mathbf{A}$, and can be written as such a linear combination with infinitely many different possible weights (x_1, x_2, x_3). Each pair of weights is of the form $(-3 - t, 5 + t, t)$.

13. The system has no solution, so $\mathbf{b}$ is not a linear combination of the columns of A.

15. $\mathbf{A} = \begin{bmatrix} 5 & -3 & 1 \\ -2 & 1 & 4 \\ 1 & 0 & -2 \end{bmatrix}$, $\mathbf{b} = \begin{bmatrix} 0 \\ 22 \\ -11 \end{bmatrix}$

17. The system has infinitely many solutions of the form $(-t, t, t)$.

19. The system has infinitely many solutions of the form $(-t/3, t)$.

21. The system has the unique solution $x_1 = x_2 = x_3 = 0$.

23. All vectors $\mathbf{b} = [b_1 \; b_2]^T$ whose entries satisfy $b_1 = b_2/2$.

25. (a) F; (b) T; (c) T; (d) F; (e) F.

27. $\mathbf{x}^{(1)} = [94.40 \; 70.40 \; 75.20]^T$, $\mathbf{x}^{(2)} = [89.52 \; 79.50 \; 70.98]^T$, $\mathbf{x}^{(3)} = [85.27 \; 87.47 \; 67.26]^T$.

Section 1.4

1. Infinitely many solutions, each of the form $(2t/11, 8t/11, t)$. Thus the solution set is the span of the vector $[2/11, 8/11, 1]^T$.

3. The span of the vector $[8/5, 1]^T$.

5. $Ax = 0$ has only the trivial solution.

7. Because A has more columns than rows, A cannot have a pivot in every column. Therefore, free variables must be present when $[A \mid 0]$ is row-reduced and nontrivial solutions exist.

9. b is not in the span of the given vectors.

11. Yes, using the weights $x_1 = -2$, $x_2 = 1$, $x_3 = 4$.

13. W is a plane through the origin in $\mathbb{R}^3$ that contains the given vectors v_1 and v_2.

15. If (x_1, x_2) satisfies $2x_1 - 3x_2 = 0$, then $x_1 = 3x_2/2$, so that the vector $x = [x_1 \ x_2]^T$ is a scalar multiple of the vector $[3 \ 2]^T$. Hence, each point on the line lies in Span$\{[3 \ 2]^T\}$.

17. (a) T; (b) F; (c) F; (d) F; (e) F.

Section 1.5

1. $Ax = b$ is consistent for every $b \in \mathbb{R}^2$ since A has a pivot in both rows.

3. $Ax = b$ is consistent for every $b \in \mathbb{R}^2$ since A has a pivot in both rows.

5. $Ax = b$ is consistent for every $b \in \mathbb{R}^3$ since A has a pivot in all three rows.

7. $Ax = b$ is not consistent for every $b \in \mathbb{R}^4$ since A does not have a pivot in row 4.

9. No. Because A has more rows than columns, it is impossible for A to have a pivot in every row.

11. b is a linear combination of the columns of A with weights $1/5$, $-6/5$.

13. b is a linear combination of the columns of A; infinitely many different weights are possible: one triple of such weights for the respective columns is $(6, -2, 0)$.

15. b is a linear combination of the columns of A with weights $x_1 = -35/11$, $x_2 = 1/11$, $x_3 = 7/11$.

17. $x = x_3[1 \ 1 \ 1]^T$.

19. $x = x_2[8/5 \ 1]^T$.

21. $x = x_3[-1 \ 1 \ 1]^T$.

23. $x = x_p + x_h = [-1 \ 1 \ 1 \ 0]^T + x_4[5 \ -3 \ -1 \ 1]^T$.

25. $x = x_p + x_h = [2 \ -3/2 \ 3/2]^T] + [0 \ 0 \ 0]^T$.

27. $Ax = b$ is always consistent.

29. Impossible. A can't have a pivot in all three rows.

31. $Ax = b$ will always be consistent. Since the described system has one free variable present, there is one non-pivot column in A. Since A has four columns,

A must have three pivot columnns and thus three pivot rows. Because of the free variable, every equation $\mathbf{Ax} = \mathbf{b}$ will have infinitely many solutions.

33. (a) F; (b) F; (c) T; (d) F; (e) T; (f) F.

35. $y = y_h + y_p = Ce^{5t} - 6/5.$

Section 1.6

1. S is linearly dependent.

3. S is linearly independent.

5. S is linearly dependent.

7. S is linearly dependent.

9. (1) no; (2) yes; (3) yes; (4) yes; (5) no; (6) yes; (7) no; (8) yes.

11. Not necessarily either.

13. S may or may not span $\mathbb{R}^4$. S cannot be linearly independent.

15. Given any nonzero vector $\mathbf{v}$, the zero vector may be written $\mathbf{0} = 0\mathbf{v}$.

19. $\{\mathbf{v}_1, \mathbf{v}_2, \mathbf{v}_3\}$ linearly independent for all real numbers k except $k = 17/7$. If $k = 17/7$, $\mathbf{v}_3$ in the span of $\{\mathbf{v}_1, \mathbf{v}_2\}$.

21. The columns of $\mathbf{A}$ are linearly dependent; the columns of $\mathbf{A}$ span $\mathbb{R}^4$. Both hold because there are four pivot columns in this 4×7 matrix.

23. (a) F; (b) T; (c) F; (d) F.

25. $c_1 = -2$ and $c_2 = 4.$

Section 1.7

1. (a) $\mathbf{B} + \mathbf{C} = \begin{bmatrix} -1 & 13 \\ 1 & 11 \\ -1 & -6 \end{bmatrix}$; (b) $\mathbf{A} + \mathbf{B}$ is undefined; (c) $-2\mathbf{A} = \begin{bmatrix} -6 & 10 & -4 \\ 2 & -10 & 8 \end{bmatrix}$;

(d) $-3\mathbf{B} + 4\mathbf{C} = \begin{bmatrix} 38 & -18 \\ -10 & -33 \\ 17 & -10 \end{bmatrix}$; (e) $\mathbf{AB} = \begin{bmatrix} -34 & -29 \\ 28 & 53 \end{bmatrix}$; (f) $\mathbf{BA} = \begin{bmatrix} -28 & 80 & -52 \\ -5 & 45 & -40 \\ -7 & 5 & 2 \end{bmatrix}$;

(g) $\mathbf{AA}$ is undefined; (h) $\mathbf{A}(\mathbf{B} + \mathbf{C}) = \begin{bmatrix} -10 & -28 \\ 10 & 66 \end{bmatrix}$; (i) $\mathbf{CA} = \begin{bmatrix} 12 & -10 & -2 \\ -3 & 5 & -2 \\ 10 & -30 & 20 \end{bmatrix}$;

(j) $\mathbf{C}(\mathbf{A} + \mathbf{B})$ is undefined; (k) $\mathbf{A}^{\mathsf{T}} + \mathbf{B} = \begin{bmatrix} -3 & 9 \\ -3 & 16 \\ -1 & -6 \end{bmatrix}$; (l) $(\mathbf{B} + \mathbf{C})^{\mathsf{T}} =$

$\begin{bmatrix} -1 & 1 & -1 \\ 13 & 11 & -6 \end{bmatrix}$; (m) $\mathbf{B}^{\mathsf{T}}\mathbf{C} = \begin{bmatrix} -38 & -6 \\ 35 & 38 \end{bmatrix}$; (n) $\mathbf{BC}^{\mathsf{T}} = \begin{bmatrix} 0 & 6 & -52 \\ 43 & -2 & -40 \\ -21 & 3 & 2 \end{bmatrix}$;

(o) $(\mathbf{AB})^{\mathrm{T}} = \begin{bmatrix} -34 & 28 \\ -29 & 53 \end{bmatrix}$; (p) $(\mathbf{BA})^{\mathrm{T}} = \begin{bmatrix} -28 & -5 & -7 \\ 80 & 45 & 5 \\ -52 & -40 & 2 \end{bmatrix}$.

3. Two square matrices of the same size can always be multiplied, and in either order. Non-square matrices can only be multiplied in both orders ($\mathbf{AB}$ and $\mathbf{BA}$) when one is $m \times n$ and the other is $n \times m$. Note that when $\mathbf{A}$ and $\mathbf{B}$ are not square, $\mathbf{AB}$ never equals $\mathbf{BA}$.

5. $\mathbf{A} = \begin{bmatrix} 1 & -2 \\ 2 & 1 \end{bmatrix}$, $\mathbf{B} = \begin{bmatrix} -3 & -4 \\ 4 & -3 \end{bmatrix}$.

7. $\mathbf{B} = \begin{bmatrix} 1/2 & 0 \\ 0 & 1/5 \end{bmatrix}$. Note that $\mathbf{BA} = \mathbf{AB}$.

9. $\mathbf{B} = \begin{bmatrix} 2 & 1 \\ 1 & 1 \end{bmatrix}$. Note that $\mathbf{BA} = \mathbf{AB}$.

11. (a) No; (b) No; (c) 1; (d) No familiar one; (e) No such matrix exists.

13. See 1.) above.

Section 1.8

1. $\mathbf{A}^{-1} = \begin{bmatrix} 1 & -\frac{1}{2} \\ -1 & 1 \end{bmatrix}$.

3. $\mathbf{A}^{-1}$ does not exist.

5. $\mathbf{A}^{-1}$ does not exist.

7. $\mathbf{Ax} = \mathbf{b}_1$ and $\mathbf{Ax} = \mathbf{b}_2$ each have infinitely many solutions, while $\mathbf{Ax} = \mathbf{b}_3$ has no solution. We see that $\mathbf{A}$ is not invertible.

9. Multiplying $\mathbf{A}$ by $\mathbf{E}$ on the left switches rows 2 and 3 in $\mathbf{A}$.

11. Multiplying $\mathbf{A}$ by $\mathbf{E}$ on the left switches multiplies row 2 by c.

13. Multiplying $\mathbf{A}$ by $\mathbf{E}$ on the left switches replaces row 3 with row 3 plus a times row 1.

15. $\mathbf{A}^{-1} = \mathbf{A}^{\mathrm{T}}$.

17. $(\mathbf{AB})^{-1} = \mathbf{B}^{-1}\mathbf{A}^{-1}$.

19. Suppose that both $\mathbf{B}$ and $\mathbf{C}$ are inverses of $\mathbf{A}$. Then $\mathbf{AB} = \mathbf{I} = \mathbf{AC}$. Since $\mathbf{A}$ is invertible, we can multiply on the left by $\mathbf{A}^{-1}$, from which it follows that $\mathbf{B} = \mathbf{C}$.

21. Yes: $\mathbf{A} = \begin{bmatrix} 1 & 1 \\ -1 & -1 \end{bmatrix}$.

23. $\mathbf{A}^n = \mathbf{PD}^n\mathbf{P}^{-1}$.

25. Every point is rotated 60° counterclockwise.

27. Every point is rotated 90° counterclockwise.

29. $\mathbf{C} = \begin{bmatrix} 2/5 & -3/10 \\ 3/10 & 2/5 \end{bmatrix}$.

31. Apply the inverse of the Markov matrix to the current population.

33. (a) F; (b) T; (c) F; (d) F; (e) T; (f) F; (g) T; (h) F.

Section 1.9

1. $\det(\mathbf{A}) = 2 \neq 0$ so $\mathbf{A}$ is invertible.

3. $\det(\mathbf{A}) = -28 \neq 0$ so $\mathbf{A}$ is invertible.

5. $\det(\mathbf{A}) = 252 \neq 0$ so $\mathbf{A}$ is invertible.

7. $\det(\mathbf{I}_n) = 1$; clearly $\mathbf{I}_n$ is invertible.

9. The matrix is invertible for all real numbers z except $z = 1, 3$.

11. $\det(\mathbf{AB}) = \det(\mathbf{A}) \cdot \det(\mathbf{B})$.

13. Since $\mathbf{A}\mathbf{A}^{-1} = \mathbf{I}$, we have $\det(\mathbf{A}\mathbf{A}^{-1}) = \det(\mathbf{I})$. Now use the property of determinants from Exercise 11 and solve for $\det(\mathbf{A})$.

15. $\det(\mathbf{A}) = 0$ since the columns are a linearly dependent set; equivalently, $\det(\mathbf{A}^{\mathrm{T}}) = 0$ since the rows of $\mathbf{A}$ are a linearly dependent set.

17. If $\mathbf{A}^2$ is not invertible, then $\mathbf{A}$ is not invertible, since $\det(\mathbf{A}^2) = \det(\mathbf{AA}) = \det(\mathbf{A})\det(\mathbf{A})$, so $\det(\mathbf{A}) = 0$ if and only if $\det(\mathbf{A}^2) = 0$.

19. $\mathbf{A}^{-1} = \frac{1}{\det(\mathbf{A})} \begin{bmatrix} d & -b \\ -c & a \end{bmatrix}$.

Section 1.10

1. $\lambda = 5, 3$ with corresponding eigenvectors $[1\ 0]^{\mathrm{T}}$, $[1\ -2]^{\mathrm{T}}$.

3. $\mathbf{A}$ does not have any real eigenvalues or eigenvectors. Its eigenvalues are $\lambda = -1 \pm 2i$.

5. $\lambda = 2$ with corresponding eigenvector $[1\ 0\ 0]^{\mathrm{T}}$.

7. $\lambda = 2$ with corresponding linearly independent eigenvectors $[1\ 0\ 0]^{\mathrm{T}}$, $[0\ 1\ 0]^{\mathrm{T}}$; $\lambda = 0$ with corresponding eigenvector $[0\ 0\ 1]^{\mathrm{T}}$.

9. $\mathbf{Ax} = [5\ 20]^{\mathrm{T}}$.

11. (a) $\lambda = -3, -3, 0$ with corresponding eigenvectors $[-1\ 1\ 0]^{\mathrm{T}}$, $[-1\ 0\ 1]^{\mathrm{T}}$, $[1\ 1\ 1]^{\mathrm{T}}$; (b) Yes.

13. (a) $\lambda = 5, 2, 2$ with corresponding eigenvectors $[1\ -1\ 1]^{\mathrm{T}}$, $[1\ 1\ 0]^{\mathrm{T}}$, $[-1\ 0\ 1]^{\mathrm{T}}$; (b) The columns of $\mathbf{P}$ are linearly independent; use the Invertible Matrix Theorem; (c) $\mathbf{AP} = \mathbf{PD}$; (d) $\mathbf{A}^{10} = \mathbf{PD}^{10}\mathbf{P}^{-1}$.

15. Hint: $\det(\mathbf{B} - \lambda\mathbf{I}) = \det(\mathbf{PAP}^{-1} - \lambda\mathbf{I}) = \det(\mathbf{PAP}^{-1} - \lambda\mathbf{PIP}^{-1}) = \det(\mathbf{P}(\mathbf{A} - \lambda\mathbf{I})\mathbf{P}^{-1}) = \det(\mathbf{P})\det(\mathbf{A} - \lambda\mathbf{I})\det(\mathbf{P}^{-1})$.

17. $D(e^{rx}) = r \cdot e^{rx}$, so taking the derivative only stretches e^{rx} by a factor of r. Thus, e^{rx} is like an eigenvector with eigenvalue r.

19. Yes; $\mathbf{v} \approx [192.41 \ \ 139.43 \ \ 77.71]^T$ (in millions).

21. (a) F; (b) T; (c) F; (d) T.

Section 1.11

1. H is not a subspace. Consider multiplying $[1 \ \ 1]^T$ by a negative scalar.

3. H is a subspace.

5. H is a subspace.

7. H is not a subspace, since the zero vector does not belong to H.

9. H is not a subspace; the zero matrix is not invertible.

11. H is a subspace.

13. H is a subspace.

15. H is a subspace.

17. For $\lambda = 1$, the corresponding eigenspace is the set of all scalar multiples of the eigenvector $[1 \ \ 1]^T$. For $\lambda = 3$, the corresponding eigenspace is the set of all scalar multiples of the eigenvector $[1 \ \ -1]^T$.

19. Because the span of a set is the set of all linear combinations of a given collection of vectors, we can always make the zero combination to get the zero vector. In addition, because any linear combination is allowed, the span of a set of vectors must be closed under scalar multiplication and closed under addition.

21. H is not a subspace of $\mathbb{R}^3$ because no values of a and b can be chosen to form the zero vector in H.

23. Col($\mathbf{A}$) is the set of all linear combinations of the columns of $\mathbf{A}$, which is equivalently the span of the columns of $\mathbf{A}$. By exercise 19, it follows that Col($\mathbf{A}$) is a subspace.

25. $\mathbf{v} = [-2 \ \ 1 \ \ 1]^T$ is not in Col($\mathbf{A}$); $\mathbf{u} = [-1 \ \ 4 \ \ -4]^T$ is in Col($\mathbf{A}$); Col($\mathbf{A}$) is the span of $\{[1 \ 3 \ -4]^T, [-2 \ 1 \ 0]^T\}$

27. Col($\mathbf{A}$), because it is simply the span of the columns of the given matrix.

29. Verify by direct substitution that $y = Ce^{3t} + 1$ is a solution to the equation. This set of all such solutions is not a subspace because the zero function is not a solution to the DE.

Section 1.12

1. A basis for H is $\{[2 \ 0 \ -1]^T\}$ and therefore H is one-dimensional.

3. A basis for H is $H = \{[2 \ 1 \ -3 \ 1]^T, [3 \ -4 \ 2 \ -1]^T\}$, so H is two-dimensional.

5. A basis for H is $H = \{[1/2\ \ 1]^T\}$; H is one-dimensional.

7. A basis for H is $\left\{ \begin{bmatrix} 1 & 0 \\ 0 & 0 \end{bmatrix}, \begin{bmatrix} 0 & 0 \\ 1 & 0 \end{bmatrix}, \begin{bmatrix} 0 & 0 \\ 0 & 1 \end{bmatrix} \right\}$, so H is three dimensional.

9. Yes, since S is a linearly independent spanning set in $\mathbb{R}^2$.

11. No, since S is not linearly independent.

13. No, a set with fewer than 4 vectors cannot span $\mathbb{R}^4$.

15. The vector space $\mathbb{P}$ of all polynomial functions is an infinite dimensional vector space because its basis has to include every power function: $1, t, t^2, t^3, \ldots, t^{100}, \ldots, t^{100000}, \ldots$. Therefore, the basis cannot have a finite number of elements.

17. $\dim(\text{Nul}(A)) + \dim(\text{Col}(A)) = n$ since the dimension of the column space of A is the number of pivot columns of A and the dimension of the null space of A is the number of non-pivot columns of A.

Section 2.2

1. (a) 2; (c) $y = e^{-2t}$.

3. $A(t) = 100$ is an equilibrium solution because it is a constant function that makes the DE true; this solution is a stable equilibrium, as seen from the direction field.

5. The direction field should show an unstable equilibrium at $P = 0$ and a stable equilibrium at $P = 25$ all solutions with initial values greater than 0 tending toward $P = 25$ as $t \to \infty$.

7. (a) i; (b) iii; (c) ii; (d) iv.

9. $y = t^2/2 + \sin t + C$.

11. $y = t^4/12 + t^2 + C_1 t + C_2$.

13. $y = \sin t - t \cos t + C$.

15. $y = -\frac{1}{2} e^{-t^2} + C$.

17. $y = t^2/2 + \sin t - \pi^2/8$.

19. $y' = t^4/12 + t^2 - \frac{13}{3} t + \frac{29}{4}$.

21. $y = \sin t - t \cos t + 2$.

23. $y = -\frac{1}{2} e^{-t^2} - \frac{1}{2}$.

25. $y = 3/2$ is a stable equilibrium.

27. $y = 1$ and $y = -1$ are stable equilibria; $y = 0$ is unstable.

29. $y = 1$ and $y = 3$ are unstable equilibria.

Section 2.3

1. linear.

3. nonlinear.

5. nonlinear.

7. $y' = Ce^{-t}$.

9. $y = Ce^{-t^2/2}$.

11. $y = C \csc t$.

13. $y = C(100 - t)^2$.

15. $y = -1/2 + t + Ce^{-2t}$.

17. $y = 2t^{-2}e^t - 2t^{-1}e^t + e^t + Ct^{-2}$.

19. $y = (t^2 + C)/(t^2 + 1)$.

21. $y = 2 + e^{-t}$.

23. $y = 10 - 5e^{-t^2/2}$.

25. $y = 1$.

27. $y = 3 - 0.03t - 0.002(100 - t)^2$.

29. $y = -1/2 + t - 1/2e^{-2t-2}$.

31. $y = 2t^{-2}e^t - 2t^{-1}e^t + e^t + (4 - e)t^{-2}$.

33. $D(f + g) = D(f) + D(g)$ and $D(cf) = cD(f)$.

Section 2.4

1. 37.73 h.

3. 129.66 min.

5. (a) $P' = 0.002P + 5$, $P(0) = 100$; (b) $P(t) = 2600e^{0.002t} - 2500$; (c) about 102 thousand walleye more.

7. (a) $A' = 1.5 - A/60$, $A(0) = 45$; (b) $A(t) \to 90$ as $t \to \infty$; (d) 65.92 min.

9. 643.76 days.

11. Use an integrating factor to show that $T = (T_0 - T_m)e^{-kt} + T_m$.

13. 13.08 h.

15. 24.76 h.

Section 2.5

1. linear, separable.

3. nonlinear, separable.

5. nonlinear, separable.

7. linear, separable.

9. linear, separable.

11. linear, separable, exact.

13. exact.

15. $y = Ce^{10t}$.

17. $y = -1/(10t + C)$.

19. $y = (1 + Ce^{2/t})/(1 - Ce^{2/t})$.

21. $y = 1 + Ct$.

23. $y = -6 + Ct/3t - 1$.

25. $y = C/(2 + t^2)$.

27. $y = -t \pm (2t^2 + C)^{1/2}$.

29. $y = 3e^{10t}$.

31. $y = -4/(40t - 41)$.

33. $y = (1 - e^{2/t-1})/(1 + e^{2/t-1})$.

35. $y = 1 + 2t$.

37. $y = (-6 + 16t)/(3t - 1)$.

39. $y = 3/(2 + t^2)$.

41. $y = -t + (2t^2 + 1)^{1/2}$.

43. Consider $y = 0$ and $y = t^2/4$. This result does not violate the noted theorem since $f(t, y) = (y)^{1/2}$ does not have a continuous partial derivative with respect to y in a rectangle containing $(0, 0)$.

Section 2.6

1. (a) $y(3) \approx y_{10} = 4.08956$; (b) $y(t) = \sqrt{8 + t^2}$.

3. (a) With $h = 0.1$, $y(1.5) \approx y_{15} = 1.56309$; with $h = 0.05$, $y(1.5) \approx y_{30} = 1.57217$.

5. (a) $y(1) \approx y_{10} = -0.76341$; (b) $y(t) = -2e^{-t^2}$.

7. $y(1) \approx y_{10} = 5.18748$; (b) $y = 2e^t$.

9. $y(1) \approx y_{10} = 3.06501$; (b) $y(t) = \sqrt[3]{24t + 1}$.

11. $y(1) \approx y_{10} = 7.56597$.

13. $y(1) \approx y_{10} = 0.77258$.

Section 2.7

1. (a) P is increases for $0 < P < A$; (c) P increases most rapidly at the instant $P = A/2$; (d) $M = (A - P_0)/P_0$.

3. (a) $P = 0$ and $P = 4$; (b) $P = 0$ is unstable, $P = 4$ is stable; (c) $P = 2$; (d) $t = 47.58$.

5. $P(t) = (6e^{5t} + 4)/(e^{5t} + 4)$, which makes sense since this is an increasing function that tends to 6 as $t \to \infty$; the equilibrium solutions of the DE are $P = 1$ (unstable) and $P = 6$ (stable).

7. $t \approx 9030.5$ s.

9. $t \approx 2849$ s.

11. (a) Because $f(t, h) = -\sqrt{h}$ does not have a continuous partial derivative with respect to h on a rectangle containing the point $(1, 0)$; (b) because we have no idea what time the tank actually emptied; (d) the solution in (c) shows that for any time $c < 1$, there is a valid solution function which represents the tank emptying at time c. This demonstrates both the nonuniqueness of the solution and the fact that the problem is ill-posed since we do not know the time the tank actually emptied.

Section 3.2

1. $\lambda = -1, 5$ with corresponding eigenvectors $[-2 \ 1]^T, [1 \ 1]^T$.

3. $\lambda = -1, 9$ with corresponding eigenvectors $[-3 \ 1]^T, [1 \ 3]^T$.

5. $\lambda = 1, 4, 0$ with corresponding eigenvectors $[-2 \ 1 \ 1]^T, [1 \ 1 \ 1]^T, [1 \ -1 \ 1]^T$.

7. $\lambda = 2, 2, 2$ with corresponding eigenvector $[1 \ 0 \ 0]^T$.

9. (a) $\mathbf{A} = \begin{bmatrix} -1 & 2 \\ -7 & 8 \end{bmatrix}$; (b) $\mathbf{x} = \mathbf{0}$ is the only constant solution; (c) $\lambda = 1, 6$ with corresponding eigenvectors $\mathbf{v} = [1 \ 1]^T, [2 \ 7]^T$; (d) $\mathbf{x}_1(t) = e^t[1 \ 1]^T$ and $\mathbf{x}_2(t) = e^{6t}[2 \ 7]^T$; (e) $\mathbf{x} = c_1 e^t[1 \ 1]^T + c_2 e^{6t}[2 \ 7]^T$; (f) $\mathbf{x} = -\frac{14}{5}e^t[1 \ 1]^T + \frac{2}{5}e^{6t}[2 \ 7]^T$; this vector function has its length grow without bound as $t \to \infty$.

11. (a) $\mathbf{A} = \begin{bmatrix} -2 & 1 \\ 0 & -2 \end{bmatrix}$; (b) $\mathbf{x} = \mathbf{0}$ is the only constant solution; (c) $\lambda = -2, -2$ with corresponding eigenvector $\mathbf{v} = [1 \ 0]^T$; (d) $\mathbf{x}_1(t) = e^{-2t}[1 \ 0]^T$; (e) $\mathbf{x} = c_1 e^{-2t}[1 \ 0]^T$; (f) There is no value of c_1 for which the solution in (e) satisfies this IVP. This tells us we must not have found the correct general solution in (e).

13. (a) $\mathbf{A} = \begin{bmatrix} -3 & 1 \\ 3 & -1 \end{bmatrix}$; (b) Any vector of form $\mathbf{x} = x_2[1 \ 3]^T$ is a constant solution to the given system, so there are infinitely many such solutions; (c) $\lambda = 0, -4$ with corresponding eigenvectors $[1 \ 3]^T, [-1 \ 1]^T$; (e) every solution is a straight line solution of form $c_1[1 \ 3]^T + c_2 e^{-4t}[-1 \ 1]^T$; (f) $\mathbf{x}(t) = \frac{3}{4}[1 \ 3]^T - \frac{9}{4}e^{-4t}[-1 \ 1]^T$, which tends to the vector $[1 \ 3]^T$ as $t \to \infty$.

15. (a) $\mathbf{A} = \begin{bmatrix} 8 & -1 & -11 \\ 18 & -3 & -19 \\ 2 & -1 & -5 \end{bmatrix}$; (b) $\mathbf{x} = \mathbf{0}$ is the only constant solution; (c) $\lambda = -4, -2, 6$ with corresponding eigenvectors $\mathbf{v} = [1\ 1\ 1]^T, [1\ -1\ 1]^T, [2\ 1\ 0]^T$; (d) $\mathbf{x}_1(t) = e^{-4t}[1\ 1\ 1]^T$, $\mathbf{x}_2(t) = e^{-2t}[1\ -1\ 1]^T$, and $\mathbf{x}_3(t) = e^{6t}[2\ 1\ 0]^T$; (e) $\mathbf{x} = c_1\mathbf{x}_1 + c_2\mathbf{x}_2 + c_3\mathbf{x}_3$; (f) $\mathbf{x} = 1e^{-4t}[1\ 1\ 1]^T$, which is a straight-line solution that approaches zero along the line through $(1, 1, 1)$.

17. $\mathbf{x}' = \mathbf{Ax}$ where $\mathbf{A} = \begin{bmatrix} 0 & 1 \\ 12 & -1 \end{bmatrix}$.

19. $\mathbf{x}' = \mathbf{Ax} + \mathbf{b}(t)$ where $\mathbf{A} = \begin{bmatrix} 0 & 1 \\ 8 & 2 \end{bmatrix}$ and $\mathbf{b}(t) = [0\ e^t]^T$.

21. $\mathbf{x}' = \mathbf{Ax}$ where $\mathbf{A} = \begin{bmatrix} 0 & 1 & 0 \\ 0 & 0 & 1 \\ -5 & 6 & 0 \end{bmatrix}$.

23. $x_1' = -\frac{7}{100}x_1 + \frac{2}{200}x_2 + 35$, $x_2' = \frac{3}{100}x_1 - \frac{12}{200}x_2 + 27$.

25. $\mathbf{x}' = \begin{bmatrix} -0.04 & 0.08 \\ 0.04 & -0.08 \end{bmatrix}\mathbf{x}$, $\mathbf{x}(0) = \begin{bmatrix} 25 \\ 150 \end{bmatrix}$.

27. Use direct substitution with $\mathbf{x}'(t) = \lambda e^{\lambda t}\mathbf{v}$ and $\mathbf{Ax} = \mathbf{A}(e^{\lambda t}\mathbf{v}) = e^{\lambda t}\mathbf{Av}$, along with the fact that $\lambda\mathbf{v} = \mathbf{Av}$.

Section 3.3

1. 4; 7.

3. 3; the given linear third-order homogeneous equation should also have a three-dimensional solution space.

5. $\mathbf{x}_1(t)$ and $\mathbf{x}_2(t)$ are linearly independent.

7. $\mathbf{x}_1(t)$, $\mathbf{x}_2(t)$, and $\mathbf{x}_3(t)$ are linearly independent.

9. For two vectors, it's equivalent to ask if they are scalar multiples of each other.

11. (a) $\mathbf{A}$ has the repeated eigenvalue $\lambda = 3$ with a single corresponding linearly independent eigenvector $\mathbf{v} = [1\ 0]^T$; (c) $\mathbf{x}(t) = c_1 e^{3t}[1\ 0]^T + c_2(te^{3t}[1\ 0]^T + e^{3t}[0\ 1]^T$; (d) $\mathbf{x}(t) = 3e^{3t}[1\ 0]^T + 2(te^{3t}[1\ 0]^T + e^{3t}[0\ 1]^T$.

13. (a) $\mathbf{A}$ has complex eigenvalues $\lambda = \pm i$ with corresponding complex eigenvectors; (c) $\mathbf{x}(t) = c_1[\cos t \quad \sin t]^T + c_2[-\sin t \quad \cos t]^T$; (d) $\mathbf{x}(t) = 3[\cos t \quad \sin t]^T + 2[-\sin t \quad \cos t]^T$.

15. $y = c_1 \cos t + c_2 \sin t$.

17. $y = c_1 + c_2 e^t + c_3 e^{-t}$.

Section 3.4

1. (a) $x(t) = c_1 e^t [1 \ 1]^T + c_2 e^{-t} [1/3 \ 1]^T]^T$; (b) The origin is a saddle point and therefore unstable.

3. (a) $x(t) = c_1 e^{-5t} [-1 \ 1]^T + c_2 e^{-t} [1 \ 1]^T]^T$; (b) The origin is a stable attracting node.

5. (a) $x(t) = c_1 [1 \ 1]^T + c_2 e^{-3t} [-2 \ 1]^T]^T$; (b) Every point of form $k[1 \ 1]^T$ is an equilibrium solution of the system. Each is stable. (c) Every nonconstant solution is a straight line because only one of the terms in $x(t)$ has an exponential function present. That term results in a straight-line solution; the added constant only shifts the line.

7. $x_1(t) = e^{4t} [-1 \ 2]^T$ and $x_2(t) = e^{-3t} [1 \ 2]^T$ are straight-line solutions; the origin is an unstable saddle point.

9. $x_1(t) = e^{0.1t} [1 \ 1]^T$ and $x_2(t) = e^{10t} [-1 \ 1]^T$ are straight line solutions; the origin is an unstable repelling node.

11. $A = \begin{bmatrix} 1/2 & -7/4 \\ -7 & 1/2 \end{bmatrix}$.

13. $x(t) = e^t/2[1 \ 1]^T + 3e^{-t}/2[1/3 \ 1]^T$.

15. $x(t) = -3e^{-5t}/2[-1 \ 1]^T - e^{-t}/2[1 \ 1]^T$.

17. $x(t) = 3[1 \ 1]^T + e^{-3t} [-2 \ 1]^T$.

19. $y = c_1 e^{-t} + c_2 e^{-5t}$.

21. $y' = c_1 e^{-t} + c_2 e^{-2t}$.

Section 3.5

1. The origin is a stable attracting node.

3. The origin is an unstable saddle point.

5. The origin is a stable center.

7. The origin is an unstable repelling node.

9. The origin is a stable center.

11. The origin is an unstable repelling node.

13. The origin is a stable attracting node.

15. (a) $x(t) = c_1 [\cos 2t \ \sin 2t]^T + c_2 [-\sin 2t \ \cos 2t]^T$; (b) the origin is a stable center; (c) none.

17. (a) $x(t) = c_1 [e^{-2t} \ 0]^T + c_2 [te^{-2t} \ e^{-2t}]^T$; (b) the origin is a stable attracting node; (c) one, along the line through (0,0) in the direction of $[1 \ 0]^T$.

19. (a) $x(t) = c_1 e^{9t} [2 \ 1]^T + c_2 e^{-2t} (t[2 \ 1]^T + [-1 \ 0]^T$; (b) the origin is an unstable repelling node; (c) one, along the line through (0,0) in the direction of $[2 \ 1]^T$.

21. $\mathbf{x}(t) = -e^{2t}[\sin 3t \;\; -\cos 3t]^{\mathrm{T}} - 3e^{2t}[\cos 3t \;\; \sin 3t]^{\mathrm{T}}$.

23. $\mathbf{x}(t) = -7/3[\sin 3t \;\; \frac{3}{5}\cos 3t + \frac{4}{5}\sin 3t]^{\mathrm{T}} - 2[\cos 3t \;\; -\frac{3}{5}\sin 3t + \frac{4}{5}\cos 3t]^{\mathrm{T}}$.

25. (a) $\mathbf{x}(t) = c_1 e^{4t}[-1 \;\; 0 \;\; 1]^{\mathrm{T}} + c_2 e^{4t}[1 \;\; 1 \;\; 0]^{\mathrm{T}} + c_3 e^{t}[1 \;\; -1 \;\; 1]^{\mathrm{T}}$; (b) the origin is an unstable repelling node; (c) there are three straight line solutions, as demonstrated in (a).

27. The characteristic polynomial for a 3×3 matrix is a cubic polynomial, and thus must have at least one real zero. This forces the matrix $\mathbf{A}$ to have at least one real eigenvalue, and with it, at least one corresponding real eigenvector, thus generating at least one straight-line solution. A 4×4 matrix may possibly have all complex eigenvalues, and thus the system may have no straight line solution. In general, any time n is odd, an $n \times n$ homogeneous system is guaranteed at least one straight-line solution.

29. $y = c_1 e^{-t} \sin 2t + c_2 e^{-t} \cos 2t$.

31. $y = c_1 e^{-7t} + c_2 e^{4t}$.

Section 3.6

1. $\mathbf{x} = c_1 e^{(-2+\sqrt{7})t}[3/(\sqrt{7}-1) \;\; 1]^{\mathrm{T}} + c_2 e^{(-2-\sqrt{7})t}[3/(-\sqrt{7}-1) \;\; 1]^{\mathrm{T}} + [-4 \;\; -3]^{\mathrm{T}}$.

3. $\mathbf{x} = c_1 e^{3t}[1 \;\; 1]^{\mathrm{T}} + c_2 e^{t}[-1 \;\; 1]^{\mathrm{T}} + \sin t[-2/5 \;\; 1/10]^{\mathrm{T}} + \cos t[-3/10 \;\; 1/5]^{\mathrm{T}}$.

5. (a) Because the forcing function $\mathbf{b}(t)$ is constant, $vx_p = [4/3 \;\; -7/3]^{\mathrm{T}}$; (b) $\mathbf{x}_h = c_1 e^{-t}[-1/2 \;\; 1]^{\mathrm{T}} + c_2 e^{3t}[1/2 \;\; 1]^{\mathrm{T}}$; (d) $vx_p = [4/3 \;\; -7/3]^{\mathrm{T}}$ is constant and thus an equilibrium solution. Since the eigenvalues have opposing signs, this equilibrium point is an unstable saddle.

7. (a) $\mathbf{x}_p = [-2/3 - 1/5e^{-2t} \;\; -1/3 - 2/5e^{-2t}]^{\mathrm{T}}$; (b) $\mathbf{x}_h = c_1 e^{3t}[1/2 \;\; 1]^{\mathrm{T}} + c_2 e^{-t}[-1/2 \;\; 1]^{\mathrm{T}}$.

9. $\mathbf{x} = c_1 e^{t}[1 \;\; 1]^{\mathrm{T}} + c_2 e^{-9t}[-1 \;\; 1]^{\mathrm{T}} + [1/9 \;\; -1/9]^{\mathrm{T}}$. 11. $\mathbf{x} = c_1 e^{-t}[1 \;\; 0]^{\mathrm{T}} + c_2 e^{t}[1/2 \;\; 1]^{\mathrm{T}} + [-13/2e^{-2t} \;\; 4/3e^{-2t}]^{\mathrm{T}}$.

13. $\mathbf{x} = c_1[\cos t \;\; \sin t]^{\mathrm{T}} + c_2[-\sin t \;\; \cos t]^{\mathrm{T}} + [2 \;\; 3]^{\mathrm{T}}$.

15. $\mathbf{x} = c_1[\cos t \;\; \sin t]^{\mathrm{T}} + c_2[-\sin t \;\; \cos t]^{\mathrm{T}} + [2 + 3e^{t}/2 \;\; 3 - e^{t}/2]^{\mathrm{T}}$.

17. $\mathbf{x} = c_1 e^{t}[1 \;\; 1]^{\mathrm{T}} + c_2 e^{-t}[1/3 \;\; 1]^{\mathrm{T}} + [4 + 3/10\sin 3t - 1/5\cos 3t \;\; 8 - 3/10\cos 3t]^{\mathrm{T}}$.

19. $\mathbf{x} = -5/2e^{-t}[1 \;\; 0]^{\mathrm{T}} + 11/18e^{t}[1/2 \;\; 1]^{\mathrm{T}} + [-13/2e^{-2t} \;\; 4/3e^{-2t}]^{\mathrm{T}}$.

21. $\mathbf{x} = -1[\cos t \;\; \sin t]^{\mathrm{T}} - 5[-\sin t \;\; \cos t]^{\mathrm{T}} + [2 + 3e^{t}/2 \;\; 3 - e^{t}/2]^{\mathrm{T}}$. 23. $\mathbf{x}_p = [a\sin 3t + b\cos 3t + c\sin 2t + d\cos 2t \;\; e\sin 3t + f\cos 3t + g\sin 2t + h\cos 2t]^{\mathrm{T}}$.

Section 3.7

1. $\mathbf{x}_p = [-17/5 \;\; 13/5]^{\mathrm{T}}$.

3. (a) $\mathbf{x}_p = [Ae^t \ \ Be^t]^T$; (b) $\mathbf{x}_h = c_1 e^t [-1 \ \ 1]^T + c_2 e^{5t} [1 \ \ 1]^T$, which includes the natural guess for $\mathbf{x}_p$, so that guess for a particular solution will fail to work; (c) $\mathbf{x}_p = [te^t \ -(t+1)e^t]^T$.

5. $\mathbf{x}_p = [-e^{-t} \ -\frac{1}{4}e^{-t} - \frac{1}{3}]^T$.

7. $\mathbf{x}_p = [-\frac{1}{5}\cos 2t + \frac{2}{5}\sin 2t \ -\frac{19}{65}\cos 2t - \frac{22}{65}\sin 2t]^T$.

9. $\mathbf{x}_p = [-11/12e^{-t} - 3/4te^{-t} \ \ 9/4te^{-t}]^T$.

11. $\mathbf{x}_p = [-t - 1 \ -t - 2]^T$.

Section 3.8

1. The IVP is $\mathbf{x}' = \mathbf{A}\mathbf{x}$ where $\mathbf{A} = \begin{bmatrix} -4/100 & 4/50 \\ 4/100 & -4/50 \end{bmatrix}$ and $\mathbf{x}(0) = \begin{bmatrix} 25 \\ 50 \end{bmatrix}$. The

solution to the IVP is $\mathbf{x}(t) = \begin{bmatrix} 50 \\ 25 \end{bmatrix} + e^{-3/25t}\begin{bmatrix} -25 \\ 25 \end{bmatrix}$, which is a straight-line
solution that tends to the stable equilibrium $(50, 25)$ as $t \to \infty$.

3. The matrix $\mathbf{A}$ in the system $\mathbf{x}' = \mathbf{A}\mathbf{x} + \mathbf{b}$ stays the same as in #2, but the system is now homogeneous of the form $\mathbf{x}' = \mathbf{A}\mathbf{x}$. As $t \to \infty$, $\mathbf{x}(t) \to \mathbf{0}$, which is consistent with the fact that the amount of salt in each tank will go to zero as time progresses.

5. The IVP is $\mathbf{x}' = \mathbf{A}\mathbf{x} + \mathbf{b}$ where $\mathbf{A} = \begin{bmatrix} -7/400 & 0 & 0 \\ 7/400 & -7/200 & 0 \\ 0 & 7/200 & -7/300 \end{bmatrix}$ and $\mathbf{b} =$

$\begin{bmatrix} 70 \\ 0 \\ 0 \end{bmatrix}$, $\mathbf{x}(0) = \begin{bmatrix} 8000 \\ 10000 \\ 0 \end{bmatrix}$. The general solution to the system is

$$\mathbf{x}(t) = c_1 e^{-7/300t}\begin{bmatrix} 0 \\ 0 \\ 1 \end{bmatrix} + c_2 e^{-7/200t}\begin{bmatrix} 0 \\ -1/3 \\ 1 \end{bmatrix} + c_3 e^{-7/400t}\begin{bmatrix} 1/6 \\ 1/6 \\ 1 \end{bmatrix} + \begin{bmatrix} 4000 \\ 2000 \\ 3000 \end{bmatrix}$$

from which we can see that our intuition is confirmed: with just one inflow putting brine at 10 g/liter into the system, eventually the concentration should stabilize throughout at a concentration of 10 percent by volume. The constants c_1 and c_2 can be determined by applying the initial conditions; $c_1 = -15000$, $c_2 = -12000$, $c_3 = 24000$.

7. (a) $y'' + 4y = 0$, $y(0) = 0.4$ and $y'(0) = 0$; (b) $\mathbf{x}' = \begin{bmatrix} 0 & 1 \\ -4 & 0 \end{bmatrix}\mathbf{x}$, $\mathbf{x}(0) = [0.4 \ 0]^T$;
(c) $\mathbf{x}(t) = [\frac{2}{5}\cos(2t) \ -\frac{4}{5}\sin(2t)]^T$, so $y = x_1 = \frac{2}{5}\cos(2t)$.

9. (a) $y'' + y' + 4y = \cos(2t)$, $y(0) = 0.3$ and $y'(0) = 0$; (b) $\mathbf{x}' = \begin{bmatrix} 0 & 1 \\ -4 & 1 \end{bmatrix}\mathbf{x}$,
$\mathbf{x}(0) = [0.3 \ 0]^T$; (c) $\mathbf{x}(t) = e^{-1/2t}[-0.0775\sin(1.936t) + 0.3\cos(1.936t) \ -$

$0.620\sin(1.936t)]^{\mathrm{T}}$, so $y = x_1 = e^{-1/2t}[-0.0775\sin(1.936t) + 0.3\cos(1.936t)$.
Thus the solution function oscillates and decays to zero as $t \to \infty$.

11. (a) $I'' + RI' + 100I = 0$, $I(0) = 100$, $I'(0) = 0$; (b) $\mathbf{x}' = \begin{bmatrix} 0 & 1 \\ -100 & -R \end{bmatrix}\mathbf{x}$,

$\mathbf{x}(0) = [100 \ 0]^{\mathrm{T}}$; (c) (i) $I = x_1 = 100\cos 10t$, (ii) $I = x_1 = e^{-8t}((400/3\sin 6t + 100\cos 6t))$, (iii) $I = x_1 = e^{-10t}(100 + 1000t)$, (iv) $I = x_1 = 400/3e^{-5t} - 100/3e^{-20t}$.

Section 4.2

1. $y = c_1 e^{4t} + c_2 e^{-3t}$.

3. $y = c_1 e^t + c_2 e^{-t}$.

5. $y = c_1 + c_2 t$.

7. $y = c_1 e^{\frac{-1+\sqrt{5}}{2}t} + c_2 e^{\frac{-1-\sqrt{5}}{2}t}$.

9. $y = 2e^t$.

11. $y = 2/3 + 1/3e^{-3t}$.

13. $y = -6e^{-t} + 4e^{-3t}$.

15. $y'' - 4y = 0$.

17. $y'' - 4y' = 0$.

19. $y'' = 0$.

21. (b) the roots of the characteristic equation are the complex numbers $r = 1 \pm 2i$; (c) the two functions are linearly independent because neither is a scalar multiple of the other; (d) $y = c_1 e^t \cos 2t + c_2 e^t \sin 2t$.

23. $y = 5e^{-t} - 3e^{-2t}$.

25. $y = 325/2e^{-t} - 125/2e^{-3t}$.

Section 4.3

1. $y = c_1 e^{4t} + c_2 te^{4t}$.

3. $y = c_1 e^{-t/2} + c_2 te^{-t/2}$.

5. $y = c_1 \cos 2t + c_2 \cos 2t$.

7. $y = c_1 e^{5t} + c_2 te^{5t}$.

9. $y = c_1 e^{-5t/2} + c_2 e^{-t}$.

11. $y = \sqrt{3}e^{-t/2}\sin\sqrt{3}t/2 + e^{-t/2}\cos\sqrt{3}t/2$.

13. $y' = 19/4e^{-2t} + 9/4e^{2t}$.

15. $y = 16/5e^{5t}\sin 5t - 3e^{5t}\cos 5t$.

17. $y = 0$.

19. (a) $y = c_1 e^{3t} + c_2 te^{3t}$; (d) $x_1 = y$.

21. The equation will have two real distinct roots when $a_1^2 - 4a_0 > 0$, one real repeated root when $a_1^2 - 4a_0 = 0$, and two distinct complex roots when $a_1^2 - 4a_0 < 0$.

23. $y = \sqrt{3}/3 \sin \sqrt{3}t + 2 \cos \sqrt{3}t$.

25. $I = 100e^{-2t} + 225te^{-2t}$.

27. $I = 25/3 \sin 3t + 100 \cos 3t$.

Section 4.4

1. $y = c_1 e^{4t} + c_2 e^{-3t} + 5/4e^{5t}$.

3. $y = c_1 e^{-t} + c_2 e^t + 11/2te^t$.

5. $y = c_1 + c_2 t + 1/12t^4 + 3/2t^2$.

7. $y = c_1 e^{-2t} + c_2 te^{-2t} + 3/8 - 1/2t + 1/4t^2$.

9. $y = c_1 \sin 2t + c_2 \cos 2t + 2e^t(\sin t + 2\cos t)$.

11. $y = -5/7e^{4t} + 41/28e^{-3t} + 5/4e^{5t}$.

13. $y = 1/4e^{-t} - 13/4e^t + 11/2te^t$.

15. $y = -2 - 2t + 1/12t^4 + 3/2t^2$.

17. $y = 37/8e^{-2t} + 51/4te^{-2t} + 3/8 - 1/2t + 1/4t^2$.

19. $y = -7/2 \sin 2t - 4\cos 2t + 2e^t(\sin t + 2\cos t)$.

21. $y = c_1 \sin t + c_2 \cos t - \cos t \cdot \ln \frac{1+\sin t}{\cos t}$.

23. $y = c_1 e^{-2t} + c_2 te^{-2t} + 1/6t^3 e^{-2t}$.

25. $y = c_1 e^t + c_2 te^t + te^t(-1 + \ln t)$.

27. $y = c_1 e^{2t} + c_2 e^{-3t} - 1/10e^{-3t}(e^{3t} - 2e^{4t} + 2\ln(e^t + 1)e^{5t} - 2te^{5t} + e^{2t} - 2e^t + 2\ln(e^t + 1))$.

29. $y = 1/2\sin 2t + 582/41 \cos 2t - 500/41 \cos \frac{21}{10}t$; y_h and y_p are each equi-oscillatory functions whose frequencies are nearly equal. When added together, they sometimes cancel each other out, leading to widely varying behavior in the amplitude of y.

31. $y = 106.28e^{-0.000625t} - 5.79e^{-3.999t} - 0.481\cos 20t + 0.096\sin 20t$.

Section 4.5

1. $y = 1/1000t \sin 5t$; there is no maximum displacement of the mass as oscillations are unbounded due to resonance.

3. $y = -1/72 \sin 6t - 1/72 \cos 6t + 1/72 e^{6t}$; the displacement is unbounded, but resonance is not present.

5. $y = -1.98 \sin 7.07t + 2 \sin 7t$; beats is present. The maximum displacement is approximately 3.98.

7. Beats are present; $y = \cos 6.93t - \cos 7t$.

9. $I = 10t \sin 10t$; resonance is present.

11. $I = -80/33 \cos 100t + 80/33 \cos 10t$; neither beats nor resonance is present.

13. $c \approx 2.5$.

Section 4.6

1. $y = c_1 e^t + c_2 e^{-t} + c_3 e^{3t}$.

3. $y = c_e^{-1/2t} + c_2 e^{-3/2t} + c_3 e^{2t}$.

5. $y = c_1 e^{-t} + c_2 t e^{-t} + c_3 t^2 e^{-t}$.

7. $y = c_1 e^t + c_2 \cos 2t + c_3 \sin 2t$.

9. $y = c_1 e^t + c_2 e^{2t} + c_3 e^{-t} + c_4 \sin t + c_5 \cos t$.

11. $y = c_1 e^{-t} + c_2 t e^{-t} + c_3 t^2 e^{-t} + + c_4 t^3 e^{-t}$.

13. $y = 1/2 + 1/4 e^{-2t} + 1/4 e^{2t}$.

15. $y = -3 e^{3t} + 8 e^{2t} - 5 e^t$.

17. $y = 5 \sin 2t$.

19. $y = t^2 + 2$.

21. $y = e^t + t e^t - t^2 e^t$.

23. $y'' - y' = 0$.

25. $y^{(5)} + y^{(4)} + 9y''' + 9y'' = 0$.

27. $y^{(4)} + y''' + 105/4 y'' + 25y' + 125/4 y = 0$.

29. $y = c_1 \cos t + c_2 \sin t + c_3 t \cos t + c_4 t \sin t + 7/32 - 1/8 t^2 \cos t$.

31. $y = c_1 e^t + c_2 e^{-t} + c_3 e^{3t} + 1$.

33. $y = c_e^{-1/2t} + c_2 e^{-3/2t} + c_3 e^{2t} - 6/325 \cos t - 17/325 \sin t$.

35. $y = c_1 e^{-t} + c_1 t e^{-t} + c_1 t^2 e^{-t} - 1/4 \cos t - 1/4 \sin t$.

37. $y = c_1 e^t + c_2 \cos 2t + c_3 \sin 2t - 1/10 e^{-t}$.

39. $y = c_1 e^t + c_2 e^{2t} + c_3 e^{-t} + c_4 \sin t + c_5 \cos t + 7/2$.

41. $y = c_1 e^{-t} + c_2 t e^{-t} + c_3 t^2 e^{-t} + c_4 t^3 e^{-t} - 4 + t - 1/4 \cos t$.

Section 5.2

1. $\lim_{r\to\infty} re^{-sr} = \lim_{r\to\infty} \frac{r}{e^{-sr}} = \lim_{r\to\infty} \frac{1}{-se^{-rs}} = 0$, where the second equality holds by an application of L'Hopital's Rule.

3. Consider applying L'Hopital's Rule n times to $\lim_{r\to\infty} \frac{r^n}{e^{-sr}}$.

5. $F(s) = \frac{2}{s^2}$.

7. $F(s) = \frac{2}{s} - \frac{1}{s^2}$.

9. $F(s) = \frac{2}{s^3} - \frac{3}{s}$

11. $F(s) = \frac{1}{s-3}$.

13. $F(s) = \frac{e^5}{s-3}$.

15. $F(s) = \frac{1}{(s-a)^2}$.

17. $F(s) = \frac{1}{s} + \frac{1}{s^2}$.

19. $F(s) = \frac{c}{s} + \frac{k}{s^2}$.

Section 5.3

1. $F(s) = 3/s - 1/(s-1)$.

3. $F(s) = (3/(s-2)) - (6/(s^2+4))$.

5. $F(s) = (4s/(s^2+25)) + (6/(s+2))$.

7. $F(s) = 2/(s+1)^3$.

9. $F(s) = (24s(s^2-1))/(s^2+1)^4$.

11. $F(s) = (20/(s^2+25)) - (6/(s+2))$.

13. $F(s) = 2/(s+1)^3$.

15. $F(s) = 2/((s+1)^2+4)$.

17. $f(t) = \cosh(2t)\sin(3t) = \frac{1}{2}(e^t+e^{-t})\sin(3t)$ so $F(s) = \frac{1}{2}(3/((s-1)^2+9)) + (3/((s+1)^2+9))$.

19. $F(s) = (8/(s+1)^3) + (7/((s+3)^1+1))$.

21. $F(s) = ((6s^2-3)/((s^2+1)^3)) - (2s/(s^2+1)^2)$.

23. $F(s) = 2(s+1)/((s+1)^2+1)^2$.

29. $\mathcal{L}[f^{(4)}(t)] = s^4\mathcal{L}[f(t)] - s^3 f(0) - s^2 f'(0) - sf''(0) - f''(0)$.

Section 5.4

1. $f(t) = u(t-1) - u(t-2)$.

3. $f(t) = t \cdot [u(t-1) - u(t-2)] + t^2 \cdot u(t-2)$.

5. $f(t) = \sin(t) \cdot [u(t) - u(t-2\pi)]$.

7. $f(t) = t \cdot [u(t) - u(t-2)] + 2 \cdot [u(t-2) - u(t-4)] + (4-t) \cdot u(t-4)$.

9. $F(s) = \frac{2e^{-s}}{s} - \frac{e^{-3s}}{+} e^{-5s}$.

11. $F(s) = 2\frac{2}{(s+3)^2+4} + e^{-8s}$.

13. $y'' + \frac{1}{2}y' + \frac{5}{2}y = \frac{3}{4}\sin(2t) \cdot u(t-4) + \frac{1}{4}\delta(t-10)$, $y(0) = 0.25$, $y'(0) = 0$.

Section 5.5

1. $y = 4 - e^{-5t}$.

3. $y = 4 - e^{-5t}$.

5. $y = (\frac{1}{4}(-1+2t)e^{2t} - \frac{3}{4})e^{-t}$.

7. $y = u(t-3)\left(-\frac{1}{64} - \frac{1}{8}t + \frac{25}{64}e^{8(t-3)}\right) - 4e^{8t}$.

9. $y = \frac{5}{3}\sin(3t)$.

11. $y = \frac{1}{3}\sin(3t) - \frac{2}{9}\cos(3t) + \frac{2}{9}$.

13. $y = \frac{5}{6}t\sin(3t)$.

15. $y = 2e^{-3t} + 6te^{-3t}$.

17. $y = \frac{1}{2}e^{-t}\sin(2t) + e^{-t}\cos(2t) - \frac{1}{5}u(t-4)\left(-1+(\frac{1}{2}\sin 2(t-4) + \cos 2(t-4))e^{-(t-4)}\right)$.

19. $y = 3/2e^{-t} + 1/2e^{3*t} + 1/12u(t-3)(-4+3e^{-(t-3)} + e^{3(t-3)}$.

21. $y = (1/5)e^{2t} - 11/5e^{-3t}$.

23. $y = 1/4(-1+2t)e^t - 3/4e^{-t}$.

25. $y(t) = 5/6t\sin 3t$.

27. (a) $y(t) = 1/36 - 1/36\cos(6t)$; (b) $y(t) = -1/5148e^{-t/2}\sin(1/2\sqrt{143}t)$ $\sqrt{143} - 1/36e^{-t/2}\cos(1/2\sqrt{143}t) + 1/36$; (c) $y(t) = -1/36e^{-6t} - 1/6te^{-6t} + 1/36$; (d) $y(t) = -1/32e^{-2t} + 1/288e^{-18t} + 1/36$.

29. (a) $y = 5/72\sin 6t - 5/12t\cos 6t$; (b) $y = 5\sqrt{143}/858e^{-t/2}\sin\sqrt{143}t/2 + 5/6e^{-t/2}\cos\sqrt{143}t/2 - 5/6\cos 6t$; (c) $y = 5/72e^{-6t} + 5/12te^{-6t} - 5/72\cos 6t$; (d) $y = 3/64e^{-2t} - 1/192e^{-18t} - 1/24\cos 6t$.

Section 5.6

1. $f(t) = (2 - 6t)e^{-3t}$.

3. $f(t) = -1/4 - t/2 + 1/4e^{2t}$.

5. $f(t) = 6/25\cos 2t + 9/50\sin 2t + 2/25(5t-3)e^{-t}$.

7. $f(t) = 1/2u(t-1)(\sin(t-1) - 2\sinh(t-1) + (t-1)\cosh(t-1))$.

9. $f(t) = 5/9u(t-\pi)(-9 + 5e^{4(t-\pi)} - e^{t-\pi}(-4 + 15(t-\pi)))$.

11. $y = 5/8\sin 2t - 1/4t\cos 2t$.

13. $y = 3/4\sin 2t - 1/2t\cos 2t + 1/2u(t-6)\sin 2(t-6) + u(t-12)\sin 2(t-12)$.

15. $y = 5/8e^{-t}\sin 2t - 1/4te^{-t}\cos 2t$.

17. $y = 1/2e^{-t}\sin 2t + 1/8e^{-t}(-2t\cos 2t + \sin 2t + (t+\pi)u(t-\pi)(1/2\sin 2t + (\pi - t)\cos 2t))$.

19. $y(t) = 17/18e^{-2t} + 4/3te^t + 5/9e^t - 1/2 - 1/3u(t-3)e^{-2(t-3)} + 1/3u(t-3)e^{t-3}$.

21. $y(t) = 1/2e^{-t}\sin 2t + 1/16e^{-t}(t+2)(-2t\cos 2t + \sin 2t) + 1/2u(t-5)e^{-t+5}\sin 2(t-5)$.

Section 6.2

1. $(0,0), (1/2, 1/4)$.

3. $(k\pi/2, j\pi/2)$, where $k = \pm 1, \pm 5, \pm 9, \ldots$ and $j = \pm 1, \pm 3, \pm 5, \ldots$.

5. The system has no equilibrium solutions.

7. $(0,0), (1/2, 1/4), (-1/2, 1/4)$.

9. $(\pm 2k\pi, 0)$, where $k = 0, 1, 2, \ldots$. At even multiples of π, the system demonstrates stable equilibria with stable centers nearby; at odd multiples of π, the system shows unstable equilibria, which correspond to the pendulum starting in a vertical position.

Section 6.3

1. $J(x_1, x_2) = \begin{bmatrix} 2x_1 & 1 \\ 1 & -2x_2 \end{bmatrix}$.   3. $J(x_1, x_2) = \begin{bmatrix} -2x_2 & 1-2x_1 \\ 4x_2 - 1 & 4x_1 \end{bmatrix}$.

5. $J(x_1, x_2, x_3) =$
$$\begin{bmatrix} -2x_1(1 + x_1^2 + x_2^2 + x_3^2)^{-2} & -2x_2(1 + x_1^2 + x_2^2 + x_3^2)^{-2} & -2x_1(1 + x_1^2 + x_2^2 + x_3^2)^{-2} \\ -2x_1 e^{-x_1^2 - x_2^2 - x_3^2} & -2x_2 e^{-x_1^2 - x_2^2 - x_3^2} & -2x_3 e^{-x_1^2 - x_2^2 - x_3^2} \\ 2 & -6x_2 & 4x_3^3 \end{bmatrix}.$$

7. $F(x_1, x_2) \approx \begin{bmatrix} 2 & 1 \\ 1 & 2 \end{bmatrix} \begin{bmatrix} x_1 - 1 \\ x_2 + 1 \end{bmatrix}$.

9. $F(x_1, x_2) \approx \begin{bmatrix} -1/2 & 0 \\ 0 & 2 \end{bmatrix} \begin{bmatrix} x_1 - 1/2 \\ x_2 - 1/4 \end{bmatrix}$

11. (a) Equilibrium solutions: $(0,0), (1/2, 1/4)$; (b) near $(0,0)$, $\begin{bmatrix} x_1' \\ x_2' \end{bmatrix} \approx \begin{bmatrix} 0 & 1 \\ -1 & 0 \end{bmatrix} \begin{bmatrix} x_1 \\ x_2 \end{bmatrix}$; near $(1/2, 1/4)$, see exercise 9; (c) the purely imaginary eigenvalues of the Jacobian matrix show that $(0,0)$ is stable since nearby trajectories are approximately elliptical.

13. (a) Equilibrium solutions: $(k\pi/2, j\pi/2)$, where $k = \pm1, \pm5, \pm9, \ldots$ and $j = \pm1, \pm3, \pm5, \ldots$; (b) near $(\pi/2), (\pi/2)$, $\begin{bmatrix} x_1' \\ x_2' \end{bmatrix} \approx \begin{bmatrix} 0 & -1 \\ 0 & 0 \end{bmatrix} \begin{bmatrix} x_1 - \pi/2 \\ x_2 - \pi/2 \end{bmatrix}$;
(c) the repeated zero eigenvalue of the Jacobian matrix does not reveal useful information; a plot of the direction field nearby shows that $(\pi/2, \pi/2)$ appears to be unstable.

15. There are no equilibrium points for this system.

17. (a) Equilibrium solutions: $(0,0)$, $(1/2, 1/4)$, $(-1/2, 1/4)$; (b) for example, near $(0,0)$, $\begin{bmatrix} x_1' \\ x_2' \end{bmatrix} \approx \begin{bmatrix} 0 & 1 \\ 1 & 0 \end{bmatrix} \begin{bmatrix} x_1 \\ x_2 \end{bmatrix}$; (c) the two real eigenvalues of the Jacobian matrix of opposing signs show that $(0,0)$ is unstable since the nearby behavior is approximately that of a saddle point.

19. Near $(0,0)$, $\begin{bmatrix} x_1' \\ x_2' \end{bmatrix} \approx \begin{bmatrix} 0 & 1 \\ -4.9 & 0 \end{bmatrix} \begin{bmatrix} x_1 \\ x_2 \end{bmatrix}$; the two purely imaginary eigenvalues of the Jacobian matrix show that $(0,0)$ is a stable center, which confirms what we expect for the pendulum. If the initial displacement and angular velocity are small, we expect the pendulum to oscillate indefinitely near its equilibrium.

Section 6.4

1. $x(1) \approx x_{10} = 0.51614$, $y(1) \approx y_{10} = 2.64423$.

3. $x(1) \approx x_{10} = 1.00781$, $y(1) \approx y_{10} = 3.04026$.

5. $x(1) \approx x_{20} = 0.66581$, $y(1) \approx y_{20} = 0.86534$.

7. $x(1) \approx x_{20} = 0.687028$, $y(1) \approx y_{20} = 0.302645$.

Section 7.2

1. (a) $y_{10} = -0.763413361$; (b) $y_{10} = -0.738106789$; (c) $y_{10} = -0.734305821$; the exact solution at $t = 1$ is $y(1) = -2e^{-1} = -0.735758882$.

3. (a) $y_{10} = 5.18748492$; (b) $y_{10} = 5.428161693$; (c) $y_{10} = 5.428161693$; the exact solution at $t = 1$ is $y(1) = 2e = 5.436563657$.

5. (a) $y_{10} = 1.396393786$; (b) $y_{10} = 1.553789505$; (c) $y_{10} = 1.543274653$; the exact solution at $t = 1$ is $y(1) = \tan 1 = 1.557407725$.

7. (a) $y_{10} = 0.875101928$; (b) $y_{10} = 0.877113041$; (c) $y_{10} = 0.877113041$.

9. (a) $y_{10} = 0.827421159$; (b) $y_{10} = 0.805202364$; (c) $y_{10} = 0.804960517$.

Section 7.3

1. (a) $y_{10} = -0.730521596$; (b) $y_{10} = -0.735762133$; the exact solution at $t = 1$ is $y(1) = -2e^{-1} = -0.735758882$.

3. (a) $y_{10} = 5.428161693$; (b) $y_{10} = 5.436559488$; the exact solution at $t = 1$ is $y(1) = 2e = 5.436563657$.

5. (a) $y_{10} = 1.53289173$; (b) $y_{10} = 1.557406443$; the exact solution at $t = 1$ is $y(1) = \tan 1 = 1.557407725$.

7. (a) $y_{10} = 0.879321827$; (b) $y_{10} = 0.881752898$.

9. (a) $y_{10} = 0.759536196$; (b) $y_{10} = 0.763163853$.

Section 7.4

1. (a) $\mathbf{x}^{(10)} = \begin{bmatrix} -0.31171756 \\ 1.45329846 \end{bmatrix}$; (b) $\mathbf{x}^{(10)} = \begin{bmatrix} -0.303502219 \\ 1.381443614 \end{bmatrix}$; (c) $\mathbf{x}(1) = \begin{bmatrix} 0.3011686789 \\ 1.381773291 \end{bmatrix}$.

3. (a) $\mathbf{x}^{(10)} = \begin{bmatrix} 1.199804688 \\ 0.600390625 \end{bmatrix}$; (b) $\mathbf{x}^{(10)} = \begin{bmatrix} 1.198181011 \\ 0.603637979 \end{bmatrix}$; (c) $\mathbf{x}(1) = \begin{bmatrix} 1.198652411 \\ 0.6026951788 \end{bmatrix}$.

5. (a) $\mathbf{x}^{(10)} = \begin{bmatrix} 1.33580647 \\ 1.74092711 \end{bmatrix}$; (b) $\mathbf{x}^{(10)} = \begin{bmatrix} 1.244809581 \\ 1.721363223 \end{bmatrix}$; (c) $\mathbf{x}(1) = \begin{bmatrix} 1.223244276 \\ 1.760866373 \end{bmatrix}$.

7. (a) $\mathbf{x}^{(10)} = \begin{bmatrix} 0.516373457 \\ 3.169684507 \end{bmatrix}$; (b) $\mathbf{x}^{(10)} = \begin{bmatrix} 0.534445981 \\ 3.490162952 \end{bmatrix}$.

9. (a) $\mathbf{x}^{(10)} = \begin{bmatrix} 1.007884422 \\ 3.043412617 \end{bmatrix}$; (b) $\mathbf{x}^{(10)} = \begin{bmatrix} 1.007920502 \\ 3.042623241 \end{bmatrix}$.

11. (a) $\mathbf{x}^{(40)} = \begin{bmatrix} 0.96546577 \\ 1.459385106 \end{bmatrix}$; (b) $\mathbf{x}^{(10)} = \begin{bmatrix} 0.965243106 \\ 1.459786552 \end{bmatrix}$.

13. (a) $\mathbf{x}^{(10)} = \begin{bmatrix} 0.694555012 \\ 0.305303136 \end{bmatrix}$; (b) $\mathbf{x}^{(10)} = \begin{bmatrix} 0.694412729 \\ 0.3051542 \end{bmatrix}$.

15. (a) $x_1' = x_2$, $x_2' = -16x_1 + 2t + 1$; (b) $x_1^{(10)} = 0.331345434$; (c) $x_1^{(10)} = 0.240361385$; (d) $x_1(1) = y(1) = 0.252002804$.

17. (a) $x_1' = x_2$, $x_2' = -16x_1^2 + 2\sin 2t$; (b) $x_1^{(10)} = 0.392559752$; (c) $x_1^{(10)} = 0.418137228$.

Section 8.2

1. $R = 1$.

3. $R = 5$.

5. $f(t) \approx 1 + t/2 - t^2/8 + t^3/16$; $R = 1$.

7. $f(t) \approx 1 - t^4 + t^8 - t^{12}$; $R = 1$.

9. $f(t) \approx 1 + t + 2t^2/3 + t^3/3$; $R = \infty$.

11. $f(t) \approx t^5 - t^9/6 + t^{13}/120 - t^{17}/5040$; $R = \infty$.

13. $f(t) \approx t - 2t^3/3 + 2t^5/15 - 4t^7/315$; $R = \infty$.

15. $f(t) \approx t - t^2 + t^3/3 - t^5/30$; $R = \infty$.

17. $f(t) \approx t^2 - t^6/3 + t^{10}/5 - t^{14}/7$; $R = 1$.

19. $t - t^3/3 + t^5/10 - t^7/42$.

21. $t - t^3/18 + t^5/600 - t^7/35280$.

23. $t - t^7/14 + t^{13}/312 - t^{19}/13680$.

Section 8.3

1. $a_0 + a_1\sqrt{2}t/\sqrt{\pi} - a_1\sqrt{2}t^3/(6\sqrt{\pi}) + a_2\sqrt{2}t^5/(40\sqrt{\pi})$.

3. $a_1 + 2a_0t - 2a_1t^2 - 4/3a_0t^3$.

5. $(a_0 + a_1) + (-a_1 - 5a_0)t + (a_1/2 + 25a_0/2)t^2 + (-a_1/6 - 125a_0/6)t^3$.

7. $(a_0 + a_1) + (-2a_1 + 3a_0)t + (2a_1 + 9/2a_0)t^2 + (-4a_1/3 + 9a_0/2)t^3$.

9. $a_0 + a_1t - \frac{1}{2}a_0t^2 - \frac{1}{6}(a_0 + a_1)t^3$.

11. $a_0 + a_1t - \frac{1}{2}a_0t^2 - \frac{2}{3}a_1t^3$.

13. $a_0 + a_1t - \frac{3}{2}a_0t^2 + \frac{3}{2}a_1t^3$.

15. $1 - \frac{1}{2}t^2 - \frac{1}{6}t^3 + \frac{1}{24}t^4$.

17. $1 - \frac{1}{2}t^2 + \frac{1}{6}t^4 - \frac{31}{720}t^6$.

Section 8.4

3. Hint: write $\ln\frac{1+t}{1-t} = \ln(1+t) - \ln(1-t)$ and use the fact that $\ln(1+t) = t - t^2/2 + t^3/3 - t^4/4 + \cdots$.

5. Using $\lambda = 2$ and the particular solution $y_p = 1$, $y = c_1 P_2(t) + c_2 Q_2(t) + y_p = c_1(\frac{3}{2}t^2 - \frac{1}{2}) + c_2((\frac{3}{2}t^2 - \frac{1}{2})\frac{1}{2}\ln\frac{1-t}{1+t} - \frac{3}{2}t) + 1$.

7. Using $\lambda = 5$ and the particular solution $y_p = \frac{t^2}{2} - \frac{1}{30}$, $y = c_1 P_5(t) + c_2 Q_5(t) + y_p$.

9. With $\lambda = \frac{-1+\sqrt{13}}{2}$, $y = 1 - \frac{1}{2!}\lambda(\lambda+1)t^2 + \frac{1}{4!}\lambda(\lambda+1)(\lambda-2)(\lambda+3)t^4 + \cdots$.

11. With $\lambda = 1/3$, $y = t - \frac{1}{3!}(\lambda-1)(\lambda+2)t^3 + \frac{1}{5!}(\lambda-1)(\lambda-3)(\lambda+2)(\lambda+4)t^5 + \cdots$.

13. Hint: use $\lambda = 4$ and the particular solution $y_p = t^2 - \frac{1}{10}$ to write the general solution $y = c_1 P_4(t) + c_2 Q_4(t) + y_p$; find c_1 and c_2.

Section 8.5

1. $H_4(t) = 16t^4 - 48t^2 + 12$ and $H_5(t) = 32t^5 - 160t^3 + 120t$.

3. Using $q = 5$, $y = c_1 + c_2 t + 15c_2 t^2 + \cdots$.

5. Using $q = 3$, $y = 2 + 10t + 6t^2 + \cdots$.

7. Using $q = 2$ and $y_p = 4t$, $y = 4t + 1 - 4(t - t^3/3 + 8t^5/5! + \cdots)$.

9. Since $y(0)$ is finite, $c_2 = 0$. With $q = 3$, $y = c_1 L_3$, and the other initial condition implies $y = \frac{6}{5}L_3(t)$.

11. $y = -\frac{78}{15}L_4(t) + t - \frac{1}{4}$.

13. $y = c_1 J_2(t) + c_2 Y_2(t)$.

15. $y = c_1 J_4(t) + c_2 Y_4(t)$.

17. $y = c_1 J_3(t)$, where $c_1 = -3/J_3(1)$.

Section 8.6

1. Using $r = 1/2$, $y = t^{1/2}(1 - t + t^2/6 + \cdots)$.

3. Hint: multiply the DE by t on both sides and let $p(t) = t - 2$, $q(t) = t$. Using $r = 3$, $y = t^3(1 - t + t^2/2 - t^3/6 + \cdots)$.

5. Using $r = 5$, $y = t^5(1 + 4t/5 + 5t^2/12 + \cdots)$.

7. Using $r = 1/2$, $y = t^{1/2}(1 - t/10 + t^2/28 + \cdots)$.

9. Using $r = \sqrt{3}$, $y = t^{\sqrt{3}}(1 - t/(1 + 2\sqrt{3}) + t^2/(4(1 + 2\sqrt{3})(1 + \sqrt{3})) + \cdots)$.

11. $r(r-1) + pr + q = 0$.

Index

CPSIA information can be obtained
at www.ICGtesting.com
Printed in the USA
LVOW04*0128281215

467942LV00004B/5/P